UMRECHNUNGSTABELLEN / CONVERSION TABLES

ENERGIE – *ENERGY**

kJ=kWs	kWh	kcal	Btu	ft · lb
1	2.78×10^{-4}	0.239	0.95	737.6
3600	1	860	3412	2.6×10^{6}
4.1868	1.163×10^{-3}	1	3.96	3100
1.054	2.929×10^{-4}	0.252	1	780
1.356×10^{-3}	0.377×10^{-6}	3.225×10^{-4}	1.282×10^{-3}	1

LEISTUNG – *POWER**

kW=kJ/s	hp	cal/s
1	1.341	238.7
0.7457	1	178

WASSERHÄRTE – *WATER HARDNESS**

German	French	American	British
1	1.786	1.041	1.25
0.56	1	0.583	0.7
0.961	1.716	1	1.201
0.8	1.429	0.832	1

RADIOAKTIVITÄT – *RADIOACTIVITY**

Energiedosis *absorbed dose*		Ionendosis *exposure*			
Rad	Gray (J/kg)	Röntgen	C/kg	Curi…	…Sv)
1	0.01	1	2.58×10^{-4}		
100	1	3876	1	2.703 …	

* Dezimalen: 0.1 (zero point one) im Englischen entspricht 0,1 (Null Komma eins) im Deutschen;
Tausender: 1,000 (one thousand) im Englischen entspricht 1.000 (eintausend) im Deutschen
– das heißt, Punkt und Komma werden genau umgekehrt verwendet !
(Diese Tabelle enthält die englische Schreibweise)

N = *nuisant* umweltgefährlicher Stoff

E = *explosive* explosionsgefährlicher Stoff

F + = *extremly flammable* hoch entzündlicher Stoff

F = *highly flammable* leicht entzündlicher Stoff

O = *oxidizing* brandfördernder Stoff

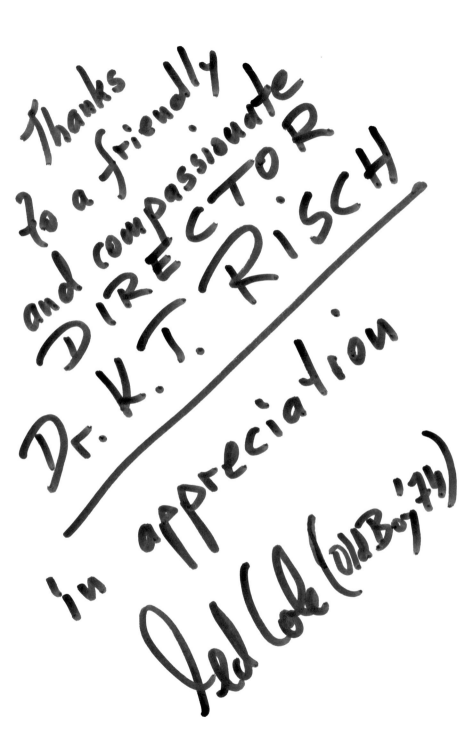

Wörterbuch Labor

Laboratory Dictionary

Theodor C. H. Cole

Wörterbuch Labor
Laboratory Dictionary

Deutsch/Englisch

English/German

Mit einem Geleitwort von
Professor Dr. Michael Wink

Dipl. rer. nat.
Theodor C. H. Cole
Heidelberg

tchcole@gmx.de

Library of Congress Control Number: 2004116528

ISBN 3-540-23419-5 Springer Berlin Heidelberg New York

This work is subject to copyright. All rights are reserved, whether the whole or part of the material is concerned, specifically the rights of translation, reprinting, reuse of illustrations, recitation, broadcasting, reproduction on microfilm or in any other way, and storage in data banks. Duplication of this publication or parts thereof is permitted only under the provisions of the German Copyright Law of September 9, 1965, in its current version, and permission for use must always be obtained from Springer. Violations are liable for prosecution under the German Copyright Law.

Springer is a part of Springer Science+Business Media
springeronline.com

© Springer-Verlag Berlin Heidelberg 2005
Printed in Germany

The use of designations, trademarks, etc. in this publication does not imply, even in the absence of a specific statement, that such names are exempt from the relevant protective laws and regulations and therefore free for general use.

Typesetting: Computer to film by author´s data
Production: Verlagsservice Heidelberg
Cover design: KünkelLopka.de
Printed on acid-free paper   2/3130XT 5 4 3 2 1 0

To "My Old School"
Gymnasium Englisches Institut, Heidelberg
at the occasion of its 60th anniversary
and to its founder
Gladys E. Fischer and her daughter Ellis Neu
and to the teachers who encouraged and believed in us

Geleitwort

Wer schon einmal versucht hat, einen Spezialtext aus dem Bereich der Zell- und Molekularbiologie, Biochemie oder Biotechnologie vom Deutschen ins Englische oder in die umgekehrte Richtung zu übersetzen, der wird irgendwann ein Wörterbuch zu Rate ziehen müssen. Doch selbst die besseren und umfangreicheren Wörterbücher, die heute verfügbar sind, kennen oft das gewünschte Wort nicht oder schlagen nicht selten eine falsche oder mehrdeutige Übersetzung vor. Vor dieser Schwierigkeit stehen insbesondere Studierende, aber auch erfahrene Wissenschaftler und Dozenten der Biologie, Molekularbiologie, Biotechnologie, Biochemie, Medizin, Pharmazie und Chemie.

Theodor C.H. Cole schließt mit dem vorliegenden „Wörterbuch Labor/Laboratory Dictionary" diese Lücke. Er ist wie kaum ein anderer Autor für diese Aufgabe gerüstet: Als Amerikaner, der viele Jahre in Deutschland gelebt und gelehrt hat, beherrscht er beide Sprachen gleich gut. Da er zudem Biologie, Chemie und Physik in Berkeley und Heidelberg studiert und als Übersetzer und Wörterbuchautor einschlägige Erfahrung erworben hat, sind ihm die Fachtermini in beiden Sprachen geläufig. Dieses Wörterbuch enthält über 12.000 der relevanten Begriffe, die in einem biologischen, biochemischen und chemischen Labor tagtäglich gebraucht werden. Aber auch viele Begriffe, die im breiteren wissenschaftlichen Kontext der im Labor durchgeführten Versuche und Experimente benötigt werden, fehlen nicht.

Dieses präzise und aktuelle Wörterbuch ergänzt die Serie der vorhandenen Wörterbücher in vorbildlicher Weise – es sollte in keinem Labor oder Büro fehlen.

Prof. Dr. Michael Wink   Universität Heidelberg

Foreword

Anyone attempting to translate the specialty literature of cell or molecular biology, biochemistry, or biotechnology from German to English or vice versa will inevitably find him/herself in need of consulting a dictionary. Even the advanced and extensive dictionaries available on the market today will often lack the desired term or propose an incorrect or ambiguous translation. According difficulties are encountered especially by students, but also by experienced scientists and educators in biology, molecular biology, biotechnology, biochemistry, medicine, pharmaceutical sciences, and chemistry.

Theodor C.H. Cole manages to fill this gap with his new "Laboratory Dictionary/ Wörterbuch Labor". More than any other author, he is suited for this task: as an American, having lived and taught in Germany for many years, he equally masters both languages. As he also studied biology, chemistry, and physics in Berkeley and Heidelberg and has gained considerable experience as a translator and author of other dictionaries, he is familiar with the corresponding specialty terminology in English and German. This dictionary contains some 12,000 relevant terms which are used daily in biological, biochemical and chemical labs. A great number of terms required in the broader scientific context of lab experiments and research are also included.

This concise and timely dictionary perfectly supplements those dictionaries currently available on the market – no lab or office should be without it.

Prof. Dr. Michael Wink   University of Heidelberg

Vorwort

Im Labor hat die englische Sprache ihren festen Platz. In den Naturwissenschaften sind Grundkenntnisse des Englischen geradezu Voraussetzung. Kataloge, Gebrauchsanweisungen, Versuchsbeschreibungen, Lehrbücher und Original-Forschungsberichte sind oft nur in Englisch vorhanden. Der Neuling im Labor wird sich über die vielen Anglizismen in der Laborsprache wundern, aber sich später vielleicht gar nicht mehr fragen, ob es überhaupt entsprechende deutsche Begriffe gibt – wenigstens nicht bis zu dem Zeitpunkt, an dem er vor dem Problem steht, die vollbrachte Laborarbeit schriftlich auf Deutsch zu verfassen.

Es gibt aber auch andere Gründe ein „Wörterbuch Labor" zu konsultieren: Die Initiative zu diesem Wörterbuch entstand bei der Vorbereitung einer Akademie für junge Nachwuchswissenschaftler zu einem Forschungsaufenthalt in den USA. Die Teilnehmer sprachen zwar gutes Alltagsenglisch, aber für die englische „Laborsprache" benötigten sie noch Zusatzkenntnisse. Ein entsprechendes Wörterbuch musste allerdings noch geschaffen werden. So entstand die Grundlage für das vorliegende Wörterbuch.

Für englischsprachige Gastwissenschaftler in Deutschland dürfte das „Wörterbuch Labor" ebenso eine nützliche Hilfe sein wie für Laboranten, Technische Assistenten, forschende und schreibende Wissenschaftler, Studenten, und Übersetzer.

Im Labor gehört der sichere und verantwortungsbewusste Umgang mit „Feuer, Luft und Wasser" – ebenso wie mit Mikroorganismen, Pflanzen und Tieren – zum Grundhandwerkszeug für erfolgreiche Forschung und Lehre. Jeder der sich zu Laboraufenthalten ins Ausland begibt sollte sich deshalb sprachlich gründlich vorbereiten, da gerade in diesem Bereich eine effiziente Kommunikation oft lebensrettend sein kann.

Für das Lesen und Verstehen von Anleitungen, Berichten und „Kochrezepten", oder für die Auswertung und Dokumentation von Forschungsergebnissen sind wir im Laboralltag eigentlich immer auf Wörterbücher angewiesen. Hier soll das vorliegende Wörterbuch Unterstützung liefern, auch wenn es die Benutzung von Großwörterbüchern mit breiterem Vokabular nicht in jedem Fall ersetzen kann.

Dieses Fachwörterbuch enthält einen Grundwortschatz zur allgemeinen Labor-Terminologie. Mit 12.000 Begriffen in jeder Sprachrichtung bietet es die wichtigsten deutschen und englischen Stichwörter zu den Bereichen:

- **Laborausstattung und ~einrichtung**
- **Laborbau, Sanitär, Elektrik, Lüftung**
- **Laborbedarf (Zubehör, Geräte, Werkzeuge, Laborglas)**
- **Basischemikalien (allgemeine Grundstoffe)**
- **Laborsicherheit (Arbeits~ und Personenschutz, Notfälle und Vorsorge)**
- **Methoden und Analytik**

Wortfelder. Um eine zusammenhängende Themenbearbeitung zu ermöglichen, die „Trefferwahrscheinlichkeit" bei der Wortsuche zu erhöhen und somit die Arbeit zu erleichtern, verwenden wir zusätzlich zur gewöhnlichen alphabetischen Ordnung ein auch in amerikanischen Wörterbüchern verwendetes Konzept der thematischen Begriffssammlung (*clusters*) unter den jeweiligen übergeordneten Hauptstichwörtern. Thematisch verwandte Begriffe werden in Wortfeldern zusammengefasst, auch wenn die einzelnen Begriffe das Hauptstichwort selbst gar nicht enthalten. Beispielsweise finden sich unter dem Hauptstichwort *Gefahrenbezeichnungen* alle entsprechenden sicherheitsrelevanten Begriffe; alle Chromatographie-Verfahren erscheinen unter dem Haupteintrag *Chromatographie*, alle Arten von *Filter*, *Kolben*, *Pipetten*, *Pinzetten*, *Pumpen*, *Schüttler*, *Ventile*, *Waagen* etc. unter dem jeweilgen Hauptstichwort - zusätzlich zu den alphabetisch geordneten Einträgen. Dies verschafft klare Übersicht und Arbeitskomfort. In anderen Wörterbüchern müsste jeder Begriff einzeln aufgesucht werden.

Die Rechtschreibung orientiert sich an der amerikanischen Schreibweise laut *Merriam Webster's Collegiate Dictionary*, 10th edn., bzw. *Wahrig Deutsches Wörterbuch*, 6. Aufl., d.h. die deutsche Rechtschreibreform wurde berücksichtigt. Die „c" Schreibweise von lateinisch und griechisch abgeleiteten deutschen Begriffen wird bevorzugt, z.B. *cyclisch* gegenüber *zyklisch*.

Danksagungen. Für fachlichen und freundschaftlichen Rat danke ich Dr. Ingrid Haußer-Siller (Universität Heidelberg), Dr. Willi Siller (Universität Heidelberg) und Dr. Dietrich Schulz (UBA, Berlin) sowie Dan Choon (Heidelberg) für seine Hilfe bei der technischen Umsetzung. Mein Dank ergeht auch an die Teilnehmer der Heidelberger „San Francisco Academy 2003" – besonders Michael Breckwoldt, cand. med. (München), Laura Michel, cand. med. (München) und Samuel Bandara, B.A. (Heidelberg) – für Inspiration und das gelungene gemeinsame Projekt.

Meiner Frau Erika Siebert-Cole gilt meine ganz besondere Wertschätzung für ihre fachliche Kompetenz und Hilfe mit der deutschen Sprache, der Durchsicht der Endversion des Manuskripts sowie für entscheidenden Zuspruch in den kritischen Phasen des Bücherschreibens.

Speziellen Dank auch an meinen Cousin Cliff Cole (Sacramento, California) für seine stets zuversichtliche moralische Unterstützung.

Den Mitarbeitern der „Planung Chemie" im Springer-Verlag bin ich für die langjährige positive Zusammenarbeit sehr verbunden. Dr. Marion Hertel war mit Erfahrung, Geduld und Zuspruch dem Projekt und dessen Autor gegenüber immer aufgeschlossen und ermöglichte in angenehmer Arbeitsatmosphäre das Zustande-kommen eines weiteren Wörterbuches – eine spannende und angenehme Zeit. Eigentlich schade, dass sie schon vorüber ist!

Heidelberg, Anfang 2005   Theodor C.H. Cole

Preface

The internationalization of science has led to the use of English as the general language of communication – a structurally simple language, with an enormous vocabulary.

Laboratory English seems to be a language of its own – at times quite incomprehensible to outsiders. "Laboratory slang" consists of neologisms, colloquialisms, contractions, acronyms, brandnames. Foreigners arriving in an English-speaking laboratory may be quite puzzled with *fleas, pigs, fuge, frige, policeman, bogie, dolly, elbows, daisy-chains, PCR, fleakers* – and *"fuzzy logic"*. Those who intend to work in laboratories abroad are well advised to consider a language preparatory course.

There is a wealth of technical laboratory terms to be learned by the non-native-speaker in the lab, if only for the reason of survival: safety precautions and proper behavior in case of emergencies require instant response and efficient communication.

The idea for writing this book arose from my involvement as a consultant to a "San Francisco Academy" for a dozen young German college students from Heidelberg. For our language preparatory course I needed to decide on a list of "essential terminology" – as no adequate book was available on the German market at the time. Now, the "Laboratory Dictionary", which is an extension of that early list, fills this gap providing the basic vocabulary to assist its users in improving their laboratory language skills.

This "Laboratory Dictionary" contains some 12,000 terms from all walks of laboratory life, emphasizing:

- **Facility engineering (electricity, plumbing, ventilation)**
- **Equipment & supplies**
- **Glassware, tools, apparatus**
- **Servicing, maintenance, repair**
- **Ordering, shipment, delivery**
- **Analytics (methods & procedures)**
- **Chemicals (general reactants)**
- **Safety (precautions & emergencies)**

The new German orthography rules have been taken into account according to *Wahrig Deutsches Wörterbuch*, 6th edn., the English orthography following the American according to *Merriam Webster's Collegiate Dictionary*, 10th edn.

We hope this dictionary may serve you as a useful tool in your research, in writing publications, and for translations.

Acknowledgements. Dr. Ingrid Haußer-Siller (Department of Dermatology, EM-Laboratory, Heidelberg University Clinic) kindly reviewed the manuscript and made valuable suggestions. Dr. Willi Siller (Biological Safety Officer, University of Heidelberg) and Dr. Dietrich Schulz (Acad.Dir., Agr.Dept., UBA, Berlin) have offered welcome advise. Dan Choon, cand. phys. (University of Heidelberg) converted the raw data into an appealing layout.

The participants of the Heidelberg "San Francisco Academy 2003" provided inspiration and impetus – and special thanks for a rewarding friendship to Michael Breckwoldt, cand. med. (Munich), Laura Michel, cand. med. (Munich), and Samuel Bandara, B.A. (Heidelberg).

Dr. Marion Hertel, Senior Editor of Chemistry at Springer, Heidelberg, has been a brilliant partner and I commend her for the encouragement and energy she devoted to implementing this project – and thanks to Jörn Mohr for editorial assistance.

Erika Siebert-Cole, M.A. has readily shared her knowledge, time, inspiration, and an intricate sense of the German language – bearing with me through quite many strenuous and euphoric moments in the bookwriting process.

Thanks to my cousin Cliff Cole, of Sacramento, California, for his steadily positive moral support.

My gratitude is wholeheartedly expressed to family and friends who helped and provided encouragement throughout this 5th dictionary project.

Heidelberg, early in 2005					Theodor C.H. Cole

Abkürzungen – Abbreviations

sg	Singular – singular
pl	Plural – plural
adv-adj	Adverb-Adjektiv – adverb-adjective
n	Nomen (Substantiv) – noun
vb	Verb – verb
f	weiblich – feminine
m	männlich – maskulin
nt	sächlich – neuter
analyt	Analytik – analytics
allg – general	allgemein – general
biot	Biotechnologie – biotechnology
centrif	Zentrifugation – centrifugation
chem	Chemie – chemistry
chromat	Chromatographie – chromatography
comp	Datenverarbeitung – computing
dest – dist	Destillation – distillation
dial	Dialyse – dialysis
ecol	Ökologie – ecology
electr	Elektrik-Elektronik – electrics-electronics
electroph	Elektrophorese – electrophoresis
gen	Genetik – genetics
geol	Geologie – geology
immun	Immunologie – immunology
lab	Labor – laboratory
math	Mathematik – mathematics
mech	Mechanik – mechanics
med	Medizin – medical science
micb	Mikrobiologie – microbiology
micros	Mikroskopie – microscopy
neuro	Neurobiologie – neurobiology
opt	Optik – optics
photo	Photographie – photography
phys	Physik – physics
physiol	Physiologie – physiology
spectr	Spektroskopie – spectroscopy
stat	Statistik – statistics
tech	Technologie – technology
vir	Virologie – virology

Deutsch – Englisch

Abbau (Zersetzung/Zerfall/ Zusammenbruch) degradation, decomposition, breakdown; (einer Apparatur) disassembly, dismantling, dismantlement, takedown
➤ **biologischer Abbau/Biodegradation** biodegradation
➤ **enzymatischer Abbau** enzymatic digestion
➤ **Stoffwechsel-Abbau** digestion; degradative metabolism, catabolism
Abbaubarkeit degradability, decomposability
➤ **biologische Abbaubarkeit** biodegradability
abbauen (zersetzen) degrade, decompose, break down; (Apparatur/Experimentiergerät) disassemble (take equipment apart)
Abbauprodukt degradation product
abbilden *opt* image; (projizieren) project
Abbildung (in einer Fachzeitschrift/Buch) figure, illustration
Abbildungsmaßstab/ Lateralvergrößerung/ Seitenverhältnis/Seitenmaßstab lateral magnification
abbinden (fest/steif werden) set
Abbrand burning, burn-up, burn-off; consumption; scalding loss; (Rückstand beim Rösten) cinder
abdampfen/abdunsten evaporate
Abdampfschale evaporating dish
Abdichtbarkeit sealability
abdichten seal off, make tight, make leakproof, insulate
Abdichtung seal, sealing; (Manschette) gasket
Abdruck (Oberflächenabdruck: *EM*) *micros* replica
➤ **genetischer Fingerabdruck/ Fingerprinting** fingerprinting, genetic fingerprinting, (DNA fingerprinting)
abfackeln flare, burn off

Abfackelung flare, flaring off
Abfall waste, trash, refuse
➤ **Sonderabfall/Sondermüll** hazardous waste
Abfall mit biologischem Gefährdungspotential biohazardous waste
Abfallbehälter trash can; waste container, litter bin
Abfallentsorgung/Abfallbeseitigung waste disposal
Abfallgesetz/ Abfallbeseitigungsgesetz (AbfG) waste disposal law, waste disposal act
Abfallvorbehandlung waste pretreatment
abfärben stain, bleed
abflammen/'flambieren' (sterilisieren) flame
Abfluss (Ausfluss) discharge, outflow, efflux, draining off; (Ablauf, z.B. am Waschbecken) drain
➤ **frei machen** unblock
➤ **verstopft** blocked, clogged, choked
Abflussbecken sink, basin
Abführmittel laxative
Abgabe/Einreichung (Ergebnisse etc.) delivery, handing in, dropoff
Abgabepuls (Pumpe) discharge stroke
Abgabetermin/Ablieferungstermin/ Deadline deadline
Abgase exhaust fumes
abgelaufen (Haltbarkeitsdatum) expired, outdated
abgeschrägt/abgekantet (Kanülenspitze/Pinzette etc.) beveled, bevelled
abgießen/dekantieren (ablassen) pour off, decant; (drain)
Abgleich equalization, adjustment, balancing, balance; alignment; tuning
abgleichen adjust, equalize; align, tune
abgraten/bördeln deburr
Abguss (an der Spüle) drain (of the sink)
➤ **in den Abguss schütten** pour s.th. down the drain

Ab

Abholung (Lieferung etc.) pickup
Abisolierzange wire stripper
abkanten/abschrägen (Metall/ Pinzetten/Kanülen/Glas etc.) bevel
Abklärflasche/Dekantiergefäß decanter
abkochen/absieden decoct
Abkochung/Absud/Dekokt decoction
abkühlen cool down, get cooler
Abkühlzeit/Abkühlphase/ Fallzeit (Autoklav)
cool-down period, cooling time
ablängen (mit Glasrohrschneider)
size, cut into discreet length
ablassen drain, discharge; (Druck reduzieren) relieve, vent
Ablasshahn/Ablaufhahn
draincock (faucet/spigot)
Ablauf drain; (Ablaufbrett/Platte an der Spüle) drainboard; (Ausfluss: Austrittstelle einer Flüssigkeit) outlet; (herausfließende Flüssigkeit) effluent
Ablaufdatum/Verfallsdatum
expiration date
ablaufen lassen drain
ableiten carry off, drain, discharge; (umleiten) deflect
➤ **zuleiten** supply, feed, pipe in, let in
Ableitung (von Flüssigkeiten)
discharge, drainage, outlet
➤ **Zuleitung** supplying, feeding, inlet
Ablenkung deflection
Ablenkungsspannung
deflection voltage
➤ **Kippspannung** sweep voltage
ablesbar readable
Ablesbarkeit (Waage) readability
Ablesefehler
reading error, false reading
Ablesegenauigkeit reading accuracy
Ablesegerät direct-reading instrument
Ablesemarke
reference point, index mark
ablesen read (off/from)
Ablesung/Ablesen (Gerät/Messwerte)
reading, readout
Abluft exhaust, exhaust air, waste air, extract air

Ablufteinrichtung/Abluftsystem
exhaust system, off-gas system
Abluftschacht exhaust duct
Abluftsystem exhaust system
abmelden deregister, sign out (schriftlich 'austragen')
abmessen measure, size
Abmessungen (Höhe/Breite/Tiefe)
dimensions (height/width/depth)
Abnahme (eines Labors nach Bau)
commissioning, certification
Abpackung pack, package
Abrauchen
fuming; (eindampfen) evaporate
Abreicherung
depletion, stripping, downgrading
abrutschen/ausrutschen slip
absaugen (Flüssigkeit) draw off, suction off, siphon off, evacuate
abschalten turn off, shut off, switch off; (Computer: herunterfahren) power down
➤ **anschalten** turn on, switch on; (Computer: hochfahren) power up
Abschaltung shutoff
➤ **automatische A.** auto-shutoff
Abschaltventil shut-off valve
Abscheider separator, precipitator, settler, trap, catcher, collector
➤ **Wasserabscheider**
water separator, water trap
abschirmen (von Strahlung)
shield (from radiation)
Abschirmung (von Strahlung)
shielding (from radiation)
Abschmelzrohr
fusion tube, melting tube
Abschnürbinde/Binde/Aderpresse/ Tourniquet tourniquet
abschöpfen skim off, scoop off/up
abschrecken turn away, repel, reject; *polym* quench
abseihen strain
absondern/abscheiden (Flüssigkeiten)
exude, secrete, discharge
Absorbanz (Extinktion)
absorbance, absorbancy (extinction: optical density)

absorbieren/aufsaugen
  soak up, absorb
Absorption absorption
Absorptionsindex
  absorbance index, absorptivity
Absorptionskoeffizient
  absorption coefficient
Absorptionsmittel/Aufsaugmittel
  absorbent, absorbant
Absorptionsspektrum
  absorption spectrum,
  dark-line spectrum
Absorptionsvermögen/
  Absorptionsfähigkeit/
  Aufnahmefähigkeit absorbency
Absperrband/Markierband
  barricade tape
Absperrhahn/Sperrhahn stopcock
Absperrung/Barriere/Sperre/Barrikade
  barrier, barricade
Absperrventil shut-off valve
Abstand (Geräte/Möbel etc.) clearance
Abstand halten! keep clear!
Abstandhalter/Abstandshalter/
  Distanzstück spacer
absteigend (DC) descending
Abstellraum/Abstellkammer
  storeroom, storage room
abstoßend repellent, repellant
Abstreicher/Rakel (Gummi) squeegee
Abstrich *med* swab; *micros* smear
  ➤ **einen Abstrich machen** *med*
    to take a swab
abtauen (Kühl-/Gefrierschrank)
  defrost
Abtransport/Entfernen
  transporting away; removal
abtrennen separate
Abtrennung separation;
  (Trennwand: räumlich) partition
Abtriebsäule/Abtreibkolonne
  dest stripping column
Abtriebsteil (Unterteil der Säule)
  dest stripping section
Abtropfbrett/Ablaufbrett drainboard
Abtropfgestell draining rack
Abtropfsieb colander
Abwärme waste heat

Abwaschwasser/Spülwasser dishwater
Abwasser wastewater, sewage
  ➤ **Rohabwasser** raw sewage
Abwasserabgabengesetz
  wastewater charges act
Abwasseraufbereitung
  sewage treatment
Abwasserkanal/Kloake
  sewer, sanitary sewer
abweichen von ... deviate from ...
Abweichung deviation;
  (Aberration) aberration
  ➤ **Standardabweichung**
    standard deviation
  ➤ **statistische Abweichung**
    statistical deviation
abwiegen (eine Teilmenge) weigh out
abwischen wipe, wipe off, wipe clean
Abzieher/Gummiwischer
  (Fensterwischer/Bodenwischer)
  squeegee
  ➤ **Fensterwischer/Fensterabzieher**
    squeegee (for windows)
  ➤ **Wasserschieber/Wasserabzieher**
    (Bodenwischer) squeegee (for floors)
Abzug/Dunstabzugshaube hood,
  fume hood, fume cupboard (*Br*)
  ➤ **begehbarer A.** walk-in hood
  ➤ **Fallstrombank**
    vertical flow workstation (hood/unit)
  ➤ **Handschuhkasten/**
    Handschuhschutzkammer
    glove box
  ➤ **Labor-Werkbank**
    laboratory/lab bench
  ➤ **Querstrombank**
    laminar flow workstation,
    laminar flow hood, laminar flow unit
  ➤ **Rauchabzug** fume hood
  ➤ **Reinraumwerkbank**
    clean-room bench
  ➤ **Saugluftabzug** forced-draft hood
  ➤ **Sicherheitswerkbank** clean bench
  ➤ **sterile Werkbank** sterile bench
Abzugschornstein exhaust stack
Abzugsöffnung/Luftschlitz vent
Abzweig *chromat* split
Abzweigventil *chromat* split valve

Ac

Acetaldehyd/Ethanal acetaldehyde, acetic aldehyde, ethanal
Acetat/Azetat (Essigsäure/Ethansäure) acetate (acetic acid/ethanoic acid)
Acetessigsäure (Acetacetat)/ 3-Oxobuttersäure acetoacetic acid (acetoacetate), acetylacetic acid, diacetic acid
Aceton (Azeton)/Propan-2-on/ 2-Propanon/Dimethylketon acetone, dimethyl ketone, 2-propanone
Achatmörser agate mortar
achromatischer Kondensor *micros* achromatic condenser, achromatic substage
achromatisches Objektiv *micros* achromatic objective
Achsenlager/Achslager/ Zapfenlager (z.B. beim Kugellager) journal
Achtkantstopfen octa-head stopper, octagonal stopper
Acidität/Azidität/Säuregrad acidity
Acridinfarbstoff acridine dye
Acrylglas acrylic glass
Adapter adapter, fitting(s)
- **Balg** bellows
- **Destilliervorstoß** receiver adapter
- **Eutervorlage/Verteilervorlage/ 'Spinne'** *dest* cow receiver adapter, 'pig', multi-limb vacuum receiver adapter (receiving adapter for three/four receiving flasks)
- **Expansionsstück (Laborglas)** expansion adapter, enlarging adapter
- **Filtervorstoß** adapter for filter funnel
- **Kern-/Gewindeadapter** cone/screwthread adapter
- **Kriechschutzadapter** *dest* anticlimb adapter
- **Krümmer (Laborglas)** bend, bent adapter
- **Nadeladapter** syringe connector
- **Reduzierstück (Laborglas/Schlauch)** reducer, reducing adapter, reduction adapter
- **Schaumbrecher-Aufsatz/ Spritzschutz-Aufsatz (Rückschlagsicherung)** *dest* antisplash adapter, splash-head adapter
- **Schlauchadapter** tubing adapter
- **Schlauch-Rohr-Verbindungsstück** pipe-to-tubing adapter
- **Septum-Adapter** septum-inlet adapter
- **Tropfenfänger** drip catcher, drip catch; splash trap, antisplash adapter (distillation apparatus); (Rückschlagschutz: Kühler/Rotationsverdampfer etc.) splash/antisplash adapter, splash-head adapter
- **Übergangsstück** adapter, connector, transition piece
- **Übergangsstück mit seitlichem Versatz** offset adapter
- **Vakuumfiltrationsvorstoß** vacuum-filtration adapter
- **Vakuumvorstoß** vacuum adapter
- **Vorlage** *dest* distillation receiver adapter, receiving flask adapter
- **Zweihalsaufsatz** two-neck (multiple) adapter

Additionsverbindung addition compound (of two compounds), additive compound (saturation of multiple bonds)
Aderendhülse *electr* wire end sleeve, wire end ferrule
Aderendhülsenzange/Crimpzange crimping pliers, crimper
Aderpresse/ Abschnürbinde/Binde/Tourniquet tourniquet
Adsorptionsmittel/Adsorbens adsorbent

Agar agar
- **Blutagar** blood agar

Agardiffusionstest
agar diffusion test

Agarnährboden agar medium

Agarose agarose

Agarplatte agar plate

Agens/Agenz (*pl* **Agentien**)
agent
- **interkalierendes Agens**
intercalating agent
- **quervernetzendes Agens**
cross linker, crosslinking agent

Aggregatzustand
physical state
- **fester Zustand**
solid state
- **flüssiger Zustand**
liquid state
- **gasförmiger Zustand**
gaseous state

Ahle awl, pricker;
(Reibahle) reamer

Airliftreaktor/pneumatischer Reaktor
(Mammutpumpenreaktor)
airlift reactor, pneumatic reactor

Akkusäure/Akkumulatorsäure
accumulator acid,
storage battery acid (electrolyte)

aktinisch actinic

Aktivkohle activated carbon

Akzeptorstamm *biochem*
(Proteinsynthese)
acceptor stem

Alarm alarm; alert
- **falscher Alarm** false alarm
- **Feueralarm** fire alarm
- **Probealarm/Probe-Notalarm**
drill, emergency drill

Alarmanlage alarm system

Alarmbereitschaft alert

alarmieren (Feuerwehr etc.) call, alert

Alarmsignal alarm signal

Alarmsirene
alarm siren, air-raid siren

Alarmstufe
emergency level, alert level

Alaun/Aluminiumsulfat alum

Aldehyd aldehyde
- **Acetaldehyd/Ethanal**
acetaldehyde,
acetic aldehyde, ethanal
- **Anisaldehyd**
anisic aldehyde, anisaldehyde
- **Formaldehyd/Methanal**
formaldehyde, methanal
- **Glutaraldehyd/Glutardialdehyd/**
Pentandial
glutaraldehyde, 1,5-pentanedione

Aliquote/aliquoter Teil
(Stoffportion als Bruchteil
einer Gesamtmenge)
aliquot

alkalibeständig/laugenbeständig
alkaliproof

Alkali-Blotting alkali blotting

alkalisch/basisch
alkaline, basic

Alkaliverätzung/Basenverätzung
alkali burn

Alkaloide alkaloids

Alkohol alcohol
- **Ethylalkohol/Ethanol/Äthanol**
(Weingeist)
ethyl alcohol, ethanol
(grain alcohol, spirit of wine)
- **Isopropylalkohol/Propan-2-ol**
isopropyl alcohol, isopropanol,
1-methyl ethanol (rubbing alcohol)
- **Methylalkohol/Methanol**
(Holzalkohol)
methanol, methyl alcohol
(wood alcohol)
- **Propylalkohol/Propan-1-ol**
n-propyl alcohol, propanol

Alkoholreihe/
aufsteigende Äthanolreihe
graded ethanol series

Allergen allergen, sensitizer

allergisch allergic

Allergisierung sensitization

Allzweck.../
Allgemeinzweck.../Mehrzweck...
all-purpose, general-purpose, utility...

Altöl waste oil, used oil

Altpapier waste paper

Altstoffe
 existing chemicals/substances
➢ **Neustoffe**
 new chemicals/substances
Aluminium (Al) aluminum
Aluminiumfolie/Alufolie
 aluminum foil
Ambulanz/Notaufnahme
 emergency room
Ames-Test Ames test
Amid amide
Amidierung amidation
Amin amine
Aminierung amination
Aminoacylierung
 aminoacylation
**Aminobuttersäure/
 γ-Aminobuttersäure (GABA)**
 gamma-aminobutyric acid (GABA)
Aminosäure amino acid
Aminozucker amino sugar
Ammoniak ammonia
Ampulle (Glasfläschchen)
 ampule, ampoule
➢ **vorgeritzte Spießampulle**
 prescored ampule/ampoule
Amt/Behörde
 agency, department; office, bureau
analog/funktionsgleich
 analogous
Analog-Digital-Wandler
 analog-to-digital converter (ADC)
Analogie analogy
analogisieren analogize
Analogon (*pl* **Analoga**)
 analog, analogue
Analysator analyzer
Analyse analysis (*pl* analyses)
analysenrein/zur Analyse
 lab reagent grade
Analysenwaage
 analytical balance
analysieren analyze
Analyt/zu analysierender Stoff
 analyte
analytisch analytic(al)
Andreaskreuz (Gefahrenzeichen)
 St. Andrew's cross

Anfangsgeschwindigkeit
 (v_0: **Enzymkinetik**)
 initial velocity (vector),
 initial rate
anfärbbar dyeable, stainable
Anfärbbarkeit
 dyeability, stainability
anfärben dye, stain
Anfärbung dyeing, staining
anfeuchten humidify, prewet
Angestellter employee
Anhäufung/Kumulation
 accumulation
Anheizzeit/Steigzeit (Autoklav)
 preheating time
Anionenaustauscher
 anion exchanger
Anisaldehyd
 anisic aldehyde, anisaldehyde
anketten (Gasflaschen etc.)
 chain (to)
➢ **mehrere Gegenstände
 aneinander ketten**
 daisy-chain
**Anlage/Einrichtung/
 Betriebseinrichtung**
 installation(s)
**Anlaufphase/Latenzphase/
 Inkubationsphase/
 Verzögerungsphase/
 Adaptationsphase/lag-Phase**
 lag phase, latent phase,
 incubation phase,
 establishment phase
Anlaufzeit/Reaktionszeit
 response time
Anleitung (Einarbeitung) instructions,
 training, guidance, directions, lead;
 (Einführung) introduction (to);
 (Gebrauchsanweisung) manual,
 instructions
anmelden register, announce oneself,
 sign in (schriftlich eintragen/
 einschreiben)
Anmeldepflicht mandatory registration,
 obligation to register
Anmeldung registration, signing in;
 (Rezeption) reception

anregen stimulate, excite
Anregung stimulation, excitation
anreichern enrich; concentrate, accumulate, fortify
Anreicherung enrichment, concentration, accumulation, fortification
Anreicherung durch Filter
  filter enrichment
Anreicherungseffekt/Gesamtwirkung
  cumulative effect
Anreicherungskultur
  enrichment culture
Ansatz (Versuchsansatz/ Versuchsaufbau)
  arrangement, set-up;
  (Charge) batch;
  (Methode) approach, method;
  (Präparat) starting material, preparation; (Versuch) attempt
Ansatzstück (Glas)
  attachment, extension (piece)
Ansatzstutzen
  (Kolben) side tubulation, side arm;
  (Schlauch) hose connection
ansäuern acidify
Ansaugpuls (Pumpe)
  suction stroke
Ansaugrohr intake pipe;
  induction pipe, suction pipe
Ansaugventil
  induction valve,
  aspirator valve
anschalten
  turn on, switch on; (Computer: hochfahren) power up
➤ **abschalten** turn off, shut off, switch off; (Computer: herunterfahren) power down
Anschlag (Endpunkt/Sperre/Stop)
  stop, limit, detent
Anschlagzettel (Gefahrgutkennzeichnung etc.)
  placard
anschließen *allg* fasten (to), connect (to/with), link (up to);
  electr connect, hook up, wire to, (make) contact

Anschluss connection;
  (*pl* Anschlüsse: Armaturen/Hähne) fixture(s), outlet(s);
  (Gas~/Strom~/Wasser~) connection, line (Leitung)
➤ **elektrische(r) Anschluss**
  electrical fixture(s),
  electricity outlet
➤ **Versorgungsanschlüsse (Wasser/Strom/Gas)**
  service fixtures, service outlets
Anschlussleitung
  electr lead, pigtail lead
Anschütz-Aufsatz Anschütz head
Anschwänzapparat/ Anschwänzvorrichtung (Fermentation)
  sparger
ansetzen (z.B. eine Lösung)
  start, prepare, mix, make, set up
Ansprechzeit (z.B. Messgerät etc.)
  response time
anspruchslos undemanding, modest, having low requirements/demands
anspruchsvoll demanding, having high requirements/demands
anstecken/infizieren infect
ansteckend/ ansteckungsfähig/infektiös
  contagious, infectious
Ansteckleuchte *micros*
  substage illuminator
Ansteckung/Infektion
  contagion, infection
Ansteckungsfähigkeit/ Infektionsvermögen
  infectivity
Ansteckungsherd/Ansteckungsquelle
  source of infection
Ansteckungskraft/Virulenz
  virulence
Anstellwinkel
  angle of attack
Antagonismus antagonism
Anteil/Hälfte/Teil moiety
Antrag application
➤ **eingereichter A.**
  submittal, submitted application

An

Antrieb/Trieb drive; Voranbringen (Fortbewegung) propulsion
Antriebskraft/Triebkraft propulsive force
Antriebssystem drive system, drive unit
Antriebswelle drive shaft
Antwort (auf Reiz) response
antworten answer; respond
Anweisung assignment, direction(s), directive, instructions; prescription, order
anwenderfreundlich user-friendly
Anzeige (an einem Gerät) display; dial, scale, reading
anzeigen display, show, read
anzeigepflichtig obligation to notify, notifiable, reportable
Anzeiger/Anzeigegerät indicator, recording instrument; monitor
Anzucht (einer Kultur) starting
anzüchten (einer Kultur) establish, start (a culture)
Anzuchtmedium starter medium (growth medium)
Anzüchtung (einer Kultur) establishing growth, starting growth
anzünden ignite, strike, start a fire
Anzünder (Gas) striker
Apertur (Blende)/Öffnung/Mündung aperture, opening, orifice
Aperturblende/Kondensorblende (Irisblende) condenser diaphragm (iris diaphragm)
Äpfelsäure (Malat) malic acid (malate)
Apiezonfett apiezon grease
Apoenzym apoenzyme
Aquarium aquarium, fishtank
Äquivalenzpunkt (Titration) end point, point of neutrality
Arachidonsäure arachidonic acid, icosatetraenoic acid
Arachinsäure/Arachidinsäure/Eicosansäure arachic acid, arachidic acid, icosanic acid
Aräometer (Densimeter/Senkwaage) areometer
Arbeitgeber employer
Arbeitsablauf sequence of operation
Arbeitsabstand *micros* working distance (objective-coverslip)
Arbeitsanweisung/Arbeitsvorschrift prescribed work procedure, prescribed operating procedure
➤ **Standard-Arbeitsanweisung** standard operating procedure (SOP)
Arbeitsbedingungen operating conditions (Geräte), working conditions (Personen)
Arbeitsbereich operating range (Geräte), work area, working range (Personen)
Arbeitsdruck working pressure (delivery pressure)
Arbeitsfläche work surface, working surface, working area
➤ **Tischoberfläche (Labortisch)** countertop, benchtop
Arbeitshygiene industrial hygiene
Arbeitskittel smock, gown
➤ **Laborkittel** frock, lab coat
Arbeitsmedizin occupational medicine
Arbeitsmethode work procedure
Arbeitsöffnung working aperture
➤ **Schutzfaktor für die Arbeitsöffnung (Werkbank)** aperture protection factor (open bench)
Arbeitspensum workload
Arbeitsplatte/Arbeitsfläche (Labor-/Werkbank) countertop, benchtop
Arbeitsplatz (Ort) workplace; (Stelle) job
➤ **Arbeitsbereich (räumlich)** workspace
Arbeitsplatzhygiene occupational hygiene

Arbeitsplatzkonzentration, zulässige/maximale (DFG: MAK)
permissible workplace exposure; *nicht identisch mit:* (*Br*) occupational exposure limit (OEL); (*US*: by ACGIH) threshold limit value (TLV)
Arbeitsplatzsicherheit
occupational safety, workplace safety
Arbeitsplatzsicherheitsvorschriften
occupational safety code
Arbeitsraum (im Inneren der Werkbank)
working space
Arbeitsrichtlinie
working guideline
Arbeitsschritt
step in a working procedure
Arbeitsschutz
occupational protection, workplace protection, safety provisions (for workers)
Arbeitsschutzanzug
coverall, boilersuit, protective suit
Arbeitsschutzkleidung
workers' protective clothing
Arbeitsschutzverordnung
workplace safety regulations
Arbeitsstoff (workplace) agent
Arbeitstagebuch logbook
Arbeitstemperatur
operating temperature
Arbeitstisch worktable
Arbeitsunfall
occupational accident
Arbeitsvertrag
contract of employment
Arbeitsvorgang work procedure
Arbeitsvorschrift/ Arbeitsanweisung
prescribed work procedure, prescribed operating procedure
➢ **für die Überwachung**
monitoring protocol
➢ **Standard-Arbeitsanweisung**
standard operating procedure (SOP)
Arbeitszeit work hours
Arbeitszyklus (Gerät)
duty cycle

arithmetisches Mittel
stat arithmetic mean
arithmetisches Wachstum
arithmetic growth
Armatur(en) (Hähne im Labor/an der Spüle etc.)
fittings, fixtures, mountings; instruments; connections
Armaturenbrett/Schalttafel
switchboard, electrical control panel; (im Fahrzeug) dashboard, dash
Aroma (Wohlgeruch) aroma, fragrance, (pleasant) odor; (Wohlgeschmack) flavor, taste (pleasant)
Aromastoff
flavoring, aromatic substance
aromatisch aromatic
Arretierhebel
stop lever, arresting lever, locking lever, blocking lever; catch, safety catch
Arretierbolzen
locking bolt, locking pin
arretieren/feststellen
arrest, stop, fix, lock in place/position; block; detent
Arretierschraube
locking screw
Arretierung *tech/mech* lock, locking device; (Klinke/Schnappverschluss) catch; (z.B. am Mikroskop) stop
Arsen (As) arsenic
Arsenwasserstoff/ Arsan/Monoarsan arsine
Artefakt artifact, artefact
artfremd (Eiweiss) foreign
Arznei/ Arzneimittel/Medizin
medicine, medication, drug
Arzneibuch
pharmacopeia
Arzneikunde/ Arzneilehre/Pharmazie pharmacy

Arzneimittel
drug, medicine, medication
➢ **nicht verschreibungspflichtiges A.**
non-prescription drug
➢ **verschreibungspflichtiges A.**
prescription drug
Arzneimittel-Rezeptbuch/ Pharmakopöe/ amtliches Arzneibuch
formulary, pharmacopoeia
Asbest asbestos
➢ **Blauasbest/Krokydolith**
blue asbestos, crocidolite
➢ **Weißasbest/Chrysotil**
white asbestos, chrysotile, Canadian asbestos
Asbestplatte asbestos board
Asbeststaublunge/ Bergflachslunge/Asbestose
asbestosis
Asbestzementplatte (Labortisch)
transite board
Asche ash
aschefrei (quantitativer Filter)
ashless (quantitative filter)
Ascorbinsäure (Ascorbat)
ascorbic acid (ascorbate)
Asparagin
asparagine, aspartamic acid
Asparaginsäure (Aspartat)
asparagic acid,
aspartic acid (aspartate)
Assemblierung/Zusammenbau
assembly
Assimilat assimilate
Assimilation
assimilation, anabolism
assimilatorisch assimilatory
assimilieren assimilate
Assoziationskoeffizient
stat coefficient of association
Atem breath
atembar inhalable
Atemgifte/Fumigantien
respiratory toxin, fumigants
Atemmaske/Atemschutzmaske
protection mask, face mask, respirator mask, respirator

Atemminutenvolumen (AMV)
minute respiratory volume
Atemschutz
breathing protection, respiratory protection
Atemschutzgerät/Atemgerät
breathing apparatus, respirator
Atemschutzmaske
protection mask, face mask, respirator mask, respirator
➢ **Feinstaubmaske**
mist (respirator) mask
➢ **Filterkartusche** filter cartridge
➢ **Fluchtgerät/Selbstretter**
emergency escape mask
➢ **Grobstaubmaske**
dust mask (respirator)
➢ **Halbmaske**
half-mask (respirator)
➢ **Operationsmaske/ chirurgische Schutzmaske**
surgical mask
➢ **Partikelfilter A.**
particulate respirator
➢ **Vollmaske**
full-mask (respirator)
➢ **Vollsicht-A.**
full-facepiece respirator
Atemschutzvollmaske/ Gesichtsmaske
full-face respirator
Atemwege respiratory system
Atemwegsverätzung
respiratory tract burn, (alkali/acid) caustic burn of the respiratory tract
Atemzentrum respiratory center
Atemzugvolumen tidal volume
Äthanol/Ethanol/Äthylalkohol/ Ethylalkohol/'Alkohol'
ethanol, ethyl alcohol, alcohol
Äther/Ether ether
ätherisches Öl
ethereal oil, essential oil
Äthylen/Ethylen ethylene
atmen breathe, respire
➢ **ausatmen** breathe out, exhale
➢ **einatmen** breathe in, inhale
Atmosphäre atmosphere

Atmung breathing, respiration
- **aerobe Atmung**
  aerobic respiration
- **anaerobe Atmung**
  anaerobic respiration
- **Ausatmung/Ausatmen/ Expiration/Exhalation**
  expiration, exhalation
- **Bauchatmung/ Zwerchfellatmung**
  abdominal breathing, diaphragmatic respiration
- **Brustatmung/Thorakalatmung**
  thoracic respiration, costal breathing
- **Einatmung/Einatmen/ Inspiration/Inhalation**
  inspiration, inhalation
- **Hautatmung**
  cutaneous respiration/breathing, integumentary respiration
- **Zellatmung**
  cellular respiration

Atmungsgift respiratory poison
Atmungsquotient/ respiratorischer Quotient
  respiratory quotient
Atom-Absorptionsspektroskopie (AAS)
  atomic absorption spectroscopy (AAS)
atomar verseucht
  radioactively contaminated
atomar/Atom... atomic
Atomemissionsdetektor (AED)
  atomic emission detector (AED)
Atom-Emissionsspektroskopie (AES)
  atomic emission spectroscopy (AES)
Atom-Fluoreszenzspektroskopie (AFS)
  atomic fluorescence spectroscopy (AFS)
Atomgewicht atomic weight
Atomisator atomizer
Atomkraft/Atomenergie
  nuclear/atomic power, nuclear/atomic energy
Atommüll nuclear waste
Atomzahl atomic number
ATP (Adenosintriphosphat)
  ATP (adenosine triphosphate)

Atropin atropine
Attenuation/Abschwächung
  attenuation
attenuieren/abschwächen
  (die Virulenz vermindern: mit herabgesetzter Virulenz)
  attenuate
Attenuierung attenuation
Attraktans (*pl* **Attraktantien**)**/ Lockmittel/Lockstoff**
  attractant
Attrape/Modell/Nachbildung
  mock-up
ätzen *vb med* cauterize;
  metall/tech/micros etch
  (*siehe:* Gefrierätzen);
  chem (korrodieren) eat into, corrode
Ätzen/Ätzung (Korrosion) corrosion;
  (Ätzverfahren) *med* cauterization;
  metall/tech/micros etching (*siehe:* Gefrierätzen)
ätzend/beizend/korrosiv *chem*
  caustic, corrosive, mordant
Ätzkali/Kaliumhydroxid KOH
  caustic potash, potassium hydroxide
Ätzkalk slaked lime
Ätzmittel *metall/tech/micros* etchant;
  (Beizmittel) *chem* caustic agent
Ätznatron/Natriumhydroxid NaOH
  caustic soda, sodium hydroxide
Audit/Prüfung (Sachverständigenprüfung)
  audit
aufarbeiten *lab/biot* work up, process
Aufarbeitung *lab/biot* work up,
  working up, processing,
  down-stream processing
Aufbau (eines Experiments)
  setup
Aufbau (Struktur)
  construction, structure,
  body plan, anatomy
Aufbau *metabol* **/Synthesestoffwechsel**
  anabolism,
  synthetic reactions/metabolism
aufbauen (Experiment) setting up
  (assemble the equipment)

Au

Aufbereitung
  processing; concentration
aufbewahren
  store, keep, save, preserve
Aufbewahrung storage
aufdampfen/bedampfen *micros*
  vacuum-metallize
Auffangbecken/Auffangbehälter
  (für Chemikalien) dunk tank
Auffanggefäß receiver,
  receiving vessel, collection vessel
Aufflackern/Auflodern/Aufflammen
  flare-up
auffüllen fill up; (nachfüllen) replenish;
  (Vorräte/Lager) restock
➤ **bis zum Rand auffüllen**
  top up/off
➤ **wiederauffüllen** refill
aufgeblasen inflated
Aufguss/Infusion infusion
Aufheller/Aufhellungsmittel
  (optischer Aufheller) *chem*
  brightener, brightening agent,
  clearant,
  clearing agent (optical brightener)
aufklären (Strukturen/
  Zusammenhänge) elucidate
Aufklärung (Strukturen/
  Zusammenhänge)
  elucidation
Aufklärungshof/Lysehof/Hof/Plaque
  plaque
Aufkleber sticker; (Etikett) label
Auflage(n) *jur* legal requirements
Auflicht/Auflichtbeleuchtung
  epiillumination,
  incident illumination
auflösen *chem* dissolve;
  opt resolve
➤ **hoch aufgelöst**
  high-resolution ...
➤ **niedrig aufgelöst**
  low-resolution ...
Auflösung *chem* dissolution;
  (optische Auflösung)
  optical resolution
Auflösungsgrenze *opt*
  limit of resolution

Auflösungsvermögen *opt*
  resolving power
Aufnahme/Annahme
  acceptance; acquisition
Aufnahme/Aufschreiben/Registration
  recording, registration
Aufnahme/Bild picture, image
➤ **mikroskopische A./**
  mikroskopisches Bild *photo*
  micrograph,
  microscopic picture/image
Aufnahme/Einnahme
  uptake/intake; ingestion
Aufnahmeleistung power input
Aufnahmezeit *vir* acquisition time
aufnehmen (aufschreiben/registrieren)
  record, register; (einnehmen/zu sich
  nehmen) take up, take in; ingest
aufputzen clean up; mop up (the floor)
aufräumen clean up, tidy up
aufreinigen purify
Aufsatz (auf ein Gerät)
  attachment, fixture; cap, top
aufsaugen/absorbieren soak up,
  absorb, take up, suck up; aspirate
Aufsaugen/Absorption
  soaking up, absorption
aufsaugend absorptive
Aufsaugmittel/Absorptionsmittel
  absorbent, absorbant
aufschlämmen *chem* suspend,
  slurry (slurrying)
Aufschlämmung
  (Suspension) suspension, slurry;
  (IR/Raman) mull
aufschließen *chem*
  dissolve, disintegrate,
  decompose, break up, digest
Aufschluss *chem*
  dissolution, disintegration,
  decomposition, digestion
➤ **Zellaufschluss**
  (Öffnen der Zellmembran)
  cell lysis
➤ **Zellfraktionierung**
  cell fractionation
➤ **Zellhomogenisierung**
  cell homogenization

aufschmelzen/schmelzen melt
Aufschrift legend; (Etikett) label;
(Brief etc.) address
➢ **mit Aufschrift (Etikett)**
labeled
Aufseher/Wächter
guard, custodian
Aufsicht/Kontrolle
supervision, control
aufspalten split
➢ **segregieren** *gen* segregate
➢ **spalten/öffnen** *chem*
crack, break down, open
➢ **verteilen** distribute
➢ **zerlegen** *chem* split
Aufspaltung splitting
➢ **Öffnen** *chem* cracking, opening
➢ **Segregation** *gen* segregation
➢ **Verteilung** distribution
➢ **Zerlegen** *chem* splitting
aufsteigend afferent, rising;
(DC) ascending
Auftauen *n* thawing
auftauen *vb* thaw
Auftrag (Auftragung) application;
(Bestellung) order
auftragen
(applizieren) *chromat* apply;
('plotten') *math/geom* plot
Auftragestab/Applikator
application rod
Auftragsbestätigung
order confirmation
Auftragsforschung
contract research
Auftragung/Applikation *chromat*
application
auftrennen/trennen/fraktionieren
separate, fractionate
Auftrennung/Trennung/Fraktionierung
separation, fractionation
Auftrieb
(in Wasser) buoyancy; (in Luft) lift
Auftrittsenergie (MS)
appearance energy
aufwachsen grow up
aufwinden *vb* coil up
Aufwinden *n* coiling

aufwischen
wipe up; mop up (the floor)
Aufwuchs/Nachkommenschaft
descendants, descendents
Aufzeichnung(en) record
➢ **Verwahrung/Verwaltung von A.**
recordkeeping
Aufzug (Personen~) elevator
Augendusche
eye-wash fountain
Augenschutzbrille goggles
Ausatemventil (am Atemschutzgerät)
exhalation valve
ausäthern/ausethern
extract with ether,
shake out with ether
ausatmen *vb*
expire, exhale, breathe out
Ausatmen/Ausatmung/
Expiration/Exhalation
expiration, exhalation
ausbalancieren balance (out)
Ausbeute/Ertrag yield
ausbeuten (Rohstoffe)
exploit
Ausblaspipette
blow-out pipet
ausbleichen/bleichen bleach;
(*passiv*, z.B. Fluoreszenzfarbstoffe)
fade
Ausbleichen/Bleichen bleaching;
(*passiv*, z.B. Fluoreszenzfarbstoffe)
fading
Ausbreitung/Propagation
spreading, expansion;
propagation, dispersal,
dissemination
Ausdauer/Dauerhaftigkeit
endurance, persistence,
hardiness, perseverance
ausdauernd (widerstandsfähig)
hardy, persistent, enduring
Ausdehnung/Erweiterung
expansion
Ausdehnung/Verlängerung
extension
Ausdruck (Drucker)
printout (from a printer)

au

ausdünnen *vb* thin
Ausdünnen/Ausdünnung
 thinning
auseinandernehmen
 (Glas-/Versuchsaufbau)
 disconnect, disassemble
ausethern extract with ether,
 shake out with ether
ausfällen/fällen precipitate
Ausfällung/Ausfällen/Fällung/Fällen
 precipitation
Ausfluss (Abfluss) *tech* discharge,
 outflow, efflux, draining off;
 med discharge, secretion, flux
Ausfuhrbestimmungen
 export regulations
ausführen/wegführen/
 ableiten (Flüssigkeit)
 discharge, drain,
 lead out, lead/carry away
ausführend/wegführend/ableitend
 (Flüssigkeit) efferent
Ausführgang/Ausführkanal
 duct, passageway
Ausgabe *tech/mech/electr* output;
 (Material/Chemikalien) issue point,
 issueing, supplies issueing;
 (Auslesen: Daten) readout
Ausgang exit;
 (Fluchtweg) egress;
 electr output
Ausgangsprodukt
 primary product, initial product
Ausgangsstoff (Ausgangsmaterial)
 starting material, basic material,
 base material, source material,
 primary material, parent material,
 raw material;
 (Reaktionsteilnehmer/Reaktand)
 reactant
Ausgangsverteilung *stat*
 initial distribution
ausgasen degas
ausgesetzt sein/exponiert sein
 to be exposed (to chemicals)
Ausgesetztsein/Gefährdung
 (durch eine Chemikalie)
 exposure

ausgießen pour out, decant
Ausgießer dispenser
Ausgießhahn tap
Ausgießring pouring ring
Ausgießschnauze
 spout, nozzle, lip, pouring lip
Ausgleichsventil relief valve
 (pressure-maintaining valve)
Ausgleichszeit/
 thermisches Nachhinken (Autoklav)
 setting time
ausglühen roast, calcine; (Glas) anneal
Ausguss (Spüle) sink; (Ansatz zum
 Ausgießen einer Flüssigkeit) spout
Ausgussstutzen (Kanister)
 nozzle (attachable/detachable)
aushärten/vulkanisieren *chem*
 (Polymere) cure, vulcanize
Aushärtung *polym* curing
Aushilfe/Hilfspersonal
 temporary worker
 (aid/helper/employee/personnel)
aushungern starve
auskreuzen/herauskreuzen *gen*
 cross out
Auskreuzen/Herauskreuzen *gen*
 outcrossing
Auslauf/Austritt (Leck) leakage;
 (Zulauf von Flüssigkeit/Gas) outlet
auslaufen (Flüssigkeit)
 leak (out), bleed
Auslaufventil plug valve
auslaugen (Boden) leach
Auslaugung (Boden) leaching
Auslese/Selektion selection
auslesen select;
 (aussortieren) sort out;
 (Daten) read out
Auslieferung delivery
ausloggen log off
auslösen (z.B. eine Reaktion) trigger,
 elicitate; initiate, actuate;
 electr trip (z.B. Sicherung)
Auslöser releaser
Auslöseschwelle *med*
 trigger threshold
Auslösung (Reaktion)
 triggering, elicitation

ausmerzen/ausrotten
 eliminate, eradicate, extirpate
Ausnahme/Sonderfall
 exception, special case
Ausnahmegenehmigung/ Sondergenehmigung
 exceptional permission, special permission
Ausräucherungsmittel
 fumigant
Ausreißer *stat* outlier
ausrotten/ausmerzen
 eradicate, eliminate, extirpate
Ausrottung/Ausmerzung *med*
 (z.B. Schädlinge) eradication, elimination, extirpation
Ausrüstung
 equipment, appliances, device; accessories, fittings; outfit
Aussalzchromatographie
 salting-out chromatography
Aussalzen *n* salting out
aussalzen *vb* salt out
ausschalten
 turn off, switch off
ausscheiden *allg* secrete; (Kristalle) precipitate; (Exkrete/Exkremente) egest, excrete
Ausscheidung *allg* secretion; (Exkretion) egestion, excretion
Ausscheidungen/Exkrete/Exkremente
 excreta, excretions
Ausschluss/Exklusion exclusion
Ausschnittszeichnung
 cutaway drawing
ausschütteln shake out
Ausschüttelung shaking out
ausschütten
 pour out, empty out; (verschütten) spill
Ausschüttung
 (z.B. Hormone/Neurotransmitter) release
Ausschwingrotor *centrif*
 swing-out rotor, swinging-bucket rotor, swing-bucket rotor

Außenanlage outside facility
Außendienstmitarbeiter
 field representative, field rep
Außenelektron
 outer electron
Außengewinde
 external thread, male thread
äußerlich/von außen/extern
 external, extrinsic
außerzellulär/extrazellulär
 extracellular
**aussetzen
 (Schadstoff/Strahlung aussetzen)**
 expose to
 (hazardous chemical/radiation)
ausspülen/ ausschwenken/nachspülen
 rinse
Ausstattung provisions, furnishings, equipment, outfit, supplies
Ausstattung/Mobiliar furnishings
Aussterben
 extinction, dying out
Ausstiegsluke (Flucht)
 escape hatch
ausstöpseln unplug, disconnect
Ausstoßen/Spritzen/Extrusion
 extrusion
ausstrahlen/verströmen/ausstoßen
 emit
ausstreichen *micb*
 (z.B. Kultur) streak, smear
ausstreuen disseminate, disperse, spread, release
Ausstreuung dissemination, dispersal, spreading, releasing
Austrich *micb* smear
Austrichkultur/Abstrichkultur *micb*
 streak culture, smear culture
Ausstrom efflux
Ausströmen/Effusion (Gas)
 effusion
**Ausströmgeschwindigkeit/ Austrittsgeschwindigkeit
 (Sicherheitswerkbank)**
 exit velocity (hood)
ausstülpen evert, evaginate, protrude, turn inside out

austarieren
(Waage: Gewicht des Behälters/
Verpackung auf Null stellen)
tare (determine weight of
container/packaging as to
substract from gross weight:
set reading to zero)
Austausch exchange
austauschbar
exchangeable
Austauschbarkeit
exchangeability
Austauschreaktion
exchange reaction
Austritt exit; release
➢ **A. bei üblichem Betrieb**
incidental release
➢ **störungsbedingter A.**
(unerwartetes Entweichen von
Prozessstoffen) accidental release
Austrittsgruppe/
Abgangsgruppe/
austretende Gruppe
leaving group,
coupling-off group
Austrittspupille *micros*
exit pupil
Austrittsspalt exit slit
austrocknen/entwässern
desiccate, dry up, dry out
Austrocknung/Entwässerung
desiccation
Austrocknungsvermeidung
desiccation avoidance
auswaschen wash out, rinse out,
flush out; (eluieren) elute
Auswaschung
(feste Bodenbestandteile in
Suspension) eluviation;
(gelöste Bodenmineralien) leaching
auswechselbar exchangeable;
(gegeneinander) interchangeable;
(ersetzbar) replaceable
auswerten (z.B. von Ergebnissen)
evaluate, analyze, interpret (results)
Auswertung (z.B. von Ergebnissen)
evaluation, analysis, interpretation

auswiegen (genau wiegen)
weigh out precisely
Auswringer/Wringer (Mop)
wringer (mop)
Auszehrphase
starvation phase
Auszeit downtime
Auszubildende(r)/Azubi
occupational trainee
(professional school &
on-the-job training)
Auszug/Extrakt extract
Autokatalyse autocatalysis
Autoklav autoclave
➢ **Abkühlzeit/Fallzeit**
cool-down period,
cooling time
➢ **Anheizzeit/Steigzeit**
preheating time, rise time
➢ **Ausgleichszeit/**
thermisches Nachhinken
setting time
autoklavierbar
autoclavable
Autoklavierbeutel
autoclave bag
autoklavieren autoclave
Autoklavier-Indikatorband
autoclave tape,
autoclave indicator tape
autolog autologous
Autolyse autolysis
Autoradiographie
autoradiography,
radioautography
Auxine auxins
Axt axe
➢ **Beil** hatchet
➢ **Brandaxt** fire axe
Azelainsäure azelaic acid
azeotrop azeotropic
azeotropes Gemisch
azeotropic mixture
azid/acid/sauer acid
Azidität/Acidität/Säuregrad
acidity
Azidose/Acidose acidosis

Ba

Backenbrecher
 jaw crusher, jaw breaker
Backhefe/Bäckerhefe
 baker's yeast
Bahre/Krankenbahre/
Trage/Krankentrage
 stretcher
Bakterie/Bakterium (*pl* **Bakterien**)
 bacterium (*pl* bacteria)
➤ **Bazillen/Bacillen (Stäbchen)**
 bacilli (*sg* bacillus) (rods)
➤ **denitrifizierende Bakterien/**
 Denitrifikanten
 denitrifying bacteria
➤ **Fäulnisbakterien**
 putrefactive bacteria
➤ **Knallgasbakterien/**
 Wasserstoffbakterien
 hydrogen bacteria
 (aerobic hydrogen-oxidizing bacteria)
➤ **Knöllchenbakterien** nodule bacteria
➤ **Kokken/Coccen (kugelig)**
 cocci (*sg* coccus) (spherical forms)
➤ **Leuchtbakterien**
 luminescent bacteria
➤ **Myxobakterien/Schleimbakterien**
 myxobacteria
➤ **Rickettsien**
 (Stäbchen- oder Kugelbakterien)
 rickettsias, rickettsiae (*sg* rickettsia)
 (rod-shaped to coccoid)
➤ **Schwefelbakterien** sulfur bacteria
➤ **Spirillen (schraubig gewunden)**
 spirilla (*sg* spirillum)
 (spiraled forms)
➤ **stickstofffixierende Bakterien**
 nitrogen-fixing bacteria
➤ **Vibrionen (meist gekrümmt)**
 vibrios (mostly comma-shaped)
➤ **wärmesuchende**
 Bakterien/thermophile Bakterien
 thermophilic bacteria
bakteriell bacterial
bakterielle Infektion bacterial infection
Bakterienflora bacterial flora
bakterienfressend bacterivorous,
 bactivorous
Bakterienkultur bacterial culture
Bakterienrasen bacterial lawn
Bakteriologie bacteriology
bakteriologisch
 bacteriologic, bacteriological
Bakteriophage/Phage
 bacteriophage, phage, bacterial virus
Bakteriose bacteriosis
bakterizid/keimtötend
 bacteriocidal, bactericidal
Ballastgruppe (*chem* **Synthese**)
 ballast group
Ballon/Ballonflasche
 (für Flüssigkeiten)
 carboy; (mit Ablaufhahn)
 bottle with faucet
 (carboy with spigot)
Bananenstecker *electr* banana plug
Band (Klebeband etc.) tape
➤ **Absperrband/Markierband**
 barricade tape
➤ **Autoklavier-Indikatorband**
 autoclave tape,
 autoclave indicator tape
➤ **Dichtungsband** sealing tape
➤ **Filamentband** filament tape
➤ **Gewindeabdichtungsband**
 thread seal tape
➤ **Klebeband**
 adhesive tape
➤ **Signalband/Warnband**
 warning tape
➤ **Teflonband** Teflon tape
Bandbreite *phys* bandwidth
Bande *electrophor/chromat* band
➤ **Hauptbande** main band
➤ **Satellitenbande** satellite band
Bandenverbreiterung *chromat*
 band broadening
Bänderungsmuster/Bandenmuster
 (von Chromosomen)
 banding pattern
Bänderungstechnik
 banding technique
Bandmaß/Messband
 tape rule, tape measure
Bank/Bibliothek/Klonbank
 library, bank (clone bank)
Bart (eines Schlüssels) bit (of a key)

Ba

Bartbildung/Signalvorlauf/
Bandenvorlauf *chromat*
  fronting, bearding
Base base
> **stickstoffhaltige Base**
  (Purine/Pyrimidine)
  nitrogenous base
Baseität/Basizität basicity
Basenanhydrid basic anhydride
Basenpaar *gen* base pair
basisch/alkalisch basic, alkaline
Basischemikalien/Grundchemikalien
  base chemicals (general reactants)
Basiseinheit base unit
Basisnährboden basal medium
Basispeak (MS) base peak
Basizität/Baseität basicity
Bauaufsichtsbehörde
  building supervisory board
Bausch/Wattebausch/
  Tupfer/Tampon
  pad, swab (cotton),
  pledget (cotton), tampon
Baustein/Bauelement
  building block, unit
Baustelle
  building site, construction site
Bauunternehmen building contractor,
  construction firm
Bauvorschriften building code,
  building regulations
Bauwerk, building, edifice, structure,
  construction
bazillär/Bazillen.../bazillenförmig/
  stäbchenförmig
  bacillary
Beamter/Beamtin
  public service officer,
  civil servant (*Br*)
beatmen apply artificial respiration
Beatmung (künstliche) artificial
  respiration
Beatmungsgerät respirator
bebrüten/brüten/inkubieren
  brood, breed, incubate
Bebrütung/Bebrüten/Inkubation
  breeding, incubation
Becher cup; *centrif* bucket

Becherglas/Zylinderglas
  (ohne Griff) beaker;
  (mit Griff/Krug) pitcher
Becherglaszange beaker tongs
Bedampfung/Bedampfen/Aufdampfen
  micros vapor blasting
Bedampfungsanlage *micros*
  vaporization apparatus
bedienen *tech/mech*
  operate, handle, work
Bedienfeld control panel
Bedienknopf/Drucktaste
  push button
Bediensteter/Angestellter
  employee;
  (staatl. B.) civil servant,
  public service officer)
Bedienung *tech/mech*
  operation; handling
Bedienungsanleitung/
  Gebrauchsanleitung (Handbuch)
  operating instructions (manual)
Bedienungspersonal
  (Arbeiter/Handwerker/Mechaniker)
  operations personnel
Bedrohung
  threat; endangerment
Befall (Schädlingsbefall)
  infestation (with pests/parasites)
> **Wiederbefall** reinfestation
befallen (Schädlingsbefall)
  infest (pests/parasites)
Befallsrate
  degree/level/rate of infestation
befeuchten
  moisten, humidify, dampen
Befeuchter damper
Befeuchtung moistening,
  humidification, dampening
Beförderung/Transport
  transport, shipment
Befund findings, result
begasen fumigate
Begasung fumigation
Begehung/Besichtigung
  (z.B. Geländebegehung)
  inspection (on-site inspection);
  (zur Abnahme) commissioning

Beglaubigung (amtlich)
(Zertifizierung) certification;
(Zertifikat) certificate
Begleitprodukt side product
begrenzender Faktor/
limitierender Faktor/
Grenzfaktor
limiting factor
Begrenzungsventil limit valve
begutachten
give an expert opinion;
review, examine, inspect, study
Begutachter expert
Begutachtung expert opinion;
examination, inspection
Begutachtungsverfahren
(wissenschaftl. Manuskripte)
peer review
Behälter/Behältnis
container (large), receptacle (small)
behandeln (behandelt)
treat (treated)
➢ **unbehandelt** untreated
behindert med handicapped
➢ **körperbehindert**
physically handicapped
Behinderung
(Hindernis) obstacle;
med handicap
Behörde agency
Beil hatchet
beimpfen/inokulieren inoculate
Beimpfung/Inokulation inoculation
Beimpfungsverfahren
inoculation method,
inoculation technique
➢ **Plattenausstrichmethode**
streak-plate method/technique
➢ **Plattengussverfahren/**
Gussplattenmethode
pour-plate method/technique
➢ **Spatelplattenverfahren**
spread-plate method/technique
Beipackzettel (package)
insert/leaflet/slip
beißend (Geruch/Geschmack) sharp,
pungent, acrid
Beißzange/Kneifzange pliers

Beitel/Stechbeitel chisel
Beize/Beizenfärbungsmittel mordant
beizen (Saatgut) dress (coat/treat with
fungicides/pesticides); (Holz) stain
Beizmittel (zur Saatgutbehandlung)
fungicide treatment,
pesticide treatment (of seeds)
Bekleidung/Kleidung
clothing, apparel
belastbar
strong, durable; loadable
belasten (belastet/verschmutzt)
contaminate(d)
Belastung
(Traglast/Last: Gewicht) weight;
(Verschmutzung) contamination
Belastungsfähigkeit/Grenze der
ökologischen Belastbarkeit/
Kapazitätsgrenze/Umweltkapazität
carrying capacity
Belastungsgrenze exposure limit
➢ **zulässige/erlaubte B.**
permissible exposure limit (PEL)
Belastungsursache strain
Belastungszustand stress
beleben (belebt) animate(d)
➢ **unbelebt**
inaminate(d), lifeless, nonliving
➢ **wiederbeleben (wiederbelebt)**
reaminate(d)
Belegexemplar voucher specimen
Belegschaft staff, employees,
personnel; *allg* force, labour force
Belehrung instruction, advice
beleuchten illuminate
Beleuchtung illumination
➢ **Auflicht/Auflichtbeleuchtung**
epiillumination,
incident illumination
➢ **Durchlicht/Durchlichtbeleuchtung**
transillumination,
transmitted light illumination
➢ **Kaltlichtbeleuchtung**
fiber optic illumination
➢ **Köhlersche Beleuchtung**
Koehler illumination
➢ **künstliche Beleuchtung**
artificial light(ing)

Be

Beleuchtungsstärke illuminance
belichten (z.B. Film/Pflanzen)
 expose
Belichtung (z.B. Film/Pflanzen)
 exposure (to light)
belüften aerate
Belüftung aeration
Benachrichtigung/Inkenntnissetzung
 notification
**Benennung/Bezeichnung/
 Namensgebung**
 naming, designation,
 nomenclature
benetzen wet; moisten
benigne/gutartig benign
Benignität/Gutartigkeit
 benignity, benign nature
Benutzer/Nutzer user
Benzin gasoline, gas, petrol (*Br*)
Benzinkanister/Kraftstoffkanister
 gasoline canister
Benzoesäure (Benzoat)
 benzoic acid (benzoate)
Benzol benzene
Beobachtungsfenster
 viewing panel
berechnen calculate
Berechnung calculation
beregnen/bewässern (künstlich)
 irrigate; (besprühen) sprinkle, spray
Beregnung/Bewässerung irrigation
**Beregnungsanlage/
 Berieselungsanlage/Sprinkler**
 sprinkler, sprinkler irrigation system
bereinigen
 clarify, clear, straighten out; adjust
Bereinigung *math/stat* adjustment
Bereitschaft
 (Gerät) standby; (Dienst) duty
Bereitschaftsstellung/Wartebetrieb
 standby mode
Bericht report
berichten report
berieseln sprinkle, spray; irrigate
Berieselung sprinkle irrigation
Berlsattel (Füllkörper) *dest*
 berl saddle (column packing)
Berlese-Apparat *ecol* Berlese funnel

Bernstein amber
Bernsteinsäure (Succinat)
 succinic acid (succinate)
**Berstscheibe/Sprengscheibe/
 Sprengring/Bruchplatte**
 bursting disk
Beruf profession;
 (Beschäftigung) occupation;
 (Arbeit/Arbeitsstelle/Job) work, job
Berufseignungstest
 vocational aptitude test
Berufsgenossenschaft
 trade cooperative association
Berufskrankheit occupational disease
Berufsrisiko occupational hazard
Berufsunfähigkeit
 working disability, disablement
Berufsverband
 professional association
 (organization)
Berufsverletzung
 occupational injury
berühren
 touch, contact; boarder
**Berührung/Kontakt
 (z.B. mit Chemikalien)**
 contact, exposure
Besatzdichte stocking density
Beschaffenheit
 (Konsistenz) consistency;
 (Zustand) state, condition;
 (Struktur) structure, constitution;
 (Eigenschaft) quality, property;
 (Art) nature, character
Beschaffung
 procuring, procurement, supply;
 (Erwerb) acquisition;
 (Kauf) purchase
beschallen/mit Schallwellen behandeln
 sonicate
Beschallung/Sonifikation/Sonikation
 sonication
Beschattung *allg* shading;
 (Schrägbedampfung bei TEM)
 shadowcasting
 (rotary shadowing in TEM)
 ➤ **Metallbeschattung**
 metallizing

Bescheinigung certification
beschichten
 line, coat, cover, laminate
Beschichtung
 lining, coat, coating,
 covering, lamination
beschicken *micb* charge, feed
Beschickungsstutzen (Kolben)
 delivery tube (flask)
Beschleunigung acceleration
Beschleunigungsphase/Anfahrphase
 acceleration phase
Beschleunigungsspannung (EM)
 micros accelerating voltage
Beschreibung description
 ➢ **technische B.** specifications, specs
beschriften mark, label
Beschriftung
 mark, label, caption, legend
Beschriftungsetikett label
Beschuss mit schnellen Atomen (MS)
 spectr fast-atom bombardment (FAB)
beseitigen/entfernen remove
Beseitigung/Entfernung removal
Besen/Kehrbesen broom
besiedeln (etablieren) settle, establish;
 micb (kolonisieren) colonize
Besiedlung (Etablierung)
 settlement, establishment;
 micb (Kolonisation/Kolonisierung)
 colonization
besprengen sprinkle
besputtern (EM) *micros* sputter
Besputtern/
 Kathodenzerstäubung (EM)
 micros sputtering
Besputterungsanlage (EM) *micros*
 sputtering unit/appliance
Bestand (Menge/Quantität) stock,
 number, quantity; stand;
 (Bevölkerung) population
beständig/resistent resistant
Beständigkeit/Resistenz resistance
Bestandsaufnahme
 (to make an) inventory
Bestandsdichte/Populationsdichte
 population density
Bestandteil component

Bestätigung/Vergewisserung
 verification
Bestätigungsprüfung
 verification assay
bestehend/existierend
 existing, existant;
 (bestehend aus) consisting of
bestellen order
Bestellung order
bestimmen *chem* determine, identify;
 (Pflanzen/Tiere) identify
Bestimmung/
 Determinierung/Determination
 identification; determination
Bestimmungen *jur* provisions
Bestimmungsbuch manual
Bestimmungsgrenze
 limit of detection
Bestimmungsschlüssel key
bestrahlen irradiate; expose
Bestrahlung irradiation; exposure
Bestrahlungsdosis
 radiation dosage, irradiation dosage
Bestrahlungsintensität/
 Bestrahlungsdichte
 irradiance, fluence rate, radiation
 intensity, radiant-flux density
betäuben/narkotisieren/anästhesieren
 stupefy, narcotize, anesthetize
betäubend/narkotisch/anästhetisch
 stupefacient, stupefying, narcotic,
 anesthetic
Betäubung/Narkose/Anästhesie
 stupefaction, narcosis, anesthesia
Betäubungsmittel/Narkosemittel/
 Anästhetikum stupefacient, narcotic,
 narcotizing agent, anesthetic,
 anesthetic agent
Betrieb/Unternehmen business,
 company, firm, enterprise
Betriebsanleitung
 operating instructions;
 (Handbuch) manual
Betriebsarzt company doctor
Betriebsdruck operating pressure
Betriebserlaubnis
 operational permission
Betriebsführung management

Be

Betriebsgeheimnis/ Geschäftsgeheimnis trade secret
Betriebsleiter operations manager, plant manager
Betriebssanitäter (company) nurse
Betriebssicherheit safety of operation
Betriebsstoffwechsel maintenance metabolism
Betriebsunfall industrial accident, accident at work
Betriebsvorschrift operating instructions
Betriebswasser/Brauchwasser (nicht trinkbares Wasser) process water, service water, industrial water (nondrinkable water)
Betrug/Schwindel/arglistige Täuschung fraud
Beugung *phys/opt* diffraction
Beugungsmuster diffraction pattern
Bevölkerung/Population population
bewachen guard
bewahren/erhalten/preservieren preserve, keep, maintain
Bewahrung/Erhaltung/Preservierung preservation
bewässern irrigate
Bewässerung irrigation
bewegen move
Bewegung motion; (Fortbewegung/Lokomotion) movement, motion, locomotion
➢ **Drehbewegung (rotierend)** spinning/rotating motion
➢ **Handbewegung** hand motion (handshaking motion)
➢ **kreisförmig-vibrierende Bewegung** vortex motion, whirlpool motion
➢ **Rüttelbewegung (hin und her/rauf und runter)** rocking motion (side-to-side/up-down)
➢ **Taumelbewegung, dreidimensionale** nutation, gyroscopic motion (threedimensional orbital & rocking motion)
➢ **Vibrationsbewegung** vibrating motion
➢ **Wippbewegung** see-saw motion, rocking motion
Bewegungsmelder/Bewegungssensor motion sensor, movement detector
Bewertung rating, evaluation; (Beurteilung) judgement; (Erfassung) assessment
Bewuchs growth, cover, stand
bewusst *psych* conscious
➢ **unbewusst** unconscious, unknowing(ly)
Bewusstheit awareness
Bewusstsein consciousness
➢ **Bewusstlosigkeit** unconsciousness
Bezettelung badging
Bezugselektrode reference electrode
Bezugstemperatur reference temperature
Bezugswert reference value
Bibliothek/Bank (Klonbank) library, bank (clone bank)
➢ **Bereichsbibliothek** unit library
➢ **Expressionsbibliothek** expression library
➢ **Institutsbibliothek** departmental library
Bibliothekar(in) librarian
biegsam flexible, pliable
Biegsamkeit flexibility, pliability; stiffness
Bienenwachs beeswax
Bilanz (Energiebilanz/Stoffwechselbilanz) balance
Bild picture, image
➢ **elektronenmikroskopisches Bild/ elektronenmikroskopische Aufnahme** electron micrograph
➢ **Endbild** *micros* final image
➢ **mikroskopisches Bild/ mikroskopische Aufnahme** microscopic image, microscopic picture, micrograph
➢ **reelles Bild** *micros* real image
➢ **virtuelles Bild** *micros* virtual image

Bilddiagramm/Begriffszeichen
pictograph (for hazard labels)
bilden (entwickeln) (z.B. Gase/Dämpfe)
generate (develop)
Bildpunkt *opt* image point;
(Rasterpunkt) pixel
Bildschirm/Monitor
display, monitor
Bildungswärme
heat of formation
Bimetallthermometer
bimetallic thermometer
bimodale Verteilung
bimodal distribution,
two-mode distribution
Bims pumice
Bimsstein pumice rock
Binde/Aderpresse/
 Abschnürbinde/Tourniquet
tourniquet
Bindefähigkeit
bonding strength
Bindemittel/Saugmaterial
 (saugfähiger Stoff)
binder, binding agent,
absorbent, absorbing agent
binden *chem* bond, link
binden/anbinden/
 zusammenbinden tether
Bindevlies strapping fabric
Bindung *chem*
bond, linkage
➢ **Atombindung** atomic bond
➢ **chemische Bindung** chemical bond
➢ **Disulfidbindung (Disulfidbrücke)**
disulfide bond, disulfide bridge
➢ **Doppelbindung** double bond
➢ **Dreifachbindung** triple bond
➢ **energiereiche Bindung**
high energy bond
➢ **glykosidische Bindung**
glycosidic bond/linkage
➢ **heteropolare Bindung**
heteropolar bond
➢ **homopolare Bindung**
homopolar bond, nonpolar bond
➢ **hydrophile Bindung**
hydrophilic bond
➢ **hydrophobe Bindung**
hydrophobic bond
➢ **Ionenbindung** ionic bond
➢ **Kohlenstoffbindung** carbon bond
➢ **konjugierte Bindung**
conjugated bond
➢ **kooperative Bindung**
cooperative binding
➢ **kovalente Bindung** covalent bond
➢ **Mehrfachbindung** multiple bond
➢ **Peptidbindung**
peptide bond, peptide linkage
Bindungsenergie
binding energy, bond energy
Bindungskurve binding curve
Bindungswinkel bond angle
Binokular binoculars
Binomialverteilung
binomial distribution
binomische Formel
binomial formula
bioanorganisch bioinorganic
Bioäquivalenz bioequivalence
Biochemie biochemistry
biochemischer Sauerstoffbedarf/
 biologischer Sauerstoffbedarf (BSB)
biochemical oxygen demand,
biological oxygen demand (BOD)
Biodegradation/biologischer Abbau
biodegradation
Bioenergetik bioenergetics
Bioethik bioethics
Biogefährdung biohazard
biogen biogenic
Bioindikator/Indikatorart/
 Zeigerart/Indikatororganismus
bioindicator, indicator species
Biolistik biolistics,
microprojectile bombardment
Biologe/Biologin biologist,
bioscientist, life scientist
Biologie/Biowissenschaften
biology, bioscience, life sciences
biologisch abbaubar biodegradable
biologisch/biotisch
biologic(al), biotic
biologische Abbaubarkeit
biodegradability

biologische Sicherheit(smaßnahmen)
biological containment
biologische Verfahrenstechnik/ Biotechnik/Bioingenieurwesen
bioengineering
biologischer Abbau/Biodegradation
biodegradation
biologischer Kampfstoff
biological warfare agent
biologischer Sauerstoffbedarf/ biochemischer Sauerstoffbedarf (BSB) biological oxygen demand, biochemical oxygen demand (BOD)
biologischer Test
bioassay, biological assay
biologisches Gleichgewicht
biological equilibrium
Biolumineszenz bioluminescence
Biomasse biomass
Bionik bionics
Biophysik biophysics
Bioreaktor (Reaktortypen *siehe:* **Reaktor)** bioreactor
Biorhythmus biorhythm
Biostatik biostatics
Biostatistik biostatistics
Biosynthese biosynthesis
Biosynthesereaktion
biosynthetic reaction (anabolic reaction)
biosynthetisch biosynthetic(al)
biosynthetisieren biosynthesize
Biotechnik/ biologische Verfahrenstechnik/ Bioingenieurwesen
bioengineering
Biotechnologie biotechnology
Biotransformation/Biokonversion
biotransformation, bioconversion
Bioverfügbarkeit bioavailability
Biowissenschaft bioscience (meist *pl* biosciences), life science (meist *pl* life sciences)
Biozid biocide
Biozön/Biozönose/Biocönose/ Lebensgemeinschaft/ Organismengemeinschaft
biocenosis, biotic community

Birnenkolben/Kjeldahl-Kolben
Kjeldahl flask
bitter bitter
Bitterkeit bitterness
Bittermandelöl
bitter almond oil
Bitterstoffe bitters
bivalent bivalent
blähen bloat
Blähschlamm bulking sludge
Blähungen/Flatulenz bloating, gas
Bläschen/Vesikel bubble, vesicle
bläschenförmig
bubble-shaped, bulliform
Blase *med* bladder
Blase (Gasblase/Luftblase/Seifenblase)
bubble
Blase/Destillierrundkolben
still pot, distilling boiler flask
blasenartig/blasenförmig
bladderlike, bladdery, vesicular
Blasen-Linker-PCR *gen*
bubble linker PCR
Blasensäulen-Reaktor
bubble column reactor
blasentreibend/blasenziehend
vesicating, vesicant
Blasenzähler
bubble counter, bubbler, gas bubbler
blasig
bullous, with blisters, vesiculate
Blattgold gold foil, gold leaf
Blausäure/Cyanwasserstoff
hydrogen cyanide, hydrocyanic acid, prussic acid
Blech sheet metal
Blechschere
sheet-metal shears, plate shears
Blei (Pb) lead
Bleiblock *rad* pig (outermost container of lead for radioactive materials)
bleich/blass pale
Bleiche/Blässe/bleiche Farbe
paleness
Bleiche/Bleichmittel bleach
bleichen/ausbleichen (*aktiv:* **weiß machen/aufhellen)** bleach

Bleicitrat (EM) lead citrate
Bleiglanz galena
Bleioxid PbO
  litharge, massicot,
  lead protooxide, lead oxide
  (yellow monoxide)
Bleioxid PbO$_2$
  lead dioxide,
  brown lead oxide,
  lead superoxide
Bleioxid Pb$_2$O
  lead oxide (yellow),
  lead suboxide
Bleioxid Pb$_3$O$_4$
  red lead oxide, red lead
Bleiring (Gewichtsring/ Stabilisierungsring/ Beschwerungsring)
  lead ring (for Erlenmeyer)
Bleistiftmarkierung
  pencil marking
Blende *opt/micros*
  (Öffnung/Apertur) aperture;
  micros (Diaphragma) diaphragm
Blendenöffnung *micros*
  diaphragm aperture
Blickfeld/Sehfeld/Gesichtsfeld
  field of view, scope of view,
  field of vision, range of vision,
  visual field
blind blind
Blindheit blindness
Blindniete blind rivet
Blindnietmutter blind rivet nut
Blindwert blank
Blindwiderstand
  reactance, relative impedance
Blitz flash (light/lightning/spark)
Blitzchromatographie/ Flash-Chromatographie
  flash-chromatography
blitzen flash
Blitzlicht flash, flashlight
Blitzlichtphotolyse
  flash photolysis
Blockhalter *micros* block holder
Blockierungsreagenz
  blocking reagent

Blockverfahren block synthesis
Blothybridisierung
  blot hybridization
blotten (klecksen/Flecken machen/ beflecken) blot
Blotten/Blotting
  blotting, blot transfer
  ➢ **Affinitäts-Blotting**
    affinity blotting
  ➢ **Alkali-Blotting**
    alkali blotting
  ➢ **Diffusionsblotting**
    capillary blotting
  ➢ **genomisches Blotting**
    genomic blotting
  ➢ **Liganden-Blotting**
    ligand blotting
  ➢ **Nassblotten** wet blotting
  ➢ **Trockenblotten** dry blotting
Blotting-Elektrophorese/ Direkttransfer-Elektrophorese
  direct blotting electrophoresis,
  direct transfer electrophoresis
blühen flower, bloom
Blut blood
  ➢ **Frischblut** fresh blood
  ➢ **Serum** (*pl* **Seren**)
    serum (*pl* sera or serums)
  ➢ **Vollblut** whole blood
Blutagar blood agar
Blutausstrich *micros* blood smear
Blutbank blood bank
Blutbild/Blutstatus/Hämatogramm
  blood count, hematogram
Blutdruck blood pressure
Bluten *n* bleeding
bluten *vb* bleed
Bluterguss/Hämatom
  bruise, hematoma
Blut-Ersatz blood substitute
Blutgerinnung blood clotting
Blutgruppe blood group
Blutgruppenbestimmung
  blood-typing
Blutgruppenunverträglichkeit
  blood group incompatibility
Blutkonserve
  stored blood, banked blood

Blutkörperchen
 blood cell,
 blood corpuscle, blood corpuscule
Blutkultur blood culture
Blutspende blood donation
Blutsperre
 arrest of blood supply
blutstillend (adstringent)
 styptic, hemostatic (astringent)
Blutvergiftung/Sepsis
 blood poisoning
Blutzellzahlbestimmung/
Blutkörperchenzählung
 blood count
blutzersetzend/hämorrhagisch
 hemorrhagic
Blutzucker blood sugar
Boden *dest/chromat* plate;
 (Erdboden) soil, ground, earth
➤ **Bodenhöhe** plate height
➤ **theoretische Böden**
 theoretical plates
Bodenabfluss/Bodenablauf
 floor drain
Bodenbestandteile soil components
Bodenkolonne *dest* plate column
Bodenkörper *chem* bottoms, deposit
 (sediment, precipitate, settlings)
Bodenpartikelgrößen soil texture
Bodenskelett
 soil skeleton (inert quartz fraction)
Bodenversalzung soil salinization
Bodenwirkungsgrad *dest*
 plate efficiency
Bodenzahl *dest/chromat*
 number of plates,
 plate number
Bogenflamme arc flame
Bogenlampe arc lamp
Bogensäge coping saw
bohren drill
Bohrer/Bohrspitze/Bohraufsatz
 bit, drill bit, drill
 (on a dental drill: bur)
➤ **Betonbohrer** concrete drill (bit)
➤ **Holzbohrer** wood drill (bit)
➤ **Metalbohrer** metal drill (bit)
➤ **Steinbohrer** rock drill (bit)

Bohrfutter drill chuck
Bohrkern/Kern *geol/paleo*
 drill core, core
Bohrmaschine drill
Bohrung (Prozess/Vorgang) drill,
 drilling, bore;
 (Ergebnis: Loch etc.) bore
Bolzenschneider bolt cutter
Bombenkalorimeter
 bomb calorimeter
Bombenrohr/
Schießrohr/Einschlussrohr
 bomb tube, Carius tube, sealing tube
Bonitur *stat* notation, scoring
Bor (B) boron
Borax/Natriumtetraborat Decahydrat
 borax, sodium tetraborate
Bördelflansch lap-joint flange
Bördelkappe (für Rollrandgläschen/
Rollrandflasche) crimp seal
Bördelkappen-Verschließzange
 cap crimper
bördeln
 bead, flange, seam, edge; crimp
Bördelrand bead, beaded rim, flange;
 (Reagenzglas/Kolben)
 deburred edge, beaded rim
Bördelzange crimping pliers
Borosilikatglas borosilicate glass
Borste bristle
bösartig/maligne malignant
Bösartigkeit/Malignität
 malignancy
Bottich vat, tub; washtub
Brackwasser
 brackish water (somewhat salty)
Brand fire, blaze; burning
Brandarten fire classification
Brandaxt fire axe
Brandbekämpfung
 fire fighting
Brandgase combustion gases
Brandgefahr
 fire risk, fire hazard
Brandgeruch burnt smell
Brandherd source of fire
Brandmauer fire wall
Brandrisiko fire hazard

Brandschutz/Brandverhütung
fire protection, fire prevention;
fire control
Brandverletzung/
Brandwunde/Verbrennung
burn, burn wound
Branntkalk caustic lime (CaO)
Brauchwasser
(**nicht trinkbares Wasser**)
process water, service water,
industrial water
(nondrinkable water)
Braunglas amber glass
Braunstein/Manganoxid
manganese dioxide
Brecheisen crowbar, jimmy
brechen/erbrechen (bei Übelkeit)
vomit
Brechung, optische/Refraktion
optical refraction
Brechungsindex/
Brechungskoeffizient/Brechzahl
refractive index, index of refraction
Brechungsvermögen refractivity
Brechungswinkel refracting angle
Brechzentrum vomiting center
Breitspektrumantibiotikum
broad-spectrum antibiotic
Brennäquivalent
fuel equivalence
brennbar
combustible, flammable
➢ **nicht brennbar**
noncombustible, nonflammable
Brennbarkeit
combustibility, flammability
Brennebene focal plane
brennen burn
➢ **anbrennen/entzünden/entflammen**
chem inflame, ignite
➢ **durchbrennen** burn through/out
➢ **rasch abbrennen (lassen)** deflagrate
➢ **verbrennen** combust, incinerate,
burn
Brenner/Flamme (Ofen)
burner, flame (oven)
➢ **Bunsenbrenner**
Bunsen burner, flame burner

➢ **Gasbrenner** gas burner
➢ **Kartuschenbrenner**
cartridge burner
➢ **Schwalbenschwanzbrenner/**
Schlitzaufsatz für Brenner
wing-tip (for burner),
burner wing top
➢ **Spiritusbrenner/Spirituslampe**
alcohol burner
➢ **Verdunstungsbrenner**
evaporation burner
Brennereihefe distiller's yeast
Brennpunkt focal point, focus
Brennweite focal length
Brennwert caloric value;
heat value, heating value
Brennwertbestimmung/Kalorimetrie
calorimetry
Brenztraubensäure (Pyruvat)
pyruvic acid (pyruvate)
Brillenträgerokular *micros*
spectacle eyepiece,
high-eyepoint ocular
Brilliantrot *micros* vital red
brodeln bubble;
(Wasser: kochen) boil;
(Wasser: sieden/leicht kochen)
simmer
Brom (Br) bromine
Broschüre/Informationsschrift
brochure, pamphlet
Bruch breakage
Bruchglas cullet, glass cullet
bruchsicher nonbreakable,
unbreakable, crashproof
Bruchstelle *gen* breakpoint
Bruchstück/Fragment fragment
Bruchstückion fragment ion
Brüden exhaust vapor,
fuel-laden vapor
Brunnenwasser well water
Brutdauer/Inkubationszeit
breeding period,
incubation period
brüten brood, breed, incubate
Brutraum
incubation room
Brutschrank incubator

BT

BTA
(biologisch-technischer Assistent)
biology lab technician,
biological lab assistant
Bücherwagen book cart
Büchner-Trichter (Schlitzsiebnutsche)
Buechner funnel
Buchse bushing
Bügel/U-Klammer/Gabelkopf
clevis bracket
Bügelmessschraube
outside micrometer
Bügelsäge hacksaw
Bügelschaft rod clevis
Bulkladung (Transport) bulk cargo
Bundesgesundheitsamt
German Federal Health Agency
Bunsenbrenner
Bunsen burner, flame burner
Bunsenstativ/Stativ
support stand, ring stand,
retort stand, stand
Bürette buret, burette (*Br*)
➢ **Wägebürette**
weight buret, weighing buret

Büro office
Bürobedarf office supplies
Büroklammer paper clip
Bürste brush
➢ **Becherglasbürste**
beaker brush
➢ **Drahtbürste** wire brush
➢ **Flaschenbürste** bottle brush
➢ **Kolbenbürste** flask brush
➢ **Laborbürste** laboratory brush
➢ **Malpinsel** paint brush
➢ **Pfeifenreiniger/Pfeifenputzer**
pipe cleaner
➢ **Pipettenbürste**
pipet brush
➢ **Reagenzglasbürste**
test-tube brush
➢ **Scheuerbürste/Schrubbbürste**
scrubbing brush, scrub brush
➢ **Spülbürste** dishwashing brush
➢ **Stahlbürste** wire brush
➢ **Trichterbürste** funnel brush
Bußgeld fine
Buttersäure/Butansäure (Butyrat)
butyric acid, butanoic acid (butyrate)

Cadmium (Cd) cadmium
Callus-Kultur/Kallus-Kultur
   callus culture
Caprinsäure/Decansäure (Caprinat/Decanat)
   capric acid, decanoic acid (caprate/decanoate)
Capronsäure/Hexansäure (Capronat/Hexanat)
   caproic acid, capronic acid, hexanoic acid (caproate/hexanoate)
Caprylsäure/Octansäure (Caprylat/Octanat)
   caprylic acid, octanoic acid (caprylate/octanoate)
Carbonsäuren/Karbonsäuren (Carbonate/Karbonate)
   carboxylic acids (carbonates)
Carrageen/Carrageenan carrageenan, carrageenin (*Irish moss* extract)
Cäsium (Cs) cesium
Cäsiumchloridgradient
   cesium chloride gradient
Catenan/Concatenat
   catenane, concatenate
Catenation/Ringbildung catenation
CBA-Papier CBA-paper
   (cyanogen bromide activated paper)
Cerotinsäure/Hexacosansäure
   cerotic acid, hexacosanoic acid
chaotrope Reihe chaotropic series
chaotrope Substanz chaotropic agent
Charge (in einem Arbeitsgang erzeugt) batch; (Produktionsmenge/-einheit) lot, unit
Chargen-Bezeichnung (Chargen-B.)
   batch number; lot number, unit number
Chelat/Komplex chelate
Chelatbildner/Komplexbildner
   chelating agent, chelator
Chelatbildung/Komplexbildung
   chelation, chelate formation
Chemie chemistry
 ➢ **Allgemeine Chemie**
   general chemistry
 ➢ **Analytische Chemie**
   analytical chemistry
 ➢ **Angewandte Chemie**
   applied chemistry
 ➢ **Anorganische Chemie**
   inorganic chemistry
 ➢ **Biochemie** biochemistry
 ➢ **Lebensmittelchemie**
   food chemistry
 ➢ **Organische Chemie**
   organic chemistry
 ➢ **Physikalische Chemie**
   physical chemistry
Chemieabfälle chemical waste
Chemiearbeiter chemical worker
Chemiefachverband chemical society
Chemiefaser artificial fiber, polyfiber
Chemieingenieur chemical engineer
Chemielaborant
   chemical lab assistant
Chemieunfall chemical accident
Chemikalie(n) chemical(s)
Chemikalienabzug
   chemical fume hood, 'hood'
Chemikalienausgabe
   chemical stockroom counter
chemikalienfest
   chemical-resistant
Chemikalienschrank
   chemical cabinet,
   chemical safety cabinet
Chemikant (chem. Facharbeiter)
   chemical worker (industry)
Chemiosmose chemiosmosis
chemiosmotische Hypothese/Theorie
   chemiosmotic hypothesis/theory
chemische Bindung chemical bond
chemische Gleichung
   chemical equation
chemischer Kampfstoff
   chemical warfare agent
chemischer Sauerstoffbedarf (CSB)
   chemical oxygen demand (COD)
Chemisorption/chemische Adsorption
   chemisorption
Chemoaffinitäts-Hypothese
   chemoaffinity hypothesis
Chemostat chemostat
Chemosynthese chemosynthesis
Chemotherapie chemotherapy

Ch

Chinasäure chinic acid, kinic acid, quinic acid (quinate)
Chinolsäure chinolic acid
chiral chiral
Chiralität chirality
Chlor (Cl) chlorine
Chlorbenzol chlorobenzene
Chlorbleiche chlorine bleach
chlorieren chlorinate
Chlorierung chlorination
chlorige Säure chlorous acid
Chloroform/Trichlormethan chloroform, trichloromethane
Chlorogensäure chlorogenic acid
Chlorophyll chlorophyll
Chlorsäure $HClO_3$ chloric acid
Cholesterin/Cholesterol cholesterol
Cholsäure (Cholat) cholic acid (cholate)
Chorisminsäure (Chorismat) chorismic acid (chorismate)
Chrom (Cr) chromium
chromaffin chromaffin, chromaffine, chromaffinic
Chromatogramm chromatogram
Chromatograph chromatograph
Chromatographie chromatography
- **Affinitätschromatographie** affinity chromatography
- **Aussalzchromatographie** salting-out chromatography
- **Ausschlusschromatographie/ Größenausschlusschromatographie** size exclusion chromatography (SEC)
- **Blitzchromatographie/Flash-Ch.** flash-chromatography
- **Dünnschichtchromatographie (DC)** thin-layer chromatography (TLC)
- **Elektrochromatographie (EC)** electrochromatography (EC)
- **enantioselektive Chromatographie** chiral chromatography
- **Festphasenchromatographie** bonded-phase chromatography
- **Flüssigkeitschromatographie** liquid chromatography (LC)
- **Gaschromatographie** gas chromatography
- **Gas-Flüssig-Chromatographie** gas-liquid chromatography
- **Gelpermeationschromatographie/ Molekularsiebchromatographie** gel permeation chromatography (GPC), molecular sieving chromatography
- **Größenausschlusschromatographie/ Ausschlusschromatographie** size exclusion chromatography (SEC)
- **Hochdruckflüssigkeitschromatographie/ Hochleistungsflüssigkeitschromatographie** high-pressure liquid chromatography, high-performance liquid chromatography (HPLC)
- **Immunaffinitätschromatographie** immunoaffinity chromatography
- **Ionenaustauschchromatographie** ion-exchange chromatography (IEX)
- **Ionenpaarchromatographie (IPC)** ion-pair chromatography (IPC)
- **Kapillarchromatographie** capillary chromatography (CC)
- **Membranchromatographie** membrane chromatography (MC)
- **Mitteldruckflüssigkeitschromatographie** medium-pressure liquid chromatography (MPLC)
- **Molekularsiebchromatographie/ Gelpermeationschromatographie/ Gelfiltration** molecular sieving chromatography, gel permeation chromatography (GPC), gel filtration
- **Normaldruck-Säulenchromatographie** gravity column chromatography
- **Papierchromatographie** paper chromatography
- **präparative Chromatographie** preparative chromatography
- **Säulenchromatographie** column chromatography

- ➤ **überkritische Fluidchromatographie/ superkritische Fluid-Chr./ Chromatographie mit überkritischen Phasen**
  supercritical fluid chromatography (SFC)
- ➤ **Umkehrphasenchromatographie**
  reversed phase chromatography, reverse-phase chromatography (RPC)
- ➤ **Verteilungschromatographie/ Flüssig-flüssig-Chr.**
  partition chromatography, liquid-liquid chromatography (LLC)
- ➤ **Zirkularchromatographie/ Rundfilterchromatographie**
  circular chromatography, circular paper chromatography

Chrombeize chromium mordant
Chromsäure H_2CrO_4
  chromic(VI) acid
Chromschwefelsäure
  chromic-sulfuric acid mixture for cleaning purposes
chronisch chronic, chronical
Cinnamonsäure/Zimtsäure (Cinnamat)
  cinnamic acid
Circulardichroismus
  circular dichroism
Citronensäure/Zitronensäure (Citrat)
  citric acid (citrate)
Coinzidenzfaktor/Koinzidenzfaktor
  coefficient of coincidence
Colinearität/Kolinearität
  colinearity
Computertomographie
  computed tomography (CT)
Coulter-Zellzählgerät
  Coulter counter, cell counter
Crotonsäure/Transbutensäure
  crotonic acid, α-butenic acid
Cutis/Haut/eigentliche Haut
  cutis, skin
Cyankali/Zyankali/Kaliumcyanid
  potassium cyanide
cyclisch/ringförmig cyclic
Cyclisierung/Ringschluss *chem*
  cyclization
Cyclus cycle
Cysteinsäure cysteic acid
Cytochemie/Zellchemie
  cytochemistry
Cytologie/Zellenlehre/Zellbiologie
  cytology, cell biology
cytolytisch cytolytic
Cytometrie cytometry
cytopathisch/zellschädigend (cytotoxisch)
  cytopathic (cytotoxic)
Cytoskelett cytoskeleton
Cytostatikum
  (meist *pl* **Cytostatika**)
  cytostatic agent, cytostatic
cytotoxisch cytotoxic
Cytotoxizität cytotoxicity

dämmen *tech* insulate
Dämmplatte insulating panel
➤ **Schalldämmplatte**
  acoustical panel/tile
Dämmstoff insulating material
Dämmung *tech* insulation
Dampf vapor
➤ **Wasserdampf**
  water vapor, steam
Dampfbad steam bath
dampfdicht/dampffest
  vaporproof, vaportight
Dampfdichte vapor density
Dampfdruck vapor pressure
Dampfdruckthermometer
  vapor pressure thermometer
dämpfen/abschwächen
  damp, dampen;
  (schlucken: Schall) deaden
Dampfentwickler/
Wasserdampfentwickler *dest*
  vaporizer, water vaporizer
Dampfkochtopf
  pressure cooker
Dampfraum-Gaschromatographie
  head-space gas chromatography
Dämpfung absorption; attenuation,
  stabilization;
  (von Schwingungen, z.B. Waage)
  damping
Darre/Darrofen
  kiln, kiln oven
  (for drying grain/lumber/tobacco)
darren kiln-dry
darstellen
  (isolieren/rein darstellen) isolate;
  chem (synthetisieren) synthesize,
  prepare
Darstellung/Synthese *chem*
  synthesis, preparation
➤ **graphische D.**
  graph, plot, chart, diagram
Daten data
Datenanalyse data analysis
➤ **explorative D.**
  explorative data analysis
➤ **konfirmatorische D.**
  confirmatory data analysis

Datenblatt/Merkblatt
  (für Chemikalien etc.) data sheet
➤ **Sicherheitsdatenblatt**
  safety data sheet
➤ **Sicherheitsdatenblätter**
  U.S.: Material Safety Data Sheet
  (MSDS)
Datenerfassung data acquisition
Datenerfassungsgerät/
Messwertschreiber/Registriergerät
  datalogger
Datenermittlung data acquisition
Datenverarbeitung data processing
Dauerbetrieb/Dauerleistung/
Non-Stop-Betrieb
  continuous run/operation/duty,
  long-term run/operation,
  permanent run/operation
Dauernutzung continuous use
Dauerpräparat *micros*
  permanent mount/slide
Daumenschraube thumbscrew
DC (Dünnschichtchromatographie)
  TLC (thin layer chromatography)
Deckanstrich finish
Deckel lid, cover, top
Deckglas *micros*
  coverslip, coverglass
Deckglaspinzette
  cover glass forceps
Deckungsgrad
  coverage percentage, coverage level
Deckungswert cover value
Dedifferenzierung/Entdifferenzierung
  dedifferentiation
Deformationsschwingung (IR)
  deformation vibration,
  bending vibration
Degeneration degeneracy
degenerieren/entarten degenerate
Dehnbarkeit expansivity
Dehydratation/Entwässerung
  dehydration
dehydratisieren/entwässern
  dehydrate
dehydrieren dehydrogenate
Dehydrierung/Dehydrogenierung
  dehydrogenation

Dekanter decanter
Dekontamination/Dekontaminierung/ Reinigung/Entseuchung
decontamination
dekontaminieren/reinigen/entseuchen
decontaminate
Demethylierung/Desmethylierung
demethylation
Demontage
disassembly, dismantling, stripping
demontieren demount, disassemble, dismantle, strip, take apart
denaturieren denature
denaturierendes Gel denaturing gel
Denaturierung denaturation, denaturing
Dephlegmation/fraktionierte D./ fraktionierte Kondensation
dephlegmation, fractional distillation
dephosphorylieren dephosphorylate
Dephosphorylierung
dephosphorylation
Depolarisation depolarization
depolarisieren depolarize
Deponie landfill
Derivat derivative
Derivatisation derivatization
derivatisieren derivatize
dermal dermal, dermic, dermatic
Desamidierung deamidation, deamidization, desamidization
Desaminierung
deamination, desamination
Desinfektion disinfection
Desinfektionsmittel disinfectant
desinfizieren (desinfizierend)
disinfect (disinfecting)
Desinfizierung/Desinfektion
disinfection
Desodorierungsmittel/ Deodorans/Desodorans/Deodorant
deodorant
Destillat distillate
Destillation distillation
➢ **Azeotropdestillation**
azeotropic distillation
➢ **Dephlegmation/fraktionierte D./ fraktionierte Kondensation**
dephlegmation, fractional distillation

➢ **Drehband-Destillation**
spinning band distillation
➢ **einfache/direkte D.**
straight-end distillation
➢ **Entspannungs-Destillation/ Flash-Destillation**
flash distillation
➢ **Extraktivdestillation/ extrahierende D.**
extractive distillation
➢ **Gleichgewichtsdestillation**
equilibrium distillation
➢ **Gleichstromdestillation**
simple distillation
➢ **Kugelrohrdestillation**
bulb-to-bulb distillation
➢ **Kurzwegdestillation**
short-path distillation
➢ **mehrfache D./Redestillation**
repeated distillation, cohobation
➢ **Nachlauf/Ablauf** tailings, tails
➢ **Reaktionsdestillation**
reaction distillation
➢ **Trägerdampfdestillation**
steam distillation
➢ **Vakuumdestillation**
vacuum distillation, reduced-pressure distillation
➢ **Vorlauf** first run, forerun
➢ **Zersetzungsdestillation**
destructive distillation
Destillationsgut
distilland, material to be distilled
Destillieraufsatz
stillhead, distillation head
Destillierblase still pot, boiler, distillation boiler flask, reboiler
Destillierbrücke stillhead
destillieren distil, distill, still
➢ **erneut d./wiederholt d.** redistil, rerun
Destilliergerät/Destillationsapparatur
distilling apparatus, still
Destillierkolben/Destillationskolben
distilling flask, distillation flask, 'pot'; (Retorte) retort
Destillierkolonne distilling column
Destillierrückstand distillation residue
Destilliervorstoß receiver adapter

De

Detektor/Fühler/Sensor (*tech:* z.B.
Temperaturfühler) sensor, detector
> **Atomemissionsdetektor (AED)**
atomic emission detector (AED)
> **Elektroneneinfangdetektor**
electron capture detector (ECD)
> **Flammenionisationsdetektor (FID)**
flame-ionization detector (FID)
> **Infrarot-Absorptionsdetektor**
infrared absorbance detector (IAD)
> **Infrarotdetektor**
infrared detector (ID)
> **Ioneneinfangdetektor**
ion trap detector (ITD)
> **Ioneneinfangdetektor (MS)**
ion trap detector (ITD)
> **massenselektiver Detektor**
mass-selective detector
> **Photoionisations-Detektor (PID)**
photo-ionization detector (PID)
> **Schnellscan-Detektor**
fast-scanning detector (FSD),
fast-scan analyzer
> **Verdampfungs-Lichtstreudetektor**
evaporative light scattering detector
(ELSD)
> **Wärmeleitfähigkeitsdetektor/
Wärmeleitfähigkeitsmesszelle**
thermal conductivity detector (TCD)
> **Widerstands-Temperatur-Detektor**
resistance temperature detector
(RTD)
Detergens/Reinigungsmittel
detergent
Dewargefäß Dewar vessel, Dewar flask
DFG (Deutsche Forschungsgemeinschaft)
'German Research Society' (German National Science Foundation)
Diagnose diagnosis
> **Differentialdiagnose**
differential diagnosis
> **pränatale Diagnose**
antenatal diagnosis,
prenatal diagnosis
> **präsymptomatische Diagnose**
presymptomatic diagnosis
Diagnostik diagnostics

Diagnostikpackung (DIN)
diagnostic kit
diagnostisch diagnostic
Diagramm (auch: Kurve) *math/graph*
diagram, plot, graph
> **Histogramm/Streifendiagramm**
histogram, strip diagram
> **Kreisdiagramm** pie chart
> **Phasendiagramm** phase diagram
> **Punktdiagramm** dot diagram
> **Röntgenbeugungsdiagramm/
Röntgenbeugungsmuster/
Röntgenbeugungsaufnahme/
Röntgendiagramm**
X-ray diffraction pattern
> **Spindeldiagramm** spindle diagram
> **Stabdiagramm**
bar diagram, bar graph
> **Strahlendiagramm** *opt* ray diagram
> **Streudiagramm** scatter diagram
(scattergram/scattergraph/scatterplot)
> **Strichdiagramm** line diagram
Dialyse dialysis
dialysieren dialyze
Diamant diamond
Diamantbohrer diamond drill
Diamantmesser diamond knife
Diamantschleifer diamond cutter
Diarrhö diarrhea
Diät diet
diät/Diät.../die Diät betreffend
dietary
Diätetik dietetics
diätetisch dietetic
Dichlordiphenyldichlorethylen (DDE)
dichlorodiphenyltrichloroethylene
(DDE)
Dichlordiphenyltrichlorethan (DDT)
dichlorodiphenyltrichloroethane
(DDT)
dicht (Masse pro Volumen) dense;
(fest verschlossen) tight, sealed tight;
(leckfrei/lecksicher) leakproof,
leaktight (sealed tight)
> **undicht/leck** leaky
Dichte (Masse pro Volumen)
density
Dichtegradient density gradient

Dichtegradientenzentrifugation
  density gradient centrifugation
Dichtigkeit tightness
Dichtkonus/Schneidring *chromat*
  ferrule
Dichtung seal, sealing; gasket
  ➤ **Abdichtung** seal, sealing
  ➤ **Gleitringdichtung (Rührer)**
    face seal
  ➤ **Gummidichtung(sring)**
    rubber gasket
  ➤ **Lippendichtung**
    (Wellendurchführung)
    lip seal, lip-type seal
  ➤ **Wellendichtung (Rotor)**
    shaft seal
Dichtungsband sealing tape
dichtungsfrei/ohne Dichtung (Pumpe)
  sealless
Dichtungskitt lute
Dichtungsmanschette gasket
Dichtungsmasse/Dichtungsmittel
  sealant, sealing compound/material
Dichtungsmuffe packing sleeve
Dichtungsmutter packing nut
Dichtungsring/Dichtungsscheibe/
  Unterlegscheibe washer
dickflüssig/zähflüssig/viskos/viskös
  viscous, viscid
Dickungsmittel thickener
Dielektrizitätskonstante
  dielectric constant
Dienst service; duty; work; (Schicht) shift
  ➤ **Spätdienst/Spätschicht** late shift
Dienst- und Treueverhältnis
  confidential employer-employee relationship,
  confidential working relationship
Dienstkleidung official dress
Dienstuniform official uniform
Dienstvergehen disciplinary offense
Dienstvertrag
  contract of employment
Dienstvorschrift service regulations,
  job regulations, official regulations
Differential-Interferenz (Nomarski)
  differential interference

Differentialdiagnose
  differential diagnosis
Differentialfärbung/Kontrastfärbung
  differential staining, contrast staining
Differentialgleichung
  differential equation
Differentialkalorimetrie
  differential scanning calorimetry (DSC)
Differentialthermoanalyse/
Differenzthermoanalyse (DTA)
  differential thermal analysis (DTA)
Differenz difference
diffundieren diffuse
Diffusionsblotting
  capillary blotting
Diffusionskoeffizient
  diffusion coefficient
Diffusionstest/Agardiffusionstest
  agar diffusion test
Diffusor diffuser
digerieren decoct, digest
  (by heat/solvents)
Digitalisiergerät digitizer
Dilatation/Ausweitung
  expansion, dilation, dilatation
dimerisieren dimerize
Dimerisierung dimerization
Dimroth-Kühler
  coil condenser (Dimroth type)
DIN (Deutsche Industrienorm)
  German Industrial Standard
Diodenarray-
  Nachweis/Diodenmatrixnachweis
  diode array detection (DAD)
Dioptrie (*Einheit*) diopter (D)
dioptrisch dioptric
diphasisch diphasic
diploid diploid
Dipolmoment dipole moment
Direkttransfer-Elektrophorese/
  Blottingelektrophorese
  direct transfer electrophoresis,
  direct blotting electrophoresis
Direktverdrängerpumpe
  positive displacement pump
Dispenserpumpe
  dispenser pump

dispergieren disperse
Dispergierung/Dispersion dispersion
Dispersion/Kolloid dispersion, colloid
Disposition/Veranlagung/Anfälligkeit
  disposition
Dissoziationsgeschwindigkeit
  dissociation rate
Dissoziationskonstante (K_i)
  dissociation constant
dissoziieren dissociate
Disulfidbindung/Disulfidbrücke
  disulfide bond, disulfide bridge,
  disulfhydryl bridge
Diurese/Harnfluss/Harnausscheidung
  diuresis
divergieren diverge
Diversität diversity
DNA/DNS (Desoxyribonucleinsäure/
  Desoxyribonukleinsäure)
  DNA (deoxyribonucleic acid)
DNA-Fingerprinting/
  genetischer Fingerabdruck
  DNA profiling, DNA fingerprinting
DNA-Fußabdruck/DNA-Footprint
  DNA footprint
DNA-Sequenzierungsautomat
  DNA sequencer
Docht wick
Donor/Spender donor
Doppelbindung double bond
Doppelblindversuch
  double blind assay,
  double-blind study
doppelbrechend
  birefringent, double-refracting
Doppelbrechung
  birefringence, double refraction
Doppeldiffusion/
  Doppelimmundiffusion
  double diffusion,
  double immunodiffusion
  (Ouchterlony technique)
Doppelkreuzung double cross
Doppelmuffe/Kreuzklemme
  clamp holder, 'boss', clamp 'boss'
  (rod clamp holder)
Doppelschicht double layer, bilayer
Doppelschleifer bench grinder

Doppelstrang *gen* double strand
doppeltwirkend double-acting
Doppelverdau *gen/biochem*
  double digest
Doppelzucker/Disaccharid
  double sugar, disaccharide
DOP-Vernebelung
  (Dioctylphthalat-Vernebelung)
  DOP smoke
  (dioctyl phthalate smoke)
dosieren dose (give a dose),
  measure out; meter, proportion
Dosieren dose, meter, proportion
Dosierpumpe dosing pump,
  proportioning pump
Dosierspender dispenser
Dosierung/Dosieren
  (im Verhältnis/anteilig)
  apportioning, proportioning
Dosierventil flow control valve;
  metering valve, proportioning valve
Dosis dose, dosage
➢ **Einzeldosis** single dose
➢ **höchste Dosis ohne beobachtete**
  Wirkung
  no observed effect level (NOEL)
➢ **letale Dosis/Letaldosis/tödliche**
  Dosis lethal dose
➢ **maximal verträgliche Dosis**
  maximum tolerated dose (MTD)
➢ **mittlere effektive Dosis (ED_{50})/**
  mittlere wirksame Dosis median
  effective dose (ED_{50})
➢ **mittlere letale Dosis (LD_{50})**
  median lethal dose (LD_{50})
➢ **Überdosis** overdose
Dosis-Wirkungskurve
  dose-response curve
Dosisäquivalent *rad* dose equivalent
Dosiseffekt dosage effect
Dosiskompensation
  dosage compensation
Draht wire
Drahtbürste wire brush
Drahtnetz *chem/lab*
  wire gauze, wire gauze screen
Drahtschere
  wire shears, wire cutters

Dr

Drahtseilschere
 wire cable shears, cable shears
Dränung/Drainage drainage
Dreck/Schmutz dirt, filth
dreckig/schmutzig
 dirty, filthy
Drehband-Destillation
 spinning band distillation
Drehbandkolonne
 spinning band column
Drehbank/Drehmaschine lathe
drehbar pivoted
Drehbewegung (rotierend)
 spinning/rotating motion
drehen/verdrehen contort
Drehgriff twist-grip
Drehhocker swivel stool
Drehkolbenzähler
 rotary-piston meter
Drehmischer
 roller wheel mixer
Drehmomentschlüssel
 torque wrench
 (torque amplifier handle)
Drehplatte (Mikrowelle) turntable
Drehpunkt/Drehzapfen/Drehbolzen
 pivot
Drehschieberpumpe
 rotary vane pump
Drehsinn/Rotationssinn
 rotational sense,
 sense of rotation
Drehstuhl swivel chair
Drehtisch *micros* rotating stage
Drehung/Torsion torsion
Drehwalze (Roller-Apparatur) roller
Drehzahl
 (UpM=Umdrehungen pro Minute)
 number of revolutions
 (rpm=revolutions per minute)
Drehzahlregelung
 rotation speed adjustment
Dreieck triangle
 ➢ **Tondreieck/Drahtdreieck**
  clay triangle, pipe clay triangle
Dreifachbindung triple bond
Dreifinger-Klemme three-finger clamp
Dreihalskolben three-neck flask

Dreiweghahn/Dreiwegehahn
 three-way cock, T-cock
Dreiwegverbindung
 three-way connection
dreiwertig trivalent
Dreiwertigkeit trivalency
Dreizack... three-prong ...
Driftröhre (TOF-MS) drift tube
Droge drug
 ➢ **Pflanzendroge** herbal drug
Drogenkunde/Pharmakognosie/
 pharmazeutische Biologie
 pharmacognosy
Drogenpflanze/Arzneipflanze
 medicinal plant
Drossel throttle, choke
Drosselklappe
 throttle valve, damper
drosseln/herunterfahren/dämpfen
 throttle, choke, slow down, dampen
Drosselventil throttle valve
Druck pressure
 ➢ **Arbeitsdruck**
  working pressure, delivery pressure
 ➢ **Betriebsdruck** operating pressure
 ➢ **Blutdruck** blood pressure
 ➢ **Dampfdruck** vapor pressure
 ➢ **Eingangsdruck (HPLC)**
  supply pressure
 ➢ **erniedrigter Druck**
  reduced pressure
 ➢ **Gegendruck** counterpressure
 ➢ **Hinterdruck** outlet pressure;
  (Arbeitsdruck: Druckausgleich)
  working pressure, delivery pressure
 ➢ **Hochdruck** high pressure
 ➢ **hydrostatischer Druck**
  hydrostatic pressure
 ➢ **Luftdruck** air pressure
 ➢ **atmosphärischer Luftdruck**
  atmospheric pressure
 ➢ **Niederdruck** low pressure
 ➢ **Normaldruck** standard pressure
 ➢ **Öffnungsdruck (Ventil)**
  breaking pressure
 ➢ **onkotischer Druck/**
  kolloidosmotischer Druck
  oncotic pressure

Dr

- **osmotischer Druck**
  osmotic pressure
- **Partialdruck** partial pressure
- **Sauerstoffpartialdruck**
  oxygen partial pressure
- **Selektionsdruck**
  selective pressure,
  selection pressure
- **Turgor/hydrostatischer Druck**
  turgor, hydrostatic pressure
- **Turgordruck** turgor pressure
- **Überdruck** positive pressure
- **Umgebungsdruck** ambient pressure
- **Unterdruck** negative pressure
- **Vordruck/Eingangsdruck
  (Hochdruck: Gasflasche)**
  initial pressure, initial compression, tank pressure, high pressure

Druckabfall pressure drop
Druckanstieg
  pressure rise, pressure increase
Druckausgleich pressure equalization
- **Dekompression** decompression

Druckbehälter pressure vessel;
  (aus Glas) glass pressure vessel
druckdicht pressure-tight
Druckentlastungseinrichtung
  pressure protection device
druckfest pressure resistant
Druckfiltration pressure filtration
Druckflasche cylinder
Druckgas
  compressed gas,
  pressurized gas
Druckgasflasche gas cylinder
Druckleistenverschluss
  zip seal, zip-lip
Druckluft compressed air
Druckluftventil pneumatic valve
Druckmesser/Manometer
  pressure gauge, pressure gage, gauge, gage
Druckminderer (Gasflasche)
  pressure regulator
**Druckminderventil/
  Druckminderungsventil/
  Druckreduzierventil**
  pressure-relief valve

**Druckpumpe/
  Saugpumpe/
  doppeltwirkende Pumpe**
  double-acting pump
Druckregelventil
  pressure control valve
Druckregler pressure regulator
Druckschlauch pressure tubing
Druckschwankung
  pressure fluctuation
druckstoßfest
  shock pressure resistant
**Druckstromtheorie/
  Druckstromhypothese**
  pressure-flow theory/hypothesis
Drucktaste/Bedienknopf push button
Druckverband *med*
  pressure bandage,
  compression dressing
Druckverlust
  loss of pressure, pressure drop
Druckverschluss compression seal
Druckverschlussbeutel
  zip storage bag, zip-lip storage bag
Dübel pin, dowel, wall plug
Duft/Geruch smell, odor, scent
- **angenehmer Duft/Geruch**
  fragrance, scent, pleasant smell
- **unangenehmer Duft/Geruch**
  unpleasant smell

duftend (angenehm) fragrant
Duftstoffe scents, odiferous substances
**Dung/Mist/
  tierische Exkremente/Tierkot**
  dung, manure
düngen fertilize, manure
Dünger/Düngemittel
  fertilizer, plant food, manure
Düngung fertilization
- **Überdüngung** overfertilization,
  excessive fertilization

Dunkelfeld *micros* dark field
Dunkelkammer *micros/photo* darkroom
Dunkelkammerlampe (Rotlichtlampe)
  safelight
Dünnschnitt thin section, microsection
- **Semidünnschnitt** semithin section
- **Ultradünnschnitt** ultrathin section

Dunstabzugshaube/Abzug
 fume hood, hood
durchbluten
 supply with blood, vascularize
Durchblutung circulation,
 blood supply, blood circulation
durchbrennen burn through/out
durchfließen
 percolate, flow through
Durchfluss
 percolation, flowing through, flux
Durchflussrate
 (D.geschwindigkeit) flow rate;
 (Verdünnungsrate) dilution rate
Durchflussreaktor (Bioreaktor)
 flow reactor
**Durchflusszytometrie/
 Durchflusscytometrie**
 flow cytometry
Durchführung
 performance, realization,
 completion, implementation
Durchgang passage, passageway;
 walkthrough; *electr* throughput
Durchgangsprüfer *electr*
 continuity tester
Durchgangsquerschnitt *tech*
 cross sectional area
Durchgeh-Reaktion
 runaway reaction
Durchlass passage, passageway,
 opening, outlet, port, conduit, duct
durchlässig/permeabel
 pervious, permeable
➢ **halbdurchlässig/semipermeabel**
 semipermeable
➢ **undurchlässig/impermeabel**
 impervious, impermeable
Durchlässigkeit/Permeabilität
 perviousness, permeability
➢ **Halbdurchlässigkeit/
 Semipermeabilität**
 semipermeability
➢ **Undurchlässigkeit/Impermeabilität**
 imperviousness, impermeability
Durchlaufgeschwindigkeit (Säule)
 chromat flow rate
 (mobile-phase velocity)

Durchlicht/Durchlichtbeleuchtung
 transillumination,
 transmitted light illumination
durchlüften (einen Raum)
 air (the room), ventilate;
 (belüften) aerate
Durchlüftung aeration
Durchmischung mixing
Durchmustern/Durchtesten screening
durchnässt/durchweicht soggy
Durchreiche service hatch
Durchsatz (Durchsatzmenge)
 throughput
Durchsatzrate throughput rate
durchscheinend
 translucent, pellucid
durchschlagen/bluten (TLC)
 bleed (spotting)
durchschneiden
 transect, cut through
Durchschnitt (Mittelmaß)
 average, mean;
 (schneiden) transection
Durchschnittsertrag average yield
Durchsickern *n* percolation
durchsickern *vb*
 percolate; seep through
Durchsuchung search
durchtränken (durchtränkt)
 soak (soaked)
Durchzug (Luft) draft, draught (*Br*)
Dusche shower
➢ **Augendusche**
 eye-wash (station/fountain)
➢ **Notdusche**
 emergency shower,
 safety shower
➢ **'Schnellflutdusche'**
 quick drench shower,
 deluge shower
Düse jet, nozzle; orifice
**Düsenumlaufreaktor/
 Strahl-Schlaufenreaktor**
 jet loop reactor
**Düsenumlaufreaktor/
 Umlaufdüsen-Reaktor**
 nozzle loop reactor,
 circulating nozzle reactor

Eb

Ebene/ebene Fläche *math/geom*
plane (flat/level surface)
➢ **Brennebene** focal plane
➢ **Sagittalebene**
(parallel zur Mittellinie)
median longitudinal plane
➢ **Schnittebene/Schnittfläche**
cutting face, cutting plane
Echtzeit real time
Edelgas inert gas, rare gas
Edelmetall precious metal
Edelstahl
high-grade steel, high-quality steel
Edelstahlschrubber
stainless-steel sponge
eichen/kalibrieren calibrate, adjust; (Maße/Gewichte) standardize, gage, gauge
Eichgerät
calibrating instrument, calibrator
Eichkurve calibration curve
Eichmarke calibrating mark
Eichmaß
calibrating standard, standard (measure)
Eichung calibration, adjustment, adjusting, standardization
Eiernährboden
egg culture medium
Eigelb/Dotter/Eidotter yolk, egg yolk
Eigengewicht
own weight; dead weight, permanent weight;
service weight, unladen weight
Eignung/Fitness
fitness, suitability
Eiklar/natives Eiweiss
native egg white
Eikultur (Hühnerei)
chicken embryo culture
Eimer bucket (plastic), pail (metal)
Eimeröffner pail opener
einarbeiten train;
(in ein Dokument etc.) work in
Einarbeitungsphase
(für Neubeschäftigte)
training period
einäschern incinerate

einatmen *vb*
breathe in, inhale
Einatmung/Einatmen/
Inspiration/Inhalation
inspiration, inhalation
einbalsamieren enbalm
einbasig monobasic
Einbau/Anschluss
installation
Einbauten
internal fittings,
built-in elements,
structural additions
Einbettautomat/Einbettungsautomat *micros* embedding machine, embedding center
einbetten *micros* embed
Einbettung *micros* embedding
Einbettungsmittel/Einschlussmittel
mountant, mounting medium
Einbettungspräparat
embedded specimen
eindämmen contain
Eindämmung containment
eindampfen (vollständig)
reduce by evaporation
(evaporate completely)
Eindampfschale evaporating dish
Eindunsten evaporation
einengen/konzentrieren
reduce, concentrate
einfachbrechend/isotrop
isotropic
Einfachzucker/
einfacher Zucker/
Monosaccharid
single sugar, monosaccharide
Einfetten/Einschmieren
lubrication, oiling
einfrieren freeze
einführen introduce; import
Einfülltrichter addition funnel
Eingabe input
➢ **Ausgabe** output
Eingang *electr* input;
(Anschluss: Gerät) port
➢ **Ausgang** *electr* output
Eingangsdruck (HPLC) supply pressure

eingeschweißt
  welded on, welded to
Eingewöhnung
  acclimation, acclimatization
Eingewöhnungsphase
  establishment phase
Einhaltung (Vorschrift)
  observance, compliance
Einhängekühler/Kühlfinger
  suspended condenser, cold finger
Einhängethermostat/
  Tauchpumpen-Wasserbad
  immersion circulator
Einheit (Maßeinheit) unit (measure)
einheitlich uniform
Einkapselung encapsulation
Einkauf/Erwerb purchase
Einkristall monocrystal
Einlagerung inclusion, intercalation
Einlasssystem inlet system
einlesen (Daten) read in; scan
➢ **auslesen** read out
einloggen log on
➢ **ausloggen** log off
Einmal.../Einweg.../Wegwerf...
  single-use, disposable
Einmalhandschuhe
  single-use gloves,
  disposable gloves
einnehmen/etwas zu sich nehmen
  ingest
einordnen/einstufen/klassifizieren
  rank, classify
einpflanzen *bot* plant; *med* implant
einrichten (Experiment etc.)
  install, set up
Einrichtung (Möbel etc.)
  furnishings
Einrichtungsgegenstände
  furnishings, pieces of equipment,
  fixtures, fittings, fitments
Einsalzen/Einsalzung *chem* salting in
Einsatz *tech* **(Gefäß etc.)** insert, inset
einsaugen suck in, draw in
Einsaugen (Rückschlag bei
  Wasserstrahlpumpe etc.)
  suck-back
einschalten turn on, switch on

Einscheibensicherheitsglas (ESG)
  tempered safety glass
Einschichtzellkultur
  monolayer cell culture
Einschiebereaktion/Insertionsreaktion
  insertion reaction
Einschlämmtechnik *chromat*
  slurry-packing technique
Einschluss inclusion
Einschlussgrad (physikalische/
  biologische Sicherheit)
  containment level
Einschlussverbindung *chem* inclusion
  compound
Einschlussverfahren *biotech*
  immurement technique
Einschnitt incision, cut; indentation
Einschnürung constriction
einseitig/unilateral unilateral
Einsetzen/Beginn (einer Reaktion)
  onset, start (of a reaction)
einspannen clamp, fix, attach; mount
Einspritzblock
  injection port, syringe port
einspritzen/injizieren inject
Einspritzer injector
Einspritzung/Injektion injection
Einspritzventil
  injection valve, syringe port
einstecken/anschließen
  electr/tech plug in
Einstellknopf adjustment knob
Einstellschraube
  adjustment screw; tuning screw
Einstellungen (eines Geräts)
  settings; adjustment
Einstichkultur/Stichkultur (Stichagar)
  micb stab culture
Einstrom influx
Einströmen ingression
Einströmgeschwindigkeit/
  Eintrittsgeschwindigkeit
  (Sicherheitswerkbank)
  inlet velocity (hood)
Einströmöffnung
  inlet, incurrent aperture
Einstufung/Kategorisierung
  categorization

Ei

Eintauchkühler (mit Kühlsonde)
 refrigerated chiller with
 immersion probe
Eintauchkultur submerged culture
Einteilung division; arrangement;
 classification; planning, scheduling;
 tech graduation, scale
eintopfen (Pflanze) pot
Eintopfreaktion *chem*
 one-pot reaction
Eintrag entry; *ecol* input
eintragen
 (z.B. Daten ins Laborbuch) enter;
 (bei Anmeldung) sign in
 ➢ **austragen (bei Abmeldung)**
 sign out
**Eintrittsgeschwindigkeit/
 Einströmgeschwindigkeit
 (Sicherheitswerkbank)**
 face velocity (not same as
 'air speed' at face of hood)
Eintrittspforte route of entry
Eintüten (Tüten/Säcke einfüllen)
 bagging
Einverständniserklärung
 agreement, consent
 ➢ **E. nach ausführlicher Aufklärung**
 informed consent
Einwaage initial weight,
 amount weighed,
 weighed amount/quantity
einwägen weigh in
Einweg.../Einmal.../Wegwerf...
 disposable
Einweghandschuhe
 disposable gloves
Einwegspritze disposable syringe
einweichen/einweichen lassen
 soak, drench, steep
einwertig/univalent/monovalent *chem*
 univalent, monovalent
Einwertigkeit/Univalenz *chem*
 univalence
einwiegen (nach Tara)
 weigh in (after setting tare)
Einwilligung/Zustimmung
 consent, agreement
 ➢ **Einhaltung** compliance

einwirken
 act, effect, contact, attack, interact
einwirken lassen (in einer Flüssigkeit)
 soak
 ➢ **reagieren lassen** let react
Einwirkung effect, action, impact
Einwirkungsdauer/Einwirkungszeit
 exposure time, duration of exposure
Einwirkzeit contact time
Einzeldosis single dose
Einzelhandel
 retail business, retail trade
Einzelhandelsgeschäft retail store
Einzelhandelspreis retail price
Einzelhändler
 retailer, retail dealer, retail vendor
Einzeller unicellular lifeform
einzellig single-celled, unicellular
einzeln/solitär single, solitary
Eis ice
 ➢ **Trockeneis (CO_2)** dry ice
 ➢ **zerstoßenes Eis** crushed ice
Eisbad ice bath, ice-bath
Eisbehälter ice bucket
Eisen (Fe) iron
 ➢ **Brecheisen** crowbar, jimmy
 ➢ **Gusseisen** cast iron
 ➢ **Tempereisen/Temperguss**
 malleable iron,
 malleable cast iron, wrought iron
Eisenkies pyrite
eisenregulierender Faktor
 iron-regulating factor (IRF)
Eisessig glacial acetic acid
Eiskernaktivität *micb* ice nucleating
 activity
Eisschnee (fürs Eisbad)
 snow, crushed ice
**Eisüberzug/überfrorene Nässe/
 gefrorener Regen**
 sleet, glaze, frozen rain
Eiweiß (Ei) egg white, egg albumen;
 (Protein) protein
 ➢ **aus Eiweiß bestehend/Eiweiß.../
 proteinartig/proteinhaltig/Protein...**
 proteinaceous
 ➢ **denaturiertes Eiweiß**
 denatured egg white

➢ **natives Eiweiß/Eiklar**
   native egg white
eiweißlos exalbuminous
Ekzem eczema
Elastizität elasticity
Elastizitätsgrenze/Dehngrenze yield strength
'Elefantenfuß'/Rollhocker (runder Trittschemel mit Rollen) (rolling) step-stool
Elektriker electrician
Elektrizität electricity
➢ **statische E.** static electricity
Elektroabscheidung electroprecipitation
Elektrode electrode
➢ **Bezugselektrode** reference electrode
➢ **ionenselektive E.** ion-selective electrode (ISE)
➢ **Quecksilbertropfelektrode** dropping mercury electrode (DME)
➢ **Tropfelektrode** dropping electrode
➢ **Wasserstoffelektrode** hydrogen electrode
Elektroencephalogramm (EEG) electroencephalogram
elektrogen electrogenic
Elektrogerät electrical appliance, electrical device
Elektroimmunodiffusion electroimmunodiffusion, counter immunoelectrophoresis
Elektrokardiogramm (EKG) electrocardiogram
Elektrolyse electrolysis
Elektrolysezelle/Elektrolysierzelle cell
Elektrolyt electrolyte
elektrolytische Dissoziation electrolytic separation
elektromotorische Kraft (EMK) electromotive force (emf/E.M.F.)
Elektron electron
➢ **Außenelektron** outer electron
➢ **Bindungselektron** binding electron
➢ **Einzelelektron** single electron
➢ **freies Elektron** free electron
➢ **gepaartes Elektron** paired electron

➢ **Rumpfelektron** inner-shell electron
➢ **ungepaartes E./einsames E.** odd electron
➢ **Valenzelektron** valence electron, valency electron
Elektronenakzeptor electron acceptor
Elektronendonor/Elektronenspender electron donor
Elektroneneinfangdetektor electron capture detector (ECD)
Elektronen-Energieverlust-Spektroskopie electron energy loss spectroscopy (EELS)
Elektronenmikroskopie electron microscopy (EM)
➢ **Höchstspannungselektronenmikroskopie** high voltage electron microscopy (HVEM)
➢ **Immun-Elektronenmikroskopie** immunoelectron microscopy (IEM)
➢ **Rasterelektronenmikroskopie (REM)** scanning electron microscopy (SEM)
➢ **Transmissionselektronenmikroskopie/Durchstrahlungselektronenmikroskopie** transmission electron microscopy (TEM)
Elektronenpaar electron pair
➢ **freies E./einsames E./nichtbindendes E.** lone pair
Elektronenraffer/Elektronenempfänger electron acceptor
Elektronenspender/Elektronendonor electron donor
Elektronen-Spinresonanz-spektroskopie (ESR)/Elektronenparamagnetische Resonanz (EPR) electron spin resonance spectroscopy (ESR), electron paramagnetic resonance (EPR)
Elektronenstoß-Ionisation electron-impact ionization (EI)

Elektronenstoß-Spektrometrie
electron-impact spectrometry (EIS)
Elektronentransport
electron transport
Elektronentransportkette
electron-transport chain
Elektronenüberträger electron carrier
Elektronenübertragung
electron transfer
elektroneutral electroneutral
(electrically silent)
elektronisch electronic
Elektroosmose/Elektroendosmose
electro-endosmosis,
electro-osmotic flow (EOF)
elektrophiler Angriff
electrophilic attack
Elektrophorese electrophoresis
> **Direkttransfer-Elektrophorese/ Blotting-Elektrophorese**
direct transfer electrophoresis,
direct blotting electrophoresis
> **Diskelektrophorese/ diskontinuierliche Elektrophorese**
disk electrophoresis
> **freie Elektrophorese**
free electrophoresis
(carrier-free electrophoresis)
> **Gegenstromelektrophorese/ Überwanderungselektrophorese**
countercurrent electrophoresis
> **Gelelektrophorese** gel electrophoresis
> **Gelgießstand/ Gelgießvorrichtung**
gel caster
> **Gelträger/Geltablett**
gel tray
> **Isotachophorese/ Gleichgeschwindigkeits-Elektrophorese**
isotachophoresis (ITP)
> **Kapillarelektrophorese** capillary electrophoresis (CE)
> **Kapillar-Zonenelektrophorese**
capillary zone electrophoresis (CZE)
> **Papierelektrophorese** paper electrophoresis

> **Puls-Feld-Gelelektrophorese**
pulsed field gel electrophoresis (PFGE)
> **Tasche/Vertiefung (Elektrophorese-Gel)** well, depression (at top of gel)
> **Trägerelektrophorese/ Elektropherografie**
carrier electrophoresis
> **Überwanderungselektrophorese/ Gegenstromelektrophorese**
countercurrent electrophoresis
> **Wechselfeld-Gelelektrophorese**
alternating field gel electrophoresis
> **Zonenelektrophorese**
zone electrophoresis
elektrophoretisch electrophoretic
elektrophoretische Mobilität
electrophoretic mobility
Elektroplaque (*pl* **Elektroplaques/** *slang:* **Elektroplaxe**)
electroplaque
elektroplatieren electroplating
Elektroporation electroporation
Elektroretinogramm (ERG)
electroretinogram
Elektrospray electrospray
elektrotonisches Potential
electrotonic potential
Element element
> **elektrochemisches E./Zelle**
cell
> **Halbelement (galvanisches)/Halbzelle**
half cell, half element
(single-electrode system)
> **Periodensystem (der Elemente)**
periodic table (of the elements)
> **Spurenelement/Mikroelement**
trace element, microelement, micronutrient
> **Thermoelement**
thermocouple
Elementarzelle unit cell
ELISA (enzymgekoppelter Immunadsorptionstest/ enzymgekoppelter Immunnachweis)
ELISA (enzyme-linked immunosorbent assay)

Ellagsäure ellagic acid, gallogen
eloxieren anodize, anodically oxidize, oxidize by anodization (electrolytic oxidation)
Eloxierung anodization, anodic oxidation
Eluat eluate
eluieren eluate
eluotrope Reihe (Lösungsmittelreihe) eluotropic series
Elutionskraft eluting strength (eluent strength)
Elutionsmittel/Eluens (Laufmittel) eluent, eluant
Elutriation/Aufstromklassierung elutriation
Emaille/Email porcelain enamel
embryotoxisch embryotoxic
Emission/Ausstoss/Ausstrahlung emission
Emissionskoeffizient emissivity coefficient (absorptivity coefficient)
emittieren/aussenden emit
Empfänger *phys/tech* receiver; (Rezeptor) receptor; (Adressat/Konsignatar) consignee
empfänglich receptive
Empfangsgerät receiver
Empfehlung recommendation
empfindbar perceptible, sensible
Empfindbarkeit sensibility, sensitiveness
empfinden/fühlen/spüren feel, sense, perceive
empfindlich (sensitiv/leicht reagierend) sensitive; (reizempfänglich) irritable, sensible; (zerbrechlich: z.B. Pflanze/Ökosystem) tender, fragile
Empfindlichkeit *photo/micros* sensitivity; (Anfälligkeit) susceptibility
Empfindung sensation, perception
empfohlener täglicher Bedarf recommended daily allowance (RDA)
empirisch empiric(al)

empirische Formel empirical formula
Emulgator emulsifier, emulsifying agent
emulgieren emulsify
Emulsion emulsion
Enantiomer enantiomere
Endbild *micros* final image
Ende/Terminus (Molekülende) terminus
endergon/energieverbrauchend endergonic
Endgruppenbestimmung end group analysis, terminal residue analysis
endotherm endothermic
Endprodukthemmung/Rückkopplungshemmung end-product inhibition, feedback inhibition
Endpunktsbestimmung end-point determination
Endpunktverdünnungsmethode (Virustitration) end-point dilution technique
endständig terminal, terminate
Energetik energetics
Energie energy
Energiebarriere energy barrier
Energiebedarf energy requirement
Energiebilanz energy balance, energy budget
Energieerhaltungssatz law of conservation of energy
Energiefluss energy flux, energy flow
Energieladung energy charge
Energieprofil energy profile
Energiequelle energy source
energiereich energy-rich
energiereiche Bindung high energy bond
energiereiche Verbindung high energy compound
Energiesparlampe energy-saving lightbulb
Energiestoffwechsel energy metabolism

En

Energieübergang/Energietransfer
energy transfer
Energieverlust-Spektroskopie
electron energy loss spectroscopy (EELS)
Energiezuführung
energy supply
Enghals ... narrow-mouthed, narrowmouthed, narrow-neck, narrownecked
Engländer/Rollgabelschlüssel
adjustable wrench
Engpass/Flaschenhals bottleneck
entarten/degenerieren degenerate
entartet/degeneriert (IR) degenerate
Entartung degeneration, degeneracy
Entartungsgrad (IR)
degree of degeneracy
Entdifferenzierung/Dedifferenzierung
dedifferentiation
Entfärbung decoloration, bleaching
entfetten degrease
Entfeuchter demister;
(Gerät) dehumidifier
entflammbar/brennbar/entzündlich
flammable, inflammable
➤ **flammbeständig/flammwidrig**
flame-resistant
➤ **nicht entflammbar/nicht brennbar**
nonflammable, incombustible
➤ **schwer entflammbar**
flameproof, flame-retardant
Entflammbarkeit/Brennbarkeit/ Entzündbarkeit
flammability
entgasen
degas, degasify, outgas
Entgasen/Entgasung
degasing, gasing-out
entgiften detoxify
Entgiftung detoxification
Entgiftungszentrale/Entgiftungsklinik
poison control center, poison control clinic
Enthalpie enthalpy
enthärten soften
Enthemmung/Disinhibition
disinhibition

Entionisierung deionizing
entkalken (ein Gerät ~) descale
Entkalkung/Dekalzifizierung
decalcification
entkernen core
entkernt (Zelle) enucleate
entkoppeln
decouple, uncouple, release
Entkoppler
uncoupler, uncoupling agent
Entkopplung
decoupling, uncoupling, release
Entladung discharge
entleeren
(ausleeren/auskippen) empty out;
(luftleer pumpen/herauspumpen) evacuate, drain, discharge
Entleeren/Entleerung (eines Gefäßes; allgemein) empty(ing) out, pour out;
(Flüssigkeit) drain, drainage;
(Gas/Luft) deflation
entlüften vent; degas
Entlüftung ventilation, venting; degassing; air extraction
Entlüftungsventil
purge valve, pressure-compensation valve, venting valve
entmischen
segregate, separate out, reseparate
Entmischung
segregation, separation, reseparation
Entnahme removal, withdrawal;
taking out; (einer Probe) sampling
Entparaffinierungsmittel *micros*
decerating agent
(for removing paraffin)
Entropie entropy
entsalzen desalinate
Entsalzung desalination
Entschädigung~/ Kompensationszahlung bei Arbeitsunfällen od. Berufskrankheiten
workman's compensation
Entschäumer/Antischaummittel
antifoam, antifoaming agent, defoamer, defrother

Entschirmung (NMR) deshielding
entschwefeln
  desulfurize, desulfur
Entschwefelung
  desulfurization, desulfuration
Entseuchung/Dekontamination/
  Dekontaminierung/Reinigung
  decontamination
entsorgen dispose of, remove
Entsorgung
  waste disposal, waste removal
  ➢ **unsachgemäße E.** improper disposal
Entsorgungsfirma/
  Entsorgungsunternehmen
  disposal firm
entspannen *physiol* relax
entspannt/relaxiert (Konformation)
  relaxed
Entspannung *physiol* relaxation
Entspannungs-Destillation/
  Flash-Destillation
  flash distillation
Entspannungsmittel/
  oberflächenaktive Substanz
  surfactant
Entwarnung all-clear
entwässern (dehydratisieren)
  dehydrate; (drainieren) drain
Entwässerung
  (Dehydratation) dehydration;
  (Drainage) drainage, draining
Entweichen (entweichen lassen)
  release; ('passiv') escape
entweichen (Gas etc.) escape
entwickeln/entstehen
  develop, emerge, unfold
Entwickler *photo* developer
Entwicklungsgenetik
  developmental genetics
Entwicklungsstadium
  (*pl* **Entwicklungsstadien**)/
  Entwicklungsphase
  developmental stage,
  developmental phase
Entwurf/Plan/Design design
entzündbar ignitable
entzünden/entflammen/anbrennen
  chem inflame, ignite
entzündet *med* inflamed
entzündlich (entflammbar/brennbar)
  chem flammable, inflammable;
  med inflammed, inflammatory
Entzündung *chem/med* inflammation
Enzym/Ferment enzyme
  ➢ **Apoenzym** apoenzyme
  ➢ **Coenzym/Koenzym** coenzyme
  ➢ **Holoenzym** holoenzyme
  ➢ **Isozym/Isoenzym**
    isozyme, isoenzyme
  ➢ **Kernenzym (RNA-Polymerase)**
    core enzyme
  ➢ **Leitenzym** tracer enzyme
  ➢ **Multienzymkomplex/**
    Multienzymsystem/Enzymkette
    multienzyme complex,
    multienzyme system
  ➢ **Proenzym/Zymogen**
    proenzyme, zymogen
  ➢ **progressiv arbeitendes Enzym**
    processive enzyme
  ➢ **Reparaturenzym** repair enzyme
  ➢ **Restriktionsenzym**
    restriction enzyme
  ➢ **Schlüsselenzym** key enzyme
  ➢ **Verdauungsenzym**
    digestive enzyme
Enzymaktivität (*katal*)
  enzyme activity (*katal*)
enzymgekoppelter
  Immunadsorptionstest/
  enzymgekoppelter Immunnachweis
  (ELISA) enzyme-linked
  immunosorbent assay (ELISA)
enzymgekoppelter
  Immunoelektrotransfer
  enzyme-linked immunotransfer blot
  (EITB)
Enzymhemmung
  enzymatic inhibition,
  repression of enzyme,
  inhibition of enzyme
Enzymimmunoassay/Enzymimmuntest
  (EMIT-Test)
  enzyme-immunoassay,
  enzyme immunassay (EIA)
Enzymkinetik enzyme kinetics

En 48

Enzymreaktion enzymatic reaction
Enzymspezifität
   enzymatic specificity,
   enzyme specificity
Epidemie epidemic
Epidemiologie epidemiology
epidemiologisch epidemiologic(al)
epidermal/Haut../die Haut betreffend
   epidermal, cutaneous
Epidermis epidermis
Epimerisierung epimerization
Epithel (*pl* **Epithelien**)
   epithelium (*pl* epithelia)
erben/ererben inherit
Erbfaktor/Gen gene
Erbgang/Vererbungsmodus
   mode of inheritance
Erbgut/Genom
   hereditary material, genome
Erbinformation
   hereditary information,
   genetic information
Erbkrankheit/erbliche Erkrankung
   hereditary disease, genetic disease,
   inherited disease, heritable disorder,
   genetic defect, genetic disorder
Erbleiden/angeborener Fehler
   inborn error
erblich/hereditär hereditary, heritable
Erbmerkmal hereditary trait
Erbrechen vomiting
   ➢ **provoziertes E.**
   induced vomiting
Erbschaden/genetischer Schaden
   genetic hazard
Erbträger/Erbsubstanz
   hereditary material
Erdbeschleunigung
   acceleration of gravity
Erde/Erdboden/Erdreich
   soil, ground, earth
Erde/Erdung *electr* ground
Erde/Welt Earth, world
Erdfehler/Erdschluss ground fault
Erdgas natural gas
Erdöl petroleum, crude oil
Erdreich/Erdboden/Erde
   soil, ground, earth

Erdschlussstrom/Fehlerstrom
   ground fault current
   (leakage current)
erfassen
   (aufnehmen) acquire, record;
   (bewerten) assess
Erfassung
   (Aufnahme von Ergebnissen etc.)
   acquiring, acquisation, recording;
   (Bewertung) assessment
Erhaltungsenergie
   maintenance energy
Erhaltungskoeffizient
   maintenance coefficient (m)
erheben *math/stat* survey
Erhebung *math/stat* survey
erhitzen heat
erholen recover
Erholung recovery
erkalten (lassen) (let) cool
Erkältung (viraler Infekt) cold
Erkennungssequenz-
   Affinitätschromatographie
   recognition site affinity
   chromatography
erkranken fall ill, get sick,
   sicken, contract a disease
Erkrankung illness, sickness,
   disease, disorder (Störung)
Erlaubnis permission
Erlenmeyer Kolben Erlenmeyer flask
ermitteln (bestimmen/herausfinden)
   determine, investigate, check;
   (finden) trace, locate,
   find out, discover
Ermittlungsergebnisse
   test results (of an investigation)
ermüden
   fatigue; tiring, become tired
Ermüdung
   fatigue, tiring
   ➢ **Materialermüdung**
   material fatigue
ernähren
   (nähren/füttern) nurture, feed;
   (sich von etwas ernähren/leben von)
   (Mensch) eat something, live on;
   (Tiere) feed on something

Ernährung/Nahrung
  food, diet, nourishment, nutrition; (Füttern) feeding, nourishing
Ernährungswissenschaft/Diätetik
  nutrition (nutrition science/nutrition studies), dietetics
Ernte harvest
ernten harvest
erregbar excitable, irritable, sensitive
Erregbarkeit
  excitability, irritability, sensitivity
erregen excite, irritate
erregend/exzitatorisch
  excitatory
Erreger (Fluoreszenzmikroskopie)
  exciter
➤ **Krankheitserreger**
  disease-causing agent, pathogen
Erregerfilter (Fluoreszenzmikroskopie)
  exciter filter
erregter Zustand/
  angeregter Zustand
  chem/med/physiol excited state
Erregung (Aufregung) arousal, excitement; (Impuls) impulse; (Irritation) excitation, irritation
Erregungsleitung
  transmission of signals, impulse propagation
erreichen/sich annähern/
  näherkommen/annähern
  (z.B. einen Wert)
  approach (*vb*) (e.g. a value)
Ersatz
  substitute, replacement
Ersatzbirnchen *electr*
  replacement bulb, replacement lamp
Ersatzname
  substitute name
Ersatzstoff
  substitute substance
Ersatzteile
  spare parts, replacement parts
Ersatztherapie
  substitution therapy

Erscheinungsbild/
  Erscheinungsform
  appearance
Erschwerniszulage extra pay (bonus/compensation) for difficult working conditions
ersetzen replace
erstarren freeze
Erste Hilfe/Erstbehandlung
  first aid
Erste-Hilfe Ausrüstung
  first-aid supplies
Erste-Hilfe-Kasten
  (Erste-Hilfe-Koffer) first-aid kit; (Medizinschrank/Medizinschränkchen) first-aid cabinet, medicine cabinet
Ersthelfer first-aider
ersticken suffocate
Ersticken suffocation
erstickend (chem.
  Gefahrenbezeichnung)
  asphyxiant
Ertrag/Ausbeute yield
Ertragsklasse/Ertragsniveau/Bonität
  yield level, quality class
Ertragskoeffizient/Ausbeutekoeffizient/
  ökonomischer Koeffizient
  yield coefficient (Y)
Ertragsminderung yield reduction
Ertragssteigerung yield increase
erwärmen heat, warm (warm up)
➤ **erhitzen** heat (heat up)
Erwärmung heating, warming
erwerben acquire
Erythrozytenschatten/Schatten
  (leeres/ausgelaugtes
  rotes Blutkörperchen)
  erythrocyte ghost
Erz ore
erzeugen produce, make
Erzeuger/Produzent
  producer; (Müll etc.) generator
Erzlaugung, mikrobielle
  microbial metal-ore leaching, microbial leaching of metal ores
ESR (Elektronenspinresonanz)
  ESR (electron spin resonance)

es

essbar edible, eatable
➢ nicht essbar inedible, uneatable
Essbarkeit edibility, edibleness
essen eat
Essen food; (Mahlzeit) meal
essentiell essential
essentielle Aminosäure
  essential amino acids
Essenz *chem/pharm* essence
➢ Fruchtessenz fruit essence
Essig vinegar
Essigsäure/Ethansäure (Acetat)
  acetic acid, ethanoic acid (acetate)
➢ 'aktivierte Essigsäure'/Acetyl-CoA
  acetyl CoA, acetyl coenzyme A
Essigsäureanhydrid
  acetic anhydride,
  ethanoic anhydride,
  acetic acid anhydride
Ether/Äther ether
➢ ausethern
  extract with ether,
  shake out with ether
➢ Petrolether/Petroläther
  petroleum ether
Etherfalle ether trap
Etikett label, tag
➢ Namensetikett
  name tag
etikettieren/markieren tag
Etikettierung
  labelling, tagging
Etui case
eutektischer Punkt
  eutectic point
Eutervorlage/Verteilervorlage/'Spinne'
  dest cow receiver adapter, 'pig',
  multi-limb vacuum receiver adapter
  (receiving adapter for three/four
  receiving flasks)
eutroph (nährstoffreich)
  eutrophic
Evakuierungsplan
  evacuation plan
Evaporator
  evaporator, concentrator
➢ Vakuum- E.
  vacuum concentrator, speedy vac

Evaporimeter/Verdunstungsmesser
  evaporimeter, evaporation gauge,
  evaporation meter
Excision/Exzision/Herausschneiden
  excision
Exclusion/Exklusion/Ausschluss
  exclusion
Exemplar/Muster/Probe
  specimen, sample
exergon/energiefreisetzend
  exergonic
Exklusion/Ausschluss
  exclusion
Exkremente excretions
Exkret/Exkretion excretion
Exkursion excursion, field trip
exogen exogenic, exogenous
exotherm exothermic
Expansionsstück (Laborglas)
  expansion adapter,
  enlarging adapter
experimentieren experiment
Expertenwissen expertise
Expiration/Ausatmen
  expiration
Explantat explant
explodieren explode
Explosion explosion
➢ Staubexplosion
  dust explosion
Explosionsgefahr
  explosion hazard
explosionsgefährlich
  (Gefahrenbezeichnungen)
  explosive (E)
explosionsgeschützt/
  explosionssicher
  explosionproof
Explosionsgrenze
  (untere=UEG/obere=OEG)
  explosion limit
explosiv explosive
Explosivstoff
  (*siehe:* Sprengstoff)
  explosive
➢ Schießstoff/Schießmittel
  low explosive
➢ Sprengstoff explosive

**exponentielle Wachstumsphase/
exponentielle Entwicklungsphase**
exponential growth phase
Exposition/Ausgesetztsein
med/chem exposure
exprimieren express
Exsikkator desiccator
Exsudat/Absonderung/Abscheidung
exudate, exudation, secretion
Extinktionskoeffizient
extinction coefficient, absorptivity
extrahieren/herauslösen
extract
Extrakt/Auszug extract
➢ **Fleischextrakt** *micb* meat extract
➢ **Hefeextrakt** yeast extract

➢ **Rohextrakt** crude extract
➢ **Zellextrakt** cell extract
➢ **zellfreier Extrakt** cell-free extract
Extraktion extraction
Extraktionshülse extraction thimble
Extraktionszange *dent* **(Zähne)**
extraction forceps
extrapolieren (hochrechnen)
extrapolate
**Extrudierdüse/Extruderdüse/
Pressdüse**
polym extrusion die
extrudieren (strangpressen)
polym extrude
Extrusion/Extrudieren/Strangpressen
polym extrusion

Fa

Fachbezeichnungen/Terminologie
terminology
Fächer fan
Fächerung/
Kompartimentierung/Unterteilung
compartmenta(liza)tion,
sectionalization, division
Fachgebiet specialty, special field,
field of specialization
Fachkenntnis/
Sachkenntnis/Expertenwissen
expertise
Fachsprache/Fachterminologie
terminology
Fackel torch
FACS (fluoreszenzaktivierte
Zelltrennung/Zellsortierung)
FACS (fluorescence-activated cell sorting)
Faden filament, thread
Fadenkreuz crosshairs
Fahrgestell
(Kistenroller/Fassroller etc.)
dolly
Fäkalien (Kot & Harn) fecal matter
(incl. urin) (*see:* Fäzes/Kot)
Faktor factor
➤ **begrenzender Faktor**
ecol limiting factor
➤ **dichteabhängiger Faktor**
ecol density-dependent factor
➤ **dichteunabhängiger Faktor**
ecol density-independent factor
➤ **Umweltfaktoren**
environmental factors
fakultativ facultative, optional
Fall *med* case
Falle trap
fallen fall
Fällen/Ausfällen/Ausfällung/
Präzipitation
chem precipitation
fällen/ausfällen/präzipitieren
chem precipitate
Fallmischer tumbler, tumbling mixer
Fallrohr downpipe
Fallstrombank
vertical flow workstation/hood/unit

Fällung/Ausfällung/Präzipitation
precipitation; (Präzipitat) precipitate
➤ **fraktionierte Fällung**
fractional precipitation
Fällungsmittel
precipitant, precipitating agent
Fällungstitration precipitation titration
Fallzahl *stat* sample size
falsch false, spurious
fälschen fake, falsify, forge, fabricate
falschpositiv (falschnegativ)
false-positive (false-negative)
Fälschung/'Erfindung'
fabrication, faking, falsification
➤ **gefälschte Daten**
fabricated data
Falte fold, plication, wrinkle
Faltenfilter folded filter
faltig folded, pleated, plicate(d)
Faradaykäfig Faraday cage
Farbanpassung color-matching
Färbbarkeit *micros* stainability
Färbegestell staining tray
Färbeglas/Färbetrog/Färbewanne
micros staining dish, staining jar, staining tray
Färbekasten staining dish
Färbemethode/Färbetechnik
staining method/technique
färben/einfärben dye, add color,
add pigment;
(kontrastieren) *tech/micros* stain
Färben/Färbung/
Einfärbung/Kontrastierung
tech/micros stain, staining
Farbensehen color vision
Farbmarker *electrophor* tracking dye
Farbstoff/Pigment dye, dyestuff;
colorant, pigment; *micros* stain;
(in Nahrungsmitteln) colors, coloring
➤ **künstliche Farbstoffe**
artificial colors, artificial coloring
➤ **natürliche Farbstoffe**
natural colors, natural coloring
➤ **Supravitalfarbstoff**
supravital dye, supravital stain
➤ **Vitalfarbstoff/Lebendfarbstoff**
vital dye, vital stain

Farbton/Tönung/Schattierung
  hue
Farbumschlag/Farbänderung
  color change
Färbung (durch Farbstoffzugabe)
  micros staining
  ➤ **Lebendfärbung/Vitalfärbung**
  vital staining
  ➤ **Supravitalfärbung** supravital staining
Färbung/Farbton/Pigmentation
  color, shade, tint, tone,
  pigmentation, coloration
Faser fiber
  ➤ **Ballaststoffe (diätätisch)**
  dietary fiber
  ➤ **Glasfaser/Faserglas**
  fiberglass
  ➤ **Hohlfaser** hollow fiber
  ➤ **Textilfaser** textile fiber
faserig/fasrig fibrous, stringy
Faserstoffplatte fiberboard
Fass barrel, drum, vat, tub, keg, tun;
  (Holzfass) cask
Fassöffner barrel opener
Fasspumpe barrel pump, drum pump
Fassschlüssel
  (zum Öffnen von Fässern)
  drum wrench
Fassung/Steckbuchse
  electr socket, receptacle
  ➤ **Gewindefassung** screw-base socket
Fassungsvermögen capacity
Fassventil (Entlüftung) drum vent
Fasten *n* fasting
fasten *vb* fast
faul/modernd
  foul, rotten, decaying, decomposing
Faulbehälter (Abwässer)
  septic tank
Fäule rot, mold, mildew, blight
faulen rot, decay, decompose,
  disintegrate;
  (im Faulturm der Kläranlage) digest
Faulgas/Klärgas (Methan)
  sludge gas, sewage gas
Fäulnis decay, rot, putrefaction
Fäulnisbakterien
  putrefactive bacteria
Fäulnisernährer/Fäulnisfresser/
  Saprovore/Saprophage
  saprophage, saprotroph, saprobiont
fäulniserregend/saprogen
  saprogenic
Faulschlamm
  (*speziell:* **ausgefaulter Klärschlamm**)
  sewage sludge
  (*esp.*: excess sludge from digester)
Faulschlamm/Sapropel
  sludge, sapropel
  ➤ **Halbfaulschlamm/**
  Grauschlamm/
  Gyttia/Gyttja
  gyttja, necron mud
Faulschlammgas sewer gas
Faulturm digester, digestor, sludge
  digester, sludge digestor
Fäustel club hammer
  ➤ **Handfäustel** mallet;
  (Gummihammer) rubber mallet
Fäzes/Kot feces; (Stuhl) human feces
Fazies facies
FCKW (Fluorchlorkohlenwasserstoffe)
  CFCs (chlorofluorocarbons/
  chlorofluorinated hydrocarbons)
Federklammer (für Kolben:
  Schüttler/Mischer)
  (four-prong) flask clamp
fegen/kehren sweep (up)
Fehlbildung malformation
fehlend lacking, missing, wanting
Fehler error, mistake; defect
  ➤ **statistischer Fehler** statistical error
  ➤ **systematischer Fehler/Bias**
  systematic error, bias
  ➤ **zufälliger Fehler/Zufallsfehler**
  random error
fehlerhaft erroneous, mistaken, flawed;
  (falsch) incorrect, wrong, false
Fehlermeldung/Falschmeldung
  false report
  ➤ **Fehleranzeige** malfunction report
fehlernährt malnourished
Fehlernährung malnutrition
Fehlerquelle source of error/mistake;
  source of trouble/defect
Fehlersuche troubleshooting

Fehlingsche Lösung Fehling's solution
Fehlzünden/Fehlzündung
   misfire, backfire
Feile file
➢ **Holzfeile** wood file
➢ **Metallfeile** metal file
➢ **Nadelfeile** needle file
Feilspäne (Metall~) filings (metal)
Feinbau/Feinstruktur
   fine structure
➢ **Ultrastruktur**
   ultrastructure
Feinchemikalien fine chemicals
Feinjustierschraube/Feintrieb
   micros fine adjustment knob
Feinjustierung/Feineinstellung *micros*
   fine adjustment,
   fine focus adjustment
Feinstaub (alveolengängig)
   fine dust, mist
Feinstruktur/Feinbau fine structure
Feinwaage precision balance
Feiung/stille Feiung/stumme Infektion
   silent infection
Fekundität fecundity
Feldblende *opt/micros* field diaphragm
Felddesorption (FD)
   field desorption (FD)
Feldionisation field ionization
Feldlinse *micros* field lens
Feldversuch/Freilanduntersuchung/
   Freilandversuch
   field study, field investigation,
   field trial
Fenster window
Fensterglas window glass
Fensterkitt glazier's putty
Fensterrahmen window frame
Fensterscheibe window pane
Fensterwischer/Fensterabzieher
   squeegee (for windows)
Ferment/Enzym enzyme
Fermenter/Gärtank
   (*siehe auch:* **Reaktor**)
   fermenter, fermentor
fermentieren/gären ferment
Fernbachkolben Fernbach flask
Fernbedienung remote control

Ferntransport long-distance transport
Fernwärme long-distance heat(ing)
Fertigarzneimittel/
   Generica/Generika generic drug
Fertigplatte *chromat* precoated plate
Fertilität/Fruchtbarkeit fertility
Ferulasäure ferulic acid
fest firm, tight; solid
fest verschlossen
   tightly closed, sealed tight
fest werden (steif werden/
   abbinden) set; (fest werden
   lassen/erstarren) solidify
Festbettreaktor (Bioreaktor)
   fixed bed reactor, solid bed reactor
Festphase solid phase, bonded phase
festsitzend/festgewachsen/
   festgeheftet/aufsitzend/sessil
   firmly attached (permanently),
   sessile
feststeckend/festgebacken
   (Schliff/Hahn)
   jammed, seized-up, stuck,
   'frozen', caked
feststellen/fixieren arrest, fixate
Feststoff solid, solid matter
Festwinkelrotor *centrif*
   fixed-angle rotor
fetales Kälberserum
   fetal calf serum (FCS)
fetotoxisch fetotoxic
Fett fat
fettartig/fetthaltig/Fett...
   fatty, adipose
Fettgießer (große 'Pipette') baster
fettig fatty
Fettigkeit/fettig-ölige Beschaffenheit
   oiliness
fettlöslich fat-soluble
Fettsäure fatty acid
➢ **einfach ungesättigte Fettsäure**
   monounsaturated fatty acid
➢ **gesättigte Fettsäure**
   saturated fatty acid
➢ **mehrfach ungesättigte Fettsäure**
   polyunsaturated fatty acid
➢ **ungesättigte Fettsäure**
   unsaturated fatty acid

Fettspeicher/Fettreserve
  fat storage, fat reserve
Fetttröpfchen/Fett-Tröpfchen
  fat droplet
feucht humid, damp, moist
Feuchte moistness, dampness
Feuchte-Orgel/Feuchtigkeitsorgel
  ecol humidity-gradient apparatus
Feuchtigkeit
  humidity, dampness, moisture
➤ **Luftfeuchtigkeit (absolute/relative)**
  (absolute/realtive) air humidity
Feuchtigkeitsmesser/Hygrometer
  hygrometer
Feuchtigkeitsschreiber/Hygrograph
  hygrograph
feuchtigkeitsundurchlässig
  moisture-proof
Feuer (*siehe auch:* **Flamm...**)
  fire
Feuer löschen
  put out a fire, quench a fire
Feueralarm fire alarm
Feueralarmanlage
  fire-alarm system
Feueralarmübung/Feuerwehrübung
  fire drill
Feuerbekämpfung fire fighting
feuerbeständig fire-resistant
feuerfest/feuersicher
  fireproof, flameproof
Feuergefahr fire hazard
feuerhemmend/flammenhemmend
  fire-retardant, flame-retardant
Feuerleiter (Nottreppe)
  fire-escape
Feuerlöschdecke fire blanket
Feuerlöscher/Feuerlöschgerät
  fire extinguisher
Feuerlöschfahrzeug
  fire engine, fire truck
Feuerlöschmittel
  fire-extinguishing agent
Feuerlöschschaum
  fire foam
Feuermelder fire alarm
feuern fire, firing
feuerpoliert fire polished

Feuerschutz fire protection,
  fire prevention; fireproofing
➤ **Sprinkleranlage (Beregnungsanlage/**
  Berieselungsanlage)
  fire sprinkler system
Feuerschutzmittel (zur Imprägnierung)
  fireproofing agent; fire retardant
Feuerschutzvorhang
  fireproofing curtain, fire curtain
Feuerschutzvorschriften fire code
Feuerschutzwand fire wall, fire barrier
feuersicher/feuerfest fireproof
Feuerwehr fire brigade, fire department
Feuerwehrmann firefighter, fireman
Feuerwehrschlauch fire hose
Feuerwehrübung fire drill
Feuerwehrvereinigung
  fire protection association
➤ **U.S.-Feuerwehrvereinigung**
  National Fire Protection Association
  (NFPA)
Feuerwiderstandsklasse
  fire resistance class
Fiberglas fiberglass
Fibroskop/Faserendoskop/
Fiberendoskop
  fiberscope
Filamentband filament tape
Filter filter
➤ **Anreicherung durch Filter**
  filter enrichment
➤ **aschefreier quantitativer Filter**
  ashless quantitative filter
➤ **Erregerfilter**
  (Fluoreszenzmikroskopie)
  exciter filter
➤ **Faltenfilter**
  folded filter, plaited filter,
  fluted filter
➤ **Filternutsche/Nutsche**
  (Büchner-Trichter) nutsch filter,
  nutsch, filter funnel, suction funnel,
  suction filter, vacuum filter
  (Buechner funnel)
➤ **HOSCH-Filter**
  (Hochleistungsschwebstofffilter)
  HEPA-filter (high-efficiency
  particulate and aerosol air filter)

Fi

- **Membranfilter** membrane filter
- **Nutsche** nutsch filter, nutsch
- **Partikelfilter** particle filter
- **Polarisationsfilter/
  'Pol-Filter'/Polarisator**
  polarizing filter, polarizer
- **Rauschfilter** noise filter
- **Rippenfilter** ribbed filter, fluted filter
- **Rundfilter**
  round filter, filter paper disk, 'circles'
- **Sperrfilter** *micros*
  selective filter, barrier filter, stopping filter, selection filter
- **Spritzenvorsatzfilter/Spritzenfilter**
  syringe filter
- **Sterilfilter** sterile filter
- **Tonfilter** ceramic filter
- **Überspannungsfilter**
  surge suppressor
- **Vakuumdrehfilter**
  rotary vacuum filter
- **Vorfilter** prefilter

Filteranreicherung
   filter enrichment
Filterblättchenmethode
   filter disk method
Filterblende (Schirm)
   filter screen
Filterhilfsmittel filter aid
Filterkerze cartridge
Filterkuchen/Filterrückstand
   filter cake, filtration residue, sludge
Filtermaske filter mask
**Filternutsche/Nutsche
(Büchner-Trichter)** nutsch, nutsch filter, suction funnel, suction filter, vacuum filter (Buchner funnel)
Filterpapier filter paper
Filterpipette filtering pipet
Filterpresse filter press
Filterpumpe filter pump
Filterrückstand/Filterkuchen
   filtration residue, filter cake, sludge
Filterstaub fly ash
Filterstopfen filter adapter
Filterträger *micros* filter holder
Filtervorstoß adapter for filter funnel
Filtrat filtrate

Filtration filtration
- **Druckfiltration** pressure filtration
- **Gelfiltration/
  Molekularsieb-chromatographie/
  Gelpermeations-Chromatographie**
  gel filtration, molecular sieving chromatography, gel permeation chromatography
- **Klärfiltration** clarifying filtration
- **Kreuzstrom-Filtration**
  cross flow filtration
- **Kuchenfiltration**
  dead-end filtration
- **Querstromfiltration**
  cross-flow filtration
- **Saugfiltration** suction filtration
- **Schwerkraftsfiltration
  (gewöhnliche F.)** gravity filtration
- **Sterilfiltration** sterile filtration
- **Ultrafiltration** ultrafiltration
- **Vakuumfiltration**
  vacuum filtration, suction filtration

filtrieren/passieren filter, pass through
Filtrierer/Filterer filter feeder
**Filtrierflasche/Filtrierkolben/
Saugflasche**
   filter flask, vacuum flask
Filtrierrate/Filtrationsrate filtering rate
Filtrierung/Filtrieren filtering, filtration
filzig felty, felt-like, tomentose
Filzstift/Filzschreiber
   felt-tip pen, felt-tipped pen
Fingerabdruck fingerprint
Fingerhut thimble
Fingerling (Schutzkappe) finger cot
**Fingerprinting/
genetischer Fingerabdruck**
   fingerprinting, genetic fingerprinting, DNA fingerprinting
Firnis varnish
**'Fisch'/Rührfisch (Magnetstab/
Magnetstäbchen/Magnetrührstab)**
   'flea', stir bar, stirrer bar, stirring bar, bar magnet
**FI-Schalter
(Fehlerstromschutzschalter)**
   leakage current circuit breaker, surge protector (fuse)

Fl

Fischer-Projektion/Fischer-Formel/ Fischer-Projektionsformel
Fischer projection, Fischer formula, Fischer projection formula
FISH (*in situ* Hybridisierung mit Fluoreszenzfarbstoffen)
FISH (fluorescence activated *in situ* hybridization)
Fisher-Verteilung/F-Verteilung/ Varianzquotientenverteilung
variance ratio distribution, F-distribution, Fisher distribution
fixieren (befestigen/fest machen) affix, attach
fixieren (mit Fixativ härten) fix
Fixiermittel/Fixativ fixative
Fixierung/Fixieren fixation
Flachbehälter/Schale tray
Flachstecker flat plug
Flachsteckhülse flat-plug socket
Flachsteckverbinder
flat-plug connector
Flachzange flat-nosed pliers
flammbeständig flame-resistant
Flamme (*siehe auch: Feuer...*) flame
➢ **Sparflamme/Zündflamme**
pilot flame, pilot light
Flammen ersticken smother the flames
Flammenemissionsspektroskopie (FES)
flame atomic emission spectroscopy (FES), flame photometry
Flammenfärbung flame coloration
flammenhemmend/feuerhemmend
flame-retardant
Flammenionisationsdetektor (FID)
flame-ionization detector (FID)
Flammenspektroskopie
flame spectroscopy
Flammensperre/ Flammenrückschlagsicherung
flame arrestor
Flammofen reverberatory furnace
Flammpunkt flash point
Flammschutzfilter flash arrestor
Flammschutzmittel
flame retardant, flame retarder
flammsicher/flammfest (schwer entflammbar) flameproof
Flansch flange
flanschen flange
Flanschverbindung flange connection, flange coupling, flanged joint
Flasche bottle
➢ **Abklärflasche/Dekantiergefäß**
decanter
➢ **Ballonflasche** carboy
➢ **Druckflasche** cylinder
➢ **Druckgasflasche** gas cylinder
➢ **Enghalsflasche**
narrow-mouthed bottle
➢ **Filtrierflasche/Filtrierkolben/ Saugflasche**
filter flask, vacuum flask
➢ **Gasflasche** gas bottle, gas cylinder, compressed-gas cylinder
➢ **Gaswaschflasche**
gas washing bottle
➢ **Gewebekulturflasche/ Zellkulturflasche**
tissue culture flask
➢ **Laborstandflasche/Standflasche**
lab bottle, laboratory bottle
➢ **Nährbodenflasche**
culture media flask
➢ **Pipettenflasche**
dropping bottle, dropper vial
➢ **Rollerflasche** roller bottle
➢ **Rollrandflasche** beaded rim bottle
➢ **Schraubflasche** screw-cap bottle
➢ **Spritzflasche**
wash bottle, squirt bottle
➢ **Sprühflasche** spray bottle
➢ **Thermoskanne/Thermosflasche**
thermos
➢ **Tropfflasche**
drop bottle, dropping bottle
➢ **Verpackungsflasche**
packaging bottle
➢ **Vierkantflasche** square bottle
➢ **Weithalsflasche** wide-mouthed bottle
➢ **Woulff'sche Flasche** Woulff bottle
Flaschenbürste tube brush (test tube brush), bottle brush (beaker/jar/cylinder brush)

Fl

Flaschendruckmanometer
cylinder pressure gauge
Flaschenhals/Engpass *stat* bottleneck
Flaschenregal bottle shelf, bottle rack
Flaschenwagen bottle cart (barrow), bottle pushcart, cylinder trolley (*Br*)
Flaschenzug pulley
Flash-Chromatographie/ Blitzchromatographie
flash chromatography
Flechtensäure lichen acid
Fleck spot, stain
Fleckenentferner spot remover
fleckig
speckled, patched, spotted, spotty
Fleischbrühe/Kochfleischbouillon
cooked-meat broth
Fleischextrakt *micb* meat extract
fleischig fleshy
Fleischwasser/ Fleischbrühe/Fleischsuppe *micb*
meat infusion
(meat digest/tryptic digest)
Fliese tile
➤ **Bodenfliese** floor tile
Fliesenfußboden tiled floor, tiling
Fließbettreaktor
fluid bed reactor
fließen flow
Fließfähigkeit/Fluidität fluidity
Fließgeschwindigkeit
flow rate
Fließgleichgewicht/ dynamisches Gleichgewicht
steady state, steady-state equilibrium
Fließgrenze/Fließpunkt yield point
Fließinjektion flow injection
Fließinjektionanalyse (FIA)
flow injection analysis (FIA)
Fließmittel *chromat*
solvent (mobile phase)
Fließmittelfront *chromat*
solvent front
Fließpunkt (Schmelzpunkt) fusion point (melting point)
Fließrichtung direction of flow
Flintglas flint glass
flockig/locker fluffy

Flockulation flocculation
Flockung flocking
florieren/gedeihen
flourish, thrive
Fluchtgerät/Selbstretter (Atemschutzgerät)
emergency escape mask
flüchtig volatile
➤ **leicht flüchtig (niedrig siedend)**
highly volatile, light
➤ **nicht flüchtig**
nonvolatile
➤ **schwer flüchtig (höhersiedend)**
less volatile, heavy
Flüchtigkeit *chem* **(von Gasen: Neigung zu verdunsten)**
volatility
Fluchtweg escape route, egress
Flügelhahnventil
butterfly valve
Flügelschraube thumbscrew
Flugstaub airborne dust;
(von Abgasen) flue dust
Fluidität/Fließfähigkeit fluidity
Fluktuation fluctuation
Fluktuationsanalyse/Rauschanalyse
fluctuation analysis, noise analysis
Fluktuationstest fluctuation test
Fluor fluorine
Fluorchlorkohlenwasserstoffe (FCKW)
chlorofluorocarbons, chlorofluorinated hydrocarbons (CFCs)
Fluoreszenz fluorescence
fluoreszenzaktivierte Zellsortierung/Zelltrennung
fluorescence-activated cell sorting (FACS)
fluoreszenzaktivierter Zellsorter/Zellsortierer
fluorescence-activated cell sorter
Fluoreszenzanalyse/Fluorimetrie
fluorescence analysis, fluorimetry
Fluoreszenzerholung nach Lichtbleichung
fluorescence photobleaching recovery, fluorescence recovery after photobleaching (FRAP)

Fluoreszenz-*in-situ*-**Hybridisation (FISH)**
fluorescence-*in-situ*-hybridization (FISH)
Fluoreszenzlöschung
fluorescence quenching
Fluoreszenzsonde/Fluoreszenzmarker
fluorescence marker
Fluoreszenzspektroskopie/ Spektrofluorimetrie
fluorescence spectroscopy
fluoreszieren fluoresce
fluoreszierend fluorescent
Fluoridierung fluoridation
fluorieren fluorinate
Fluorkohlenwasserstoff
fluorinated hydrocarbon
Fluoroschwefelsäure/Fluorsulfonsäure
fluorosulfonic acid
Fluorwasserstoff/ Fluoran/Hydrogenfluorid
hydrogen fluoride
Fluorwasserstoffsäure/Flusssäure
hydrofluoric acid, phthoric acid
Flur/Korridor hallway, hall, corridor
Fluss (Fließen) flow; (Licht/Energie; Volumen pro Zeit pro Querschnitt) flux
 ➢ **diffuser Fluss** diffuse flux
flüssig fluid, liquid
Flüssigextrakt/ flüssiger Extrakt/Fluidextrakt
fluid extract
Flüssiggas liquid gas, liquefied gas; (verflüssigtes Erdgas) liquefied natural gas (LNG)
Flüssigkeit fluid, liquid
Flüssigkeit ablassen drain
Flüssigkeitschromatographie
liquid chromatography (LC)
Flüssigkristallanzeige
liquid crystal display
Flussmittel/Schmelzmittel/Zuschlag
flux, fusion reagent
Flussrate fluence
Flussregler flow regulator
fluten flood, flush; inundate
fokussieren focus (focussing)

Fokussierung focussing
Folie foil
Folienschweißgerät
wrapfoil heat sealer
folieren foliate (coat s.th. with foil)
Folierung foliation
Folsäure (Folat)/ Pteroylglutaminsäure
folic acid (folate), pteroylglutamic acid
Förderband conveyor belt
Förderleistung (Pumpe) flow rate
Förderpumpe feed pump
Forensik/forensische Medizin/ Gerichtsmedizin/Rechtsmedizin
forensics, forensic medicine
forensisch/gerichtsmedizinisch
forensic
Formaldehyd/Methanal
formaldehyde, methanal
Formänderung/ Verformung/Deformation
deformation
Formel formula
 ➢ **Fischer-Projektion/Fischer-Formel/ Fischer-Projektionsformel**
Fischer projection, Fischer formula, Fischer projection formula
 ➢ **Haworth-Projektion/Haworth-Formel**
Haworth projection, Haworth formula
 ➢ **Ionenformel** ionic formula
 ➢ **Kettenformel**
chain formula, open-chain formula
 ➢ **Molekularformel/Molekülformel**
molecular formula
 ➢ **Ringformel** ring formula
 ➢ **Strukturformel** structural formula
 ➢ **Summenformel/Elementarformel/ Verhältnisformel/empirische Formel**
empirical formula
 ➢ **Verhältnisformel**
empirical formula
Formelsammlung formulary
Forscher researcher, research scientist, research worker, investigator
Forschung research; trial, experimentation, investigation

Forschungsabteilung
research department
Forschungsauftrag
research assignment/contract
Forschungsbeirat
Research Advisory Committee
Forschungsfinanzierung
research funding
Forschungsgelder research funds
Forschungslabor
research laboratory
Forschungsprogramm
research program
**Forschungsvorhaben/
Forschungsprojekt**
research project
Fortbewegung/Bewegung/Lokomotion
movement, motion, locomotion
Fortbildung advanced training
Fortbildungskurs
advanced-training course/workshop
fortleiten/weiterleiten (Nervenimpuls)
propagate
**Fortleitung/Weiterleitung
(Nervenimpuls)** propagation
fortpflanzen/vermehren/reproduzieren
propagate, reproduce
**Fortpflanzung/
Vermehrung/Reproduktion**
propagation, reproduction
> **geschlechtliche/sexuelle F.**
sexual reproduction
> **ungeschlechtliche/vegetative F.**
asexual/vegetative reproduction
**fortpflanzungsfähig/
fruchtbar/fertil** fertile
**Fortpflanzungsfähigkeit/
Fruchtbarkeit/Fertilität**
fertility
**fortpflanzungsgefährdend/
reproduktionstoxisch**
toxic to reproduction (T)
Fortpflanzungszelle reproductive cell
fossile Brennstoffe fossil fuels
fossilisieren/versteinern fossilize
fossilisiert/versteinert fossilized
Fossilisierung/Versteinerung
fossilization

Fotolabor photographic laboratory
Fotopapier photographic paper
Fotoplatte photographic plate
Fotovervielfacher photomultipier
Fotowiderstand
photoresist, photoresistor
Fracht freight, load, cargo, goods;
(Flüssigkeit/Abwasser) load, freight
Frachtbrief bill of lading;
(als Formular) lading form
Frachtcontainer freight container
Frachtgut/Ladung cargo
Frachtkessel cargo tank
Frachtliste/Frachtdokument/Manifest
manifest
Frachtpapiere shipping papers
Fragmentierungsmuster
fragmentation pattern
Fraktion fraction
fraktionieren fractionate
Fraktioniersäule
fractionating column
Fraktionierung fractionation
Fraktionssammler fraction collector
Fraßhemmer/fraßverhinderndes Mittel
antifeeding agent,
antifeeding compound,
feeding deterrent
frei schwebend
free-floating, pendulous
Freiheitsgrad *stat*
degree of freedom (df)
**Freilanduntersuchung/
Freilandversuch/Feldversuch/
vor-Ort-Untersuchung**
field study, field investigation,
field trial
freilebend free-living
freisetzen (Wärme/Energie/Gase etc.)
liberate, release, set free
Freisetzung release;
(Sekretion) secretion
> **absichtliche F.** deliberate release
Freisetzungsexperiment
deliberate release experiment,
environmental release experiment
Fremdkörper foreign body/matter/
substance; contaminant

Frequenz/Häufigkeit frequency
fressen feed (on something),
  ingest (etwas zu sich nehmen)
Frischgewicht
  (*sensu stricto*: **Frischmasse**)
  fresh weight
  (*sensu stricto*: fresh mass)
Frischhaltefolie
  cling wrap, cling foil
Fritte frit
➢ **Glasfritte**
  fritted glass filter
fritten/sintern frit, sinter
Frontscheibe (Sicherheitswerkbank)
  sash
Frost frost, rime frost, white frost
frostbeständig/frostresistent
  frost-resistant, frost hardy
frostempfindlich frost-tender,
  susceptible to frost
frostresistent/frostbeständig
  frost-resistant
Frostschutzmittel cryoprotectant
frostsicher frostproof
fruchtbar/fertil fertile, fecund
fruchtbar machen/befruchten
  fertilize, fecundate
➢ **unfruchtbar/steril**
  infertile, sterile
Fruchtbarkeit/Fertilität
  fertility; (Fekundität) fecundity
➢ **Unfruchtbarkeit/Sterilität**
  infertility, sterility
Fruchtgeschmack fruity taste
Fruchtmark/Obstpulpe/Fruchtmus
  fruit pulp
Fruchtwasser/
  Amnionwasser/Amnionflüssigkeit
  amniotic fluid
Fruchtwasserpunktion/
  Amniozentese/Amnionpunktion
  amniocentesis
Fruchtzucker/Fruktose
  fruit sugar, fructose
Frühbeet (Mistbeet/Treibbeet: beheizt)
  forcing bed, hotbed;
  (Anzuchtkasten: unbeheizt)
  cold frame

Fruktose/Fructose (Fruchtzucker)
  fructose (fruit sugar)
Fuge/Naht/Verwachsungslinie
  seam, suture, raphe
Fugendichtungsmasse
  seam sealant, joint filler
Fühler/Sensor/Detektor
  (*tech*: z.B. Temperaturfühler)
  sensor, detector
Fühlerlehre
  feeler gage, feeler gauge (*Br*)
Führungsbuchse (Rührwelle etc.)
  bushing, guide bushing
Fukose/Fucose/6-Desoxygalaktose
  fucose, 6-deoxygalactose
Füllkörper (für Destillierkolonnen)
  column packing
➢ **Raschig-Ring (Glasring)**
  Raschig ring
➢ **Sattelkörper (Berlsättel)**
  saddle (berl saddles)
➢ **Spirale** spiral
➢ **Wendel** helice
Füllkörperkolonne
  packed distillation column
Füllstand
  (z.B. Flüssigkeit eines Gefäßes)
  fill level
Füllstoff
  (auch: Füllmaterial/Verpackung)
  filler
Fumarsäure (Fumarat)
  fumaric acid (fumarate)
Fundort/Lage site, location
fünfwertig pentavalent
Fungizid fungicide
Funke spark
Funkenspektrum spark spectrum
Funktion function
➢ **Verteilungsfunktion**
  distribution function
➢ **Wahrscheinlichkeitsfunktion**
  likelihood function
Funktionalität functionality
funktionelle Gruppe
  functional group
Funktionseinheit/Modul
  functional unit, module

fu

funktionsgleich/analog analogous
Funktionsstörung malfunction; *med* functional disorder
Funktionszustand working order, operating condition
Furan furan
Fürsorgepflicht obligation to provide welfare services
Fuselöl fusel oil
Fußabdruckmethode footprinting

Fußboden floor; ground
➤ **monolithischer F. (Labor: Stein/Beton aus einem Guß)** monolithic floor
Fußmatte mat, step mat, foot mat; (Abstreifer: vor der Tür) doormat
Futter feed
füttern *vb* feed
Futterpflanze fodder, forage (plant)
Fütterung feeding

Gabelschlüssel/Maulschlüssel
  open-end wrench,
  open-end spanner (*Br*)
Gabelstapler/Hubstapler forklift
Galaktosamin galactosamine
Galaktose galactose
Galakturonsäure galacturonic acid
Galle/Gallflüssigkeit bile
Gallensalze bile salts
gallertartig/gelartig/gelatinös
  gelatinous, gel-like
Gallerte/Gelatine jelly, gelatin, gel
Gallussäure gallic acid
Gamasche (Schutzkleidung: Bein/Fuß)
  (bis zum Knie) gaiter;
  (Fuß/Schuhe) spat
Gamet/Keimzelle/Geschlechtszelle
  gamete, sex cell
Gang/Flur/Korridor
  aisle, corridor
Ganghöhe (*DNA-Helix:*** Anzahl Basenpaare pro Windung)**
  pitch (DNA: helix periodicity)
Gangunterschied *opt*
  path difference
Gänsehaut gooseflesh,
  goose pimples, goose bumps
Ganzwäsche washdown
Garantie (Hersteller~) warranty
Garbe (Licht/Funke etc.)
  sheaf, bundle
Gärbottich fermentation tank
gären/fermentieren ferment
➤ **obergärig** top fermenting
➤ **untergärig** bottom fermenting
Gärmittel/Gärstoff/Treibmittel
  leavening
Gärröhrchen/Einhorn-Kölbchen
  fermentation tube, bubbler
Gärtassenreaktor tray reactor
Gartenschere pruners, pruning shears,
  (*Br*) secateurs
Gartenschlauch garden hose
Gärung/Fermentation fermentation
Gas gas
➤ **Druckgas**
  compressed gas, pressurized gas
➤ **Edelgas** inert gas, rare gas
➤ **Erdgas** natural gas
➤ **Faulgas/Klärgas (Methan)**
  sludge gas, sewage gas
➤ **Faulschlammgas** sewer gas
➤ **Flüssiggas**
  liquid gas, liquefied gas
➤ **Generatorgas** producer gas
➤ **Lachgas (Distickstoffoxid/Dinitrogenoxid)**
  laughing gas, nitrous oxide
➤ **Prüfgas** tracer gas, probe gas
➤ **Rauchgase** flue gases, fumes
➤ **Reizgas** irritant gas
➤ **Schutzgas**
  protective gas,
  shielding gas (in welding)
➤ **Spülgas** purge gas
➤ **Trägergas, Schleppgas (GC)**
  carrier gas (an inert gas)
➤ **Vergleichsgas (GC)**
  reference gas
➤ **Wassergas** water gas
Gasabscheider gas separator
Gasanzünder gas lighter
Gasaustausch
  gas exchange, gaseous interchange,
  exchange of gases
Gasaustritt/Gasausgang/Gasabgang (aus Geräten) gas outlet
Gasblase gas bubble
Gasbrenner gas burner
Gaschromatographie
  gas chromatography
Gasdetektor/Gasspürgerät
  gas detector, gas leak detector
gasdicht gasproof
Gasdruckreduzierventil/ Druckminderventil/ Druckminderungsventil/ Reduzierventil (für Gasflaschen)
  pressure-relief valve (gas regulator,
  gas cylinder pressure regulator)
Gasdurchflusszähler/ Gasströmungsmesser
  gas flowmeter
Gasentladungsröhre
  gas-discharge tube
Gasentwicklung evolution of gas

Gasflasche gas bottle, gas cylinder, compressed-gas cylinder
➢ **Gasdruckreduzierventil/ Druckminderventil/ Druckminderungsventil/ Reduzierventil (für Gasflaschen)** pressure-relief valve (gas regulator, gas cylinder pressure regulator)
Gasflaschen-Transportkarren gas bottle cart, gas cylinder trolley (*Br*)
gasförmig gaseous
gasförmiger Zustand gaseous state
Gashahn gas cock, gas tap
Gaskocher gas burner
Gaskonstante gas constant
Gasleitung (Erdgasleitung) gas line (natural gas line)
Gasmaske gas mask
Gasmessflasche gas measuring bottle
Gasprobenrohr/ Gassammelrohr/Gasmaus gas collecting tube, gas sampling bulb, gas sampling tube
Gasraum/Dampfraum/Headspace headspace
Gasreiniger gas purifier
Gasreinigung gas cleaning, pas purification
Gassammelrohr/ Gas-Probenrohr/Gasmaus gas sampling bulb, gas sampling tube
Gasspürgerät gas leak detector
Gasthermometer gas thermometer
gasundurchlässig gastight, impervious to gas
Gasvergiftung gas poisoning
Gaswaage gas balance; dasymeter
Gaswächter/Gaswarngerät gas detector, gas monitor
Gaswäsche gas scrubbing
Gaswaschflasche gas washing bottle
Gaszählrohr gas counter
Gaszufuhr gas supply
Gaszustand gaseous state
Gattung genus (*pl* genera)

Gattungsname genus name, generic name
Gauß-Kurve/Gauß'sche Kurve *stat* Gaussian curve
Gauß-Verteilung/Normalverteilung/ Gauß'sche Normalverteilung *stat* Gaussian distribution (Gaussian curve/normal probability curve)
Gaze gauze
GC (Gaschromatographie) GC (gas chromatography)
geädert veined, venulous
gebändert/breit gestreift banded, fasciate
Gebäudeevakuierungsplan building evacuation plan
Gebäudereinigungspersonal building cleaners
Gebeine/sterbliche Hülle corpse
Gebinde bundle, bunch, lashing, packaging (larger quantities of items fastened together)
Gebläse (Föhn) blower, fan
Gebläselampe blowtorch
gebrauchsfertig ready-to-use
gedeihen/florieren thrive, flourish
geeicht/kalibriert calibrated; (standardisiert) standardized
geerdet grounded, earthed (*Br*)
Gefahr/Gefährdung/Risiko danger, hazard, risk, chance
➢ **biologische Gefahr/ biologisches Risiko** biohazard
➢ **akute G.** immediate danger, imminent danger
➢ **außer G.** out of danger, safe, secure
➢ **drohende G.** imminent danger
➢ **Gefahr am Arbeitsplatz** occupational hazard
➢ **höchste G.** extreme danger
➢ **öffentliche G.** public danger
gefährden endanger, imperil
gefährdet endangered, in danger, at risk

Gefährdung endangerment, imperilment; (Ausgesetztsein) exposure
Gefahrenbereich/Gefahrenzone danger area, danger zone
Gefahrenbezeichnungen/ Gefährlichkeitsmerkmale hazard warnings
➢ **ätzend** corrosive (C)
➢ **brandfördernd** oxidizing (O), pyrophoric
➢ **entzündlich** flammable (R10)
➢ **erbgutverändernd/mutagen** mutagenic (T)
➢ **erstickend** asphyxiant
➢ **explosionsgefährlich** explosive (E)
➢ **fortpflanzungsgefährdend/ reproduktionstoxisch** toxic to reproduction (T)
➢ **gefährlicher Stoff** hazardous material
➢ **gesundheitsschädlich** harmful, nocent (Xn)
➢ **giftig** toxic (T)
➢ **hochentzündlich** extremely flammable (F+)
➢ **krebserzeugend/ karzinogen/kanzerogen** carcinogenic (Xn)
➢ **leicht entzündlich** highly flammable (F)
➢ **mindergiftig** moderately toxic
➢ **mutagen** mutagenic
➢ **onkogen** oncogenic
➢ **radioaktiv** radioactive
➢ **reizend** irritant (Xi)
➢ **sehr giftig** extremely toxic (T+)
➢ **sensibilisierend** sensitizing
➢ **teratogen** teratogenic
➢ **toxisch** (*siehe auch dort*) toxic
➢ **tränend (Tränen hervorrufend)** lachrymatory
➢ **umweltgefährlich** dangerous for the environment (N = nuisant)

Gefahrencode/Gefahrenkennziffer hazard code
Gefahrendiamant hazard diamond
Gefahrenherd source of danger; troublespot
Gefahrenklasse danger class, category of risk, class of risk
Gefahrenquelle hazard, source of danger
➢ **biologische G.** biohazard
Gefahrenstoffklasse hazardous material class
Gefahrenstoffverordnung hazardous materials regulations
Gefahrenstufe/ Gefahrenklasse/Risikostufe hazard rating, hazard class, hazard level
Gefahrensymbol/Gefahrenwarnsymbol hazard icon, hazard symbol, hazard sign, hazard warning symbol
Gefahrenwarnzeichen hazard warning sign, hazard sign, warning sign, danger signal
Gefahrenzone danger zone
Gefahrenzulage danger allowance, hazard bonus
Gefahrgut/Gefahrgüter (gefährliche Frachtgüter) dangerous goods, hazardous materials
Gefahrgutbestimmungen hazardous materials regulations
Gefahrguttransport transport of dangerous goods, transport of hazardous materials
gefährlich dangerous; (gesundheitsgefährdend) hazardous; (riskant) dangerous, hazardous, risky
➢ **ungefährlich** not dangerous; (nicht gesundheitsgefährdend) nonhazardous
Gefahrstoff dangerous substance, hazardous substance/material
➢ **biologischer G.** biohazard, biohazardous substance

Ge

Gefahrstoffschrank
  hazardous materials safety cabinet
Gefahrzettel hazard label
Gefälle/Gradient *chem* gradient
gefaltet folded, pleated, plicate
Gefäß vessel; (Behälter) container;
  (Trachee) trachea
**Gefäßklemme/
  Arterienklemme/Venenklemme**
  hemostatic forceps, artery clamp
gefleckt spotted, mottled
gefliest (mit Fliesen ausgelegt)
  tiled
gefrierätzen freeze-etch
Gefrierätzung freeze-etching
➤ **Tiefenätzung** deep etching
Gefrierbruch *micros* freeze-fracture,
  freeze-fracturing, cryofracture
gefrieren freeze
➤ **schnellgefrieren** quickfreeze
Gefrierfach freezer compartment;
  (vom Kühlschrank)
  freezing compartment,
  freezer (of the refrigerator)
**Gefrierkonservierung/
  Kryokonservierung**
  freeze preservation, cryopreservation
Gefrierlagerung freeze storage
Gefriermikrotom
  freezing microtome, cryomicrotome
Gefrierpunkt freezing point
Gefrierpunktserniedrigung
  freezing point depression
Gefrierschnitt *micros*
  cryosection, frozen section
Gefrierschrank upright freezer;
  (Gefriertruhe) chest freezer
Gefrierschutz cryoprotection
Gefrierschutzmittel
  cryoprotectant
gefriertrocknen/lyophilisieren
  freeze-dry, lyophilize
Gefriertrocknung/Lyophilisierung
  freeze-drying, lyophilization
Gefriertruhe chest freezer
Gefühl feeling, sensation
Gegendruck counterpressure
gegenfärben *micros* counterstain

Gegenfärbung *micros*
  counterstain, counterstaining
Gegengewicht
  counterbalance, counterpoise
Gegengift/Gegenmittel/Antidot
  antidote, antitoxin,
  antivenin (tierische Gifte)
Gegenion counterion
Gegenkraft/Rückwirkungskraft
  reactive force
Gegenreaktion counterreaction
Gegenschattierung
  countershading
Gegenselektion/Gegenauslese
  counterselection
Gegenstrom
  countercurrent
Gegenstromextraktion
  countercurrent extraction
Gegenstromverteilung
  countercurrent distribution
gegliedert/unterteilt divided
Gehalt
  salary; (akademisch) stipend;
  (Lohn) wage(s); (Bezahlung) pay
gehärtet (Metall) tempered
Gehäuse
  housing; shell, case, casing
Geheimhaltungsvereinbarung
  secrecy agreement
Gehör hearing;
  (Hörfähigkeit) sense of hearing
Gehörschutz
  hearing protection
Gehörschützer
  ear muffs, hearing protectors
**Gehörschutzstöpsel/
  Ohrenstöpsel**
  ear plugs
Gehrungsschneidlade miter box
Geiger-Müller-Zähler
  Geiger-Müller counter
Geiger-Zähler
  Geiger counter
gekachelt tiled
geklärt cleared
gekoppelte Reaktion
  coupled reaction

Gel gel
> **denaturierendes Gel**
  denaturing gel
> **hochkant angeordnetes Plattengel**
  slab gel
> **horizontal angeordnetes Plattengel**
  flat bed gel, horizontal gel
> **natives Gel** native gel
> **Sammelgel** stacking gel
> **Trenngel** running gel, separating gel
Geländeaufnahme
  topographic survey
Geländekartierung
  topographic mapping
gelartig/gallertartig/gelatinös
  gelatinous, gel-like
Gelatine gelatin, gelatine
Gelbildner gelatinizing agent
Gelbbrennen
  pickling, dipping (metal etching)
Gelbbrennsäure/Scheidewasser (konz. Salpetersäure)
  aquafortis
  (nitric acid used in metal etching)
Gelee jelly
Gelelektrophorese
  gel electrophoresis
> **Feldinversions-Gelelektrophorese**
  field inversion gel electrophoresis (FIGE)
> **Gradienten-Gelelektrophorese**
  gradient gel electrophoresis
> **Pulsfeld-Gelelektrophorese**
  pulsed field gel electrophoresis (PFGE)
> **Temperaturgradienten-Gelelektrophorese**
  temperature gradient gel electrophoresis
> **Verschluss-Scheibe**
  gate (gel-casting)
> **Wechselfeld-Gelelektrophorese**
  alternating field gel electrophoresis
Gelenk *tech* joint; (Scharnier) hinge; articulation
Gelenkkopf swivel head; ball of a joint
Gelenkkupplung ball-joint connection
Gelenkschraube hinged bolt

Gelenkverbindung
  hinged joint, swivel joint, articulated joint
Gelenkwelle
  cardan shaft, universal joint shaft
Gelenkzapfen pivot pin; hinge pin
Gelfiltration/ Molekularsiebchromatographie/ Gelpermeationschromatographie
  gel filtration,
  molecular sieving chromatography,
  gel permeation chromatography (GPC)
Gelgießstand/ Gelgießvorrichtung *electrophor*
  gel caster
Gelieren *n* gelation
gelieren *vb* gel
Geliermittel gelling agent
Gelierpunkt gelling point
Gelkamm *electrophor* gel comb
Gelkammer *electrophor* gel chamber
gelöst (lösen) dissolved
gelöster Stoff solute
Gelpräzipitationstest/ Immunodiffusionstest
  immunodiffusion
Gelretardationsexperiment
  mobility shift experiment
Gelretentionsanalyse
  gel retention analysis,
  band shift assay
Gelretentionstest gel retention assay, electrophoretic mobility shift assay (EMSA)
Gel-Sol-Übergang gel-sol-transition
Gelträger/Geltablett *electrophor*
  gel tray
gemäßigt temperate, moderate
Gemeinschaftseinrichtung
  communal installation
Gemeinschaftsraum/Pausenraum
  lounge
Gemenge/Mischung mixture
Genauigkeit
  precision, accuracy
Gendiagnostik/ Bestimmung des Genotyps
  genetic diagnostics, genotyping

genehmigungsbedürftig
permit required,
requiring official permit
Genehmigungsbescheid
notice of approval
genehmigungspflichtig
subject to approval,
requiring permission/authorization
Genehmigungsverfahren
authorization procedure
Generation generation
➤ **Filialgeneration/Tochtergeneration**
filial generation
Generationsdauer generation period
Generationszeit (Verdopplungszeit)
generation time (doubling time)
Generatorgas producer gas
Generica/Generika/
Fertigarzneimittel
generic drug
Genetik/Vererbungslehre
genetics (study of inheritance)
genetische Beratung
genetic counsel(l)ing
genetischer Fingerabdruck/
DNA-Fingerprinting
DNA profiling, DNA fingerprinting
genetischer Suchtest genetic screening
genießbar/essbar comestible, eatable,
edible
➤ **ungenießbar/nicht essbar**
uneatable, inedible
genießbar/schmackhaft
palatable
➤ **ungenießbar/nicht schmackhaft**
unpalatable
Genkarte gene map, genetic map
Genkartierung
gene mapping, genetic mapping
Genmanipulation
gene manipulation
Genom genome
Gentechnik/
Gentechnologie/Genmanipulation
genetic engineering, gene technology
gentechnisch verändert
genetically engineered,
genetically modified

gentechnisch veränderter
Mikroorganismus
genetically modified microorganism
(GMM)
gentechnisch veränderter Organismus
(GVO)
genetically engineered organism,
genetically modified organism
(GMO)
Gentechnologie/
Gentechnik/Genmanipulation
gene technology, *sensu lato:* genetic
engineering (: Gentechnik)
Gentherapie gene therapy, gene surgery
Gentisinsäure gentisic acid
gepökelt/eingesalzen
cured, pickled;
corned (e.g. corned beef)
Geraniumsäure geranic acid
Geranylacetat geranyl acetate
Gerät/Anlage/Apparat instrument,
equipment, set, apparatus; appliance
➤ **Ablesegerät**
direct-reading instrument
➤ **Anzeigegerät** indicator,
recording instrument; monitor
➤ **Atemschutzgerät/Atemgerät**
breathing apparatus, respirator
➤ **Datenerfassungsgerät/**
Messwertschreiber/Registriergerät
datalogger
➤ **Elektrogerät**
electric appliance
➤ **Kontrollgerät**
controlling instrument,
control instrument,
monitoring instrument;
(Anzeige: monitor)
➤ **Laborgerät** *allg*
laboratory/lab equipment
➤ **Ladegerät** charger
➤ **Messgerät**
measuring apparatus,
measuring instrument;
gage, gauge (*Br*)
➤ **Netzgerät/Netzteil**
power supply unit;
(Adapter) adapter

- **Prüfgerät/Prüfer/Nachweisgerät**
  tester, testing device, checking instrument; detector
- **Regelgerät** control unit
- **Sichtgerät** visualizer, visual indicator, viewing unit, display unit
- **Steuergerät**
  control unit, control gear, controller
- **Untersuchungsgerät**
  testing equipment/apparatus
- **Vorschaltgerät** *electr* ballast unit; (Starter: Leuchtstoffröhren) starter

Gerätefehler instrumental error
Geräteraum equipment room
Gerätesonde equipment probe
geräuscharm low-noise
Geräuschpegel noise level
gerben tan
Gerben tanning
Gerbsäure (Tannat)
  tannic acid (tannate)
gerbsäurehaltig/gerbstoffhaltig
  tanniferous
Gerbstoff tanning agent, tannin
Gerichtsmedizin/Rechtsmedizin/ Forensik/forensische Medizin
  forensics, forensic medicine
geriffelt (z.B. Schlauchadapter)
  barbed, fluted, serrated
  (e.g. tubing adapters)
gerinnen/koagulieren set; curdle, coagulate; (Milch) curdle; (Blut) clot
Gerinnsel (z.B. Blut)
  clot (e.g. blood clot)
Gerinnung clotting
Gerinnungsfaktor clotting factor
- **Blutgerinnungsfaktor**
  blood clotting factor

Geruch *allg* smell, scent, odor
- **angenehmer Geruch/Duft**
  pleasant smell, fragrance, scent, odor
- **stechender Geruch** pungency
- **unangenehmer Geruch**
  unpleasant smell

geruchlos odorless, scentless
Geruchssinn/olfaktorischer Sinn
  olfactory sense

Geruchsstoff
  (angenehmer G.) fragrance, perfume (stronger scent); (unangenehmer/abweisender G.) repugnant substance
Gerüst
  scaffolding, framework, stroma, reticulum
Gesamthärte (Wasser) total hardness
Gesamtkeimzahl *micb*
  total germ count, total cell count
Gesamtvergrößerung *micros*
  total magnification, overall magnification
gesättigt (sättigen) saturated (saturate)
- **ungesättigt** unsaturated

Geschirr dishes
- **Glasgeschirr** glassware

Geschirr spülen
  wash/clean the dishes
Geschirrablage (Spüle/Spültisch)
  dishboard
Geschirrhandtuch dish towel
Geschirrlappen dishcloth
Geschirrspülbürste
  dishwashing brush
Geschlecht
  (männlich/weiblich/neutral)
  sex (male/female/neuter), gender
geschlechtlich sexual
- **ungeschlechtlich/asexuell**
  asexual

Geschlechtszelle/Keimzelle/Gamet
  sex cell, gamete
Geschmack taste
Geschmackssinn sense of taste, gustatory sense/sensation
Geschmackstoff(e)
  flavor, flavoring
- **künstliche(r) G.**
  artificial flavor, artificial flavoring
- **natürliche(r) G.**
  natural flavor, natural flavoring

geschmolzen melted
geschützt (schützen)
  protected (protect)

Ge

Geschwindigkeit
speed; velocity (vector); rate
geschwindigkeitsbegrenzende(r) Schritt/Reaktion
rate-limiting step/reaction
geschwindigkeitsbestimmende(r) Schritt/Reaktion
rate-determining step/reaction
Geschwindigkeitskonstante (Enzymkinetik) rate constant
geschwollen (schwellen)
turgid, swollen (swell)
Geschwollenheit/Turgidität
turgidity
Gesetz law, act, statute;
(siehe auch bei: Verordnung)
➢ **Abfallgesetz/ Abfallbeseitigungs-gesetz (AbfG)**
Federal Waste Disposal Act
➢ **Abwasserabgabengesetz (AbwAG)**
Wastewater Charges Act
➢ **Arbeitsschutzgesetz**
Industrial Safety Law;
Factory Act (*Br*)
➢ **Bundes-Imissionsschutzgesetz (BImSchG)**
Law on Immission Control,
Federal Law on Air Pollution Control
➢ **Bundes-Seuchengesetz (BSeuchG)**
Federal Law on Epidemics,
Epidemics Control Act
➢ **Bürgerliches Gesetzbuch**
Civil Code
➢ **Chemikaliengesetz (ChemG)**
Federal Chemical Law
➢ **Embryonenschutzgesetz (ESchG)**
Law on Embryonic Research,
Embryonic Research Act
➢ **Gentechnikgesetz (GenTG)**
Law on Genetic Engineering
➢ **Infektionsschutzgesetz (IfSG)**
Law for the Protection Against Contagious Disease
➢ **Pflanzenschutzgesetz (PflSchG)**
Federal Law on Pesticide Usage,
Pesticide Regulation Act
➢ **Tierschutzgesetz (TierSG)**
Law for the Protection of Animals
➢ **Tierseuchengesetz (TierSG)**
Federal Law on Epizootic Diseases
➢ **U.S. Gesetz zur Kontrolle toxischer Substanzen (Gefahrstoffe)**
Toxic Substances Control Act (TSCA)
➢ **Wasserhaushaltsgesetz (WHG)**
Water Resources Policy Act
gesetzeswidrig
against the law, illegal, unlawful
Gesichtsfeld/Sehfeld/Blickfeld
field of vision, field of view, scope of view, range of vision, visual field
Gesichtsfeldblende/Okularblende
micros ocular diaphragm, eyepiece diaphragm, eyepiece field stop
Gesichtsmaske face mask
Gesichtsschutz/Gesichtsschirm
faceshield
Gesichtssinn vision, eyesight
gespornt spurred
Gestalt
shape, form, appearance, contour
gestapelt (stapeln) stacked (stack)
gestaucht/zusammengezogen
compressed, contracted
Gestell (Sammlung/Aufbewahrung etc.)
rack
gesund healthy
➢ **ungesund**
unhealthy, detrimental to one's health
Gesundheit health
Gesundheitsattest health certificate
gesundheitsbedrohend
health-threatening
Gesundheitserziehung
health education
Gesundheitsfürsorge
health care, medical welfare
Gesundheitsrisiko health hazard
gesundheitsschädlich
harmful, detrimental to one's health;
(Xn) harmful, nocent
gesundheitswidrig
unhealthy, harmful

Gesundheitszeugnis (ärtzliches Attest)
health certificate
Gesundheitszustand health,
state of health, physical condition
geteilt divided, parted, partite
(divided into parts)
➢ **ungeteilt** undivided, not divided
Getreidemehl *grob:* meal, *fein:* flour
Getriebe (Motor) *tech*
transmission (of gearing)
Gewächshaus/Treibhaus
greenhouse, hothouse, forcing house
gewaltsam öffnen force open
Gewässergüte/Wassergüte
water quality
Gewebe fabric, cloth, tissue;
(Zellassoziation) tissue
Gewebeabstoßung tissue rejection
Gewebeband/Textilband (einfach)
cloth tape
➢ **Panzerband/Gewebeklebeband**
(Universalband/Vielzweckband)
duct tape (polycoated cloth tape)
Gewebekultur tissue culture
Gewebekulturflasche/Zellkulturflasche
tissue culture flask
Gewebelehre/Histologie histology
Gewebeschutzsalbe/
Arbeitsschutzsalbe/Schutzcreme
barrier cream
Gewerbe trade, business, occupation
➢ **Beruf/Erwerbstätigkeit**
profession
Gewerbeaufsicht (staatl. Behörde)
trade & industrial supervision
(federal agency)
Gewerbeordnung (GewO)
Industrial Trade Law,
Industrial Code
Gewicht weight;
(Wägemasse) weight (*actually:* mass)
➢ **Atomgewicht** atomic weight
➢ **Bruttogewicht** gross weight
➢ **Frischgewicht** (*sensu stricto*:
Frischmasse)
fresh weight
(*sensu stricto*: fresh mass)
➢ **Lebendgewicht** live weight

➢ **Molekulargewicht/**
relative Molekülmasse (M_r)
molecular weight,
relative molecular mass
➢ **Nettogewicht** net weight
➢ **spezifisches Gewicht**
specific gravity
➢ **Tara**
(Gewicht des Behälters/
der Verpackung)
tare (weight of container/packaging)
➢ **Trockengewicht**
(*sensu stricto*: **Trockenmasse**)
dry weight (*sensu stricto*: dry mass)
Gewichtsanalyse/Gravimetrie
gravimetry, gravimetric analysis
Gewichtsring/Stabilisierungsring/
Beschwerungsring/Bleiring
(für Erlenmeyerkolben)
lead ring (for Erlenmeyer)
Gewinde
(Schrauben/Bolzen etc.) thread;
(Spirale) spiral, coil
➢ **Außengewinde**
external thread, male thread
➢ **Britisches Standard Gewinde**
British Standard Pipe (BSP)
thread/fittings
➢ **Innengewinde**
internal thread, female thread
➢ **U.S. Rohrgewindestandard**
National Pipe Taper (NPT)
➢ **UNF-Feingewinde**
Unified Fine Thread (UNF)
Gewindeabdichtungsband
thread seal tape
Gewindebohrer tap (tool for forming
an internal screw thread)
Gewindefassung screw-base socket
Gewindestutzen
threaded socket (connector/nozzle)
gewöhnen/anpassen
habituate, get used to, adapt
Gewöhnung/Anpassung
habituation, habit-formation,
adaptation
GFC (Gas-Flüssig-Chromatographie)
GLC (gas-liquid chromatography)

gi

gießen
  pour, irrigate, water (plants etc.)
Gießform mold, mould (*Br*)
Gießschnauze (an Gefäß)
  pouring spout
Gift/Toxin poison, toxin
➢ **Summationsgift/kumulatives G.**
  cumulative poison
➢ **Tiergift** venom
giftig (Tiere) venomous
giftig/toxisch poisonous, toxic
Giftigkeit/Toxizität
  poisonousness, toxicity
Giftinformationszentrale
  poison information center
Giftmüll
  toxic waste, poisonous waste
Giftschrank poison cabinet
Giftstoffe
  poisonous materials,
  poisonous substances
Gips (für Gipsverband) *med*
  plaster of Paris (POP)
Gips CaSO$_4$ × 2H$_2$O gypsum (selenite)
Gipsplatte (Deckenbeschalung)
  gypsum board (ceiling)
Gipsschiene *med* plaster splint
Gipsverband *med* plaster cast
Gitter screen, wire-screen, grate;
  lattice; *micros* (Netz/Gitternetz/
  Probenträgernetz für
  Elektronenmikroskopie) grid
Gitterenergie lattice energy
Gitterspannung *electr* grid voltage
Gitterstichprobenverfahren *stat*
  lattice sampling, grid sampling
Glanz luster
Glanzkohle/Anthrazit
  hard coal, anthracite
Glanzpapier
  (glanzbeschichtetes Papier)
  glazed paper
Glas glass
➢ **anschlagen/Ecke abschlagen**
  chip, chipping
➢ **Borosilikatglas** borosilicate glass
➢ **Einscheibensicherheitsglas (ESG)**
  tempered safety glass

➢ **Fensterglas** window glass
➢ **Flintglas** flint glass
➢ **gehärtet** toughened
➢ **Hartglas**
  tempered glass, resistance glass
➢ **Milchglas** milk glass
➢ **Quarzglas** quartz glass
➢ **Rippenglas/**
  geripptes Glas/geriffeltes Glas
  ribbed glass
➢ **Schutzglas/Sicherheitsglas**
  safety glass, laminated glass
➢ **Sicherheitsglas** safety glass
➢ **Sinterglas** fritted glass
➢ **Verbundsicherheitsglas**
  laminated safety glass
➢ **Wasserglas M$_2$O×(SiO$_2$)$_x$**
  water glass, soluble glass
glasartig/glasig
  glasslike, glassy, vitreous
Glasbehälter glass vessel
Glasbläser glassblower
Glasbläserei
  glassblower's workshop
  ('glass shop')
Glasbruch
  glass scrap, shattered glass,
  broken glass
Gläschen/
Glasfläschchen/Phiole vial
Glaser glazier (one who sets glass)
Glaserei
  (Handwerk) glasswork, glazing;
  (Werkstatt) glazier's workshop,
  glass shop
gläsern/aus Glas
  glassy, made out of glass, vitreous
Glasfaser/Faserglas fiberglass
Glasfaseroptik/Fiberoptik/Faseroptik
  fiber optics
Glasfritte fritted glass filter
Glasgeschirr glassware, glasswork
Glashahn glass stopcock
Glashersteller glassmaker
Glashomogenisator ('Potter'; Dounce)
  glass homogenizer (Potter-Elvehjem
  homogenizer; Dounce homogenizer)
Glaskeramik glass ceramics

Glasperle/Glaskügelchen
 glass bead
Glasplatte sheet of glass
Glasrohr/Glasröhre/Glasröhrchen
 glass tube, glass tubing (:Glasrohre)
Glasrohrschneider
 glass tubing cutter;
 (Zange) glass-tube cutting pliers
Glasrührstab glass stirring rod
Glasscheibe
 sheet of glass, pane
Glasschneider glass cutter
Glasschreiber/Glasmarker
 glass marker
Glassplitter bits of broken glass
Glasstab glass rod
Glasstößel/Glaspistill (Homogenisator)
 glass pestle
Glastemperatur/
Glasübergangstemperatur (T_g)
 polym glass transition temperature
Glasübergang *polym* glass transition
Glaswaren/Glassachen
 glassware, glasswork
Glaszylinder glass cylinder
gleich/identisch
 (völlig gleich/ein und dasselbe)
 equal, same, identical
 ➤ **ungleich/nicht identisch/anders**
 unequal, different, nonidentical
gleichartig
 (sehr ähnlich) very similar;
 (verwandt/kongenial) congenial
gleichbleibender Zustand/
stationärer Zustand
 steady state
gleichen *math* equate
 ➤ **sich gleichen/gleichartig sein**
 resemble
gleichförmig uniform
Gleichförmigkeit uniformity
gleichgestaltet
 similar-structured
Gleichgewicht
 balance, equilibrium
 ➤ **Fließgleichgewicht/**
 dynamisches Gleichgewicht
 steady state, steady-state equilibrium
 ➤ **Ionengleichgewicht**
 ion equilibrium, ionic steady state
 ➤ **natürliches Gleichgewicht**
 (Naturhaushalt)
 natural balance
 ➤ **ökologisches Gleichgewicht**
 ecological balance,
 ecological equilibrium
 ➤ **Säure-Basen-Gleichgewicht**
 acid-base balance
 ➤ **Ungleichgewicht**
 imbalance, disequilibrium
Gleichgewichtsdestillation
 equilibrium distillation
Gleichgewichtsdialyse
 equilibrium dialysis
Gleichgewichtskonstante
 equilibrium constant
Gleichgewichtspotential
 equilibrium potential
Gleichgewichtszentrifugation
 equilibrium centrifugation,
 equilibrium centrifuging
Gleichgewichtszustand
 equilibrium state
gleichrichten rectify
Gleichrichter rectifier
Gleichrichtung rectification
Gleichstrom direct current (DC)
Gleichstromdestillation
 simple distillation
Gleichung equation
 ➤ **'eingerichtete' G.** balanced equation
 ➤ **chemische G.** chemical equation
 ➤ **Gleichung** *x***ten Grades**
 equation of the xth order
gleichzählig/isomer isomerous
Gleitmittel/Schmiermittel
 lubricant
Gleitringdichtung (Rührer)
 face seal
Gleitwinkel *aer*
 glide angle, gliding angle
Gliedermaßstab folding rule
gliedern/einteilen divide;
 (klassifizieren) classify
 ➤ **untergliedern/unterteilen**
 subdivide

Gl

Gliederung/Einteilung
division
➢ **Untergliederung/Unterteilung**
subdivision
Gliederung/Klassifikation
classification
Glockenkurve (Gauß'sche Kurve)
bell-shaped curve (Gaussian curve)
Glockentrichter
(Fülltrichter für Dialyse)
thistle tube funnel,
thistle top funnel tube
Glucarsäure
glucaric acid, saccharic acid
Gluconsäure (Gluconat)
gluconic acid (gluconate),
dextronic acid
Glucuronsäure (Glukuronat)
glucuronic acid (glucuronate)
Glühbirne/Glühlampe
light bulb, lightbulb,
incandescent lamp
➢ **Glühbirnchen**
miniature lamp/bulb
Glühofen annealing furnace
Glühröhrchen combustion tube
Glühschälchen incineration dish
Glühverlust
loss on ignition, ignition loss
Glühwendel (z.B. Glühbirne) filament
Glukosamin/Glucosamin
glucosamine
Glukose/Glucose (Traubenzucker)
glucose (grape sugar)
Glukosurie/Glycosurie
glucosuria, glycosuria
Glutamin glutamine
Glutaminsäure (Glutamat)/
2-Aminoglutarsäure
glutamic acid (glutamate),
2-aminoglutaric acid
Glutaraldehyd/Glutardialdehyd/
Pentandial
glutaraldehyde, 1,5-pentanedione
Glutarsäure glutaric acid
Glutathion glutathione
Glycin/Glyzin/Glykokoll
glycine, glycocoll
Glycyrrhetinsäure
glycyrrhetinic acid
Glykokoll/Glycin/Glyzin
glycocoll, glycine
Glykol/Glycol/Ethylenglykol
glycol, ethylene glycol,
1,2-ethanediol
Glykolaldehyd/Hydroxyacetaldehyd
glycol aldehyde, glycolal,
hydroxyaldehyde
Glykolsäure (Glykolat)
glycolic acid (glycolate)
glykosidische Bindung
glycosidic bond,
glycosidic linkage
Glyoxalsäure (Glyoxalat)
glyoxalic acid (glyoxalate)
Glyoxylsäure (Glyoxylat)
glyoxylic acid (glyoxylate)
Glyzerin/Glycerin/Propantriol
glycerol, glycerin, 1,2,3-propanetriol
Glyzerinaldehyd/Glycerinaldehyd
glyceraldehyde, dihydroxypropanal
Glyzin/Glycin/Glykokoll
glycine, glycocoll
Gold (Au) gold
➢ **Blattgold**
gold foil, gold leaf
➢ **vergolden** gilding
Gold(I)... aurus
Gold(III)... auric
Goldmarkierung
gold-labelling
Goldsäure auric acid
Golgi-Anfärbemethode
Golgi staining method
Gooch-Tiegel Gooch crucible
graduiert/
mit einer Gradeinteilung versehen
graduated
Grafitofen graphite furnace
Gram-Färbung
Gram stain, Gram's method
Grammäquivalent
gram equivalent
gramnegativ gram-negative
grampositiv gram-positive
granulär granular

Gravimetrie/Gewichtsanalyse
  gravimetry, gravimetric analysis
Greifzange grippers
➤ **Haltezange/Gripzange** Vise-Grip®
➤ **Haltezange/Klasper**
  grasping claws, clasper(s), clasps
Grenzdifferenz *stat*
  least significant difference,
  critical difference
Grenzfaktor/begrenzender
  Faktor/limitierender Faktor *ecol*
  limiting factor
Grenzfläche interface
Grenzflächenspannung
  surface tension
Grenzfrequenz corner frequency
Grenzkonzentration
  limiting concentration
Grenzschicht boundary layer
Grenzwert/Schwellenwert
  limit, limiting value;
  physio/med liminal value
Griff grip, handle; (klammernd) clutch;
  (zupackend/festhaltend) grip, grasp
Griffe (z.B. Tragegriffe)
  grips; handgrips
Grind/Schorf scab
Gripzange/Haltezange Vise-Grip®
grobfaserig coarse-grained
Grobjustierschraube/Grobtrieb *micros*
  coarse adjustment knob
Grobjustierung/Grobeinstellung
  (Grobtrieb) *micros*
  coarse adjustment,
  coarse focus adjustment
Großlieferung
  bulk delivery, bulk shipment
Großpackmittel
  intermediate bulk container (IBC)
Großpackung bulk package
Großverbraucher bulk consumer
Grundbaustein basic building block
Grundkörper (Strukturformel)
  parent compound,
  parent molecule (backbone)
Grundlage base, foundation
Grundlagenforschung
  basic research

Grundnahrungsmittel
  staple food, basic food
Grundstoff/Rohstoff
  base material, starting material,
  raw material
Grundsubstanz/Grundgerüst/Matrix
  base material,
  ground substance, matrix
Grundwasser groundwater
Grundzustand ground state
Gruppe group, assemblage;
  gen cluster
Gruppenleiter (Forschung/Labor)
  group leader, principal investigator
Guano guano
Guanylsäure (Guanylat)
  guanylic acid (guanylate)
Guar-Gummi/Guarmehl
  guar gum, guar flour
Guar-Samen-Mehl
  guar meal, guar seed meal
Gulonsäure (Gulonat)
  gulonic acid (gulonate)
Gummi *tech* (Kautschuk) rubber;
  (Lebensmittel etc.) gum
➤ **Schaumgummi**
  foam rubber, plastic foam, foam
Gummi arabicum/
  Arabisches Gummi/Acacia Gummi
  gum arabic, acacia gum
Gummiband/Gummi
  rubber band, elastic (*Br*)
Gummidichtung(sring)
  rubber gasket
Gummihammer rubber mallet
Gummiharz resinous gum
Gummihütchen (Pipettierhütchen)
  rubber nipple (pipeting nipple)
Gummimanschette (für Laborglas)
  rubber sleeve
Gummiring
  rubber ring (e.g. flask support)
Gummischaber/Gummiwischer
  (zum Loslösen von festgebackenen
  Rückständen im Kolben)
  policeman, rubber policeman
  (scraper rod with rubber or
  Teflon tip)

Gu

Gummischlauch
rubber tubing
Gummistiefel
rubber boots
Gummistopfen/Gummistöpsel
rubber stopper, rubber bung (*Br*)
**Gummiwischer/
Gummischaber
(zum Loslösen von festge-
backenen Rückständen im Kolben)**
policeman, rubber policeman
(scraper rod with rubber
or Teflon tip)
Gürtel/Gurt/Cingulum
girdle, cingulum
Gusseisen cast iron
Gussmessing cast brass
**Gussplattenmethode/
Plattengussverfahren** *micb*
pour-plate method

Gutachten/Expertise
expert opinion, expertise
➢ **medizinisches G.**
medical certificate
Gutachter expert; (Berater) consultant
gutartig/benigne benign
➢ **bösartig/maligne** malignant
Gutartigkeit/Benignität
benignity, benign nature
➢ **Bösartigkeit/Malignität**
malignancy, malignant nature
Gute Arbeitspraxis
Good Work Practices (GWP)
Gute Industriepraxis (Produktqualität)
Good Manufacturing Practice (GMP)
Gute Laborpraxis
Good Laboratory Practice (GLP)
Güter articles; goods; freight
Güterzugwagen (Gefahrguttransport)
railcar

Haargefäß/Kapillare capillary
Haarnetz hair net
Haarschutzhaube bouffant cap
haften (kleben) adhere, stick, cling
Haftkleber contact adhesive
haftpflichtig liable
Haftpflichtversicherung
 liability insurance
Haftung adhesion, adhesive power; *chem* adsorption; (Verantwortung) responsibility; *jur* liability; warranty, guarantee
Haftwasser film water, retained water
Hahn (Leitungen/Behälter/Kanister)
 spigot, tap, cock, stopcock
➢ **Ablasshahn/Ablaufhahn** draincock
➢ **Absperrhahn/Sperrhahn** stopcock
➢ **Ausgießhahn** tap
➢ **Dreiweghahn/Dreiwegehahn**
 three-way cock,
 T-cock, three-way tap
➢ **Einweghahn** single-way cock
➢ **feststecken/festgebacken**
 jammed, stuck, 'frozen', caked
➢ **Gashahn** gas cock, gas tap
➢ **Glashahn** glass stopcock
➢ **Küken** key, plug
➢ **Quetschhahn** pinchcock
➢ **Wasserhahn** faucet
➢ **Zapfhahn/Fasshahn** spigot
➢ **Zylinder** barrel (stopcock barrel)
Hahnfett tap grease
Hahnküken key, stopcock key, plug
Hakenklemme (Stativ) hook clamp
Halbacetal hemiacetal
halbdurchlässig/semipermeabel
 semipermeable
Halbdurchlässigkeit/Semipermeabilität
 semipermeability
Halbedelmetall
 semiprecious metal
Halbelement (galvanisches)/
 Halbzelle half cell, half element
 (single-electrode system)
halbieren halve
Halblebenszeit (Enzyme) half-life
Halbleiter semiconductor
Halbmetalle semimetals

Halbmikroansatz semimicro batch
Halbmikroverfahren/
 Halbmikromethode
 semimicro procedure/method
Halbpfeil (in chem. Reaktionsgleichungen)
 fish hook
Halbsättigungskonstante/
 Michaeliskonstante (K_M)
 Michaelis constant,
 Michaelis-Menten constant
halbsynthetisch semisynthetic
Halbwertsbreite *math/stat*
 full width at half-maximum (fwhm),
 half intensity width
Halbwertszeit half-life
Halbzelle/galvanisches Halbelement
 half-cell, half element
 (single-electrode system)
Halbzellenpotential
 half-cell potential
Halbzeug
 semifinished product
Hälfte/Anteil/Teil moiety
Hals/Tubusträger *micros* neck
haltbar storable, durable, lasting
Haltbarkeit
 storability, durability, shelf life
Haltebolzen fixing bolt
Halterung
 (holding) fixture, mounting, support
Häm heme
Hammer hammer
➢ **Fäustel/Handfäustel**
 club hammer, mallet
➢ **Gummihammer (Fäustel)**
 rubber mallet
➢ **Klauenhammer/Splitthammer**
 claw hammer
➢ **Schlosserhammer**
 fitter's hammer,
 locksmith's hammer
➢ **Schlosserhammer mit Kugelfinne**
 ball pane hammer,
 ball peen hammer,
 ball pein hammer
➢ **Vorschlaghammer**
 sledge hammer

Ha

Handbedienung (Gerät)
manual operation
Handbesen/Handfeger
household brush
Handdesinfektion
disinfection of hands
Handel trade, business
➤ **Einzelhandel**
retail business, retail trade
➤ **Großhandel**
wholesale business,
wholesale trade
handelsüblich trade, commercial
(commonly available)
handgearbeitet (Glas etc.)
handtooled
Handhabung/
Hantieren/Gebrauch/Umgang
handling
Händler
dealer; seller; commercial vendor
➤ **Einzelhändler**
retailer, retail dealer, retail vendor
➤ **Großhändler**
wholesaler, wholesale vendor
Händlerkatalog
supplier catalog,
distributor catalog
Händlerrabatt dealer discount
Handpumpe hand pump
Handsäge handsaw
Handschuhe gloves
➤ **Arbeitshandschuhe**
work gloves
➤ **Ärmelschoner/Stulpen**
sleeve gauntlets
➤ **Baumwollhandschuhe**
cotton gloves
➤ **Einweg-/Einmalhandschuhe**
disposable gloves,
single-use gloves
➤ **Fingerling** finger cot
➤ **Hitzehandschuhe**
heat defier gloves,
heat-resistant gloves
➤ **Hoch-Hitzehandschuhe/**
Ofenhandschuhe
oven gloves

➤ **Isolierhandschuhe**
insulated gloves
➤ **Kälteschutzhandschuhe**
cold-resistant gloves
➤ **medizinische Handschuhe/**
OP-Handschuhe
medical gloves
➤ **Reinraumhandschuhe**
cleanroom gloves
➤ **Säureschutzhandschuhe**
acid gloves,
acid-resistant gloves
➤ **Schnittschutz-Handschuhe**
cut-resistant gloves
➤ **Schutzhandschuhe**
protective gloves, gauntlets
➤ **Tiefkühlhandschuhe/**
Kryo-Handschuhe
deep-freeze gloves
Handschuhinnenfutter
glove liners
Handschuhkasten/
Handschuhschutzkammer
glove box, dry-box
Handtuch towel
➤ **Geschirrhandtuch** dish towel
➤ **Küchenhandtuch** kitchen towel
➤ **Papierhandtuch** paper towel
Handtuchhalter/Handtuchständer
towel rack
Handwerker
craftsman (practicing a handicraft),
workman
Handwerker/Arbeiter
operations worker, worker
Hardy-Weinberg-Gesetz
(Hardy-Weinberg-Gleichgewicht)
Hardy-Weinberg law
(Hardy-Weinberg equilibrium)
Harn/Urin urine
➤ **Primärharn/Glomerulusfiltrat**
glomerular ultrafiltrate
➤ **Sekundärharn**
secondary urine
Harnen/Harnlassen/Urinieren/Miktion
urination, micturition
harnen/urinieren/miktuieren
urinate, micturate

Harnfluss/Harnausscheidung/Diurese
diuresis
Harnsäure (Urat) uric acid (urate)
Harnstoff (Ureid) urea (ureide)
Härte hardness, toughness
➤ **bleibende Härte**
permanent hardness
➤ **Gesamthärte** total hardness
➤ **vorrübergehende Härte**
temporary hardness
➤ **Wasserhärte** water hardness
härten harden; *polym* (aushärten) cure;
(von Stahl) temper
Härten hardening;
(Aushärten) *polym* curing;
(von Stahl) tempering
Härter/Aushärtungskatalysator
polym curing agent
Härtezeit/Abbindezeit
polym curing period
Hartglas tempered glass
Hartpapier
laminated paper
Harz resin
➤ **Gummiharz**
resinous gum
➤ **Ionenaustauscherharz**
ion-exchange resin
➤ **Kunstharz**
artificial resin, synthetic resin
➤ **selbsthärtend (Harze/Polymere)**
self-curing
➤ **Terpentinharz**
pitch (resin from conifers)
harzabsondernd resiniferous
harzig resinous
Harzsäure resin acids
häufig frequent, abundant
Häufigkeit/Frequenz
frequency (of occurrence),
abundance
➤ **relative H.** *stat*
frequency ratio
Häufigkeitshistogramm
frequency histogram
Häufigkeitsverteilung *stat*
frequency distribution (FD)
Häufungsgrad/Häufigkeitsgrad kurtosis

Hauptassoziation chief association
Hauptbande *chromat/electrophor*
main band
Hauptplatine mother board
Hauptsatz
(1./2.Hauptsatz der Thermodynamik)
first/second law of thermodynamics
Haushalt household
➤ **Naturhaushalt**
(natürliches Gleichgewicht)
natural balance
➤ **Stoffwechsel/Metabolismus**
metabolism
➤ **Wasserhaushalt/Wasserregime**
water regime
Haushaltsmüll/Haushaltsabfälle
household waste/trash
Haushaltsrolle/Küchenrolle/
Tücherrolle/Küchentücher/
Haushaltstücher
kitchen tissue
(kitchen paper towels)
Hausmeister/Hausverwalter
caretaker, janitor, custodian
➤ **Wachpersonal/Aufsichtspersonal**
custodial personnel,
security personnel
(Belegschaft: staff)
Hausverwaltung
property management,
custodian, management
➤ **Büro des Hausmeisters**
caretaker's office, custodian's office
Haut... (dermal) dermal, dermic,
dermatic; (die Haut betreffend)
epidermal, cutaneous
Haut skin; hide, peel; integument
➤ **Kutis/Cutis (eigentliche Haut;**
Epidermis & Dermis)
skin, cutis
➤ **Lederhaut/Korium/Corium/Dermis**
cutis vera, true skin, corium, dermis
➤ **Oberhaut/Epidermis** epidermis
➤ **Schleimhaut/Schleimhautepithel**
mucous membrane, mucosa
➤ **Unterhaut/Unterhautbindegewebe/**
Subcutis/Tela subcutanea
subcutis

Ha

Hautatmung
cutaneous respiration/breathing, integumentary respiration
Hautausschlag
rash, skin rash, skin eruptions
Hautpflege skin care
Hautpflegemittel
skin care product
hautreizend skin-irritant
Hautreizung skin irritation
Hautsalbe skin ointment
Haworth-Projektion/Haworth-Formel
Haworth projection, Haworth formula
Hebebühne
hoist, lifting platform
Hebelmechanismus
leverage mechanism
Heber/Hebevorrichtung/Hebebock
jack
Hebestativ/Hebebühne (fürs Labor)
laboratory jack, lab-jack
Heckenschere
hedge clippers, hedge trimmers
Hefe yeast
➤ **Backhefe/Bäckerhefe** baker's yeast
➤ **Bierhefe/Brauhefe** brewers' yeast
➤ **Brennereihefe** distiller's yeast
➤ **hochvergärende Hefe ('Staubhefe')** top yeast
➤ **Mineralhefe**
mineral accumulating yeast
➤ **niedrigvergärende Hefe ('Bruchhefe')** bottom yeast
➤ **Spalthefe** fission yeast (*Saccharomyces pombe*)
➤ **Stellhefe/Anstellhefe/Impfhefe**
pitching yeast
➤ **Trockenhefe**
dried yeast
Hefeextrakt yeast extract
Hefter (Büro~) stapler
heftig (Reaktion etc.) vigorous
Heftklammer staple
Heftpflaster (Streifen) *med*
band-aid (adhesive strip), sticking plaster, patch
heilen cure, heal

Heilpflanze/Arzneipflanze
medicinal plant
Heilung cure, healing
heimisch local, endemic
Heißluft hot air
Heißluftgebläse/ Labortrockner/Föhn
hot-air gun
Heißluftpistole heat gun
Heißwassertrichter
hot-water funnel (double-wall funnel)
Heizbad heating bath
Heizband/Heizbandage
heating tape, heating cord
heizen heat
Heizhaube/Heizmantel/Heizpilz
heating mantle
Heizkörper radiator
Heizplatte (Kochplatte)
hot plate
➤ **Doppelkochplatte**
double-burner hot plate
➤ **Einfachkochplatte**
single-burner hot plate
➤ **Magnetrührer mit Heizplatte**
stirring hot plate
Heizschlange heating coil
Heizung heater, heating system
Heizwendel heating coil
Helix/Spirale (*pl* Helices)
helix (*pl* helices or helixes), spiral
Hellfeld *micros* bright field
Hellkeimung
light-induced germination (photodormancy)
Helm helmet
➤ **Schutzhelm**
safety helmet; hard hat, hardhat
hemizygot hemizygous
hemizyklisch/hemicyclisch
hemicyclic
hemmen inhibit
hemmend/inhibierend/inhibitorisch
inhibitory
Hemmkonzentration
inhibitory concentration

Hemmstoff inhibitor
Hemmung/Inhibition inhibition
➢ **irreversible Hemmung**
  irreversible inhibition
➢ **kompetitive Hemmung/
  Konkurrenzhemmung**
  competitive inhibition
➢ **nichtkompetitive Hemmung**
  noncompetitive inhibition
➢ **reversible Hemmung**
  reversible inhibition
➢ **Suizidhemmung**
  suicide inhibition
➢ **unkompetitive Hemmung**
  uncompetitive inhibition
Hemmzone
  inhibition zone
Heparreaktion/Heparprobe
  hepar reaction, hepar test
Herabregulation
  down regulation
Heraufregulation
  up regulation
Herauskreuzen/Auskreuzen
  outcrossing
herausragen emerge;
  (hervorstehen) protrude, stand out
Herausschneiden/Excision/Exzision
  excision
herausschneiden/exzidieren
  excise
Herbar herbarium
**Herbizid/
  Unkrautvernichtungsmittel/
  Unkrautbekämpfungsmittel**
  herbicide, weed killer
Herkunft/Abstammung
  origin, descent,
  provenance (Provenienz)
Hersteller/Produzent
  manufacturer, producer
Herstellerangaben
  manufacturer's specifications
Herstellerfirma
  manufacturer,
  manufacturing company/firm
Herstellerkatalog
  manufacturer catalog

Herstellung/Produktion
  manufacture, manufacturing,
  preparation, production
Herstellungskosten
  production costs,
  manufacturing costs
Herstellungsverfahren
  preparation process/procedure,
  manufacturing process/procedure
herunterfahren (Reaktor/Computer)
  power down
**hervorkommen/
  herauskommen/auftauchen**
  emerge
**heterogen/ungleichartig/
  verschiedenartig/andersartig**
  heterogeneous
  (consisting of dissimilar parts)
heterogen/unterschiedlicher Herkunft
  heterogenous (of different origin)
**Heterogenie/
  unterschiedlicher Herkunft**
  heterogeny
**Heterogenität/Ungleichartigkeit/
  Verschiedenartigkeit/Andersartigkeit**
  heterogeneity
heterolog heterologous
Heteropolymer heteropolymer
heterotroph heterotroph, heterotrophic
heterotypisch heterotypic
heterozyklisch/heterocyclisch
  heterocyclic
Hilfe help, aid, assistance, support,
  rescue operation
**Hilfseinrichtung (Apparat der nicht
  direkt mit dem Produkt in Berührung
  kommt)** ancillary unit of equipment
Hilfsstoff/Adjuvans
  auxiliary drug, adjuvant
Hill-Gleichung (Hill-Auftragung)
  Hill equation (Hill plot)
Hinterdruck outlet pressure;
  (Arbeitsdruck: mit Druckausgleich)
  working pressure, delivery pressure
Hirschhornsalz/Ammoniumcarbonat
  hartshorn salt, ammonium carbonate
Histamin histamine
Histidin histidine

Hi

Histogramm/Streifendiagramm *stat*
histogram, strip diagram
Hitze heat
➢ **erhitzen** heat
➢ **Überhitzung**
overheating, superheating
Hitzebehandlung/Backen
heat treatment, baking
hitzebeständig/hitzestabil
heat-resistant, heat-stable
Hitzeentwicklung heat evolution
hitzemeidend/thermophob
thermophobic
Hitzeschock heat shock
Hitzeschockreaktion
heat shock reaction,
heat shock response
hitzestabil/hitzebeständig
heat-stable, heat-resistant
hitzeverträglich heat-tolerant
Hitzschlag heatstroke
hochauflösend/hochaufgelöst
high-resolution ...
Hochdruck/Bluthochdruck
hypertension
Hochdruck-Steckverbindung
compression fitting
**Hochdruckflüssigkeits-
chromatographie/
Hochleistungschromatographie**
high-pressure liquid chromatography,
high performance liquid
chromatography (HPLC)
Hochdurchsatz
high-throughput
hochentzündlich
highly ignitable
hochfahren (Reaktor/Computer)
power up
Hochfeldverschiebung
high-field shift
hochmolekular high-molecular
Hochofen blast furnace
Höchsterträge
maximum yield
**Hof/Lysehof/
Aufklärungshof/Plaque**
plaque

**Hofmeistersche Reihe/
lyotrope Reihe**
Hofmeister series, lyotropic series
Höhle/Kammer/Ventrikel
cavity, chamber, ventricle
Hohlfaser hollow fiber
Hohlleiter (z.B. an Mikrowelle)
wave guide
Hohlraum/Höhlung/Lumen
cavity, lumen; airspace
Hohlspiegel concave mirror
Hohlstopfen/Hohlglasstopfen
hollow stopper
Höhlung crypt, cavity, cave
Hohlwelle (Rührer)
hollow impeller shaft
Holoenzym holoenzyme
Holzessig
wood vinegar, pyroligneous acid
Holzfäule wood rot
Holzgeist
wood spirit, wood alcohol,
pyroligneous spirit,
pyroligneous alcohol
(chiefly: methanol)
Holzkohle charcoal
Holzteer wood tar
holzverarbeitende Industrie
timber industry
Holzwirtschaft
lumber industry, timber industry
Holzwolle wood-wool
holzzersetzend
decomposing wood, xylophilous
Holzzucker/Xylose
wood sugar, xylose
homogen (einheitlich/gleichartig)
homogeneous (having same kind of
constituents); (gleicher Herkunft)
homogenous (of same origin)
Homogenisation homogenization
Homogenisator homogenizer
homogenisieren homogenize
Homogenisierung homogenization
**Homogenität/
Einheitlichkeit/Gleichartigkeit**
homogeneity
(with same kind of constituents)

Homogentisinsäure
  homogentisic acid
homoiosmotisch
  homoiosmotic, homeosmotic
homolog/ursprungsgleich
  homologous
Homologie homology
homologisieren homologize
homonym *adv/adj*
  homonymous, homonymic
Homopolymer homopolymer
Homoserin homoserine
Hörbarkeit audibility
Hörgrenze
  hearing limit, auditory limit, limit of audibility
horizontal angeordnetes Plattengel
  horizontal gel, flat bed gel
Hormon hormone *(siehe auch unter individuellen Begriffen)*
hormonal/hormonell hormonal
Hörschwelle
  hearing threshold, auditory threshold
Hörvermögen/Gehör
  audition
HOSCH-Filter (Hochleistungsschwebstofffilter)
  HEPA-filter (high-efficiency particulate and aerosol air filter)
Hubstapler/Gabelstapler
  forklift
Hubwagen
  lifting truck, jacklift
Hülle (z.B. Wasser)
  envelope, jacket; (Mantel) body covering, vesture, vestiture
Hülse/Ring socket, ferrule
Hülse/Schliffhülse ('Futteral'/Einsteckstutzen)
  socket (female: ground-glass joint)
humifizieren humify
Humifizierung/ Humifikation/Humusbildung
  humification
Huminsäure humic acid
Huminstoffe humic substances

Humus *geol* humus
Hunger hunger
Hungern *n mich* starvation
hungern *vb mich* starve
hungrig hungry
Hutmutter acorn nut
Hyaluronsäure hyaluronic acid
hybrid/durch Kreuzung erzeugt
  hybrid, crossbred
Hybride hybrid, crossbreed
hybridisieren hybridize
Hybridisierungsinkubator
  hybridization incubator
Hydrat hydrate
Hydratation/ Hydratisierung/ Solvation (Wassereinlagerung/ Wasseranlagerung)
  hydration, solvation
Hydrathülle/ Wasserhülle/Hydratationsschale
  hydration shell
Hydratwasser water of hydration
hydraulisch vorgesteuert (Ventil)
  pilot-operated (valve)
hydrieren/hydrogenieren hydrogenate
Hydrierung (Wasserstoffanlagerung)
  hydrogenation
hydrisch hydric
Hydrokultur hydroponics
  (soil-less culture/solution culture)
Hydrologie hydrology
Hydrolyse/Wasserspaltung
  hydrolysis
hydrolytisch/wasserspaltend
  hydrolytic
hydrophil (wasseranziehend/wasserlöslich)
  hydrophilic (water-attracting/water-soluble)
Hydrophilie (Wasserlöslichkeit)
  hydrophilicity (water-attraction/water-solubility)
hydrophob (wasserabweisend/wasserabstoßend /nicht wasserlöslich)
  hydrophobic (water-repelling/water-insoluble)

hy

hydrophobe Bindung
  hydrophobic bond
Hydrophobie (Wasserabweisung/ Wasserunlöslichkeit)
  hydrophobicity
  (water-insolubility)
hydrostatischer Druck
  hydrostatic pressure
Hydroxyapatit hydroxyapatite
Hydroxylierung hydroxylation
Hydroxyprolin hydroxyproline
Hygiene hygiene
➢ **Arbeitshygiene**
  industrial hygiene
➢ **Arbeitsplatzhygiene**
  occupational hygiene

Hygienebedingungen
  hygienic conditions
Hygienemaßnahme
  sanitary measure
hygienisch hygienic
hygroskopisch hygroscopic
Hyperchromizität
  hyperchromicity,
  hyperchromic effect,
  hyperchromic shift
Hypersensibilität/Allergie
  hypersensitivity, allergy
Hypothese hypothesis
hypothetisch
  hypothetic, hypothetical
Hypoxie/Sauerstoffmangel hypoxia

identisch identical
identisch aufgrund von Zufällen
  identity by state (IBS)
ikosaedrisch *vir* icosahedral
imbibieren/hydratieren
  imbibe, hydrate
Imbibition/Hydratation
  imbibition, hydration
IMDG (Intl. Maritime Dangerous Goods Code) Internat. Code für die Beförderung von gefährlichen Gütern mit Seeschiffen
Imidazol imidazole
Iminosäure imino acid
Immission (Belastung durch Luftschadstoffe) exposure level of air pollutants; (Einwirkung) immission, injection, admission, introduction
immobil/fixiert/bewegungslos
  immobile, fixed, motionless
Immobilisation immobilization
immobilisieren
  immobilize (to make immobile)
Immobilität/Bewegungslosigkeit
  immobility, motionlessness
immortalisierte Zelle
  immortalized cell
immun immune
Immunadsorptionstest, enzymgekoppelter (ELISA)
  enzyme-linked immunosorbent assay (ELISA)
Immunaffinitätschromatographie
  immunoaffinity chromatography
Immunantwort immune response
Immundefekt
  immune deficiency
➢ **erworbenes Immunschwächesyndrom**
  acquired immune deficiency syndrome (AIDS)
➢ **schwerer kombinierter Immundefekt**
  severe combined immune deficiency (SCID)
Immundiffusion immunodiffusion
➢ **Doppelimmundiffusion**
  double diffusion, double immunodiffusion
➢ **doppelte radiale Immundiffusion (Ouchterlony-Methode)**
  double radial immunodiffusion (DRI) (Ouchterlony technique)
➢ **einfache Immundiffusion/ lineare Immundiffusion (Oudin-Methode)**
  single immunodiffusion (Oudin test)
➢ **einfache radiale Immundiffusion (Mancini-Methode)**
  single radial immunodiffusion (SRI) (Mancini technique)
➢ **Identität** identity
➢ **radiale Immundiffusion**
  radial immunodiffusion (RID)
➢ **Teilidentität/ partielle Übereinstimmung**
  partial identity
➢ **Verschiedenheit (Nicht-Identität)**
  nonidentity
Immun-Elektronenmikroskopie (IEM)
  immunoelectron microscopy (IEM)
Immunelektrophorese
  immunoelectrophoresis
➢ **Kreuzimmunelektrophorese**
  crossed immunoelectrophoresis, two-dimensional immunoelectrophoresis
➢ **Linienimmunelektrophorese**
  immunoelectrophoresis
➢ **Raketenimmunelektrophorese**
  rocket immunoelectrophoresis
➢ **Tandem-Kreuzimmunelektrophorese**
  charge-shift immunoelectrophoresis
➢ **Überwanderungsimmun- elektrophorese/ Überwanderungselektrophorese**
  countercurrent immunoelectrophoresis, counterelectrophoresis
Immunfluoreszenzchromatographie
  immunofluorescence chromatography
Immunfluoreszenzmikroskopie
  immunofluorescence microscopy
Immungenetik
  immunogenetics
immunisieren/impfen
  immunize, vaccinate

Immunisierung/Impfung
 immunization, vaccination
Immunität immunity
Immunkrankheit/Immunopathie
 immunopathy
Immunoblot/Western-Blot
 immunoblot, Western blot
Immunogold-Silberfärbung (IGSS)
 immunogold-silver staining (IGSS)
Immunologie immunology
immunoradiometrischer Assay
 immunoradiometric assay (IRMA)
Immunpräzipitation
 immunoprecipitation
Immunprophylaxe
 immunoprophylaxis
Immunreaktion immune reaction
Immunschwäche
 immune deficiency,
 immunodeficiency
**Immunschwächesyndrom/
Immunmangel-Syndrom**
 immune deficiency syndrome
➤ **erworbenes
 Immunschwächesyndrom**
 acquired immune deficiency
 syndrome (AIDS)
➤ **schwerer kombinierter Immundefekt**
 severe combined immune deficiency
 (SCID)
Immunsuppression
 immunosuppression,
 immune suppression
Immuntoleranz immune tolerance,
 immunological tolerance
**Immunüberwachung/
 immunologische Überwachung**
 immunosurveillance,
 immunologic(al) surveillance
impermeabel/undurchlässig
 impermeable, impervious
**Impermeabilität/
 Undurchlässigkeit**
 impermeability,
 imperviousness
Impfdraht inoculating wire
impfen *med* inoculate, vaccinate;
 micb inoculate, seed

**Impfen/Impfung/Vakzination
 (Immunisierung)**
 inoculation, vaccination
Impfnadel inoculating needle
Impföse/Impfschlinge
 inoculating loop
Impfstoff/Inokulum/Inokulat/Vakzine
 inoculum, vaccine
**Impfung/Inokulation/Vakzination
 (Immunisierung)** inoculation,
 vaccination (immunization)
Implosion implosion
in situ **Hybridisierung**
 in situ hybridization
➤ **FISH (***in situ* **Hybridisierung mit
 Fluoreszenzfarbstoffen)**
 FISH (fluorescence activated
 in situ hybridization)
*in vitro***-Verpackung**
 in vitro packaging
inaktiv inactive
Inbetriebnahme
 putting into operation,
 startup, starting-up;
 (offizielle Übergabe einer Anlage
 etc.) commissioning
Inbusschlüssel Allen wrench
Inbusschraube
 socket screw, socket-head screw
Indikan/Indoxylsulfat
 indican, indoxyl sulfate
**Indikatororganismus/
 Indikatorart/Bioindikator**
 bioindicator
Indolessigsäure indolyl acetic acid,
 indoleacetic acid (IAA)
Induktionsofen
 induction furnace,
 inductance furnace
induktiv gekoppeltes Plasma
 inductively coupled plasma (ICP)
Industriegase/technische Gase
 industrial gases,
 manufactured gases
Industriemüll/Industrieabfall
 industrial waste
induzierbar inducible
induzieren induce

Infektion/Ansteckung
infection
➢ **abortive Infektion**
abortive infection
➢ **anhaltende/persistierende Infektion**
persisting Infektion
➢ **Doppelinfektion**
double infection
➢ **latente Infektion**
latent infection
➢ **lytische Infektion**
lytic infection
➢ **produktive Infektion**
productive infection
➢ **Schmierinfektion**
smear infection
➢ **stumme Infektion/stille Feiung**
silent infection
➢ **Superinfektion/Überinfektion**
superinfection
➢ **unvollständige Infektion**
incomplete infection
Infektionsdosis
infectious dose
(ID_{50} = 50% infectious dose)
Infektionskrankheit
infectious disease
**Infektionsvermögen/
 Ansteckungsfähigkeit**
infectivity
infektiös/ansteckend infectious
infektiöser Abfall infectious waste
infizieren/anstecken infect
**Infrarot-Spektroskopie/
 IR-Spektroskopie**
infrared spectroscopy
inhibitorisch/hemmend inhibitory
Injektion/Spritze injection, shot
Injektionsnadel syringe needle
➢ **abnehmbare N.**
removable needle (syringe needle)
➢ **geklebte N.**
cemented needle (syringe needle)
Injektionsspritze hypodermic syringe
injizieren/spritzen inject, shoot
Inkohlung *paleo/geol*
carbonization, coalification
inkompatibel incompatible

Inkompatibilität incompatibility
Inkubation (Bebrütung/Bebrüten)
incubation
Inkubationsschüttler
shaking incubator,
incubating shaker,
incubator shaker
Inkubationszeit incubation period
inkubieren/brood/breed
incubate, brüten, bebrüten
Innengewinde
internal thread, female thread
innerlich/von innen/intern
internal, intrinsic
Inokulation/Einimpfung/Impfung
inoculation
inokulieren/einimpfen/impfen
inoculate
Inosit/Inositol inositol
Inprozesskontrolle
in-process verification
Insektenbekämpfungsmittel/Insektizid
insecticide
Insektenkunde/Entomologie
entomology
Insektenplage insect pest
**Insektenvernichtungsmittel/
 Insektizid** insecticide
inserieren (inseriert)
insert (inserted)
Inspektions-Logbuch inspection log
Inspiration/Einatmen inspiration
inspirieren/einatmen inspire
instabil unstable (instable)
**Installation(en)/Installierung/
 Einbau** installations
Instandhaltung/Wartung
maintenance, servicing
Instandhaltungskosten
maintenance costs
Instandsetzung/Reparatur
repair, restoration;
(überholen) overhaul,
reconditioning
Instrumentenanzeige
instrument display;
(abgelesener Wert)
instrument reading

in

interdisziplinäre Forschung
  interdisciplinary research
Interferenzassay
  interference assay
Interferenz-Mikroskopie
  interference microscopy
interkalierendes Agens
  intercalation agent,
  intercalating agent
Internationale Maßeinheit/SI Einheit
  International Unit (IU), SI unit
  (*fr:* Système Internationale)
interpolieren interpolate
Intervall interval
Intervallskala *stat* interval scale
Inventar inventory; stock
inventarisieren make an inventory
Inventur/Bestandsaufnahme
  inventory
  ➢ **eine I./B. machen**
  to take inventory,
  to make an inventory
invers inverted
Invertzucker invert sugar
Iod (I) iodine
Iodessigsäure iodoacetic acid
iodieren
  (mit Jod/Jodsalzen versehen)
  iodize
Iodierung (mit Jod reagieren/
  substituieren) iodination;
  (mit Jod/Jodsalzen versehen)
  iodization
Iodsalz iodized salt
Iodwasserstoffsäure
  hydroiodic acid, hydrogen iodide
Iodzahl
  iodine number, iodine value
Ion ion
  ➢ **Bruchstückion** fragment ion
  ➢ **Gegenion** counterion
  ➢ **Molekülion (MS)** molecular ion
  ➢ **Mutterion/Ausgangsion (MS)**
  parent ion
  ➢ **Radikalion** radical ion
  ➢ **Tochterion** daughter ion
  ➢ **Zwitterion** zwitterion
  (*not translated!*)

Ionenaustauscher
  ion exchanger
  ➢ **Anionenaustauscher**
  (starker/schwacher)
  anion exchanger
  (*strong*: SAX/*weak*: WAX)
  ➢ **Kationenaustauscher**
  (starker/schwacher)
  cation exchanger
  (*strong*: SCX/*weak*: WCX)
Ionenaustauscherharz
  ion-exchange resin
Ionenbindung ionic bond
Ioneneinfangdetektor (MS)
  ion trap detector (ITD)
Ionen-Fallen-Spektrometrie
  ion trap spectrometry
Ionenformel ionic formula
Ionengleichgewicht
  ion equilibrium,
  ionic steady state
Ionenkanal (Membrankanal)
  ion channel (membrane channel)
Ionenkopplung ionic coupling
Ionenleitfähigkeit
  ionic conductivity
Ionenpaar ion pair
Ionenpore ion pore
Ionenprodukt ion product
Ionenpumpe ion pump
Ionenradius ionic radius
Ionenschleuse gated ion channel
Ionenspray ion spray
Ionenstärke ionic strength
Ionenstrom
  ionic current, ion current
Ionentransport ion transport
Ionisation ionization
Ionisationskammer
  ionization chamber
ionisch ionic
ionisieren ionize
ionisierende Strahlen/
  ionisierende Strahlung
  ionizing radiation
Ionophor ionophore
Ionophorese/Iontophorese
  ionophoresis

Irisblende *micros*
  iris diaphragm
IRMA
  (immunoradiometrischer Assay)
  immunoradiometric assay (IRMA)
isoelektrische Fokussierung/
Isoelektrofokussierung
  isoelectric focusing
isoelektrischer Punkt
  isoelectric point
Isolationsmedium *micb*
  isolation medium
Isolierband
  insulating tape, duct tape
➢ **Elektro-Isolierband**
  electric tape, insulating tape, friction tape
isolieren/abtrennen isolate, separate
Isolierhandschuhe
  insulated gloves
isomer *adv/adj* isomeric
Isomer *n* isomer
Isomeratzucker/Isomerose
  high fructose corn syrup
Isomerie
  isomerism, isomery
Isomerisation isomerization
isomerisieren isomerize
isopyknische Zentrifugation
  isopycnic centrifugation
isosmotisch isosmotic
Isotachophorese/
Gleichgeschwindigkeits-
Elektrophorese
  isotachophoresis (ITP)
Isothiocyansäure
  isothiocyanic acid
Isotonie isotonicity
isotonisch isotonic
Isotopenverdünnung
  isotopic dilution
Isotopenversuch isotope assay
Isovaleriansäure isovaleric acid
Istwert
  actual value, effective value
➢ **Sollwert**
  nominal value, rated value, desired value, set point

Jo

Jod (*siehe:* **Iod**) **(I)**
iodine
justierbar
adjustable
justieren adjust; (fokussieren:
Scharfeinstellung des Mikroskops:
fein/grob) focus (*fine/coarse*);
(eichen) calibrate

Justierschraube/Justierknopf
adjusting screw, adjustment knob;
(Triebknopf *micros*)
focus adjustment knob
Justierung adjustment; (Fokussierung:
Scharfeinstellung des Mikroskops:
fein/grob) focus adjustment, focus
(*fine/coarse*)

Kabel cable
Kabelbinder/Spannband
  cable tie(s),
  wrap-it tie(s), wrap-it tie cable
  ➢ **Spannzange**
    tensioning tool, tensioning gun
    (for cable ties/wrap-it-ties)
Kabelöse/Kabelschuh
  cable lug
Kabelschuh/Ansatz/Öhr *electr*
  lug
Kabeltrommel cable drum
Kabelverbinder cable connector
Kachel tile
Kadaver/Tierleiche
  cadaver, carcass, corpse
Kaffeesäure caffeic acid
Käfig cage
kalibrieren calibrate
Kalibrierung calibration
Kalilauge/Kaliumhydroxidlösung
  potassium hydroxide solution
Kalium (K) potassium
Kaliumcyanid/Cyankali/Zyankali
  potassium cyanide
Kaliumpermanganat
  potassium permanganate
Kalk lime
  ➢ **Ätzkalk/Löschkalk/
    gelöschter Kalk**
    slaked lime [$Ca(OH)_2$]
  ➢ **Branntkalk** caustic lime (CaO)
  ➢ **entkalken (ein Gerät ~)**
    descale
  ➢ **verkalken (verkalkt)**
    calcify (calcified)
Kalkablagerung
  lime(stone) deposit
**Kalkeinlagerung/
  Verkalkung/Calcifikation**
  calcification
kalken lime, calcify
kalkig/kalkartig/kalkhaltig
  limy, limey, calcareous
Kalk-Soda-Glas
  soda-lime glass
Kalkspat calcite
Kalkstein limestone

Kalkung liming
Kallus/Callus callus
  ➢ **Wundkallus/Wundcallus/
    Wundgewebe/Wundholz** *bot*
    wound tissue, callus
Kallus-Kultur/Callus-Kultur
  callus culture
Kalorie calorie
Kalorimeter calorimeter
**Kalorimeterbombe/
  Bombenkalorimeter/
  Verbrennungsbombe**
  bomb calorimeter
Kalorimetrie calorimetry
  ➢ **Bombenkalorimeter**
    bomb calorimeter
  ➢ **Leistungskompensations-DSC**
    power-compensated DSC
  ➢ **Differentialkalorimetrie**
    differential scanning calorimetry
    (DSC)
  ➢ **Raster-Kalorimetrie**
    scanning calorimetry
Kalottenmodell *chem*
  space-filling model
Kälteakku/Kühlakku cooling pack
kälteempfindlich/kältesensitiv
  cold-sensitive
Kältemittel/Kühlflüssigkeit/Kühlmittel
  coolant (allg/direkt); refrigerant
Kälteraum/Kühlraum
  cold room ('walk-in refrigerator')
Kälteresistenz cold resistance
Kälteschaden/Kälteschädigung
  chilling damage/injury
Kälteschock cold shock
Kälte-Spray cold spray
**Kältethermostat/
  Kühlthermostat/
  Umwälzkühler**
  refrigerated circulating bath
Kältetoleranz cold hardiness
**Kalthaus/Frigidarium
  (kühles Gewächshaus)**
  cold house
Kaltlagerung cold storage
Kaltlichtbeleuchtung
  fiber optic illumination

Ka

Kaltschweißen cold welding
Kaltziehen *metal* cold-draw
kalzinieren calcine
Kalzinierung calcination
Kalzium/Calcium (Ca) calcium
Kammer *electrophor* chamber, tank
Kammerjäger (Schädlingsbekämpfung) exterminator
Kanal (zum Weiterleiten von Flüssigkeiten) canal, duct, tube; *neuro* (Membrankanal) channel, membrane channel
Kanalisation
  sewage system, sewer
Kanister (Behälter)
  jug (container);
  carboy (Ballonflasche),
  canister
Kanüle cannula
kanzerogen/karzinogen/ carcinogen/krebserzeugend
  carcinogenic
Kapazität capacity
➢ **elektrische K.** capacitance (C)
Kapazitätsfaktor/Verteilungsverhältnis
  capacity factor
Kapazitätsgrenze/ Grenze der ökologischen Belastbarkeit/ Tragfähigkeit (Ökosystem)
  carrying capacity
Kapazitätskontrollsystem, limitiertes
  limited capacity control system (LCCS)
kapazitiver Strom
  capacitative current
Kapillare/Haargefäß
  capillary; *tech* capillary tube
Kapillarelektrophorese
  capillary electrophoresis
Kapillarpipette
  capillary pipet, capillary pipette
Kapillarrohr/Kapillarröhrchen
  capillary tube/tubing
Kapillarsäule (Trennkapillare: GC)
  capillary column;
  (offene) open tubular column

Kappe (Verschluss/Deckel)
  cap, top, lid
Kapuze hood; (für Labor: Haarschutzhaube) bouffant cap
Karabinerhaken
  spring hook, snap hook
Karbonisation carbonization
Karenzzeit waiting period
Karobgummi/Johannisbrotkernmehl
  carob gum, locust bean gum
Karotinoide/Carotinoide carotinoids
Karte
  (Landkarte/Stadtplan etc.) map;
  (Tafel/Schaubild/Tabelle) chart
kartieren map;
  (grafisch darstellen: Kurven..) plot;
  (skizzieren) chart
Kartierung mapping, plotting
Karton/Kartonpapier (feste Pappe)
  cardboard, paperboard, fiberboard
Kartusche cartridge
Kartuschenbrenner
  cartridge burner
Karzinogen *n* carcinogen
karzinogen/carcinogen/ kanzerogen/krebserzeugend
  carcinogenic
Karzinom carcinoma
kaschieren
  (Bücher etc.) laminate;
  (Textilien) bond
Kassette/Patrone
  cartridge, cassette
Kasten/Kiste box, crate
kastrieren castrate, geld, neuter
Katalysator catalyst
Katalyse catalysis
katalysieren catalyze
katalytisch catalytic, catalytical
katalytische Einheit/ Einheit der Enzymaktivität (*katal*)
  catalytical unit,
  unit of enzyme activity (*katal*)
Kation cation
Kationenaustauscher
  cation exchanger
kauen/zerkauen chew, masticate

Kauforder/Bestellung
  purchase order
Kausche thimble
Kautschuk
  caoutchouc, rubber, india rubber
Kegelhülse conical socket
Kegelventil
  cone valve,
  mushroom valve,
  pocketed valve
Kehrbesen broom
kehren/fegen sweep (up)
Kehrschaufel/Kehrblech
  dustpan
Keil wedge, peg
Keim (Mikroorganismus) germ;
  (Keimling/Embryo) germ, embryo
keimen germinate, sprout
Keimfähigkeit
  germinability
keimfrei/steril germ-free, sterile
keimtötend/bakterizid
  bacteriocidal, bactericidal
Keimung germination
Keimzahl
  (Anzahl von Mikroorganismen)
  cell count, germ count;
  (Samenkeimung)
  germination percentage
Kelle trowel
Kelter fruit/juice press
  (e.g. for making juice)
keltern press (fruit/grapes)
Kenngröße/Parameter
  parameter; *math* dimensionless
  group/quantity/number
Kennwert characteristic value
Kennzahl basic number,
  characteristic number;
  (Chiffre) key, cipher;
  (Kennziffer) *stat* index number,
  indicator; (statistische Maßzahl)
  statistic, statistic value
Kennzeichen/
  Abzeichen/Marke/Banderole
  badge
Kennzeichen für Fahrzeuge/Container
  placard

Kennzeichnung marking, labeling
Kennzeichnungspflicht
  labeling requirement
keratinisieren (verhornen)
  keratinize (cornify)
Kerbe indentation, notch;
  (Schlitz/Bruchstelle) nick
kerbig/gekerbt notched, nicked
Kern/Zentrum (Mark/Core) core, center
  ➢ **Bohrkern** *geol* drill core
  ➢ **Schliffkern (Steckerteil)**
  cone (male: ground-glass joint)
kernmagnetische Resonanz/
  Kernspinresonanz
  nuclear magnetic resonance (NMR)
kernmagnetische
  Resonanzspektroskopie/
  Kernspinresonanz-Spektroskopie
  nuclear magnetic resonance
  spectroscopy, NMR spectroscopy
Kernnährelemente macronutrients
Kernseife (feste Natronseife)
  curd soap (domestic soap)
Kernspinresonanz/
  kernmagnetische Resonanz
  nuclear magnetic resonance (NMR)
Kernspinresonanz-Spektroskopie/
  kernmagnetische
  Resonanzspektroskopie
  nuclear magnetic resonance
  spectroscopy, NMR spectroscopy
Kernspintomographie (KST)/
  Magnetresonanztomographie (MRT)
  magnetic resonance imaging (MRI),
  nuclear magnetic resonance imaging
Kerntransplantation
  nuclear transfer,
  nuclear transplantation
Kerosin kerosene
Kescher/Käscher
  (Fangnetz für Fische)
  landing net, aquatic net
  (collecting net for fish)
Kesselstein boiler scale, incrustation
Kesselstein entfernen descale
Kesselwagen (Chemikalientransport)
  tank car,
  tank truck (Schiene: rail tank car)

Ke

Ketoaldehyd
  ketoaldehyde, aldehyde ketone
Keton ketone
  ➢ **Aceton (Azeton)/
  Propan-2-on/2-Propanon/
  Dimethylketon**
  acetone, 2-propanone,
  dimethyl ketone
Ketonkörper
  ketone body (acetone body)
Ketosäure keto acid
Kette (verzweigte/unverzweigte)
  chain (branched/unbranched)
Kettenform *chem*
  chain form, open-chain form
Kettenformel
  chain formula,
  open-chain formula
Kettenklammer chain clamp
Kettenlänge chain length
Kettenreaktion chain reaction
Kettensäge chain saw
Kienspan
  chip of pinewood, pinewood chip
Kies gravel
Kieselerde diatomaceous earth
Kieselgel/Silicagel silica gel
Kieselgur
  kieselguhr (loose/porous diatomite;
  diatomaceous/infusorial earth)
Kieselsäure H_4SiO_4 silicic acid
kieselsäurehaltig siliceous
Kieselstein pebble
**Kimwipes (Kimberley-Clark
  Reinraum Wischtücher)**
  Kimwipes (Kimberley-Clark
  cleanroom wipes)
Kinetik (nullter/erster/zweiter Ordnung)
  (zero-/first-/second-order...) kinetics
  ➢ **Reaktionskinetik** reaction kinetics
  ➢ **Reassoziationskinetik**
  reassociation kinetics
Kipphebel
  tumbler; lever
Kipphebelschalter
  tumbler switch, knife switch
Kippschalter
  toggle switch, rocker

**Kippscher Apparat/
  'Kipp'/Gasentwickler**
  Kipp generator
Kitt/Kittsubstanz
  allg adhesive, cement;
  (Fensterkitt etc.) putty
Kittel coat, gown; frock
  ➢ **Arbeitskittel/Overall** overall
  ➢ **Laborkittel/Labormantel**
  laboratory coat, labcoat
  ➢ **Schutzkittel/Schutzmantel**
  protective coat,
  protective gown
Kittmesser putty knife
klaffen/offen stehen
  gape
Klammer clamp, clip
  ➢ **Büroklammer** paper clip
  ➢ **Halteklammer (Büro)**
  binder clip
  ➢ **Heftklammer** staple
  ➢ **Kettenklammer** chain clamp
  ➢ **Objekttisch-Klammer** *micros*
  stage clip
  ➢ **Schliffklammer/Schliffklemme
  (Schliffsicherung)**
  joint clip, joint clamp,
  ground-joint clip,
  ground-joint clamp
  ➢ **Schliffklammer/Schliffklemme
  (Schliffsicherung)**
  joint clip, ground-joint clip,
  ground-joint clamp
Kläranlage (kommunal) sewage
  treatment plant; (industriell)
  waste-water purification plant
Klärbecken/Absetzbecken
  settling tank
klären (z.B. absetzen/entfernen von
  Schwebstoffen aus einer Flüssigkeit)
  clear, clarify, purify;
  (filtrieren) filtrate
Klärflasche purge
Klärgas/Faulgas (Methan)
  sludge gas
Klarglas clear glass
Klärschlamm (Faulschlamm)
  sludge, sewage sludge

**Klarsichtfolie
(Einwickelfolie/**auch**: Haushaltsfolie)**
film wrap (transparent film/foil),
cling wrap
**Klärung (z.B. absetzen/entfernen
von Schwebstoffen aus einer
Flüssigkeit)**
clarification, purification
➢ **Abwasseraufbereitung**
sewage treatment
➢ **Filtrierung/Filtration** filtration
Klärwerk/Kläranlage (Abwasser)
sewage treatment plant
**Klassenhäufigkeit/
Besetzungszahl/
absolute Häufigkeit** *stat*
class frequency, cell frequency
klassieren (nach Korngröße)
screen, size
Klassierung *stat* grouping of classes
klassifizieren classify
Klassifizierung/Klassifikation
classifying, classification
Klebeband adhesive tape
➢ **Elektro-Isolierband**
insulating tape, electric tape,
friction tape
➢ **Gewebeband/Textilband (einfach)**
cloth tape
➢ **Gewebeklebeband/Panzerband
(Universalband/Vielzweckband)**
duct tape (polycoated cloth tape)
➢ **Isolierband**
insulating tape, duct tape
➢ **Kreppband**
masking tape
➢ **Verpackungsklebeband**
packaging tape
kleben
stick, adhere; paste; cement
➢ **leimen** glue
klebend adhesive
Kleber/Klebstoff/Leim
adhesive, glue, gum;
paste; cement
➢ **Mehrkomponentenkleber**
multicomponent adhesive/cement
Klebestreifen adhesive tape

klebrig/glutinös
sticky, glutinous, viscid
Klebstoff/Kleber
adhesive; (Leim) glue, gum
Kleiderbügel coat hanger
Kleiderordnung dress code
Kleie bran
Kleinanwendung
small-scale application
Klemme clamp; clip
Klemme (Kegelschliffsicherung)
clip (for ground joint)
➢ **Arterienklemme**
artery forceps, artery clamp
➢ **Bürettenklemme** buret clamp
➢ **Doppelmuffe/Kreuzklemme**
clamp holder, 'boss', clamp 'boss'
(rod clamp holder)
➢ **Gefäßklemme/Arterienklemme/
Venenklemme**
hemostatic forceps, artery clamp
➢ **Hakenklemme (Stativ)** hook clamp
➢ **Kettenklammer** chain clamp
➢ **Klemme mit runden Backen**
round jaw clamp
➢ **Krokodilklemme** alligator clip
➢ **Lüsterklemme** luster terminal
(insulating screw joint)
➢ **Schlauchklemme/Quetschhahn**
tubing clamp, pinchcock clamp,
pinch clamp
➢ **Schraubklemme** pinch clamp
➢ **Spannungsklemme**
voltage clamp
➢ **Tupferklemme** sponge forceps
➢ **Verlängerungsklemme**
extension clamp
Klemmpinzette/Umkehrpinzette
reverse-action tweezers
(self-locking tweezers)
Klempner/Installateur plumber
Klettverschluss (Haken & Flausch)
Velcro, Velcro fastener,
hook and loop fastener
Klinge blade
Klinikmüll clinical waste
klinisch getestet/geprüft
clinically tested

kl

klonen/klonieren clone
Kloning/Klonierung cloning
Klumpen clump; lump; *chem* (Kruste: fest verbackender Niederschlag) cake
klumpen clump; lump; *chem* (zusammenbacken: Präzipitat) cake
Knallgas (2×H$_2$ + O$_2$) oxyhydrogen (gas), detonating gas
Knarre ratchet
➢ **Hebelknarre** lever ratchet
➢ **Umschaltknarre** change-over ratchet
kneifen pinch
Kneifzange pliers, nippers (*Br*), cutting pliers, pincers
Knochen bone
Knochenmehl bone meal
Knochensäge bone saw
Knochenzange bone-cutting forceps
knöchern/Knochen... bony
Knopf button; (Regler) control
Knopfzelle (Batterie) coin cell, button cell (button battery)
Knorpel cartilage
knorpelig cartilaginous
Koazervat coacervate
Kobalt/Cobalt (Co) cobalt
Kochblutagar/Schokoladenagar chocolate agar
köcheln (auf kleiner Flamme) simmer (boil gently)
kochen cook, boil
Kochfleischbouillon/Fleischbrühe cooked-meat broth
Koch's Postulat/Koch'sches Postulat Koch's postulate
Kochsalz (NaCl) table salt
Kochsalzlösung saline
➢ **physiologische K.** saline, physiological saline solution
kodieren/codieren encode, code
Koffein/Thein caffeine, theine

Kohle coal
➢ **Anthrazit/Kohlenblende** anthracite, hard coal
➢ **Glanzbraunkohle/ subbituminöse Kohle** subbituminous coal
➢ **Steinkohle/bituminöse Kohle** bituminous coal
➢ **Weichbraunkohle & Mattbraunkohle/ Lignit** lignite
Kohlebürste (Motor) *tech* carbon brush
Kohlendioxid carbon dioxide
kohlendioxidliebend/kapnophil capnophilic
Kohlenhydrat carbohydrate
Kohlenmonoxid carbon monoxide
Kohlensäure (Karbonat/Carbonat) carbonic acid (carbonate)
Kohlenstoff carbon
Kohlenstoffbindung carbon bond
Kohlenstoffquelle carbon source
Kohlenstoffverbindung carbon compound
Kohlenwasserstoff hydrocarbon
➢ **chlorierter Kohlenwasserstoff** chlorinated hydrocarbon
➢ **Fluorchlorkohlenwasserstoffe (FCKW)** chlorofluorocarbons, chlorofluorinated hydrocarbons (CFCs)
➢ **Fluorkohlenwasserstoff** fluorinated hydrocarbon
Köhlersche Beleuchtung *micros* Koehler illumination
Kojisäure kojic acid
Kolben *chem* flask; *tech* (Stempel/Schieber: Spritze etc.) piston, plunger (e.g. of syringe)
➢ **Birnenkolben/Kjeldahl-Kolben** Kjeldahl flask
➢ **Destillierkolben/Destillationskolben** distilling flask, retort
➢ **Dreihalskolben** three-neck flask
➢ **Enghalskolben** narrow-mouthed flask, narrow-necked flask
➢ **Erlenmeyer Kolben** Erlenmeyer flask

Ko

- **Extraktionskolben** extraction flask
- **Fernbachkolben** Fernbach flask
- **Filtrierkolben/ Filtrierflasche/Saugflasche** filter flask, filtering flask, vacuum flask
- **Kulturkolben** culture flask
- **Messkolben** volumetric flask
- **Rotationsverdampferkolben** rotary evaporator flask
- **Rundkolben/Siedegefäß** round-bottomed flask, round-bottom flask, boiling flask with round bottom
- **Säbelkolben/Sichelkolben** saber flask, sickle flask, sausage flask
- **Schliffkolben** ground-jointed flask
- **Schüttelkolben** shake flask
- **Schwanenhalskolben** swan-necked flask, S-necked flask, gooseneck flask
- **Seitenhalskolben** sidearm flask
- **Spitzkolben** pear-shaped flask (small/pointed)
- **Stehkolben/Siedegefäß** Florence boiling flask, Florence flask (boiling flask with flat bottom)
- **Verdampferkolben** evaporating flask
- **Weithalskolben** wide-mouthed flask, wide-necked flask
- **Zweihalskolben** two-neck flask

Kolbenhubpipette/Mikroliterpipette micropipet, pipettor
Kolbenklemme flask clamp
Kolbenpumpe piston pump, reciprocating pump
Kolbenwischer/Gummiwischer (zum mechanischen Loslösen von Rückständen im Glaskolben) policeman (glass/plastic or metal rod with rubber or Teflon tip)
kolieren filter, percolate, strain
Kollektorblende/Leuchtfeldblende field diaphragm
Kollektorlinse collector lens, collecting lens
Kollimationsblende/Spaltblende *micros* collimating slit
Kollimator collimator
Kollision collision
Kollodium collodion
Kollodiumwolle collodion cotton
kolonial/koloniebildend colonial, colony-forming
Kolonie colony
koloniebildend/kolonial colony-forming, colonial
Kolonne/Turm (Bioreaktor) column
Kolophonium colophony, rosin
Kombizange combination pliers, linesman pliers; (verstellbar) slip-joint pliers
Kompartimentierung compartmentalization, compartmentation
kompatibel/verträglich compatible
Kompatibilität/Verträglichkeit compatibility
Kompensationspunkt compensation point
kompetitiv competitive
Komplettmedium complete medium, rich medium
Komplexbildner/Chelatbildner complexing agent, chelating agent, chelator
Komplexbildung/Chelatbildung complexing, chelation, chelate formation
komplexieren chelate
Komplexität complexity
Kondensat condensate
Kondensation condensation
Kondensationspunkt condensing point
Kondensationsreaktion/ Dehydrierungsreaktion condensation reaction, dehydration reaction
Kondensator *opt* condenser; *electr* capacitor
kondensieren condense
Kondensorblende/Aperturblende condenser diaphragm (iris diaphragm)

Ko

Kondensortrieb *micros*
  condenser adjustment knob,
  substage adjustment knob
konditionieren *med/chromat* condition
konditioniertes Medium
  conditioned medium
Konditionierung *med/chromat*
  conditioning
Konfektionierung
  (ready-made/industrial)
  manufacture/manufacturing
Konfidenzgrenze/
Vertrauensgrenze/
Mutungsgrenze *stat*
  confidence limit
Konfidenzintervall/
Vertrauensintervall/
Vertrauensbereich *stat*
  confidence interval
Konfidenzniveau/
Konfidenzwahrscheinlichkeit *stat*
  confidence level
Konformation
  conformation
➢ **Knäuelkonformation/**
Schleifenkonformation *gen*
  coil conformation, loop conformation
➢ **relaxiert/entspannt**
  relaxed (conformation)
➢ **Repulsionskonformation** *gen*
  repulsion conformation
➢ **Ringform**
  ring form, ring conformation
➢ **Schleifenkonformation/**
Knäuelkonformation *gen*
  loop conformation, coil conformation
➢ **Sesselform (Cycloalkane)** *chem*
  chair conformation
➢ **Wannenform (Cycloalkane)** *chem*
  boat conformation
kongenial/verwandt/gleichartig
  congenial
Königswasser aqua regia
konjugierte Bindung *chem*
  conjugated bond
Konkurrent/Mitbewerber competitor
Konkurrenz/Kompetition/Wettbewerb
  competition

konkurrieren/in Wettstreit stehen
  compete
konservieren/präservieren/
haltbar machen/erhalten
  conserve, preserve; store, keep
Konservierung preservation; storage
Konservierungsstoff preservative
Konsistenz/Beschaffenheit
  consistency
konsistieren/beschaffen sein consist
Konsument/Verbraucher consumer
Kontagiosität contagiousness
Kontakt contact; *electr* lead
Kontaktallergen contact allergen
Kontaktinfektion contact infection
Kontaktpestizid contact pesticide
Kontaktrisiko (Gefahr bei Berühren)
  contact hazard
Kontamination/Verunreinigung
  contamination
kontaminieren/verunreinigen
  contaminate
kontrahieren/zusammenziehen
  contract
Kontrastfärbung/Differentialfärbung
  contrast staining, differential staining
kontrastieren contrast; *tech/micros*
  (färben/einfärben) stain
Kontrastierung/
Färben/Färbung/Einfärbung
  tech/micros stain, staining
Kontrollbereich/
kontrollierter Bereich
  controlled area
Kontrolle control, check; inspection;
  (Überwachung/Beaufsichtigung)
  supervision
Kontrollgerät
  controlling instrument,
  control instrument,
  monitoring instrument;
  (Anzeige: monitor)
kontrollieren control, check, inspect;
  (überwachen/beaufsichtigen)
  supervise
Konvektionsofen convection oven;
  (mit natürlicher Luftumwälzung)
  gravity c.o.

Kr

Konzentration concentration
> **Arbeitsplatzkonzentration, zulässige**
  permissible workplace exposure
> **Grenzkonzentration**
  limiting concentration
> **Hemmkonzentration**
  inhibitory concentration
> **Hemmkonzentration, minimale (MHK)**
  minimal inhibitory concentration, minimum inhibitory concentration (MIC)
> **MAK-Wert (maximale Arbeitsplatz-Konzentration)**
  maximum permissible workplace concentration, maximum permissible exposure
> **mittlere letale Konzentration (LC_{50})**
  median lethal concentration (LC_{50})
> **Osmolarität/ osmotische Konzentration**
  osmolarity, osmotic concentration

Konzentrationsgefälle/ Konzentrationsgradient
concentration gradient
konzentrieren concentrate
Kooperation/Zusammenarbeit
cooperation, collaboration
kooperative Bindung
cooperative binding
Kooperativität cooperativity
kooperieren/zusammenarbeiten
cooperate, collaborate
Koordination coordination
koordinieren coordinate
Kopf (Fettmolekül) head
Kopfbedeckung head cover
Kopienzahl copy number
Kopiergerät copying machine
koppeln/ verbinden/aneinander festmachen
couple, join, link
Kopplung coupling; linkage
> **chemische K.**
  chemical coupling
> **geminale K. (NMR)**
  geminal coupling

Korb basket
> **zusammenlegbarer Korb/Faltkorb**
  collapsible basket

Korkbohrer cork-borer
Korkring cork ring
Korn grain, granule, particle
Körner (Werkzeug) center punch
Korngröße particle size, grain size
Körnigkeit granulation
Kornklasse grain-size class
Körnung grain
Körper body, soma
körperbehindert
physically handicapped
Körperflüssigkeit body fluid
körperliche Arbeit physical work
Körpertemperatur
body temperature
Korrosion corrosion
korrosionsbeständig
corrosionproof; stainless
Korrosionsmittel/Ätzmittel
corrosive
korrosiv/korrodierend/ zerfressend/angreifend/ätzend
corrosive
Kost/Essen/Speise/Nahrung/Diät
diet, food, feed, nutrition
Kosten-Nutzen-Analyse
cost-benefit analysis
Kot/Fäkalien feces
Kraftfahrzeug motor vehicle
Kraftmikroskopie force microscopy
> **Rasterkraftmikroskopie**
  atomic force microscopy (AFM)

krank sick, ill, diseased
krank schreiben certify s.o. as ill; (krank geschrieben sein) to be on sick-leave
Krankenversicherung health insurance
krankhaft/pathologisch pathological
Krankheit disease, illness; sickness
> **ansteckende Krankheit/ infektiöse Krankheit**
  contagious disease, infectious disease
> **Erbkrankheit**
  inheritable disease

> **erbliche Erkrankung/Erbkrankheit**
hereditary disease, genetic disease,
inherited disease, heritable disorder
> **Strahlenkrankheit**
radiation sickness
> **übertragbare Krankheit**
transmissible disease,
communicable disease
> **Zivilisationskrankheiten**
diseases of civilization
('affluent peoples' diseases')

krankheitserregend/pathogen
disease-causing, pathogenic
Krankheitserreger
disease-causing agent, pathogen
Krankheitsüberträger
transmitter of disease
**Krankheitsursache/
Ätiologie**
etiology
**Krankheitsverursacher
(Wirkstoff/Agens/Mittel)**
etiological agent
Krankmeldung notification of illness
(to one's employer)
Kratzer (Gerät zum abkratzen) scraper
Krebs (malignes Karzinom)
cancer
(malignant neoplasm/carcinoma)
krebsartig cancerous
krebserregend/karzinogen/carcinogen
carcinogenic
krebserzeugend/onkogen/oncogen
cancer causing, oncogenic,
oncogenous
Krebsrisiko cancer risk
krebsverdächtige Substanz
cancer suspect agent,
suspected carcinogen
Kreisdiagramm pie chart
Kreiselpumpe/Zentrifugalpumpe
impeller pump, centrifugal pump
Kreisschüttler/Rundschüttler
circular shaker, orbital shaker,
rotary shaker
Kreppband masking tape
kreuzen/züchten
cross, crossbreed, breed, interbreed
Kreuzklemme/Doppelmuffe
clamp holder, 'boss', clamp 'boss'
(rod clamp holder)
Kreuzkontamination
cross-contamination
Kreuzprobe *immun* cross-matching
Kreuzschlüssel
spider wrench, spider spanner (*Br*)
**Kreuzschraubenzieher/
Kreuzschlitzschraubenzieher**
Phillips®-head screwdriver;
Phillips® screwdriver
Kreuzstrom-Filtration
cross flow filtration
Kreuztisch *micros* mechanical stage
Kreuzung/Züchtung
crossing, cross, crossbre(e)d, breed,
crossbreeding, interbreeding;
(Kreuzungsprodukt) cross, breed
> **aus der Kreuzung entfernt oder
nicht verwandter Individuen
gezüchtet** outbred
> **Dihybridkreuzung**
dihybrid cross
> **Doppelkreuzung** double cross
> **Drei-Faktor-Kreuzung**
three-point testcross
> **Einfachkreuzung** single cross
> **Herauskreuzen/Auskreuzen**
outcrossing
> **Monohybridkreuzung**
monohybrid cross
> **nicht verwandte Individuen kreuzen**
outbreed
> **Testkreuzung** testcross
> **Überbrückungskreuzung**
bridging cross
Kriechschutzadapter *dest*
anticlimb adapter
Kriechstrom
leakage current, creepage
Kristall crystal
Kriställchen small crystal; crystallite
Kristallisation crystallization
**Kristallisationskern/
Kristallisationskeim**
crystallization nucleus
kristallisieren crystallize

Kristallographie crystallography
Kristallstruktur
  crystal structure, crystalline structure
Kristallwasser
  crystal water, water of crystallization
kritischer Punkt critical point
Kritisch-Punkt-Trocknung
  critical point drying (CPD)
Krokodilklemme
  alligator clip,
  alligator connector clip
Kronenkorken crown cap
Krug/Kanne/Kännchen
  jug; (mit Griff) pitcher
Krümel scraps, shavings
Krümmer (gebogenes Rohrstück)/ Winkelrohr/Winkelstück (Glas/Metal etc. zur Verbindung)
  ell, elbow, elbow fitting, bend,
  bent tube, angle connector
krustenbildend encrusting
Kryostat cryostat
Kryostatschnitt *micros* cryostat section
Kryo-Ultramikrotomie
  cryoultramicrotomy
Küchenhandtuch kitchen towel
Küchenrolle/Haushaltsrolle/ Tücherrolle/Küchentücher/ Haushaltstücher
  kitchen tissue (kitchen paper towels)
Kugelbettreaktor (Bioreaktor)
  bead-bed reactor
Kugelgelenk
  ball-and-socket joint,
  spheroid joint
Kugelkühler Allihn condenser
Kugellager ball bearing
➢ **Achsenlager/Achslager/Zapfenlager (z.B. beim Kugellager)**
  journal
➢ **Laufring (beim Kugellager)**
  race
Kugelrohrdestillation
  bulb-to-bulb distillation
Kugel-Stab-Modell/Stab-Kugel-Modell
  chem ball-and-stick model,
  stick-and-ball model
Kugelventil ball valve

Kühlakku/Kälteakku
  cooling pack, cooling unit
Kühlbox cooler
kühlen cool, chill, refrigerate
➢ **abkühlen**
  cool down, get cooler
➢ **gefrieren** freeze
➢ **in den Kühlschrank stellen**
  refrigerate
➢ **tiefkühlen/tiefgefrieren**
  deep-freeze
➢ **unterkühlen** supercool
Kühler
  condenser
➢ **Dimroth-Kühler**
  coil condenser (Dimroth type)
➢ **Einhängekühler/Kühlfinger**
  suspended condenser, cold finger
➢ **Intensivkühler**
  jacketed coil condenser
➢ **Kugelkühler** Allihn condenser
➢ **Liebigkühler** Liebig condenser
➢ **Luftkühler** air condenser
➢ **Rückflusskühler**
  reflux condenser
➢ **Schlangenkühler**
  coil distillate condenser,
  coil condenser,
  coiled-tube condenser
➢ **Vigreux-Kolonne**
  Vigreux column
Kühlfach/Gefrierfach (im Kühlschrank)
  freezer compartment,
  freezing compartment, freezer
Kühlfalle cold trap, cryogenic trap
Kühlfinger *dest*
  cold finger (finger-type condenser)
Kühlflüssigkeit/Kühlmittel coolant
  (*allg*/direkt); refrigerant
Kühlhaus cold store
Kühlmantel condenser jacket
Kühlraum (Gefrierraum)
  cold room ('walk-in refrigerator'),
  cold-storage room, cold store,
  'freezer'; (Kühlkammer/Kühlhaus)
  cold storage, deep freeze
Kühlschlange
  cooling coil, condensing coil

Kü

Kühlschmierstoff/Kühlschmiermittel
coolant (lubricant)
Kühlschrank
refrigerator, fridge; icebox
➢ **Gefrierfach**
freezing compartment
➢ **Tiefkühlschrank**
deep freezer, 'cryo'
Kühltruhe/Gefriertruhe
chest freezer;
(Gefrierschrank) upright freezer
➢ **Tiefkühltruhe**
deep-freeze, deep freezer
Kühlwasser coolant, cooling water
kultivierbar cultivatible, arable
kultivieren
agr cultivate; *micb* culture, culturing
Kultur culture
➢ **Anreicherungskultur**
enrichment culture
➢ **Ausstrichkultur** streak culture
➢ **Blutkultur** blood culture
➢ **Dauerkultur** long-term culture
➢ **diskontinuierliche Kultur/
Batch-Kultur/Satzkultur**
batch culture
➢ **Eikultur**
chicken embryo culture
➢ **Einstichkultur/Stichkultur
(Stichagar)** stab culture
➢ **Eintauchkultur**
submerged culture
➢ **Erhaltungskultur**
maintenance culture
➢ **Gewebekultur** tissue culture
➢ **kontinuierliche Kultur**
continuous culture,
maintenance culture
➢ **Mischkultur** mixed culture
➢ **Oberflächenkultur** surface culture
➢ **Perfusionskultur** perfusion culture
➢ **Plattenausstrichmethode**
streak-plate method/technique
➢ **Plattengussverfahren/
Gussplattenmethode**
pour-plate method/technique
➢ **Reinkultur**
pure culture, axenic culture
➢ **Rollerflaschenkultur**
roller tube culture
➢ **Satzkultur/Batch-Kultur/
diskontinuierliche Kultur**
batch culture
➢ **Schrägkultur (Schrägagar)**
slant culture, slope culture
➢ **Schüttelkultur** shake culture
➢ **Spatelplattenverfahren**
spread-plate method/technique
➢ **Stammkultur**
stem culture, stock culture
➢ **statische Kultur** static culture
➢ **Stichkultur/Einstichkultur
(Stichagar)** stab culture
➢ **Submerskultur** submerged culture
➢ **Synchronkultur**
synchronous culture
➢ **Verdünnungs-Schüttelkultur**
dilution shake culture
➢ **Zellkultur** cell culture
Kulturbeutel toilet bag
Kulturflasche culture bottle
Kulturkolben culture flask
Kulturmedium/Medium/Nährmedium
medium, culture medium
➢ **Anreicherungsmedium**
enrichment medium
➢ **Differenzierungsmedium**
differential medium
➢ **Elektivmedium/Selektivmedium**
selective medium
➢ **Komplettmedium/Vollmedium**
complete medium
➢ **komplexes Medium**
complex medium
➢ **Mangelmedium** deficiency medium
➢ **Minimalmedium** minimal medium
➢ **Selektivmedium** selective medium
➢ **synthetisches Medium
(chem. definiertes Medium)**
defined medium
➢ **Vollmedium**
complete medium
Kulturpflanze
crop plant, cultivated plant
Kulturröhrchen culture tube
Kulturschale culture dish

Kü

Kunde customer
Kundendienst/Kundenbetreuung customer service
Kunstharz artificial resin, synthetic resin
künstlich artificial
Kupfer (Cu) copper
Kupfer(I) ... cuprous ...
Kupfer(I)-oxid cuprous oxide
Kupfer(II) ... cupric ...
Kupfer(II)-oxid cupric oxide, black copper
Kupferdrahtnetz copper grid mesh
Kupferglanz Cu_2S chalcocite
Kupfernetz *micros* copper grid
Kupferspäne/Kupferfeilspäne copper filings
Kupfersulfat/Kupfervitriol copper sulfate, copper vitriol, cupric sulfate
Kupplung *tech/mech* clutch, coupling, coupler, attachment; (Verbinder: z.B. Schlauch) fitting, coupler
➢ **Gelenkkupplung** ball-joint connection
➢ **Schlauchkupplung** tubing connection, tube coupling
➢ **Schnellkupplung** quick-disconnect fitting
➢ **starre Kupplung** fixed coupling
➢ **Stecker/männliche Kupplung** (male) insert; male
➢ **weibliche Kupplung/Körper** body, (female) fitting; female
Kupplungsreaktion *chem* coupling reaction
Kurort health resort
Kurzhalstrichter/Kurzstieltrichter short-stem funnel, short-stemmed funnel
kurzkettig short-chain
kurzschließen short-circuit
Kurzschluss short circuit, short-circuiting, short; (Sicherung 'rausfliegen' lassen) blow/kick a fuse
Kurzwegdestillation/ Molekulardestillation short-path distillation, flash distillation
Küvette (für Spektrometer) cuvette, spectrophotometer tube
Küvettenhalter *analyt* cell holder

La

Labor (*pl* **Labors**)/
Laboratorium (*pl* **Laboratorien**)
laboratory, lab
➢ **Forschungslabor**
research laboratory/lab
➢ **Fotolabor**
photographic laboratory/lab
➢ **Gute Laborpraxis**
Good Laboratory Practice (GLP)
➢ **im Labormaßstab**
laboratory-scale, lab-scale
➢ **Lernlabor/Lehrlabor**
teaching laboratory,
educational laboratory
➢ **Sicherheitslabor/
Sicherheitsraum/
Sicherheitsbereich (S1-S4)**
biohazard containment (laboratory)
(classified into biosafety containment classes)
➢ **Tierlabor** animal laboratory/lab
Labor-Anstandsregeln
laboratory/lab courtesy
Laborant(in) laboratory/lab worker
Laborarbeiter laboratory/lab worker
Laborarbeitstisch
laboratory/lab bench
**Laborassistent(in)/
technische(r) Assistent(in)**
technical lab assistant,
laboratory/lab technician
Laboratorium (siehe Labor)
laboratory, lab
Laboraufzeichnungen
laboratory/lab notes,
laboratory/lab documentation
Laborbank
laboratory/lab counter
Laborbedarf
labware, laboratory/lab supplies
Laborbedingungen
laboratory/lab conditions
Laborbefund
laboratory findings, laboratory results
Laborbericht laboratory/lab report
Laborbürste laboratory/lab brush
Laborchemikalie
laboratory/lab chemical

Labordiagnostik
laboratory/lab diagnostics
Laboreinheit laboratory/lab unit
Laboreinrichtung/Laborausstattung
laboratory/lab facilities
**Laboretikette/
Laborgepflogenheiten/
Laborbenimmregeln/Labor'knigge'**
lab etiquette
Laborgehilfe laboratory/lab aide
Laborgerät laboratory/lab equipment
Laborhocker lab stool
Laborjournal/Protokollheft
laboratory/lab notebook
Laborkakerlake laboratory/lab roach
Laborkaugummi lab chewing gum
(sticks to glass, metal, wood! ☺)
Laborkittel/Labormantel
laboratory coat, labcoat
Laborleiter laboratory/lab head
Labormaßstab laboratory/lab scale
Labormöbel laboratory/lab furniture
Laborpersonal
laboratory/lab personnel
Laborplatz/Laborarbeitsplatz
laboratory/lab space,
laboratory/lab working space
Laborpraxis: Gute Laborpraxis
Good Laboratory Practice (GLP)
Laborprotokoll
laboratory/lab protocol
Laborreagens
laboratory/lab reagent, bench reagent
Laborreinigung
laboratory/lab cleanup
Laborschale laboratory/lab tray
Laborschürze laboratory/lab apron
Laborschutzplatte (Keramikplatte)
laboratory protection plate
Laborsicherheit laboratory/lab safety
Laborsicherheitsbeauftragter
laboratory safety officer
Laborsicherheitsstufe
physical containment (level)
Laborstandard
laboratory/lab standard
Laborstandflasche/Standflasche
laboratory/lab bottle

Labortagebuch
lab diary, lab manual, log book
Labortechnik
laboratory/lab technique
labortechnisch/im Labormaßstab
laboratory-/lab-scale
Labortisch/Labor-Werkbank
laboratory/lab table,
laboratory/lab bench,
laboratory/lab workbench
Labortrakt/Laboratoriumstrakt
laboratory/lab suite
Labortratsch laboratory/lab gossip
Labortrockner/
Heißluftgebläse/Föhn
hot-air gun
Laborverfahren
laboratory/lab procedure
Laborversuch/Labortest
laboratory/lab experiment,
laboratory/lab test
Laborwaage
laboratory/lab balance,
laboratory/lab scales
Laborwagen/Laborschiebewagen
laboratory cart,
lab pushcart (*Br* trolley)
Laborzange tongs
Laborzeile bench row
Lachgas/
Distickstoffoxid/Dinitrogenoxid
laughing gas, nitrous oxide
Lack/Firnis/Farblack
lacquer; (Lasur) varnish
lackieren
varnish, lacquer;
repaint; refinish; coat
Ladegerät charger
Ladeverzeichnis/
Ladungsdokument
(Warenverzeichnis)
manifest document
Ladung/elektrische Ladung
charge, electrical charge
Ladungstrennung *electr*
charge separation
Lage (Position: in Bezug) position;
(Ort) location

Lager (Lagerraum/Warenlager)
stockroom, storage room, repository;
(Gebäude) warehouse; (Vorrat)
stock, store, supplies;
(Achsen~/Rührer etc.) bearing(s)
➤ **Kugellager** ball bearing
➤ **Lagerbüchse (Kugellager)**
journal box
➤ **Laufring (des Kugellagers)**
race
Lagerbestand stock, store, supplies
Lagerhalter/Lagerist
stockkeeper; stockman;
supplies manager
Lagerhaltung
stockkeeping, storekeeping;
warehousing
Lagerhülse (Glasaussatz)
stirrer bearing
Lagerkapazität storage capacity
lagern (Holz) season, store
Lagertank storage tank
Lagerung (Waren/Gerät/Chemikalien)
storage, warehousing
Lagerverwalter
stockroom manager
lag-Phase/Adaptationsphase/
Anlaufphase/Latenzphase/
Inkubationsphase
lag phase, incubation phase,
latent phase, establishment phase
Laktamid/Lactamid/Milchsäureamid
lactamide
Laktat (Milchsäure)
lactate (lactic acid)
Laktatgärung/Milchsäuregärung
lactic acid fermentation,
lactic fermentation
Laktation lactation
Laktose/Lactose (Milchzucker)
lactose (milk sugar)
laminare Strömung/Schichtströmung
laminar flow
Laminat laminate (laminated plastic)
langkettig long-chain
langlebig long-lived, long-living
Langlebigkeit longevity
länglich oblong

langsam wachsend slow-growing
Längskonstante length constant
Längsschnitt longisection,
  longitudinal section, long section
Langzeitversuch
  long-term experiment
Lanzette lancet
Lärm noise
Lärmschutz noise protection
Läsion/
  Schädigung/Verletzung/Störung
  lesion
Last (Beladung) load; (Gewicht)
  weight; tech/mech (Traglast) load;
  (Belastung) burden
Lasur/Lack/Lackfirnis
  varnish
latent/verborgen/unsichtbar/versteckt
  latent
Latenz latency
Latenzphase/
  Adaptationsphase/Anlaufphase/
  Inkubationsphase/lag-Phase
  latent phase, incubation phase,
  establishment phase, lag phase
Latenzzeit (Inkubationszeit)
  latency period, latent period
  (incubation period)
lateral/seitlich lateral
Lateralvergrößerung/Seitenverhältnis/
  Seitenmaßstab/Abbildungsmaßstab
  micros lateral magnification
Latte (aus Holz) lath, plank
Latthammer
  carpenter's roofing hammer
Laubsäge scroll saw
Laufmittel/Elutionsmittel/
  Fließmittel/Eluent (mobile Phase)
  solvent, mobile solvent, eluent,
  eluant (mobile phase)
Laufmittelfront solvent front
Laufring (beim Kugellager) race
Laufzeit (Vertrag) term;
  (Gerät/Lebenszeit) life, service life;
  (Gerät: für eine 'Runde') cycle time,
  running time
Lauge *chem* lye;
  (Bodenauslaugung) leachate

laugenbeständig/alkalibeständig
  alkaliproof
Laut/Ton sound, noise
Lautstärke volume, loudness
lauwarm lukewarm
Lävan levan
Lävulinsäure levulinic acid
LD$_{50}$ (mittlere letale Dosis)
  LD$_{50}$ (median lethal dose)
LDL (Lipoproteinfraktion niedriger
  Dichte)
  LDL (low density lipoprotein)
Leben *n* life
leben *vb* live
lebend alive, living; biological, biotic
Lebendbeobachtung
  live observation
Lebendfärbung/Vitalfärbung
  vital staining
Lebendgewicht live weight
lebendig alive
Lebendimpfstoff/Lebendvakzine
  live vaccine
Lebendkeimzahl live germ count
Lebendkultur
  live culture, living culture
Lebensdauer life span;
  (Laufzeit: Gerät etc.) service life;
  (Nutzungsdauer) *tech/mech* working
  life
lebensfähig viable
Lebensfähigkeit viability
Lebensgefahr
  danger of life, life threat
➤ **Vorsicht, Lebensgefahr!**
  caution, danger!
lebensgefährlich
  life-threatening
Lebensgröße life size
Lebensmittel foodstuff, nutrients
Lebensmittelchemie food chemistry
lebensmittelecht
  suitable for use in contact with food
Lebensmittelkonservierungsstoff
  food preservative
Lebensmittelkontrolle/
  Lebensmittelprüfung
  food quality control

Lebensmittelüberwachung/ Lebensmittelkontrolle food inspection
Lebensmittelvergiftung food poisoning
Lebensmittelzusatzstoff food additive
lebenswichtig/lebensnotwendig/vital essential for life, vital
Lebenszeit lifetime
leberschädigend/hepatotoxisch hepatotoxic
Lebertran cod-liver oil
Lebewesen/Organismus lifeform, organism
leblos/tot lifeless, inanimate, dead
Leck/Leckage leak, leakage
Leckagerate leak rate
lecken lick; (auslaufen) leak
leer empty; void
leerlaufen/trockenlaufen run dry
legieren alloy
Legierung alloy
Lehranstalt educational facility/institution
Lehrling apprentice, trainee (on-the-job)
Leiche/Kadaver (auch: Tierleiche) corpse, carcass, cadaver
Leichengeruch cadaverous smell
Leichenschau inspection of corpse, postmortem examination
Leichenstarre/Totenstarre rigor mortis
Leichnam body, dead body, corpse
leicht entzündlich highly flammable
leichtgewicht(ig) lightweight
leichtlöslich easily soluble, readily soluble
Leim glue
➢ **Holzleim** wood glue
Leinwand (Projektions~) screen (projection)
Leiste ledge, lath, border, strip; (dünne L.) slat; molding
Leistung achievement, performance; *phys/electr* power

Leistungsaudit/Leistungsprüfung/ Tauglichkeitsprüfung performance audit
Leistungsbereich performance range
Leistungskompensations- Differentialkalorimetrie power-compensated differential scanning calorimetry (PC-DSC)
Leistungskriterien (Geräte etc.) performance criteria
Leistungsregelung power control
Leistungszahl performance value, performance coefficient
leiten (Elektrizität/Flüssigkeiten) conduct, transport, translocate, lead
Leitenzym tracer enzyme
Leiter ladder; *electr* conductor; (Führungskraft: Vorgesetzter/'Chef') leader, head ('boss')
➢ **Stehleiter/Treppenleiter/Trittleiter** stepladder, steps
Leitfaden/Handbuch guide; manual, handbook
leitfähig conductive
Leitfähigkeit conductivity; (G) *neuro* conductance
Leitfähigkeitsmessgerät conductivity meter
Leitfähigkeitstitration/ konduktometrische Titration/ Konduktometrie conductometric titration
Leitlinie guideline
Leitnuklid tracer nuclide
Leitung conduction, conductance, transport, translocation; (Rohre/ Kabel für Wasser/Strom/Gas) line
Leitungswasser tap water
Lernlabor teaching laboratory, educational laboratory
letal/tödlich lethal, deadly
➢ **balanciert letal** balanced lethal
➢ **bedingt letal/konditional letal** conditional lethal

letale Dosis lethal dose
Letalität lethality
Leuchte
  lamp, illuminator;
  micros illuminator
> **Kaltlichtbeleuchtung**
  fiber optic illumination
leuchten
  shine, light; glow; burn
Leuchtfarbe
  luminous paint;
  (Farbstoff) fluorescent dye
Leuchtfeldblende/Kollektorblende
  micros field diaphragm
Leuchtkraft luminosity
Leuchtprobe flame test
Leuchtschirm
  luminescent screen
Leuchtstoff/Luminophor ('Phosphor')
  luminophore (phosphor)
Leuchtstoffröhre/
  Leuchtstofflampe ('Neonröhre')
  fluorescent tube
Leuchttest/Leuchtprobe
  flame test
Libelle (Glasröhrchen der Wasserwaage)
  bubble tube (slightly bowed glass tube/vial in spirit level)
Licht light
> **Auflicht/Auflichtbeleuchtung**
  epiillumination,
  incident illumination
> **ausgestrahltes Licht**
  emergent light
> **Blitzlicht** flash, flashlight
> **Durchlicht/**
  Durchlichtbeleuchtung
  transillumination,
  transmitted light illumination
> **einfallendes Licht**
  incident light
> **linear polarisiertes Licht**
  plane-polarized light
> **polarisiertes Licht**
  polarized light
> **Streulicht**
  scattered light, stray light

lichtbeständig/lichtecht
  photostable, light-fast, nonfading
Lichtbeständigkeit photostability
Lichtbleichung photobleaching
Lichtblitz flash
Lichtbogenofen arc furnace
Lichtbogenspektrum arc spectrum
lichtbrechend refractive
Lichtbrechung optical refraction
lichtdurchlässig
  translucent, transparent
Lichtdurchlässigkeit
  light permeability
Lichtechtheit/Lichtbeständigkeit
  light fastness
lichtempfindlich (leicht reagierend)
  light-sensitive, photosensitive,
  sensitive to light
Lichtempfindlichkeit
  light sensitivty, sensivity to light,
  photosensitivity
Lichtleiter (Kaltlicht)
  light pipe (fiberoptics)
> **Schwanenhals** gooseneck
Lichtmikroskop
  light microscope
  (compound microscope)
Lichtpunkt point of light
Lichtquelle light source
Lichtreiz light stimulus
Lichtschranke
  light barrier
Lichtschutzmittel (Sonnenschutzcreme)
  light-stability agent
  (sunscreen lotion)
lichtstark bright, luminous
Lichtstärke/Lichtintensität
  luminosity, light intensity
Lichtstrahl/Lichtbündel
  beam of light
Lichtstreuung light scattering
Lichtwahrnehmung
  photoperception
Liebigkühler Liebig condenser
Lieferant/Vertrieb supplier,
  distributor (Firma: supply house);
  (Vertragslieferant) contractor

lieferbar
on stock; available
➢ **ausstehende Lieferung wird nachgeliefert (sobald wieder auf Lager)**
on backorder
➢ **derzeit nicht lieferbar**
temporarily out of stock
➢ **nicht lieferbar** out of stock
Lieferbedingungen terms of delivery (terms and conditions of sale)
Lieferdruck
delivery pressure, discharge pressure
Lieferfrist
term of delivery, time of delivery
Lieferkosten
cost of delivery, shipment costs
Lieferschein delivery note, note/bill/confirmation of delivery
Lieferumfang
extent/scope of supply (delivery); package contents
Lieferung
supply, shipment, delivery, consignment
➢ **Großlieferung**
bulk shipment, bulk delivery
Lieferverzug delay in delivery
Lieferzeit
lead time; time frame of delivery
Ligament/Band ligament
Ligand ligand
Liganden-Blotting ligand blotting
Ligation/Verknüpfung
ligation
ligationsvermittelte Polymerasekettenreaktion
ligation-mediated PCR
Lignifizierung lignification
Lignocerinsäure/Tetracosansäure
lignoceric acid, tetracosanoic acid
Ligroin ligroin, petroleum spirit
Lineweaver-Burk-Diagramm
Lineweaver-Burk plot, double-reciprocal plot
Linienstichprobenverfahren *stat/ecol*
line transect method

linksgängig left-handed
linkshändig left-handed, sinistral
Linolensäure linolenic acid
Linolsäure linolic acid, linoleic acid
Linse lens (*also:* lense)
Linsenpapier/Linsenreinigungspapier *micros*
lens tissue, lens paper
Lipid lipid
Lipiddoppelschicht (biol. Membran)
lipid bilayer
Lipofektion lipofection
Liponsäure/Dithiooctansäure/Thioctsäure/Thioctansäure (Liponat)
lipoic acid (lipoate), thioctic acid
lipophil lipophilic
Lipoprotein hoher Dichte
high density lipoprotein (HDL)
Lipoprotein mittlerer Dichte
intermediate density lipoprotein (IDL)
Lipoprotein niedriger Dichte
low density lipoprotein (LDL)
Lipoprotein sehr niedriger Dichte
very low density lipoprotein (VLDL)
Lipoteichonsäure
lipoteichoic acid
Lippendichtung (Wellendurchführung)
lip seal, lip-type seal, lip gasket
Litocholsäure litocholic acid
Lizenz licence (or license)
Lizentinhaber
licensee, licence holder
Lochbodenkaskadenreaktor/Siebbodenkaskadenreaktor
sieve plate reactor
löcherig/perforiert perforated
Lochplatte *gen/micb* well plate
Lockmittel/Lockstoff/Attraktans
attractant
Lod-Wert lod score
('logarithm of the odds ratio')
Löffel spoon, scoop
logarithmische Phase
logarithmic phase (log-phase)
Logarithmuspapier/Logarithmenpapier
log paper

Lo

Logistikdienstleister/Spedition
shipper, freight company,
shipping company
**Lognormalverteilung/
logarithmische Normalverteilung**
lognormal distribution,
logarithmic normal distribution
Lokalanästhetikum
local anesthetic
**Lokomotion/Bewegung
(Ortsveränderung)**
locomotion
Löschdecke/Feuerlöschdecke
fire blanket
löschen (Feuer)
extinguish, put out
Löschgerät/Feuerlöscher
fire extinguisher
Löschmittel/Feuerlöschmittel
fire-extinguishing agent
Löschpapier
bibulous paper (for blotting dry)
Lösemittel/Lösungsmittel solvent
Lösemittelbeständigkeit
solvent resistance
Lösemittelfront solvent front
Lösemittelrückgewinnung solvent recovery
lösen *mech*
detach, separate, disconnect;
chem (in einem Lösungsmittel)
dissolve; *math* solve
löslich soluble
➢ **kaum löslich/wenig löslich**
sparingly soluble,
barely soluble
➢ **leichtlöslich**
easily soluble, readily soluble
➢ **schwerlöslich**
of low solubility
➢ **unlöslich** insoluble
Löslichkeit solubility
➢ **Unlöslichkeit**
insolubility
Löslichkeitspotential
solute potential
Löslichkeitsprodukt
solubility product

**Löslichkeitsvermittler/
Lösungsvermittler**
solubilizer, solutizer
Löslichkeitsvermittlung
solubilization
Lösung solution
➢ **Fehlingsche Lösung**
Fehling's solution
➢ **Gebrauchslösung/
gebrauchsfertige L./Fertiglösung**
ready-to-use solution, test solution
➢ **gesättigte Lösung**
saturated solution
➢ **Kochsalzlösung** saline
➢ **Maßlösung** volumetric solution
(a standard analytical solution)
➢ **Nährlösung**
nutrient solution, culture solution
➢ **physiologische Kochsalzlösung**
saline, physiological saline solution
➢ **Pufferlösung** buffer solution
➢ **Reagenzlösung** reagent solution
➢ **Ringerlösung/Ringer-Lösung**
Ringer's solution
➢ **Stammlösung/Vorratslösung**
stock solution
➢ **Standardlösung**
standard solution
➢ **Untersuchungslösung**
test solution,
solution to be analyzed
➢ **verdünnte Lösung** dilute solution
➢ **wässrige Lösung**
aqueous solution
Lösungsmittel/Lösemittel solvent
Lösungsmittelfront solvent front
Lösungsmittelrückgewinnung
solvent recovery
**Lösungsvermittler/
Löslichkeitsvermittler**
solubilizer, solutizer
Lösungswärme/Lösungsenthalpie
heat of solution,
heat of dissolution
Lot/Lötmittel/Lötmetall solder
Lötdraht soldering wire
Lötflussmittel
soldering flux, solder flux

Lötkolben soldering iron
Lötöse soldering lug
Lötpistole soldering gun
Lötrohrprobe
 blowpipe assay/test
Lötsäure soldering acid
Lötwasser
 soldering fluid, soldering liquid
Lücke gap
Luer T-Stück Luer tee
Luerhülse
 female Luer hub (lock)
Luerkern
 male Luer hub (lock)
Luerlock/Luerverschluss
 Luer lock
Luerspitze Luer tip
Luft air
➢ **Abluft**
 exhaust, exhaust air,
 waste air, extract air
➢ **Druckluft** compressed air
➢ **flüssige Luft** liquid air
➢ **Heißluft** hot air
➢ **Luft ablassen**
 (Gas ablassen/herauslassen)
 deflate
➢ **Pressluft**
 compressed air, pressurized air
➢ **Umluft**
 forced air, recirculating air;
 air circulation
➢ **Zugluft** draft
➢ **Zuluft** input air
Luftausschluss/Luftabschluss
 exclusion of air (air-tight)
Luftbad air bath
Luftblase air bubble;
➢ **Luftbläschen** small air bubble
luftdicht airtight, airproof
Luftdruck air pressure
➢ **atmosphärischer L.**
 atmospheric pressure
Luftdruckmessgerät/Barometer
 barometer
Lufteinlassventil
 air inlet valve, air bleed
Lufteintrittsgeschwindigkeit/
Einströmgeschwindigkeit
(Sicherheitswerkbank)
 face velocity (not same as 'air speed'
 at face of hood)
luftempfindlich air-sensitive
lüften air, ventilate, aerate
Luftentfeuchter (Gerät) air dryer
Lüfter fan, blower, ventilator
Luftfeuchtigkeit air humidity;
 atmospheric moisture
Luftfeuchtigkeitsmessgerät/
Feuchtigkeitsmesser/Hygrometer
 hygrometer
Luftfilter air filter
Luftführung air flow
➢ **vertikale Luftführung**
 (Vertikalflow-Biobench)
 vertical air flow (clean bench with
 vertical air curtain)
Luftgeschwindigkeit air speed;
 (ausgedrückt als Vektor) air velocity
luftgetragen airborne
Luftgrenzwert
 air threshold value,
 atmospheric threshold value
Luftkammer (Schacht: z.B. Abzug)
 plenum (*pl* plena)
Luftkanal
 air duct, air conduit, airway
Luftkapazität air capacity
Luftkapillare air capillary
Luftkühler
 air cooler, air condenser
Luftpolster-Folie
 air-cushion foil
luftreaktiv air reactive
Luftrückführung
 air recirculation
Luftsauerstoff
 atmospheric oxygen
Luftschadstoff air pollutant
Luftschleuse airlock
Luftstickstoff
 atmospheric nitrogen
Luftstrahl air jet
Luftstrom/Luftströmung
 air current, airflow,
 current of air, air stream

Lu

Luftströmung/
Luftgeschwindigkeit
(Sicherheitswerkbank)
air speed
Luftumwälzung
air circulation
Lüftung/Ventilation
ventilation;
(Belüftung) aeration
Lüftungsanlage
ventilation system, vent
Lüftungskanal
air duct
Lüftungsrohr
ventilating pipe, vent pipe
Lüftungsschacht/Luftschacht
air shaft, air duct,
ventilating shaft/duct,
vent shaft/duct
Luftventil air valve
Luftverflüssigung
liquefaction of air
Luftverschmutzung/
Luftverunreinigung
air pollution

Luftvorhang/Luftschranke
(z.B. an Vertikalflow-Biobench)
air curtain, air barrier
Luftzirkulation air circulation
Luftzufuhr air supply
Luftzug draft (*Br* draught)
Luke hatch, door, window;
(Dachfenster) skylight
Lumineszenz luminescence
➢ **Biolumineszenz**
bioluminescence
Lungenödem pulmonary edema
Lupe/Vergrößerungsglas lens,
magnifying glass
Lüster metallic luster
Lüsterklemme luster terminal
(insulating screw joint)
Lyophilisierung/Gefriertrocknung
lyophilization, freeze-drying
Lysat lysate
Lyse lysis
Lysehof/Aufklärungshof/Hof/Plaque
lytic plaque, plaque
Lysergsäure lysergic acid
lysieren lyse

Magensaft/Magenflüssigkeit
stomach juice, gastric juice
Magensäure stomach acid
Magenspülung
gastric lavage, gastric irrigation
Magenstein/Magensteinchen/
Hummerstein/Gastrolith
gastrolith
Magische Säure (HSO_3F/SbF_5)
magic acid
Magnesia/Magnesiumoxid
magnesia, mangesium oxide
Magnesium (Mg) magnesium
Magnetfeld magnetic field
Magnetresonanztomographie (MRT)/
Kernspintomographie (KST)
magnetic resonance imaging (MRI), nuclear magnetic resonance imaging
Magnetrührer magnetic stirrer
Magnetrührer mit Heizplatte
stirring hot plate
Magnetstab/Magnetstäbchen/
Magnetrührstab/'Fisch'/Rühr'fisch'
stir bar, stirrer bar, stirring bar, bar magnet, 'flea'
Magnetstabentferner
(zum 'Angeln' von Magnetstäbchen)
stirring bar retriever, 'flea' extractor
Magnetventil solenoid valve
Mahlbecher (Mühle) grinding jar
mahlen/zerkleinern
grind, crush, pulverize
Mahlkugeln (Mühle) grinding balls
Maische mash
Maisquellwasser cornsteep liquor
Makromolekül macromolecule
makroskopisch macroscopic
MAK-Wert (maximale
Arbeitsplatz-Konzentration)
maximum permissible workplace concentration, maximum permissible exposure
Maleinsäure (Maleat)
maleic acid (maleate)
Maler-Krepp masking tape
maligne/bösartig malignant
➤ **benigne/gutartig** benign
Malignität/Bösartigkeit malignancy

Malonsäure (Malonat)
malonic acid (malonate)
Malpinsel paint brush
Maltose (Malzzucker)
maltose (malt sugar)
Malz malt
Malzzucker/Maltose
malt sugar, maltose
Mandelsäure/Phenylglykolsäure
mandelic acid, phenylglycolic acid, amygdalic acid
Mangan (Mn) manganese
Mangel/Defizienz deficiency
Mangelerscheinung/
Defizienzerscheinung/
Mangelsymptom
deficiency symptom
Mangelmedium deficiency medium
mangelnd/Mangel../defizient
deficient, lacking
Mannit mannitol
Mannuronsäure mannuronic acid
Manschette adapter; *mech* sleeve, collar; (Tropfschutz: Wicklung/ Ummantelung) jacket (insulation)
➤ **Filtermanschette/Guko**
filter adapter, Guko
➤ **für Schliffverbindungen**
sleeve, joint sleeve
Marienglas (Gips) foliated gypsum, selenite, spectacle stone
Mark medulla, pith, core
Marke (Ware/Handel) brand
Markenbezeichnung/Warenzeichen
brand name, trade name
Marker/Markersubstanz
(genetischer/radioaktiver)
marker (genetic/radioactive)
Markierband/Absperrband
barricade tape
markieren/etikettieren tag; *chem* label; (kennzeichnen) mark, brand, earmark
Markierstift marker
➤ **wischfester/wasserfester M.**
permanent marker (water-resistant), sharpie

ma

markiertes Molekül tagged molecule
Markierung label(l)ing
➢ **Immunmarkierung**
immunolabeling
➢ **radioaktive M.** radiolabeling
Marshsche Probe Marsh test
Maschensieb mesh screen
maschig meshy
Maschinist/Bediener/Durchführender
operator
Maserung/Fladerung *allg*
figure, design;
(Faserorientierung) grain
Maske mask
➢ **Atemschutzmaske**
protection mask,
respirator mask, respirator
➢ **Ausatemventil** exhalation valve
➢ **Feinstaubmaske** dust-mist mask
➢ **Filtermaske** filter mask
➢ **Gasmaske** gas mask
➢ **Operationsmaske/**
chirurgische Schutzmaske
surgical mask
➢ **Staubschutzmaske (Partikelfilternde**
Masken) (DIN FFP)
dust mask, particulate respirator
(U.S. safety levels N/R/P according
to regulation 42 CFR 84)
Maß measure
Maßanalyse/Volumetrie/
volumetrische Analyse
volumetric analysis
Masse mass; (Fülle) bulk
➢ **Biomasse** biomass
➢ **'Frischmasse' (Frischgewicht)**
'fresh mass' (fresh weight)
➢ **Molekülmasse ('Molekulargewicht')**
molecular mass ('molecular weight')
➢ **Molmasse/**
molare Masse ('Molgewicht')
molar mass ('molar weight')
➢ **relative Molekülmasse/**
Molekulargewicht (M_r)
relative molecular mass,
molecular weight (M_r)
➢ **Trockenmasse/Trockensubstanz**
dry mass, dry matter

Masse-Ladungsverhältnis m/z **(MS)**
mass-to-charge ratio
Massenanteil (Massenbruch)
mass fraction
Massenerhaltungssatz
law of conservation of matter
Massenfilter mass filter
Massenspektrometer
mass spectrometer, mass spec
Massenspektrometrie (MS)
mass spectrometry (MS)
Massenströmung (Wasser)
mass flow, bulk flow
Massenübergang/
Massentransfer/Stoffübergang
mass transfer
Massenvermehrung
mass reproduction,
mass spread, outbreak
Massenwirkungsgesetz
law of mass action
Massenwirkungskonstante
mass action constant
Maßkorrelationskoeffizient/
Produkt-Moment-
Korrelationskoeffizient
product-moment correlation
coefficient
Maßlöffel weighing spoon
Maßlösung volumetric solution
(a standard analytical solution)
Maßstab scale
➢ **Großmaßstab** large scale
➢ **Halbmikromaßstab**
semimicro scale
➢ **Kleinmaßstab**
small scale
➢ **Labormaßstab**
lab scale, laboratory scale
➢ **Mikromaßstab** micro scale
➢ **Pilotmaßstab** pilot scale
Maßstabsvergrößerung
scale-up, scaling up
Maßstabzahl *micros*
initial magnification
Material material; (Zubehör) supplies
Materialermüdung
material fatigue

Materialfehler
defect in material,
flaw in material
Materialmangel
material shortage
Matrix matrix
Matrize *biochem* template
Matrizenstrang/Mutterstrang *gen*
template strand
Mattrand-Objektträger
frosted-end slide
Maul (Öffnung am Schraubenschlüssel) bit
Maximalgeschwindigkeit
(V_{max} **Enzymkinetik/Wachstum**)
maximum rate
Mazeration maceration
mazerieren macerate
Mechaniker mechanic
Medianwert/Zentralwert *stat*
median value
Medikament/Medizin/Droge
medicine, medicament, drug
- **zielgerichtete 'Konstruktion' neuer Medikamente am Computer**
  drug design
- **frei erhältliches M. (nicht verschreibungspflichtig)**
  over-the-counter drug
- **verschreibungspflichtiges M.**
  prescription drug

Medium/Kulturmedium/Nährmedium
medium, culture medium,
nutrient medium
- **Anreicherungsmedium**
  enrichment medium
- **Basisnährmedium**
  basal medium
- **Differenzierungsmedium**
  differential medium
- **Eiermedium/Eiernährmedium**
  egg medium
- **Elektivmedium/Selektivmedium**
  selective medium
- **Erhaltungsmedium**
  maintenance medium
- **Komplettmedium**
  complete medium, rich medium
- **komplexes Medium**
  complex medium
- **konditioniertes Medium**
  conditioned medium
- **Mangelmedium** deficiency medium
- **Minimalmedium** minimal medium
- **Selektivmedium/Elektivmedium**
  selective medium
- **synthetisches Medium (chemisch definiertes Medium)**
  defined medium
- **Testmedium/Prüfmedium (zur Diagnose)** test medium
- **Vollmedium/Komplettmedium**
  rich medium, complete medium

Medizin medicine;
(Medikament/Droge) medicine, drug
- **Biomedizin** biomedicine
- **Defensivmedizin**
  defensive medicine
- **Forensik/forensische Medizin/ Gerichtsmedizin/Rechtsmedizin**
  forensics, forensic medicine
- **Präventivmedizin**
  preventive medicine
- **Umweltmedizin**
  environmental medicine
- **Veterinärmedizin/ Tiermedizin/Tierheilkunde**
  veterinary medicine,
  veterinary science
- **vorhersagende Medizin**
  predictive medicine

Mediziner doctor, physician
Medizinerkittel
physician's white coat,
white coat
medizinische Untersuchung/ ärztliche U.
medical examination, medical exam,
physical examination, physical
Medizinstudent medical student
Meerwasser seawater, saltwater
Megaphon/Megafon bull horn
Mehl flour
- **Blutmehl** blood meal
- **Guarmehl/Guar-Gummi**
  guar gum, guar flour

- Guar-Samen-Mehl guar meal, guar seed meal
- Johannisbrotkernmehl/Karobgummi locust bean gum, carob gum
- Knochenmehl bone meal
- Sägemehl sawdust

mehlig mealy, farinaceous
Mehrfachbindung *chem* multiple bond
Mehrfachsteckdose/Steckdosenleiste outlet strip
mehrjährig/ausdauernd perennial
Mehrkomponentenkleber multicomponent adhesive/cement, multiple-component adhesive/cement
Mehrschichtfolie multilayer film
Mehrweg... reusable ...
Mehrzweckzange utility pliers
Meißel chisel
Meldepflicht mandatory report, compulsory registration, obligation to register
meldepflichtig reportable (by law), subject to registration
Melder/Messfühler/Sensor detector, sensor
- Feuermelder fire alarm

Membran membrane
- Außenmembran outer membrane
- Elementarmembran/Doppelmembran unit membrane, double membrane
- Kernmembran nuclear membrane
- Plasmamembran/Zellmembran/Ektoplast/Plasmalemma plasma membrane, (outer) cell membrane, unit membrane, ectoplast, plasmalemma
- Schleimhaut/Schleimhautepithel mucous membrane, mucosa
- Zellmembran/Plasmamembran/Ektoplast/Plasmalemma (outer) cell membrane, plasma membrane, unit membrane, ectoplast, plasmalemma

Membrandruckminderer diaphragm pressure regulator
Membrandurchfluss membrane flux
Membranfilter membrane filter
Membranfluss membrane flow
membrangebunden membrane-bound
Membrankapazität membrane capacitance
Membranlängskonstante (Raumkonstante) membrane length constant (space constant)
Membranleitfähigkeit membrane conductance
membranös membraneous
Membranpinzette membrane forceps
Membranpumpe diaphragm pump
Membranreaktor (Bioreaktor) membrane reactor
Membranventil diaphragm valve
Menge (Anzahl) quantity, amount, number
Mengen... bulk
Mengenverhältnis quantitative ratio, relative proportions
Mengenrabatt quantity discount
Meniskus meniscus
menschlich (den Menschen betreffend) human; (wie ein guter Mensch handelnd/hilfsbereit/selbstlos) humane
Mensur (Messbehälter: z.B. auch Reagierkelch) graduated cylinder, graduate
Merkblatt leaflet, notice, instructions; (Datenblatt: für Chemikalien etc.) data sheet
Merkmal/Eigenschaft trait, characteristic, feature
Mesomerie mesomerism
Messader *electr* pilot wire
messbar measurable

Messbecher measuring cup
➢ **Mensur**
graduate, graduated cylinder
Messbereich range of measurement
messen (abmessen) measure;
(prüfen) test; (ablesen) read, record
Messer
knife; (Klinge) blade;
(Messgerät: Zähler) meter;
measuring instrument
➢ **Amputiermesser** amputating knife
➢ **Diamantmesser** diamond knife
➢ **Fleischmesser** fleshing knife
➢ **Kabelmesser** cable stripping knife
➢ **Kittmesser** putty knife
➢ **Klappmesser** jack knife
➢ **Knorpelmesser** cartilage knife
➢ **Sicherheitsmesser** safety cutter
➢ **Spachtelmesser/Kittmesser**
putty knife
➢ **Taschenmesser** pocket knife
Messerhalter *micros* knife holder
Messerschalter *electr* knife switch
Messfehler
error in measurement,
measuring mistake
Messfühler/Sensor/Sonde *lab*
sensor, probe
Messgas measuring gas; sample gas
Messgenauigkeit
accuracy/precision of measurement,
measurement precision
Messgerät
measuring apparatus,
measuring instrument;
(Lehre) gage, gauge (Br)
➢ **Zähler** meter
Messglied *math* (**Größe**)
measuring unit, measuring device
Messgröße quantity to be measured
Messing brass
➢ **Gussmessing** cast brass
Messkolben volumetric flask
Messpipette
graduated pipette, measuring pipet
Messschaufel measuring scoop
Messschieber (*siehe:* **Schublehre**)
caliper gage (gauge *Br*)

Messtechnik
metrology; measurement techniques,
measuring techniques; test methods
➢ **Mess- und Regeltechnik**
instrumentation and control
messtechnisch metrological
Messung measurement, test,
testing, reading, recording
Messverfahren measuring procedure
Messwert measured value
Messzylinder graduated cylinder
Metall metal
➢ **Buntmetall** nonferrous metal
➢ **Edelmetall** precious metal
➢ **Halbedelmetall** semiprecious metal
➢ **Halbmetalle** semimetals
➢ **Schwermetall** heavy metal
➢ **Übergangsmetall** transition metal
Metallaufdampfung *micros*
metal deposition
Metallbelag metallization
Metallgewinnung, elektrolytische
(**Elektrometallurgie**)
electrowinning
Metallglanz metallic lustre
metallisch metallic
metallische Bindung metallic bond
Metallkunde (Metallurgie)
metal science (metallurgy)
Metalllegierung metal alloy
Metallsäge metal-cutting saw
Metallurgie/Hüttenkunde metallurgy
(science & technology of metals)
Methan methane
Methode method
methylieren methylate
Methylierung/Methylieren methylation
metrische Skala metric scale
Mevalonsäure (Mevalonat)
mevalonic acid (mevalonate)
Micelle micelle
Micellierung micellation
Michaeliskonstante/
Halbsättigungskonstante (K_M)
Michaelis constant,
Michaelis-Menten constant
Michaelis-Menten-Gleichung
Michaelis-Menten equation

Mikrobe/Mikroorganismus
  microbe, microorganism
mikrobiell microbial
Mikroinjektion microinjection
Mikromanipulation
  micromanipulation
Mikromanipulator micromanipulator
Mikrometerschraube *micros*
  micrometer screw, fine-adjustment,
  fine-adjustment knob
Mikroorganismus
  (*pl* **Mikrorganismen**)/**Mikrobe**
  microorganism, microbe
Mikropinzette micro-forceps
➤ **anatomische M.**
  microdissecting forceps,
  microdissection forceps
Mikropipette micropipet
Mikropipettenspitze micropipet tip
Mikropräparat
  prepared microscope slide
Mikroskop
  microscope
➤ **Kursmikroskop**
  course microscope
➤ **Polarisationsmikroskop**
  polarizing microscope
➤ **Präpariermikroskop**
  dissecting microscope
➤ **Stereomikroskop**
  stereo microscope
➤ **Umkehrmikroskop/
  Inversmikroskop**
  inverted microscope
➤ **zusammengesetztes Mikroskop**
  compound microscope
Mikroskopie
  microscopy
➤ **Dunkelfeld-Mikroskopie**
  darkfield microscopy
➤ **Hellfeld-Mikroskopie**
  brightfield microscopy
➤ **Hochspannungselektronenmikroskopie**
  high voltage electron microscopy
  (HVEM)
➤ **Immun-Elektronenmikroskopie**
  immunoelectron microscopy
➤ **Interferenzmikroskopie**
  interference microscopy
➤ **konfokale Laser-Scanning
  Mikroskopie**
  confocal laser scanning microscopy
➤ **Kraftmikroskopie**
  force microscopy (FM)
➤ **Lichtmikroskopie**
  light microscopy
  (compound microscope)
➤ **Phasenkontrastmikroskopie**
  phase contrast microscopy
➤ **Polarisationsmikroskopie**
  polarizing microscopy
➤ **Rasterelektronenmikroskopie (REM)**
  scanning electron microscopy (SEM)
➤ **Rasterkraftmikroskopie**
  atomic force microscopy (AFM)
➤ **Rastertunnelmikroskopie (RTM)**
  scanning tunneling microscopy
  (STM)
➤ **Transmissionselektronen-
  mikroskopie/Durchstrahlungs-
  elektronenmikroskopie**
  transmission electron microscopy
  (TEM)
Mikroskopieren *n*
  examination under a microscope,
  usage of a microscope
mikroskopieren *vb*
  examine under a microscope,
  use a microscope
Mikroskopierleuchte
  microscope illuminator
Mikroskopierverfahren
  microscopic procedure
Mikroskopierzubehör
  microscopy accessories
mikroskopisch
  microscopic, microscopical
**mikroskopische Aufnahme/
  mikroskopisches Bild**
  micrograph, microscopic image
mikroskopisches Präparat
  microscopical preparation/mount
Mikroskopzubehör
  microscope accessories
Mikrosonde microprobe

Mikrotom microtome
➤ **Gefriermikrotom**
  freezing microtome, cryomicrotome
➤ **Kryo-Ultramikrotom**
  cryoultramicrotome
➤ **Rotationsmikrotom** rotary microtome
➤ **Schlittenmikrotom** sliding microtome
➤ **Ultramikrotom** ultramicrotome
Mikrotomie microtomy
Mikrotommesser microtome blade
Mikrotom-Präparatehalter/ Objekthalter (Spannkopf)
  microtome chuck
Mikroträger microcarrier
Mikroumwelt micro-environment
Mikroverfahren microprocedure
Mikrowellen-Synthese
  microwave synthesis
Mikrowellenofen/ Mikrowellengerät
  microwave oven
Milchglas milk glass
milchig/opak milky, opaque
Milchsaft/Latex latex
Milchsäure (Laktat)
  lactic acid (lactate)
Milchsäureamid/Laktamid/Lactamid
  lactamide
Milchsäuregärung/Laktatgärung
  lactic acid fermentation,
  lactic fermentation
➤ **heterofermentative Milchsäuregärung**
  heterolactic fermentation
➤ **homofermentative Milchsäuregärung**
  homolactic fermentation
Milchzucker/Laktose
  milk sugar, lactose
Millimeterpapier
  graph paper, metric graph paper
Mindestzündenergie
  minimum ignition energy
Mineral (*pl* **Mineralien**) mineral(s)
Mineraldünger
  mineral fertilizer, inorganic fertilizer
Mineralisation/Mineralisierung
  mineralization

Mineralöl mineral oil
Mineralquelle mineral spring
Mineralstoffe/Mineralien minerals
Mineralwasser mineral water
Minimalmedium minimal medium
Miniprep/Minipräparation
  miniprep, minipreparation
mischbar miscible
➤ **unvermischbar** immiscible
Mischbettfilter/ Mischbettionenaustauscher
  mixed-bed filter,
  mixed-bed ion exchanger
Mischer/Mixer mixer
➤ **Drehmischer**
  roller wheel mixer
➤ **Fallmischer**
  tumbler, tumbling mixer
➤ **Mixette/Küchenmaschine (Vortex)**
  blender (vortex)
➤ **Schaufelmischer**
  blade mixer
➤ **Trommelmischer**
  barrel mixer, drum mixer
➤ **Überkopfmischer**
  mixer/shaker with spinning/rotating
  motion (vertically rotating 360°)
➤ **Vortexmischer/ Vortexschüttler/Vortexer**
  vortex shaker, vortex
Mischregel (Mischungskreuz)
  dilution rule
Mischtrommel mixing drum
Mischung mixture
Mischungsverhältnis mixing ratio
Mischzylinder volumetric flask
Missachtung/Vergehen (einer Vorschrift)
  violation
Missbildungen verursachend/ teratogen teratogenic
Mitarbeiter
  (Kollege/Arbeitskollege) colleague,
  co-worker, fellow-worker,
  collaborator;
  (Betriebszugehöriger) employee,
  staff member
Mitfällung coprecipitation

Mitteilungspflicht
  duty to inform,
  obligation to provide information
Mittel/Durchschnittswert
  (*siehe auch:* **Mittelwert**)
  mean, average
**Mittelwert/Mittel/
  arithmetisches Mittel/
  Durchschnittswert** *stat*
  mean value, mean,
  arithmetic mean, average
  ➢ **bereinigter Mittelwert/korrigierter
    Mittelwert** adjusted mean
  ➢ **Elternmittelwert** midparent value
  ➢ **Quadratmittel** quadratic mean
  ➢ **Regression zum Mittelwert**
    regression to the mean
Mittelwertbildung averaging
**Mixer/Mixette/
  Mischer/Küchenmaschine (Vortex)**
  mixer, blender (vortex)
mixotrope Reihe mixotropic series
Modalwert *stat* modal value
Modellbau model building
Modellierknete modeling clay
Moder (Schimmel) mould, mildew
moderig/faulend/verfaulend
  rotting, decaying, putrefying,
  decomposing; (Geruch) mouldy,
  putrid, musty
modern/vermodern/faulen/verfaulen
  rot, decay, putrefy, decompose
Modul/Funktionseinheit module
Modus/Art und Weise/Modalwert
  mode
Mohrsches Salz Mohr's salt,
  ammonium iron(II) sulfate
  hexahydrate
  (ferrous ammonium sulfate)
**molare Masse/
  Molmasse ('Molgewicht')**
  molar mass ('molar weight')
Molekül molecule
Molekularbiologie molecular biology
Molekularformel/Molekülformel
  molecular formula
Molekulargenetik
  molecular genetics

Molekulargewicht
  (*siehe auch:* **Molmasse**)
  molecular weight
  ➢ **relative Molmasse (M_r)**
    relative molecular mass (M_r)
Molekularleck molecular leak
Molekularsieb/Molekülsieb
  molecular sieve
Molekülion (MS) molecular ion
Molekülmasse ('Molekulargewicht')
  molecular mass ('molecular weight')
Molekülpeak molecular peak
Molenbruch/Stoffmengenanteil
  mole fraction
**Molmasse/molare Masse
  ('Molgewicht')**
  molar mass ('molar weight')
  ➢ **Durchschnitts-Molmasse (M_w)
    (gewichtsmittlere Molmasse/
    Gewichtsmittel des
    Molekulargewichts)**
    weight average molecular mass
  ➢ **relative Molmasse (M_r)**
    relative molecular mass
  ➢ **zahlenmittlere Molmasse (M_n)
    (Zahlenmittel des Molekular-
    gewichts)**
    number average molecular mass
Molmassenverteilung
  molecular-weight distribution
Molvolumen molar volume
Molybdän (Mo) molybdenum
Mop/Aufwischer mop
  ➢ **Auswringer/Wringer**
    mop wringer
**Morbidität
  (Häufigkeit der Erkrankungen)**
  morbidity
Mörser/Reibschale mortar
  ➢ **Achatmörser** agate mortar
  ➢ **Aluminiunoxid-Mörser**
    alumina mortar
  ➢ **Apotheker-Mörser** apothecary mortar
  ➢ **Glasmörser** glass mortar
  ➢ **Pistill** (*zu Mörser*) pestle
  ➢ **Porzellanmörser** porcelain mortar
Mortalität/Sterblichkeit/Sterberate
  mortality

MS (Massenspektroskopie)
MS (mass spectroscopy)
MSQ-Schätzung (Methode der kleinsten Quadrate)
LSE (least squares estimation)
MTA (medizinisch-technische(r) AssistentIn)
medical technician, medical assistant (*auch:* Sprechstundenhilfe: doctor's assistant)
MTLA (medizinisch-technische(r) LaborassistentIn)
medical lab technician, medical lab assistant
Muffe (Flanschstück) muff; (Stativ) clamp holder, 'boss', clamp 'boss' (rod clamp holder); (Röhrenleitung) faucet
Muffelofen muffle furnace
Muffenverbindung (Rohr) spigot
Mühle
(*allg*) mill, (*grob*) crusher, (*mittel*) grinder, (*fein*) pulverizer
➤ **Analysenmühle** analytical mill
➤ **Handmühle** hand mill
➤ **Kaffeemühle**
coffee mill, coffee grinder
➤ **Kugelmühle**
ball mill, bead mill
➤ **Mahlbecher** grinding jar
➤ **Mahlkugeln** grinding balls
➤ **Mischmühle** mixer mill
➤ **Mörsermühle** mortar grinder mill
➤ **Pulverisiermühle** pulverizer
➤ **Rotormühle**
centrifugal grinding mill
➤ **Scheibenmühle** plate mill, disk mill
➤ **Schneidmühle**
cutting-grinding mill, shearing machine
➤ **Schwing-Kugelmühle**
bead mill (shaking motion)
➤ **Tellermühle** disk mill
➤ **Trommelmühle**
drum mill, tube mill, barrel mill
➤ **Zentrifugalmühle/ Fliehkraftmühle**
centrifugal grinding mill

mühselig/schwer/arbeitsam
laborious
Mulde depression, basin
muldenförmig trough-shaped
Mull (Gaze) cheesecloth (gauze)
Mullbinde/Gazebinde
gauze bandage
Müll/Abfall waste;
trash, rubbish, refuse, garbage
➤ **Atommüll** nuclear waste
➤ **Chemieabfälle**
chemical waste
➤ **Giftmüll**
toxic waste, poisonous waste
➤ **Haushaltsmüll/Haushaltsabfälle**
household waste/trash
➤ **Industriemüll/Industrieabfall**
industrial waste
➤ **Klinikmüll** clinical waste
➤ **kommunaler Müll**
municipal solid waste (MSW)
➤ **Problemabfall** hazardous waste
➤ **radioaktive Abfälle**
radioactive waste, nuclear waste
➤ **Sondermüll/Sonderabfall**
hazardous waste
Müllabfuhr waste collection
Müllbeutel/Müllsack
trash bag, waste bag
Mülldeponie/Müllplatz/ Müllabladeplatz/Müllkippe
waste disposal site, waste dump; (Müllgrube: geordnet) landfill, sanitary landfill
Mülleimer
garbage can, dustbin (*Br*)
Müllmann sanitation worker
Müllschacht
garbage/waste chute
Mülltonne
waste container, garbage can, dustbin (*Br*)
Mülltrennung/Abfalltrennung
waste separation
Müllverbrennungsanlage
waste incineration plant, incinerator
Müllvermeidung waste avoidance

Müllverwertungsanlage
(waste) recycling plant
Müllwiederverwertung
waste recycling
Mulm/Fäule rot, decaying matter, mold
**Multienzymkomplex/
Multienzymsystem/Enzymkette**
multienzyme complex,
multienzyme system
Multimeter/Universalmessgerät
multimeter
Multiplett-Signal (NMR)
multiplet signal
Mund/Öffnung mouth, opening, orifice
Mundschutz mask, face mask,
protection mask (Atemschutzmaske)
Mundspatel/Zungenspatel
tongue depressor
Mundspiegel mouth mirror
Mundspülung mouth wash
Mundstück/Schnauze spout
Muraminsäure
muramic acid
Muster
(Vorlage/Modell) pattern, sample,
model; specimen;
(Musterung/Zeichnung) pattern,
design; (Probe) sample
**Mutabilität/
Mutierbarkeit/
Mutationsfähigkeit**
mutability
Mutagen/mutagene Substanz
mutagen

**mutagen/
mutationsauslösend/
erbgutverändernd**
mutagenic
Mutagenität
mutagenicity
Mutante mutant
Mutarotation mutarotation
Mutation mutation
Mutationsrate mutation rate
**Mutierbarkeit/
Mutationsfähigkeit/
Mutabilität**
mutability
mutieren mutate
Mutter (und Schraube) *tech*
nut (and bolt)
➢ **Flügelmutter** wing nut
➢ **Hutmutter** acorn nut
➢ **Rändelmutter** knurled nut
Mutterion/Ausgangsion (MS)
parent ion
Mutterlauge mother liquor
Muttersubstanz
parent substance
Mutterzelle mother cell
Myelom myeloma
Mykoplasma (*pl* **Mykoplasmen)**
mycoplasma (*pl* myoplasmas)
Mykose mycosis
**Myristinsäure/Tetradecansäure
(Myristat)**
myristic acid, tetradecanoic acid
(myristate/tetradecanate)

Nachbehandlung aftertreatment
nachfüllbar refillable
Nachhaltigkeit sustained yield
Nachklärbecken
  secondary settling tank
Nachlauf/Ablauf *dest/chromat*
  tailings, tails
nachprüfen check, control
Nachreifen after-ripening
Nachschlagewerk reference book
Nachteil disadvantage
Nachuntersuchung *med*
  posttreatment examination,
  follow-up (exam),
  reexamination after treatment
nachvollziehbar
  duplicable, duplicatable
nachvollziehen duplicate
nachwachsen regenerate, regrow,
  grow back, reestablish
Nachweis detection, proof
nachweisen detect, prove
Nachweisgerät/
  Suchgerät/Prüfgerät
  detector
Nachweisgrenze
  detection limit,
  limit of detection (LOD)
Nachweismethode detection method
Nacktmaus nude mouse
Nadel needle;
  (Kanüle/Hohlnadel: Spritze)
  hypodermic needle
➤ **chirurgische N.**
  suture needle
Nadeladapter syringe connector
Nadelfeile needle file
Nadelventil/
  Nadelreduzierventil
  (Gasflasche/Hähne)
  needle valve
Nagel nail
Nagelzieher nail extractor
näherkommen/annähern/
  sich annähern/erreichen *math/stat*
  approach (e.g. a value)
Näherung *math* approximation
Nähragar nutrient agar

Nährboden/
  Nährmedium/Kulturmedium/
  Medium/Substrat (*siehe auch:*
  Medium/Kulturmedium)
  nutrient medium (solid and liquid),
  culture medium, substrate
Nährbodenflasche
  culture media flask
Nährbouillon/Nährbrühe
  nutrient broth
nahrhaft/nährend/nutritiv
  nutritious, nutritive
Nährlösung
  nutrient solution, culture solution
Nährmedium/Kulturmedium/Medium
  nutrient medium, culture medium
➤ **Anreicherungsmedium**
  enrichment medium
➤ **Basisnährmedium** basal medium
➤ **Differenzierungsmedium**
  differential medium
➤ **Eiermedium/Eiernährmedium**
  egg medium
➤ **Elektivmedium/Selektivmedium**
  selective medium
➤ **Erhaltungsmedium**
  maintenance medium
➤ **komplexes Medium**
  complex medium
➤ **konditioniertes Medium**
  conditioned medium
➤ **Mangelmedium**
  deficiency medium
➤ **Minimalmedium**
  minimal medium
➤ **Selektivmedium/Elektivmedium**
  selective medium
➤ **synthetisches Medium**
  (chemisch definiertes Medium)
  defined medium
➤ **Testmedium/Prüfmedium**
  (zur Diagnose)
  test medium
➤ **Vollmedium/Komplettmedium**
  rich medium, complete medium
Nährsalz nutrient salt
Nährstoff nutrient
Nährstoffmangel nutritional deficit

nährstoffreich/eutroph
nutrient-rich, eutroph, eutrophic
Nahrung (Essen/Fressen) food,
feed; (Nährstoff) nutrient;
(Ernährung) nutrition
Nahrungsaufnahme
ingestion, food intake
Nahrungsbedarf (*pl*
Nahrungsbedürfnisse)
nutritional requirements
Nahrungsmangel
nutrient deficiency, food shortage
Nahrungsmenge food quantity
Nahrungsmittelkonservierung
food preservation
Nahrungsmittelvergiftung
food poisoning
Nahrungsquelle
food source, nutrient source
Nährwert food value, nutritive value
Nährwert-Tabelle
nutrient table,
food composition table
Name name, term; (ungeschützter N. einer Substanz) generic name
Namensetikett/Namensschildchen
name tag
Narbe/Wundnarbe/Cicatricula
scar, cicatrix, cicatrice
Narkose anesthesia
➤ **Vollnarkose**
general anesthesia
Nasenschleimhaut
olfactory epithelium, nasal mucosa
Nassblotten wet blotting
Nassfäule wet rot
Nasspräparat (Frischpräparat/ Lebendpräparat/Nativpräparat)
wet mount
nativ (nicht-denaturiert)
native (not denatured)
Natrium (Na) sodium
Natriumdodecylsulfat
sodium dodecyl sulfate (SDS)
Natriumhydroxid NaOH
sodium hydroxide
Natriumhypochlorit NaOCl
sodium hypochlorite

Natron/Natriumhydrogencarbonat/ Natriumbicarbonat baking soda,
sodium hydrogencarbonate
Natronkalk soda lime
Natronlauge/Natriumhydroxidlösung
sodium hydroxide solution
naturfern/künstlich/synthetisch
man-made, artificial, synthetic
Naturforscher research scientist,
natural scientist
naturidentisch (synthetisch)
synthetic (having same chemical structure as the natural equivalent)
Naturkunde/Biologie
life science, biology
natürlich natural
➤ **unnatürlich** unnatural
naturnah near-natural
Naturschutz
environmental protection, nature protection/conservation/preservation
Naturstoff natural product
Naturstoffchemie
natural product chemistry
Naturwissenschaften
natural sciences, science
Naturwissenschaftler(in)
natural scientist, scientist
naturwissenschaftlich scientific
Nebel fog; (fein) mist
➤ **leichter Nebel** mist
nebelig foggy
➤ **leicht nebelig** misty
Nebelkammer *phys* cloud chamber
Nebengruppenmetall/Übergangsmetall
transition metal
Nebenprodukt by-product, residual product, side product
Nebenreaktion side reaction
Nebenwirkung(en) side effect(s)
Negativkontrastierung *micros*
negative staining,
negative contrasting
Neigung
inclination; slope, slant, dip; gradient
Neigungswinkel inclination
Nekrose necrosis
nekrotisch necrotic

Nennleistung
  power output, rated power output
Nennmasse/Nominalmasse
  nominal mass
Nennstrom
  rated output,
  rated amperage output
Nennvolumen nominal volume
Nennwert/Nominalwert face value
**'Neonröhre'/Leuchtstoffröhre/
Leuchtstofflampe**
  fluorescent tube
Nerv *neuro* nerve
Netz *electr* (Versorgungs~) network,
  power network;
  (Verteilungs~) grid, power grid
Netzanschluss mains connection (*Br*),
  power supply (electric hookup)
Netzgerät/Netzteil
  power supply unit; (Adapter) adapter
Netzkabel
  mains cable (*Br*), power cable
Netzschalter power switch
Netzstecker power plug
Netzteil/Netzgerät power supply unit;
  (Adapter) adapter
Neuordnung/Neusortierung *gen*
  reassortment
Neuraminsäure neuraminic acid
neurotoxisch neurotoxic
Neustoffe
  new chemicals/substances
➢ **Altstoffe**
  existing chemicals/substances
Neusynthese/*de-novo*-**Synthese**
  de-novo synthesis
Neutronenaktivierungsanalyse (NAA)
  neutron activation analysis (NAA)
**Neutronenbeugung/
Neutronendiffraktometrie**
  neutron diffraction
Neutronenstreuung
  neutron scattering
Newton'sche Flüssigkeit
  Newtonian fluid
nichtessentiell nonessential
Nickel (Ni) nickel
niedermolekular low-molecular

Niederschlag *meteo* precipitation;
  (Sediment/Präzipitat) *chem* deposit,
  sediment, precipitate
Niederschlagsmesser rain gauge
Niedervoltleuchte
  low-voltage lamp/illuminator
  (spotlight)
Nikotinsäure/Nicotinsäure (Nikotinat)
  nicotinic acid (nicotinate), niacin
**NIOSH (National Institute for
Occupational Safety and Health)**
  U.S. Institut für Sicherheit und
  Gesundheit am Arbeitsplatz
Nitrat nitrate
nitrieren nitrify
Nitrierung nitration, nitrification
Nitrifikation/Nitrifizierung nitrification
Nitrobenzol nitrobenzene
Nitroglycerin/Glycerintrinitrat
  nitroglycerin, glycerol trinitrate
Niveauschalter level switch
nivellieren leveling
Nominalskala *stat* nominal scale
Nonius vernier
Norm norm, standard; (Regel) rule
Normaldruck/Normdruck
  standard pressure
Normaldruck-Säulenchromatographie
  gravity column chromatography
Normalmaß standard measure
Normalschliff (NS)
  standard taper (S.T.)
Normalverteilung *stat*
  normal distribution
Normalwasserstoffelektrode
  standard hydrogen electrode
Normalwert standard
normen (normieren) standardize
Normierung standardization
Normschliffglas (Kegelschliff)
  standard-taper glassware
Normtemperatur (0 °C)
  standard temperature
Normung standardization
**Normzustand (Normtemperatur 0 °C &
Normdruck 1 bar)**
  STP (s.t.p./NTP) (standard
  temperature & pressure)

Nosokomialinfektion/
  nosokomiale Infektion/
  Krankenhausinfektion
  nosocomial infection,
  hospital-acquired infection
Notabschaltung emergency shutdown
Notaggregat standby unit
Notaufnahme/Unfallstation
  (Krankenhaus)
  emergency ward (clinic)
Notausgang emergency exit
Notdienst/Hilfsdienst
  emergency service
Notdusche emergency shower;
  ('Schnellflutdusche')
  quick drench shower, deluge shower
Notfall emergency
Notfalleinsatz emergency response
Notfalleinsatzplan
  emergency response plan (ERP)
Notfalleinsatztruppe
  emergency response team
Notfall-Evakuierungsplan/
  Notfall-Fluchtplan
  emergency evacuation plan
Notfall-Fluchtweg
  emergency evacuation route,
  emergency escape route
Notfallvorkehrungen
  emergency provisions
Nothilfe first aid
Notruf (Notfallnummer)
  emergency call (emergency number)
Notschacht (Flucht~/Rettungsschacht)
  escape shaft
Notstromaggregat
  emergency generator,
  standby generator
nüchtern (ohne Nahrung)
  with an empty stomach

Nüchternheit *med/physio*
  emptiness (of stomach); soberness
Nucleinsäure/Nukleinsäure
  nucleic acid
nucleophiler Angriff *chem*
  nucleophilic attack
Nukleinsäure/Nucleinsäure
  nucleic acid
nukleophiler Angriff
  nucleophilic attack
Null zero
➢ **auf Null stellen** zero
Nullabgleich
  zero adjustment, null balance
Nullabgleichmethode null method
Null-Anzeige zero reading
Nullpunkteinstellung
  zero-point adjustment,
  zero-point setting
nullwertig zero-valent, nonvalent
Nuss/
  Stecknuss/Steckschlüsseleinsatz
  socket, chuck
Nutsche/Filternutsche
  nutsch, nutsch filter,
  suction filter, vacuum filter
nützen benefit
Nutzen benefit, use;
  (Vorteil) advantage;
  (Anwendung) application
nutzen utilize, use;
  (anwenden) apply
nützlich beneficial, useful
➢ **schädlich**
  harmful, causing damage
Nützling/Nutzart
  beneficial species,
  beneficient species
➢ **Schädling/Ungeziefer** pest
Nutzung utilization, use

Oberfläche surface
Oberflächenabfluss surface runoff
oberflächenaktiv surface-active
oberflächenaktive Substanz/ Entspannungsmittel
surfactant
Oberflächenkultur *micb*
surface culture
Oberflächenmarkierung
surface labeling
Oberflächenspannung/ Grenzflächenspannung
surface tension
Oberflächen-Volumen-Verhältnis
surface-to-volume ratio
oberflächlich
on the surface, superficial
Oberphase (flüssig-flüssig)
upper phase
Oberschwingung (IR) overtone
Oberseite upperside, upper surface
➢ **Unterseite**
underside, undersurface
Objektiv *micros* objective
➢ **achromatisches Objektiv** *micros*
achromatic objective
Objektivrevolver/Revolver *micros*
nosepiece, nosepiece turret
➢ **Zweifachrevolver** double nosepiece
➢ **Dreifachrevolver** triple nosepiece
➢ **Vierfachrevolver**
quadruple nosepiece
➢ **Fünffachrevolver**
quintuple nosepiece
Objektmikrometer *micros*
stage micrometer
Objekttisch *micros*
stage, microscope stage
Objekttisch-Klammer *micros* stage clip
Objektträger (microscope) slide;
(mit Vertiefung) microscope
depression slide, concavity slide,
cavity slide
Objektträgerbeschriftungsetikett
microscope slide label
Ofen oven, furnace
➢ **Flammofen** reverberatory furnace
➢ **Glühofen** annealing furnace

➢ **Hochofen** blast furnace
➢ **Hybridisierungsofen**
hybridization oven
➢ **Induktionsofen**
induction furnace,
inductance furnace
➢ **Konvektionsofen** convection oven
➢ **Lichtbogenofen** arc furnace
➢ **Mikrowellenofen** microwave oven
➢ **Muffelofen** muffle furnace
➢ **Röstofen**
roasting furnace,
roasting oven, roaster
➢ **Schmelzofen** smelting furnace
➢ **Tiegelofen** crucible furnace
➢ **Trockenofen** drying oven
➢ **Verbrennungsofen**
combustion furnace
➢ **Wärmeofen** heating oven,
heating furnace (more intense)
Ofentrocknung
oven drying, kiln drying, kilning
öffnen open
➢ **gewaltsam öffnen** force open
Öffnung/Mund/Mündung
opening, aperture, orifice, mouth,
perforation, entrance
Öffnungsdauer (Membrankanal)
life-time
Öffnungsdruck (Ventil)
breaking pressure
Öffnungswinkel *micros*
angular aperture
Öffnungszeit/Offenzeit *neuro* open time
Ohnmacht unconsciousness, faint;
blackout (short)
ohnmächtig werden
faint, become unconscious,
pass out, black out
Ohrenstöpsel earplugs
Öko-Audit/Umweltaudit
environmental audit
Ökobilanz life cycle assessment,
life cycle analysis (LCA)
ökologisch ecological
ökologisches Gleichgewicht
ecological balance,
ecological equilibrium

Ok

Okular *micros* ocular, eyepiece
- **Binokular** binoculars
- **Brillenträgerokular** spectacle eyepiece, high-eyepoint ocular
- **Trinokularaufsatz/Tritubus** trinocular head
- **Zeigerokular** pointer eyepiece

Okularblende/ Gesichtsfeldblende des Okulars ocular diaphragm, eyepiece diaphragm, eyepiece field stop

Okularlinse/Augenlinse ocular lens
Okularmikrometer ocular micrometer
Öl oil
- **Altöl** waste oil, used oil
- **ätherisches Öl** essential oil, ethereal oil
- **Baumwollsaatöl** cotton oil
- **Behenöl** ben oil, benne oil
- **Distelöl/Safloröl** safflower oil
- **Erdnussöl** peanut oil
- **Erdöl** crude oil, petroleum
- **Fuselöl** fusel oil
- **Kokosöl** coconut oil
- **Kürbiskernöl** pumpkinseed oil
- **Lebertran** cod-liver oil
- **Leinöl** linseed oil
- **Maisöl** corn oil
- **Mineralöl** mineral oil
- **natives Öl** virgin oil (olive)
- **Olivenkernöl** olive kernel oil
- **Olivenöl** olive oil
- **Palmöl** palm oil
- **Pflanzenöl** vegetable oil
- **Rizinusöl** castor oil, ricinus oil
- **Schmieröl** lubricating oil
- **Senföl** mustard oil
- **Sesamöl** sesame oil
- **Sojaöl** soybean oil
- **Sonnenblumenöl** sunflower seed oil
- **Speise-Rapsöl/Rüböl** canola oil (rapeseed oil)
- **Walratöl** sperm oil (whale)

Ölabscheidepipette baster
Ölbad oil bath
ölig oily

oligomer *adj/adv* oligomerous
Oligomer *n* oligomer
Oligonucleotid/Oligonukleotid oligonucleotide
Oligosaccharid oligosaccharide
Olive (meist geriffelter Ansatzstutzen: Schlauch-/Kolbenverbindungsstück) barbed hose connection (flask: side tubulation/side arm)
Ölkatastrophe *ecol* oil spill
Ölpest/Ölverschmutzung oil pollution
Ölteppich *ecol* oil slick
Ölverschmutzung/Ölpest oil pollution
Ölzeug oilskin(s)
onkogen/oncogen/krebserzeugend oncogenic, oncogenous
Onkogenität oncogenicity
Onkologie oncology
onkotischer Druck/ kolloidosmotischer Druck oncotic pressure
OP-Besteck *med* surgical instruments
Operationsmaske/ chirurgische Schutzmaske surgical mask
Opfer victim; (Verletzter/Verwundeter) casualty
Optik optics
optische Dichte/Absorption optical density, absorbance
optische Spezifität optical specificity
Ordentlichkeit/Aufräumen neatness (in cleaning-up)
Ordinalskala *stat* ordinal scale
Ordnung order
- **Gleichung** x**ter Ordnung** equation of the xth order

Ordnungsstatistik order statistics
organisch organic
organische Chemie/'Organik' organic chemistry
organische Substanz/ organisches Material organic matter
Organismus organism, lifeform

Orientierung/Orientierungsverhalten
orientation, orientational behavior
Orotsäure orotic acid
Orsatblase Orsat rubber expansion bag
orten locate
Öse/Metallöse grommet
osmiophil
(färbbar mit Osmiumfarbstoffen)
osmiophilic
Osmiumsäure osmic acid
Osmiumtetroxid osmium tetraoxide
Osmolalität osmolality
Osmolarität/osmotische Konzentration
osmolarity, osmotic concentration
Osmose osmosis
osmotisch osmotic
osmotischer Druck osmotic pressure
osmotischer Schock osmotic shock

Oszillator (IR) oscillator
Oszillometrie/
oszillometrische Titration/
Hochfrequenztitration
oscillometry,
high-frequency titration
Overall (Einteiler) overalls
Oxidation oxidation
Oxidationsmittel
oxidizing agent, oxidant, oxidizer
oxidativ oxidative
oxidieren oxidize
oxidierend oxidizing
Oxoglutarsäure (Oxoglutarat)
oxoglutaric acid (oxoglutarate)
Ozon ozone
Ozonisierung ozonization
Ozonolyse ozonolysis

Pa

Packmaterial packaging material
Packpapier brown paper, kraft;
  (Einpackpapier) wrapping paper
Packung package
> **Großpackung** bulk package
Packungsbeilage package insert
Paket package; (Post) parcel
Paketdienst parcel service
Palette pallet
Palladium (Pd) palladium
**Panzerband/Gewebeband/
  Gewebeklebeband/
  Duct Gewebeklebeband/
  Universalband/Vielzweckband**
  duct tape (polycoated cloth tape)
Panzerglas bulletproof glass
PAP-Färbung/Papanicolaou-Färbung
  PAP stain, Papanicolaou's stain
Papier paper
> **Altpapier** waste paper
> **Bastelpapier** construction paper
> **Briefpapier** stationery;
  (mit Briefkopf) letter-head
> **Einpackpapier** wrapping paper
> **Filterpapier** filter paper
> **Fotopapier** photographic paper
> **Glanzpapier**
  (glanzbeschichtetes Papier)
  glazed paper
> **Hartpapier** laminated paper
> **Kartonpapier/Pappe** cardboard
> **Lackmuspapier** litmus paper
> **Linsenreinigungspapier** lens paper
> **Logarithmuspapier/
  Logarithmenpapier**
  log paper
> **Löschpapier**
  bibulous paper (for blotting dry)
> **Millimeterpapier**
  graph paper, metric graph paper
> **Packpapier** brown paper, kraft;
  (Einpackpapier) wrapping paper
> **Pauspapier** tracing paper
> **Pergamentpapier** parchment paper
> **Pergamin**
  (durchsichtiges festes Papier)
  glassine paper, glassine
> **satiniertes Papier** glazed paper

> **Saugpapier ('Löschpapier')**
  absorbent paper, bibulous paper
> **Schreibpapier**
  bond paper, stationery
> **Seidenpapier**
  tissue paper (wrapping paper)
> **Umweltschutzpapier** recycled paper
> **Wachspapier** wax paper
> **Wägepapier** weighing paper
Papierhandtuch paper towel
Papierholz pulpwood
Papiertaschentuch tissue, paper tissue
Pappe/Pappdeckel/Karton
  cardboard, pasteboard
Pappkarton cardboard box
Parameter parameter
parasitär/parasitisch/schmarotzend
  parasitic
parasitieren/schmarotzen parasitize
Parasitismus/Schmarotzertum
  parasitism
Partialdruck partial pressure
Partialverdau partial digest
Partikelfilter particle filter
Parzelle plot
**PAS-Anfärbung
  (Periodsäure/Schiff-Reagens)**
  PAS stain (periodic acid-Schiff stain)
Passage/Subkultivierung
  passage, subculture
**passend/gut passend
  (z.B. Verschluss/Stopfen etc.)**
  fitted/well fitted
Passstück fitting
Passteil(e) (Zubehör) fitting(s)
Pasteur-Effekt Pasteur effect
pasteurisieren pasteurize
Pasteurisierung/Pasteurisieren
  pasteurizing, pasteurization
Pasteurpipette Pasteur pipet
Patch-Clamp Verfahren
  patch clamp technique
  (patch = Flicken)
Patching/Verklumpung patching
pathogen/krankheitserregend
  pathogenic (causing or capable of
  causing disease)
Pathogenität pathogenicity

Pathologie/
 Lehre von den Krankheiten
 pathology
pathologisch/krankhaft
 pathological
 (altered or caused by disease)
Patrone cartridge
PCR (Polymerasekettenreaktion)
 PCR (polymerase chain reaction)
➢ **Blasen-Linker-PCR**
 bubble linker PCR
➢ **differentieller Display**
 (Form der RT-PCR)
 differential display
 (Form of RT-PCR)
➢ **DOP-PCR**
 (PCR mit degeneriertem
 Oligonucleotidprimer)
 DOP-PCR (degenerate
 oligonucleotide primer PCR)
➢ **inverse Polymerasekettenreaktion**
 inverse PCR
➢ **IRP (inselspezifische PCR)**
 IRP (island rescue PCR)
➢ **ligationsvermittelte**
 Polymerasekettenreaktion
 ligation-mediated PCR
➢ **RACE-PCR**
 (schnelle Vervielfältigung von cDNA-
 Enden)-PCR
 RACE-PCR (rapid amplification of
 cDNA ends)-PCR
➢ **RT-PCR**
 (PCR mit reverser Transcriptase)
 RT-PCR (reverse transcriptase-PCR)
Pektinsäure (Pektat)
 pectic acid (pectate)
Peleusball (Pipettierball) *lab*
 safety pipet filler,
 safety pipet ball
Peptidbindung
 peptide bond, peptide linkage
Peptonwasser peptone water
Perameisensäure performic acid
Perchlorsäure perchloric acid
perforieren (perforiert/löcherig)
 perforate(d)
Perfusionskultur perfusion culture

Pergamin
 (durchsichtiges festes Papier)
 glassine paper, glassine
Periodensystem (der Elemente)
 periodic table (of the elements)
periodisch periodic(al)
Periodizität periodicity
Periodsäure/Schiff-Reagens
 (PAS-Anfärbung)
 periodic acid-Schiff stain (PAS stain)
Peristaltik peristalsis
peristaltisch peristaltic
Perlit/Perlstein perlite
Perlmutt/Perlmutter
 nacre, mother-of-pearl
permeabel/durchlässig
 permeable, pervious
➢ **impermeabel/undurchlässig**
 impermeable, impervious
➢ **semipermeabel/halbdurchlässig**
 semipermeable
Permeabilität/Durchlässigkeit
 permeability
Permissivität
 permissivity, permissive conditions
persistente Infektion/
 anhaltende Infektion
 persisting infection
Persistenz/Beharrlichkeit/Ausdauer
 persistence
persistieren/verharren/ausdauern
 persist
Perubalsam
 Peruvian balsam, balsam of Peru
Perzeption/Wahrnehmung perception
perzipieren/sinnlich wahrnehmen
 perceive
Pestizid/
 Schädlingsbekämpfungsmittel/
 Biozid
 pesticide, biocide
➢ **Algenbekämpfungsmittel/Algizid**
 algicide
➢ **Insektenbekämpfungsmittel/**
 Insektizid insecticide
➢ **Kontaktpestizid** contact pesticide
➢ **Milbenbekämpfungsmittel/Akarizid**
 acaricide

> **Nematodenbekämpfungsmittel/
  Nematizid** nematicide
> **Schneckenbekämpfungsmittel/
  Molluskizid** molluscicide

Pestizidanreicherung
  pesticide accumulation
Pestizidresistenz pesticide resistance
Pestizidrückstand
  pesticide residue
**PET
  (Positronenemissionstomographie)**
  PET (positron emission tomography)
Petrischale Petri dish
Petrolether/Petroläther
  petroleum ether
Petrolatum/Vaseline
  petroleum jelly, vaseline
Pfanne pan
> **Schliffpfanne**
  socket (female: spherical joint)
Pfeifenreiniger/Pfeifenputzer
  pipe cleaner
Pflanzendroge herbal drug
Pflanzenfarbstoff plant pigment
Pflanzeninhaltsstoff
  plant chemical, phytochemical
Pflanzenöl (diätetisch) vegetable oil
pflanzenschädlich/phytotoxisch
  phytotoxic
Pflanzenschädling plant pest
Pflanzenschutz plant protection
Pflanzenschutzmittel
  plant-protective agent, pesticide
Pflaster/Heftpflaster (Streifen) *med*
  band-aid (adhesive strip),
  sticking plaster, patch
Pflichtuntersuchung
  mandatory investigation
Pflichtverletzung breach of duty
pfropfen *vb* graft
Pharmakognosie pharmacognosy
Pharmakologie pharmacology
**Pharmakopöe/
  Arzneimittel-Rezeptbuch/
  amtliches Arzneibuch**
  pharmacopoeia, formulary
Pharmareferent
  pharmaceutical sales representative

Pharmaunternehmen
  pharmaceutical company
pharmazeutisch pharmaceutical
Pharmazie/Arzneilehre/Arzneikunde
  pharmacy
Phase
  chem (nicht mischbare Flüssigkeiten)
  phase, layer; *electr* conductor
> **obere/untere P.**
  upper/lower phase,
  upper/lower layer
> **gebundene Phase** *chromat*
  bonded phase
> **stationäre Phase** *chromat*
  stationary phase, adsorbent
Phasendiagramm phase diagram
Phasengrenze phase boundary
Phasenkontrast phase contrast
Phasenkontrastmikroskop
  phase contrast microscope
Phasenring phase ring, phase annulus
Phasenübergang phase transition
Phasenübergangstemperatur
  phase transition temperature
Phasenveränderung phase variation
Phosgen phosgene
Phosphat phosphate
Phosphatidsäure phosphatidic acid
Phosphodiesterbindung
  phosphodiester bond
Phosphor (P) phosphorus
**phosphorhaltig/
  phosphorig/Phosphor...** *adj/adv*
  phosphorous
phosphorige Säure phosphorous acid
Phosphorsäure phosphoric acid
photoallergen photoallergenic
**Photoatmung/
  Lichtatmung/Photorespiration**
  photorespiration
Photoionisations-Detektor (PID)
  photo-ionization detector (PID)
Photonenstromdichte
  photosynthetic photon flux (PPF)
Photosensibilisierung
  photosensibilization
Photosynthese photosynthesis
photosynthetisch photosynthetic

photosynthetisch aktive Strahlung
  photosynthetically active radiation (PAR)
photosynthetisieren photosynthesize
Phthalsäure phthalic acid
Physik physics
> **Experimentalphysik**
  experimental physics
> **Kernphysik** nuclear physics
> **Teilchenphysik** particle physics
> **Theoretische Physik**
  theoretical physics
> **Umweltphysik**
  environmental physics
physikalische Karte
  physical map
physikalische Sicherheit(smaßnahmen)
  physical containment
Physiologe physiologist
Physiologie physiology
physiologisch physiological
Phytansäure phytanic acid
Phytinsäure phytic acid
Pigment pigment
Pigmentierung pigmentation
Pikrinsäure picric acid
Pilotanlage pilot plant
Pilotmaßstab/ Pilotanlagen-Größe
  pilot scale
Pilzbefall fungal infestation
Pilzbekämpfungsmittel/Fungizid
  fungicide
Pilzvergiftung
  mushroom poisoning, mycetism
Pimelinsäure pimelic acid
Pinzette tweezers, forceps
  (*syn* pincers, tongs)
> **Arterienklemme**
  artery forceps,
  artery clamp (hemostat)
> **Deckglaspinzette**
  cover glass forceps
> **Gewebepinzette** tissue forceps
> **Knorpelpinzette** cartilage forceps
> **Membranpinzette**
  membrane forceps

> **Mikropinzette, anatomische/ Splitterpinzette**
  microdissection forceps,
  microdissecting forceps
> **Präparierpinzette/ Sezierpinzette/anatomische Pinzette**
  dissection tweezers,
  dissecting forceps
> **Präzisionspinzette**
  high-precision tweezers
> **Probennahmepinzette**
  specimen tweezers
> **Sezierpinzette/ Präparierpinzette/ anatomische Pinzette**
  dissection tweezers,
  dissecting forceps
> **Spitzpinzette**
  sharp-point tweezers,
  fine-tip tweezers
> **Uhrmacherpinzette**
  watchmaker forceps,
  jeweler's forceps
> **Umkehrpinzette/Klemmpinzette**
  reverse-action tweezers
  (self-locking tweezers)
pinzieren/entspitzen
  pinch off, tip
Pipette pipet, pipette (*Br*); pipettor
> **Ausblaspipette** blow-out pipet
> **Filterpipette** filtering pipet
> **Kapillarpipette** capillary pipet
> **Messpipette**
  graduated pipet, measuring pipet
> **Mikroliterpipette (Kolbenhubpipette)**
  micropipet, pipettor
> **Pasteurpipette** Pasteur pipet
> **Saugkolbenpipette**
  piston-type pipet
> **Saugpipette**
  suction pipet (patch pipet)
> **serologische Pipette**
  serological pipet
> **Tropfpipette/Tropfglas**
  dropper, dropping pipet
> **Vollpipette/volumetrische Pipette**
  transfer pipet, volumetric pipet

Pi

Pipettenflasche
dropping bottle, dropper vial
Pipettensauger
pipet filler, pipet aspirator
Pipettenspitze pipet tip
Pipettenständer
pipette rack, pipette support;
(für Mikropipetten) pipettor stand
Pipettierball/Pipettierbällchen
pipet bulb, rubber bulb
➢ **Peleusball**
safety pipet filler, safety pipet ball
pipettieren pipet
Pipettierhilfe
pipet aid, pipetting aid, pipet helper
Pipettierhütchen/
Pipettenhütchen/Gummihütchen
pipeting nipple, rubber nipple,
teat (*Br*)
Pipettierpumpe pipet pump
Pistill (*zu Mörser*) pestle
Placebo/Plazebo/Scheinarznei
placebo
Plan-Hohlspiegel/Plankonkav
plano-concave mirror
Planschliff (glatte Enden)
flat-flange ground joint,
flat-ground joint,
plane-ground joint
Planspiegel plane mirror, plano-mirror
Plaque plaque (*siehe:* Zahnbelag;
siehe: Lysehof/Aufklärungshof)
➢ **klarer Plaque** clear plaque
Plaque-Test plaque assay
Plasmensäure plasmenic acid
Plastikfolie (Frischhaltefolie)
plastic wrap (household wrap)
Plastilin plasticine
Plastination plastination
➢ **Ganzkörperplastination**
whole mount plastination
Plastizität plasticity
Platin (Pt) platinum
Platine *elektr/tech* board,
(printed) circuit board;
mech blank; mounting plate
Plattenausstrichmethode
streak-plate method/technique

Plattengel slab gel
Plattengussverfahren/
Gussplattenmethode
pour-plate method/technique
Platten-Test plate assay
Plattenverfahren *micb*
(Kultur) late assay, plating;
disk assay (for antibiotics)
Plattenzählverfahren *micb* plate count
Plattform platform
Plattformwagen/-karren
platform truck; (kleines/rundes
Gestell: Kistenroller/Fassroller etc.)
dolly
Plattierung/Plattieren *micb*
plating (plating out)
➢ **Replikaplattierung**
replica-plating
Plattierungseffizienz
efficiency of plating
Platzbeschränkung/Platznot
(z.B. im Labor)
space restrictions
Plotter/Kurvenzeichner plotter
Plus-~/Minus-Verbindung
(Rohrverbindungen etc.) male/female
joint; *electr* plus/minus connection
Pol pole; *electr* (Plus~/Minuspol:
Anschlussklemme) terminal
polar polar
➢ **unpolar** apolar
Polarisationsfilter/
'Pol-Filter'/Polarisator
polarizing filter, polarizer
Polarisationsmikroskop
polarizing microscope
polarisiertes Licht
polarized light
➢ **linear p. L.**
plane-polarized light
Polyacrylamid polyacrylamide
Polyadenylierung *gen* polyadenylation
Polydispersitätsindex (PDI)
polydispersity index (PDI)
Polymer polymer
Polymerasekettenreaktion
polymerase chain reaction (PCR)
Polymerisat polymerization product

Polymerisationsgrad
degree of polymerization
poolen/vereinigen/zusammenbringen
pool, combine, accumulate
Population/Bevölkerung population
Populationskurve/Bevölkerungskurve
population curve
Porenweite (Filter/Gitter etc.)
pore size, mesh size
porig/porös/durchlässig porous
porös/porig/durchlässig porous
Porosität/Durchlässigkeit porosity
Portionierung portioning
Porzellanschale porcelain dish
Positronenemissionstomographie (PET)
positron emission tomography (PET)
Posten/Partie (Waren) lot
Potential potential
Potentialdifferenz/Spannung
potential difference, voltage
potentiell potential
Pottasche/Kaliumcarbonat
potash, potassium carbonate
'Potter' (Glashomogenisator)
Potter-Elvehjem homogenizer
(glass homogenizer)
Prädisposition/Veranlagung
predisposition
Präkursor/Vorläufer precursor
prall/schwellend/turgeszent turgescent
Prallblech/Prallplatte/Ablenkplatte (Strombrecher z.B. an Rührer von Bioreaktoren) baffle plate
pränatale Diagnostik
prenatal diagnostics
Präparat preparation
(*Lebewesen*: preserved specimen)
➢ **Dauerpräparat** *micros*
permanent mount
➢ **mikroskopisches Präparat**
microscopical preparation,
microscopic mount
➢ **Nasspräparat (Frischpräparat/ Lebendpräparat/Nativpräparat)**
wet mount
➢ **Quetschpräparat** *micros*
squash (mount)
➢ **Schabepräparat** *micros*
scraping (mount)
➢ **Totalpräparat** whole mount
Präparation *anat* dissection
präparativ preparative
Präparator/Tierpräparator taxidermist
Präparierbesteck
dissecting instruments
(dissecting set)
präparieren
allg prepare; *anat* dissect;
micros mount
Präpariernadel
dissecting needle, probe
Präparierpinzette/ Sezierpinzette/ anatomische Pinzette dissection
tweezers, dissecting forceps
Präparierschale dissecting dish,
dissecting pan
präsymptomatische Diagnostik
presymptomatic diagnostics
Prävalenz prevalence, prevalency
Prävention prevention
Präzipitat/ Niederschlag/Sediment/Fällung
deposit, sediment, precipitate
Präzipitation/Fällung
precipitation
präzipitieren/fällen/ausfällen
precipitate
präzis/genau precise, exact
Präzision/Genauigkeit
precision, exactness
preisgünstige Produktion
cost-efficient production
Prephensäure (Prephenat)
prephenic acid
Pressling
pressed piece/article/item; pellet;
polym molding, molded piece
Pressluft
compressed air, pressurized air
Pressluftatmer
compressed air breathing apparatus
Pressspan flakeboard
Prisma prism
Proband/Propositus propositus

Pr

Probe/
Versuch/Untersuchung/Test/Prüfung
assay, test, trial, examination, exam,
investigation; *chem* proof, check (die
Probe machen); (Teilmenge eines zu
untersuchenden Stoffes)
chem/med/micb sample
➢ **Probensubstanz/**
Untersuchungsmaterial
assay material, test material,
examination material
Probealarm/Probe-Notalarm
drill, emergency drill
Probefläschchen/Probegläschen
sample vial, specimen vial;
(größer:) specimen jar;
(mit Schraubverschluss)
screw-cap vial
Probelauf
trial run ('experimental experiment')
Probenehmer/Probenentnahmegerät
sampler
Probengeber dispenser
Probenkonzentrator
sample concentrator
Probennahme/Probeentnahme
sample-taking, taking a sample
Probennahmepinzette
specimen tweezers
Probennahmevorrichtung
sampling device
Probenverwaltung sample custody
Probenvorbereitung
sample preparation
probieren/versuchen try, attempt
Probierstein touchstone
Problemabfall hazardous waste
Problemfall problematic case;
(verzwickte Lage) quandary
Produkt product
Produkthemmung product inhibition
Produktivität productivity
Produktreinheit product purity
Produzent/Erzeuger/Hersteller
producer
produzieren/erzeugen/herstellen
produce, manufacture, make
Prognose prognosis

projizieren/abbilden project
Proliferation proliferation
proliferieren proliferate
propagieren propagate
Propellerpumpe vane-type pump
prophylaktisch prophylactic
Prophylaxe prophylaxis
Propionaldehyd
propionic aldehyde, propionaldehyde
Propionsäure (Propionat)
propionic acid (propionate)
proportionaler Schwellenwert
proportional truncation
Prostansäure prostanoic acid
Protein/Eiweiß protein
proteinartig/
proteinhaltig/Protein.../
aus Eiweiß bestehend/Eiweiß...
proteinaceous
proteolytisch/eiweißspaltend
proteolytic
Protokoll/Aufzeichnungen
protocol, record, minutes
Protokollheft/Laborjournal
laboratory notebook
Protokollierung recordkeeping
Protonengradient proton gradient
protonenmotorische Kraft
proton motive force
Protonenpumpe proton pump
Protonensonde proton microprobe
proximal/ursprungsnah proximal
Prozentsatz/prozentualer Anteil
percentage
prozessieren/weiterverarbeiten process
Prozessierung/Verarbeitung processing
Prozesssteuerung/Prozess-Kontrolle
process control
Prüfbarkeit testability
Prüfbericht test report
Prüfdaten test data
prüfen/untersuchen/
testen/probieren/analysieren
investigate, examine, test, try,
assay, analyze
Prüfgas probe gas, tracer gas;
(Kalibrierung) calibration gas;
(zu prüfendes Gas) test gas

Prüfgerät/Prüfer
  tester, testing device/apparatus, checking instrument
Prüflabor testing laboratory
Prüfmittel testing device
Prüfprotokoll
  case report form (CRF), case record form
Prüfsumme check sum
Prüfung/
  Untersuchung/Test/Probe/Analyse
  investigation, examination (exam), test, trial, assay, analysis
Prüfverfahren
  testing procedure; audit procedure
Psychrometer
  (ein Luftfeuchtigkeitsmessgerät)
  psychrometer, wet-and-dry-bulb hygrometer
PTT (Nachweis verkürzter Proteine)
  PTT (protein truncation test)
Puffer buffer
Pufferkapazität
  buffering capacity
Pufferlösung buffer solution
puffern buffer
Pufferung buffering
Pufferzone buffer zone
Puls pulse
Puls-Feld-Gelelektrophorese/
  Wechselfeld-Gelelektrophorese
  pulsed field gel electrophoresis (PFGE)
pulsieren
  pulsate, throb, beat
Pulsmarkierung
  pulse labeling, pulse chase
Pulspolarografie
  pulse polarography
➢ **differentielle Pulspolarografie**
  differential pulse polarography (DPP)
Pulver/Puder pulver, powder
pulverisieren pulverize
Pulverisierung pulverization
Pulverspatel
  powder spatula

Pumpe pump
➢ **Abgabepuls** discharge stroke
➢ **Absaugpumpe/Saugpumpe**
  aspirator pump, vacuum pump
➢ **Ansaughöhe** suction head
➢ **Ansaugpuls** suction stroke
➢ **Ansaugtiefe** suction lift
➢ **Balgpumpe** bellows pump
➢ **Direktverdrängerpumpe**
  positive displacement pump
➢ **Dispenserpumpe**
  dispenser pump, dispensing pump
➢ **Dosierpumpe** dosing pump, proportioning pump, metering pump
➢ **Drehkolbenpumpe**
  rotary piston pump
➢ **Drehschieberpumpe**
  rotary vane pump
➢ **Druckpumpe/**
  doppeltwirkende Pumpe
  double-acting pump
➢ **Fasspumpe** barrel pump, drum pump
➢ **Filterpumpe** filter pump
➢ **Förderhöhe** discharge head
➢ **Förderleistung/Saugvermögen**
  flow rate
➢ **Förderpumpe** feed pump
➢ **Gesamtförderhöhe** total static head
➢ **Handpumpe** hand pump
➢ **Ionenpumpe** ion pump
➢ **Kolbenpumpe**
  piston pump, reciprocating pump
➢ **Kreiselpumpe/Zentrifugalpumpe**
  impeller pump, centrifugal pump
➢ **Laufradpumpe** impeller pump
➢ **manuelle Vakuumpumpe** hand-operated vacuum pump
➢ **Mehrkanal-Pumpe**
  multichannel pump
➢ **Membranpumpe**
  diaphragm pump
➢ **peristaltische Pumpe**
  peristaltic pump
➢ **Pipettierpumpe** pipet pump
➢ **Propellerpumpe**
  vane-type pump
➢ **Protonenpumpe**
  proton pump

Pu

- Quetschpumpe
  (Handpumpe für Fässer)
  squeeze-bulb pump
  (hand pump for barrels)
- Saugpumpe/Vakuumpumpe
  suction pump, aspirator pump, vacuum pump
- Schlauchpumpe tubing pump;
  (größere Durchmesser) hose pump
- Schneckenantriebspumpe
  progressing cavity pump
- selbstansaugend prime
- Spritzenpumpe syringe pump
- Umwälzpumpe
  circulation pump
- Vakuumpumpe
  vacuum pump
- Verdrängungspumpe/
  Kolbenpumpe (HPLC)
  displacement pump
- Wärmepumpe heat pump
- Wasserpumpe water pump
- Wasserstrahlpumpe
  water pump, filter pump, vacuum filter pump
- Zahnradpumpe gear pump
- Zentrifugalpumpe/Kreiselpumpe
  centrifugal pump, impeller pump

Pumpenantrieb pump drive
Pumpenkopf pump head
Pumpenöl pump oil
Pumpenzange/Wasserpumpenzange
water pump pliers, slip-joint adjustable water pump pliers (adjustable-joint pliers)
Punktdiagramm dot diagram
punktieren puncture, tap
Punktion puncture (needle biopsy)
Pupillenerweiterung
pupil dilatation
putzen clean, cleanse
Putzkolonne/Reinigungstrupp
cleaning squad(ron)
Putzmittel cleaning agent
Putzschwamm
cleaning pad, scrubber, sponge
Putztuch/Putzlappen
(cleaning) rag, cloth
Putzzeug cleaning utensils
Pyknometerflasche
specific gravity bottle
Pyrethrinsäure pyrethric acid
Pyridin pyridine
Pyrolyse/Thermolyse
pyrolysis, thermolysis
Pyrometer pyrometer
- **Pyropter/optisches Pyrometer**
  optical pyrometer

Pyrometrie pyrometry
Pyrrol pyrrole

Qu

Quaddel welt (weal)
Quadratmethode *ecol*
  quadrat method, quadrat sampling
Qualitätsbeurteilung/
Qualitätsbewertung
  quality assessment
Qualitätsfaktor/Bewertungsfaktor
  quality factor
Qualitätskennzeichen quality indicator
Qualitätskontrolle/
Qualitätsprüfung/
Qualitätsüberwachung
  quality control (QC)
Qualitätsmerkmal sign of quality
Qualitätssicherung
  quality assurance (QA)
Qualitätssicherungshandbuch
(EU-CEN)
  quality manual
Qualitätszertifikat
  certificate of performance
Qualm smoke
quantifizieren *med/chem*
  quantify, quantitate
Quantifizierung *med/chem*
  quantification, quantitation
Quantil/Fraktil *stat* quantile, fractile
Quantität quantity
Quarantäne quarantine
Quartil/Viertelswert *stat* quartile
Quarzglas quartz glass
Quarzgut
  (milchig-trübes Quarzglas)
  fused quartz
Quarzthermometer
  quartz thermometer
Quecksilber (Hg) mercury
Quecksilber-(I)/einwertiges Q.
  mercurous, mercury(I) ...
Quecksilber-(I)-chlorid/Kalomel
  mercurous chloride, calomel,
  mercury subchloride
Quecksilber-(II)/zweiwertiges Q.
  mercuric, mercury(II) ...
Quecksilber-(II)-chlorid/Sublimat
  mercuric chloride, sublimate,
  mercury dichloride,
  corrosive mercury chloride

Quecksilberdampflampe
  mercury vapor lamp
Quecksilberfalle
  mercury trap,
  mercury well
Quecksilberthermometer
  mercury-in-glass thermometer
Quecksilbertropfelektrode
  dropping mercury electrode (DME)
Quecksilbervergiftung/
Merkurialismus
  mercury poisoning
Quelle
  source; origin
quellen (Wasseraufnahme)
  soak, steep
➤ **anschwellen** swell
➤ **hervorquellen** emanate
Quellwasser springwater
Querschnitt cross section
Querstrombank
  laminar flow workstation,
  laminar flow hood,
  laminar flow unit
Querstromfiltration
  cross-flow filtration
quervernetzendes Agens
  cross linker, crosslinking agent
quervernetzt cross-linked
Quervernetzung
  cross-link, cross-linking;
  cross-linkage
quetschen squeeze, pinch
Quetschhahn
  pinchcock
➤ **Schraubquetschhahn**
  screw compression pinchcock
Quetschpräparat *micros*
  squash (mount)
Quetschpumpe
  (Handpumpe für Fässer)
  squeeze-bulb pump
  (hand pump for barrels)
Quetschventil
  pinch valve
Quittung/Erhaltsbestätigung
  receipt
Quotient/Verhältnis ratio, relation

R

R-Sätze (Gefahrenhinweise)
R phrases (Risk phrases)
RF-Wert *chromat* R_F-value
(retention factor; ratio of fronts)
Racemat/racemische Verbindung
racemate
Racemisierung racemization
Radikal radical
➢ **freies Radikal** free radical
Radikalfänger radical scavenger
Radikalion radical ion
radioactiv (Atomzerfall)
radioactive (nuclear disintegration)
radioaktive Abfälle
radioactive waste, nuclear waste
radioaktive Markierung radiolabelling
radioaktiver Marker radioactive marker
Radioaktivität radioactivity
Radio-Allergo-Sorbent Test
radioallergosorbent test (RAST)
Radioimmunassay/Radioimmunoassay
radioimmunoassay
Radioimmunelektrophorese
radioimmunoelectrophoresis
Radiokarbonmethode/
Radiokohlenstoffmethode/
Radiokohlenstoffdatierung
radiocarbon method
Radionuklid/Radionuclid radionuclide
Rakel/Rakelmesser/
Schabeisen/Abstreichmesser
scraper, wiper blade, spreading
knife, coating knife, doctor knife
Rand edge, margin;
(eines Gefäßes) rim, edge
Rändelmutter knurled nut
randomisieren *stat* randomize
Randomisierung *stat* randomization
Randverteilung *stat* marginal
distribution
Rangkorrelationskoeffizient *stat* rank
correlation coefficient
Rangmaßzahlen *stat* rank statistics,
rank order statistics
Rangordnung/
Rangfolge/Stufenfolge/Hierarchie
order of rank, ranking, hierarchy
ranzig rancid

Rasen *micb/bact* lawn
Rasenkultur lawn culture
Rasierklinge razor blade
Raspel rasp; (Haushaltsraspel) grater
Raster grid, screen, raster
Rasterelektronenmikroskop (REM)
scanning electron microscope (SEM)
Raster-Kalorimetrie
scanning calorimetry
Rasterkartierung *ecol/biogeo* frame
raster mapping, grid mapping
Rasterkraftmikroskopie
atomic force microscopy (AFM)
Rastermethode grid method
rastern scan, screen
Rasterstichprobenerhebung *ecol*
grid sampling
Rastertunnelmikroskopie
scanning tunneling microscopy
(STM)
Rasteruntersuchung/
Reihenuntersuchung *med*
screening
Ratsche/Rätsch ratchet (ratchet
wrench)
Ratschen-Klemme/Ratschen-
Absperrklemme (Schlauchklemme)
ratchet clamp
Rauch (sichtbar) smoke;
(Dämpfe/meist schädlich) fume
Rauchabzug
(Raumentlüftung) fume extraction;
(Abzug) fume hood;
(Abzugskanal) flue, flue duct
rauchend (Säure) fuming (acid)
Rauchentwicklung
smoke generation;
development of smoke
Rauchfang chimney, flue
rauchfrei *adj* nonsmoking
Rauchgase
(sichtbarer Qualm) smoke gas;
(Abluft aus Feuerung mit
Schwebstoffen) flue gases;
(Dämpfe) fumes
Rauchmelder smoke detector
Rauchschranke/
Rauchschutzwand smoke barrier

Rauchschwaden
clouds of smoke/fumes
Rauchverbot
ban on smoking, smoking ban
Rauchvergiftung smoke poisoning
Rauchzug flue, flue duct
Raum (Länge-Breite-Höhe)
room, compartment; space;
(Gebiet/Gegend/Region/Zone)
area, region, zone, territory;
(Platz) place
➢ **Kühlraum/Gefrierraum**
cold-storage room,
cold store, 'freezer'
➢ **Lebensraum/Lebenszone/Biotop**
life zone, biotope
➢ **Reinraum**
clean room (*auch:* Reinstraum)
➢ **Sicherheitsraum/**
Sicherheitslabor (S1-S4)
biohazard containment (laboratory)
(classified into biosafety containment classes)
Raumheizung space heating
Rauminhalt (Volumen)
capacity (volume)
räumlich spatial, of space;
(dreidimensional) three-dimensional
Räumlichkeit(en) premises, location
Raumstruktur/räumliche Struktur
three-dimensional structure,
spatial structure
Raumtemperatur
room temperature,
ambient temperature
Rauschanalyse/
Fluktuationsanalyse
noise analysis, fluctuation analysis
Rauschen *tech/electro/neuro* noise
Rauschfilter noise filter
Rauschminderung noise reduction
Rauschmittel/Rauschgift/Rauschdroge
psychoactive/psychotropic drug
Rauschthermometer
noise thermometer
Reagenz
(jetzt: Reagens, *pl* **Reagentien)**
reagent; (Reaktand) reactant

Reagenzglas test tube, glass tube, assay tube
Reagenzglasbefruchtung/
In-vitro-Fertilisation
in-vitro fertilization (IVF)
Reagenzglasbürste test tube brush
Reagenzglashalter test tube holder
Reagenzglasständer/
Reagenzglasgestell
test tube rack
Reagenzienflasche reagent bottle
Reagenzlösung reagent solution
reagieren react
Reaktand/Reaktionsteilnehmer/
Ausgangsstoff
reactant
Reaktion *chem* reaction; *ethol*
(bedingte/unbedingte R.) response
(conditioned/unconditioned r.)
➢ **Austauschreaktion**
exchange reaction
➢ **Biosynthesereaktion**
biosynthetic reaction
(anabolic reaction)
➢ **Dunkelreaktion** dark reaction
➢ **Durchgeh-Reaktion** runaway reaction
➢ **Einschiebereaktion/**
Insertionsreaktion
insertion reaction
➢ **Eintopfreaktion** one-pot reaction
➢ **Enzymreaktion** enzymatic reaction
➢ **Gegenreaktion** counterreaction
➢ **gekoppelte Reaktion** coupled reaction
➢ **geschwindigkeitsbegrenzende(r)**
Schritt/Reaktion
rate-limiting step/reaction
➢ **geschwindigkeitsbestimmende(r)**
Schritt/Reaktion
rate-determining step/reaction
➢ **heftige Reaktion**
vigorous reaction, violent reaction
➢ **Immunreaktion** immune reaction
➢ **Kettenreaktion** chain reaction
➢ **Kondensationsreaktion/**
Dehydrierungsreaktion
condensation reaction,
dehydration reaction

- **Kupplungsreaktion** *chem*
  coupling reaction
- **Nebenreaktion** *chem* side reaction
- **nullter/erster/zweiter.. Ordnung**
  zero-order/first-order/second-order..
- **Polymerasekettenreaktion (PCR)**
  polymerase chain reaction (PCR)
- **Redoxreaktion**
  redox reaction,
  reduction-oxidation reaction,
  oxidation-reduction reaction
- **sequentielle Reaktion/Kettenreaktion**
  sequential reaction, chain reaction
- **Teilreaktion** partial reaction;
  (electrode potentials:) half-reaction
- **Verdrängungsreaktion** *biochem*
  displacement reaction
- **Zweisubstratreaktion/**
  Bisubstratreaktion
  bisubstrate reaction

Reaktionsfolge
reaction sequence, reaction pathway
Reaktionsgefäß reaction vessel
Reaktionsgeschwindigkeit/
Reaktionsrate
reaction rate
Reaktionsgleichung chemical equation
Reaktionskette reaction pathway
Reaktionskinetik reaction kinetics
Reaktionsnorm norm of reaction
Reaktionswärme/Wärmetönung
heat of reaction
Reaktionszwischenprodukt
reaction intermediate
Reaktor/Bioreaktor *biot*
reactor, bioreactor
- **Airliftreaktor/**
  pneumatischer Reaktor
  airlift reactor, pneumatic reactor
- **Blasensäulen-Reaktor**
  bubble column reactor
- **Druckumlaufreaktor**
  pressure cycle reactor
- **Durchflussreaktor** flow reactor
- **Düsenumlaufreaktor/**
  Umlaufdüsen-Reaktor
  nozzle loop reactor,
  circulating nozzle reactor
- **Fedbatch-Reaktor/**
  Fed-Batch-Reaktor/Zulaufreaktor
  fedbatch reactor, fed-batch reactor
- **Festbettreaktor** fixed bed reactor,
  solid bed reactor
- **Festphasenreaktor**
  solid phase reactor
- **Filmreaktor** film reactor
- **Fließbettreaktor/**
  Wirbelschichtreaktor/
  Wirbelbettreaktor
  fluidized bed reactor,
  moving bed reactor
- **Füllkörperreaktor/Packbettreaktor**
  packed bed reactor
- **Gärtassenreaktor** tray reactor
- **Kugelbettreaktor** bead-bed reactor
- **Lochbodenkaskadenreaktor/**
  Siebbodenkaskadenreaktor
  sieve plate reactor
- **Mammutpumpenreaktor/**
  Airliftreaktor
  airlift reactor
- **Mammutschlaufenreaktor**
  airlift loop reactor
- **Membranreaktor** membrane reactor
- **Packbettreaktor/Füllkörperreaktor**
  packed bed reactor
- **Pfropfenströmungsreaktor/**
  Kolbenströmungsreaktor
  plug-flow reactor
- **Rohrschlaufenreaktor**
  tubular loop reactor
- **Rührkammerreaktor**
  fermentation chamber reactor,
  compartment reactor, cascade
  reactor, stirred tray reactor
- **Rührkaskadenreaktor**
  stirred cascade reactor
- **Rührkesselreaktor**
  stirred-tank reactor
- **Rührschlaufenreaktor/**
  Umwurfreaktor
  stirred loop reactor
- **Säulenreaktor/Turmreaktor**
  column reactor
- **Schlaufenradreaktor**
  paddle wheel reactor

Re

- **Schlaufenreaktor/Umlaufreaktor**
  loop reactor
- **Siebbodenkaskadenreaktor/
  Lochbodenkaskadenreaktor**
  sieve plate reactor
- **Strahlreaktor** jet reactor
- **Strahlschlaufenreaktor/
  Strahl-Schlaufenreaktor**
  jet loop reactor
- **Tauchflächenreaktor**
  immersing surface reactor
- **Tauchkanalreaktor**
  immersed slot reactor
- **Tauchstrahlreaktor**
  plunging jet reactor, deep jet reactor, immersing jet reactor
- **Tropfkörperreaktor/Rieselfilmreaktor**
  trickling filter reactor
- **Turmreaktor/Säulenreaktor**
  column reactor
- **Umlaufdüsen-Reaktor/
  Düsenumlaufreaktor**
  nozzle loop reactor, circulating nozzle reactor
- **Umlaufreaktor/Umwälzreaktor/
  Schlaufenreaktor**
  loop reactor, circulating reactor, recycle reactor
- **Umwurfreaktor/
  Rührschlaufenreaktor**
  stirred loop reactor
- **Wirbelschichtreaktor/
  Wirbelbettreaktor/Fließbettreaktor**
  fluidized bed reactor, moving bed reactor
- **Zulaufreaktor/
  Fedbatch-Reaktor/
  Fed-Batch-Reaktor**
  fedbatch reactor, fed-batch reactor

Reassoziationskinetik
  reassociation kinetics
Rechen rake; grid, screen; (der Kläranlage) grate, bar screen
Rechenschieber slide rule
Rechnung (Waren~)/Faktura
  bill, invoice
Rechnungsprüfung auditing
rechtsgängig right-handed
rechtshändig right-handed, dextral
**Rechtsmedizin/Gerichtsmedizin/
Forensik/forensische Medizin**
  forensics, forensic medicine
Redestillation/mehrfache Destillation
  repeated distillation, cohobation
**redestillieren/umdestillieren
(nochmal destillieren)**
  redistill
Redoxpaar redox couple
Redoxpotential redox potential
Redoxreaktion
  redox reaction,
  reduction-oxidation reaction,
  oxidation-reduction reaction
Reduktion reduction
Reduktionsmittel reducing agent
Redundanz redundancy
Reduzenten *ecol* reducers
reduzieren reduce
Reduzierstück (Laborglas/Schlauch)
  reducer, reducing adapter, reduction adapter
**Reduzierventil/
Druckreduzierventil/
Druckminderventil/
Druckminderungsventil
(für Gasflaschen)**
  pressure-relief valve, gas regulator
reelles Bild *micros* real image
Referenzstamm *micb* reference strain
Reflexhammer percussion hammer, plexor, plessor, percussor
Refraktion/Brechung refraction
Refraktometer refractometer
Regel rule
Regelgerät control unit
Regelglied control element, control unit
Regelgröße controlled variable, controlled condition
Regelkreis feedback system, feedback control system
regelmäßig regular
- **unregelmäßig** irregular
regeln/kontrollieren regulate, control
Regelspannung control voltage
Regelstrecke
  control system of a process

Re

Regeltechnik/Regelungstechnik
control technology
➢ **Mess- und Regeltechnik**
instrumentation and control
Regelung (Regulierung/Kontrolle)
control, regulation;
(Vereinbarung) arrangement
Regelungsprozess
regulatory procedure
regenerieren regenerate
Regenerierung/Regeneration
regeneration
Regenmesser pluviometer, rain gauge
Regenwasser rainwater
Regler regulator; (Schalter/Knopf)
control, adjustment knob/button
Regressionsanalyse *stat*
regression analysis
Regressionskoeffizient *stat*
regression coefficient,
coefficient of regression
regressiv/zurückbildend/
zurückentwickelnd
regressive
Regulationsmechanismen
regulatory mechanisms
Regulierungsbehörde
regulatory agency
Rehydratation/Rehydratisierung
rehydration
Reibe (Reibeisen) grater
Reibschale/Mörser (*siehe auch dort*)
mortar
Reichweite (Strahlung) range
reif mature, ripe
➢ **unreif** unripe, immature
Reif/Raureif rime, hoarfrost, white frost
Reife maturity, ripeness
➢ **Unreife** immaturity, immatureness
Reifen *n* maturing, ripening
reifen *vb* mature, ripen
Reifeteilung/Reduktionsteilung/Meiose
reduction division, meiosis
Reifung maturation
Reihe row; series
➢ **Alkoholreihe/**
aufsteigende Äthanolreihe
graded ethanol series
➢ **chaotrope Reihe** chaotropic series
➢ **eluotrope Reihe**
(Lösungsmittelreihe)
eluotropic series
➢ **Hofmeistersche Reihe/**
lyotrope Reihe
Hofmeister series, lyotropic series
➢ **mixotrope Reihe** mixotropic series
➢ **Transformationsreihe**
transformation series
➢ **Versuchsreihe** experimental series
rein (pur) neat, pure; (sauber) clean;
(ohne Zusatz) pure
Reinheit (ohne Zusätze) purity
Reinheitsgrade, chemische
purity grades, chemical grades
➢ **chemisch rein**
chemically pure (CP);
laboratory (lab)
➢ **pro Analysis (pro analysi = p.a.)**
reagent, reagent-grade, analytical
reagent (AR), analytical grade
➢ **reinst (purissimum, puriss.)** pure
➢ **roh (crudum, crd.)** crude
➢ **technisch** technical
Reinigbarkeit cleanability
reinigen (säubern) cleanse, clean up,
tidy (up); (aufbereiten) clean, purify
Reinigung (Saubermachen) cleaning,
cleansing; (Dekontamination/
Dekontaminierung/Entseuchung)
decontamination;
(Reindarstellung) purification
➢ **Reinigung ohne Zerlegung**
von Bauteilen
cleaning in place (CIP)
Reinigungskraft (Reinigungskräfte)/
Reinigungspersonal
cleaner(s), cleaning personnel
Reinigungsmittel
cleanser, cleaning agent;
(Detergens) detergent
Reinigungsmöglichkeit(en) cleanability
Reinigungstuch (Papier)
cleansing tissue
Reinigungsverfahren (Aufreinigung)
purification procedure,
purification technique

Reinkultur pure culture, axenic culture
Reinlichkeit cleanliness
Reinraum/Reinstraum
   clean room, cleanroom
Reinraumhandschuhe
   cleanroom gloves
Reinraumwerkbank cleanroom bench
reinst *lab/chem*
   highly pure (superpure/ultrapure)
Reinstoff pure substance
Reißnagel tack, thumb tack
Reißverschluss zipper
Reiz/Stimulus
   irritation, stimulus
   ➢ **Außenreiz** external stimulus
   ➢ **Lichtreiz** light stimulus
   ➢ **Schlüsselreiz/Auslösereiz**
   key stimulus,
   sign stimulus (release stimulus)
reizbar irritable
Reizbarkeit irritability
reizempfänglich irritable, excitable, sensitive
reizen (anregen/stimulieren) excite, stimulate; (irritieren)
   med/physio/chem irritate
Reizgas irritant gas
Reizschwelle stimulus threshold
Reizumwandlung stimulus transduction
Reizung/Stimulation
   irritation, stimulation
Rekombinante (Zelle)
   recombinant (cell)
rekombinieren recombine
rekonstituieren reconstitute
Rekonstitution reconstitution
Rektifikation rectification
rekultivieren recultivate, replant
Relais *electr* relay
relaxiert/entspannt
   relaxed (conformation)
renaturieren renature
Renaturierung renaturation, renaturing; *gen* (Annealing/Reannealing) annealing, reannealing, reassociation (of DNA)
Reparatur repair, restoration
reparieren repair, fix, mend, restore

Repellens (*pl* **Repellentien**) repellent
Replikaplattierung replica-plating
reprimieren/unterdrücken/hemmen
   gen/med/tech repress, control, suppress, subdue
Reprimierung/
   Unterdrückung/Hemmung
   repression, control, suppression
Reproduzierbarkeit reproducibility
reproduzieren reproduce
Reservestoff reserve material, storage material, food reserve
resistent resistant
Resistenz resistance
resorbieren resorb
Resorption resorption
Ressource/Rohstoffquelle resource
Rest rest, residue;
   (Rückstand) residue
   ➢ **unveränderter Rest/invarianter Rest**
   math invariant residue
   ➢ **variabler Rest** *math* variable residue
Restfeuchte residual dampness,
   (H_2O) residual humidity
restituieren/wiederherstellen restitute
Restitution/Wiederherstellung
   restitution
Restriktionsenzym restriction enzyme
Retention hold-up, retention
Retentionsfaktor *chromat*
   retention factor
Retentionszeit/
   Verweildauer/Aufenthaltszeit
   retention time
Retinal retinal, retinene
Retinsäure retinic acid
Retorte retort
Rettung rescue, help
Rettungsdienst
   rescue service, lifesaving service
Rettungshubschrauber
   rescue helicopter
Rettungswagen/Sanitätswagen
   ambulance
reversibel/umkehrbar reversible
Reversibilität/Umkehrbarkeit
   reversibility
Reversion/Umkehrung reversion

Reversosmose/Umkehrosmose
reverse osmosis
Reversphase/Umkehrphase
reversed phase
**Revertase/Umkehrtranskriptase/
reverse Transkriptase**
reverse transcriptase
Revolver/Objektivrevolver *micros*
nosepiece, nosepiece turret
Revolverlochzange
revolving punch pliers
rezent/gegenwärtig/heute lebend
recent, contemporary, extant
Rezeptor/Empfänger receptor
Rezeptor-Ausdünnungsregulation
receptor-down regulation
Rezeptur *pharm* formula
Reziprokschüttler reciprocating shaker
**Ribonucleinsäure/Ribonukleinsäure
(RNA/RNS)**
ribonucleic acid (RNA)
Ribosonde/RNA-Sonde riboprobe
Richtigkeit *stat* (Genauigkeit)
correctness, exactness, accuracy;
(Qualitätskontrolle) trueness
Richtlinie(n) guideline(s); rules of
conduct, EU *jur:* Directive(s);
instructions, directions, regulations,
rules, policy, standards
➢ **allgemeine R.** general policy
➢ **einheitliche R.**
uniform rules/standards
➢ **internationale R.**
international standards
Richtwert
(Näherungszahl) approximate value;
(Richtzahl) index number,
index figure, guiding figure
riechbar smellable,
perceptible to one's sense of smell
riechen smell
Riechschwelle/Geruchsschwellenwert
odor threshold, olfactory threshold
Riegel/Verriegelung (elektr. Sicherung)
interlock, fail safe circuit
Rieselfelder (Abwasser-Kläranlage)
sewage fields, sewage farm
Rieselfilm falling liquid film

Rieselfilmreaktor/Tropfkörperreaktor
trickling filter reactor
rieseln trickle
**Riffelung/Riefen
(z.B. Schlauchverbinder)**
flutings (e.g. tube connections)
Ringbildung/Catenation catenation
Ringblende *micros* disk diaphragm
(annular aperture)
Ringerlösung/Ringer-Lösung
Ringer's solution
Ringform *chem*
ring form, ring conformation
Ringformel ring formula
ringförmig/cyclisch (zyklisch)
annular, cyclic
Ringkabelschuh cable lug, terminal
Ringmarke (Laborglas etc.) graduation
Ringschluss *chem*
(Ringbildung) ring closure,
ring formation, cyclization;
(Zirkularisierung) circularization
Ringschlüssel
box wrench, box/ring spanner (*Br*)
Ringspaltung *chem* ring cleavage
Ringstruktur ring structure
**Rippenglas/
geripptes Glas/geriffeltes Glas**
ribbed glass
Risiko (*pl* **Risiken)/Gefahr**
risk, danger; hazard
➢ **Berufsrisiko** occupational hazard
➢ **Brandrisiko** fire hazard
➢ **Gesundheitsrisiko** health hazard
➢ **Kontaktrisiko (Gefahr bei Berühren)**
contact hazard
➢ **Krebsrisiko** cancer risk
➢ **Wiederholungsrisiko** recurrence risk
Risikoabschätzung risk assessment
Riss/Fissur/Furche/Einschnitt
fissure; (Spalte) crevice
➢ **Sternriss (im Glas)** star-crack
**RNA/RNS (Ribonucleinsäure/
Ribonukleinsäure)**
RNA (ribonucleic acid)
roh raw, crude
Rohabwasser raw sewage
Rohextrakt crude extract

Rohöl crude oil, petroleum
Rohprodukt (unaufgereinigt)
 crude product
Rohr/Röhre pipe, tube;
 (Rohre/Rohrleitungen) pipes,
 plumbing
Röhrchen vial, tube
➤ **Gärröhrchen/Einhorn-Kölbchen**
 fermentation tube
➤ **Glasröhrchen**
 glass tube, glass tubing (: Glasrohre)
➤ **Kulturröhrchen** culture tube
➤ **Siederöhrchen** ebullition tube
➤ **Trockenröhrchen** drying tube
➤ **Zentrifugenröhrchen**
 centrifuge tube
➤ **Zündröhrchen/Glühröhrchen**
 ignition tube
Rohrleitung conduit, pipe, duct, tube;
 (-en in einem Gebäude) plumbing
Rohrleitungssystem
 (Lüftung) ductwork, airduct system;
 (Wasser) plumbing system
Rohrofen tube furnace
Rohrschelle pipe clamp, pipe clip
Rohrverbinder/Rohrverbindung(en)
 pipe fitting(s), fittings
Rohrzange pipe wrench (rib-lock
 pliers/adjustable-joint pliers)
Rohrzucker/Rübenzucker/
Saccharose/Sukrose/Sucrose
 cane sugar, beet sugar,
 table sugar, sucrose
Rohschlamm raw sludge
Rohstoff raw material, resource
Rohstoffquelle/Ressource
 resource
Rohzucker raw sugar,
 crude sugar (unrefined sugar)
rollen roll
Rollen
 (zum Schieben: Laborwagen etc.)
 casters, castors
➤ **Lenkrollen/Schwenkrollen/**
 Schwenklaufrollen
 swivel casters
Rollerflasche roller bottle
Rollerflaschenkultur roller tube culture

Rollfüße/Laufrollen/Rollen (Wagen)
 casters, castors
Rollgabelschlüssel/'Engländer'
 adjustable wrench
Rollhocker/'Elefantenfuß'
 (runder Trittschemel mit Rollen)
 (rolling) step-stool
Rollrand (Glas: Ampullen etc.)
 beaded rim
Rollrandgläschen/Rollrandflasche
 (mit Bördelkappenverschluss)
 crimp-seal vial;
 allg beaded rim bottle
➤ **Bördelkappe** crimp seal
➤ **Verschließzange für Bördelkappen**
 cap crimper
Rollstuhl wheelchair
rollstuhlgerecht wheelchair accessible
Röntgenabsorptionsspektroskopie
 X-ray absorption spectroscopy
Röntgenbeugung X-ray diffraction
Röntgenbeugungsdiagramm/
Röntgenbeugungsaufnahme/
Röntgendiagramm
 X-ray diffraction pattern
Röntgenbeugungsmethode
 X-ray diffraction method
Röntgenbeugungsmuster
 X-ray diffraction pattern
Röntgenemissionsspektroskopie
 X-ray emission spectroscopy
Röntgenfluoreszenzspektroskopie
(RFS)
 X-ray fluorescence spectroscopy
 (XFS)
Röntgenkleinwinkelstreuung
 small-angle X-ray scattering (SAXS)
Röntgenkristallographie
 X-ray crystallography
Röntgenmikroskopie X-ray microscopy
Röntgenstrahl X-ray
Röntgenstrahl-Mikroanalyse
 X-ray microanalysis
Röntgenstrukturanalyse
 X-ray structural analysis,
 X-ray structure analysis
Röntgenweitwinkelstreuung
 wide-angle X-ray scattering (WAXS)

Rost rust
rösten/rötten (Flachsrösten) retting
**Rostentferner/Rostlöser/
Rostentfernungsmittel/
Entrostungsmittel**
  rust remover, rust-removing agent
Röstofen roasting furnace, roaster
Rostschutzmittel rust inhibitor,
  antirust agent, anticorrosive agent
Rotationsbewegung rotational motion
Rotationsmikrotom rotary microtome
Rotationssinn/Drehsinn
  rotational sense, sense of rotation
Rotationsverdampfer
  rotary evaporator,
  rotary film evaporator (*Br*), 'rovap'
Rote Liste Red Data Book
Rotor rotor
➤ **Ausschwingrotor** *centrif*
  swing-out rotor,
  swinging-bucket rotor,
  swing-bucket rotor
➤ **Festwinkelrotor** *centrif*
  fixed-angle rotor
➤ **Trommelrotor** *centrif*
  drum rotor, drum-type rotor
➤ **Vertikalrotor** *centrif* vertical rotor
➤ **Winkelrotor** *centrif*
  angle rotor, angle head rotor
rötten/rösten (Flachsrösten) retting
**Rübenzucker/Rohrzucker/
Sukrose/Sucrose**
  beet sugar, cane sugar,
  table sugar, sucrose
rückbilden degenerate, regress
Rückbildung degeneration, regression
Rückextraktion/Strippen
  back extraction, stripping
Rückfluss reflux
Rückflusskühler reflux condenser
**Rückflusssperre/
Rücklaufsperre/Rückstauventil**
  backflow prevention,
  backstop (valve)
Rückführbarkeit/Rückverfolgbarkeit
  traceability
**rückgebildet/abortiv/
  rudimentär/verkümmert** abortive

Rückgewinnung recovery, reclamation
Rückhaltevermögen
  retainment capacity,
  retainability, retention efficiency
Rückkopplung
  feedback
➤ **Rückkopplungshemmung/
  Endprodukthemmung/
  negative Rückkopplung/**
  feedback inhibition,
  end-product inhibition
Rückkopplungsschleife
  feedback loop
Rückkreuzung backcrossing, backcross
Rücklauf/Rückfluss/Reflux reflux
Rücklaufschlamm/Belebtschlamm
  activated sludge
Rücklaufsperre backstop
Rückmutation
  back-mutation, reverse mutation
Rückschlagschutz bump tube
Rückschlagventil
  backstop valve, check valve
rückseitig/dorsal dorsal
Rücksendung (einer Ware) return
Rückspülen/Rückspülung *chromat*
  backflushing
Rückstand residue; (abgesetzte
  Teilchen) bottoms, heel
Rückstauschutz
  backdraft preventer/protection
Rückstoß recoil (return motion)
Rückstoßstrahlung recoil radiation
Rückstrahlvermögen/Albedo albedo
Rückstreuung backscatter
Rückströmsperre
  backflow preventer/protection,
  backstop
Rücktitration back titration
Rückverfolgbarkeit trackability
Rudiment
  rudiment (*sensu lato*: vestige)
rudimentär rudimentary (*sensu lato*:
  vestigial);
  (abortiv/rückgebildet/verkümmert)
  abortive
ruhen rest, lie dormant
ruhend resting, quiescent, dormant

Rührbehälter/Rührkessel
agitator vessel
rühren stir, agitate; (umrühren) stir; (umwirbeln) swirl
Rührer/Rührwerk
stirrer, impeller, agitator
➢ **Ankerrührer**
anchor impeller
➢ **Axialrührer mit profilierten Blättern**
profiled axial flow impeller
➢ **Blattrührer**
blade impeller,
flat-blade paddle impeller
➢ **exzentrisch angeordneter Rührer**
off-center impeller
➢ **Gitterrührer** gate impeller
➢ **Hohlrührer** hollow stirrer
➢ **Kreuzbalkenrührer**
crossbeam impeller
➢ **Kreuzblattrührer**
four flat-blade paddle impeller
➢ **Magnetrührer** magnetic stirrer
➢ **Mehrstufen-Impuls-Gegenstrom (MIG) Rührer**
multistage impulse countercurrent impeller
➢ **Propellerrührer**
propeller impeller
➢ **Rotor-Stator-Rührsystem**
rotor-stator impeller,
Rushton-turbine impeller
➢ **Schaufelrührer/Paddelrührer**
paddle stirrer, paddle impeller
➢ **Scheibenrührer/Impellerrührer**
flat-blade impeller
➢ **Scheibenturbinenrührer**
disk turbine impeller
➢ **Schneckenrührer**
screw impeller
➢ **Schrägblattrührer**
pitched blade impeller,
pitched-blade fan impeller,
pitched-blade paddle impeller,
inclined paddle impeller
➢ **Schraubenrührer**
marine screw impeller
➢ **Schraubenspindelrührer**
pitch screw impeller
➢ **Schraubenspindelrührer mit unterschiedlicher Steigung**
variable pitch screw impeller
➢ **selbstansaugender Rührer mit Hohlwelle**
self-inducting impeller with hollow impeller shaft
➢ **Stator-Rotor-Rührsystem**
stator-rotor impeller,
Rushton-turbine impeller
➢ **Turbinenrührer**
turbine impeller
➢ **Wendelrührer**
helical ribbon impeller
➢ **zweistufiger Rührer**
two-stage impeller
Rührerblatt stirrer blade
Rührerlager (Rührwelle)
stirrer bearing
Rührerschaft/Rührerwelle
stirrer shaft
Rührerwelle impeller shaft
Rührfisch/'Fisch'/
Rührstab/Rührstäbchen/
Magnetrührstab/
Magnetrührstäbchen
stirring bar, stir bar, 'flea'
Rührgerät/Mixer stirrer, mixer
Rührhülse stirrer gland
Rührkessel/Rührbehälter
agitator vessel
Rührstab (Glasstab)
stirring rod
Rührstab/Rührstäbchen/
Magnetrührstab/
Magnetrührstäbchen/
Rührfisch/'Fisch'
stirring bar, stirrer bar, stir bar, 'flea'
Rührstabentferner/
Magnetrührstabentferner
stirring bar extractor,
stirring bar retriever,
'flea' extractor
Rührverschluss stirrer seal
Rührwelle stirrer shaft
Rührwerk impeller
Rumpfelektron
inner electron, inner-shell electron

Rundfilter round filter,
  filter paper disk, 'circles'
Rundkolben/Siedegefäß
  round-bottomed flask,
  round-bottom flask,
  boiling flask with round bottom
Rundlochplatte
  dot blot, spot blot
**Rundschüttler/
  Kreisschüttler**
  circular shaker,
  orbital shaker,
  rotary shaker
rußend/rußig
  smoking, forming soot, sooty

rutschen skid
➤ **nicht-rutschend/Antirutsch ...
  (Gerät auf Unterlage)**
  nonskid, skid-proof
rutschfest/rutschsicher
  slip resistant; nonskid,
  skid-proof, antiskid
rutschig slippery
**Rüttelbewegung
  (schnell hin und her/rauf-runter)**
  rocking motion
  (side-to-side/up-down)
rütteln shake, vibrate
Rüttelsieb shaking screen
Rüttler vibrator

S-Sätze (Sicherheitsratschläge)
S phrases (Safety phrases)
Säbelkolben/Sichelkolben
saber flask, sickle flask, sausage flask
Saccharimeter saccharimeter
Saccharose/Sucrose (Rübenzucker/Rohrzucker)
sucrose (beet sugar/cane sugar)
Sachkundiger expert, authority
Sachverständigengutachten/Expertise
expertise, expert opinion
Sachverständiger
expert, specialist, authority
Sackkammer/Sackraum (Staubabscheider)
baghouse (fabric filter dust collector)
Sackkarre
hand cart, barrow (*Br* trolley)
Säen/Aussäen/Aussaat
sowing, seed sowing
säen/aussäen/einsäen sow
saftig juicy
Säge saw
➢ **Bogensäge** coping saw
➢ **Bügelsäge** hacksaw
➢ **Gehrungssäge** miter-box saw
➢ **Handsäge** handsaw
➢ **Kettensäge** chain saw
➢ **Laubsäge (Blatt <2 mm)/ Dekupiersäge (Blatt >2 mm)**
scroll saw, jigsaw, fretsaw
➢ **Metallsäge** metal-cutting saw
➢ **Schweifsäge (Maschine)**
jigsaw machine
➢ **Stichsäge**
compass saw (with open handle), pad saw
Sägeblatt saw blade
Sägemehl sawdust
Sagittalebene (parallel zur Mittellinie)
median longitudinal plane
Sagittalschnitt
sagittal section, median longisection
Salbentopf/Medikamententopf (Apotheke) gallipot
Salicylsäure (Salicylat)
salicic acid (salicylate)

Salinität/Salzgehalt salinity, saltiness
Salmiak/Ammoniumchlorid
ammonium chloride
(sal ammoniac, salmiac)
Salmiakgeist/Ammoniumhydroxid (Ammoniaklösung)
ammonium hydroxide,
ammonia water (ammonia solution)
Salpetersäure nitric acid
salpetrige Säure nitrous acid
Salve *neuro* burst
Salz salt
➢ **Bittersalz/Magnesiumsulfat**
Epsom salts, epsomite,
magnesium sulfate
➢ **Blutlaugensalz/ Kaliumhexacyanoferrat** prussiat
➢ **Doppelsalz** double salt
➢ **Gallensalze** bile salts
➢ **Hirschhornsalz/Ammoniumcarbonat**
hartshorn salt, ammonium carbonate
➢ **Jodsalz** iodized salt
➢ **Kochsalz (NaCl)**
table salt, common salt
➢ **Komplexsalz** complex salt
➢ **Meersalz** sea salt
➢ **Mohrsches Salz**
Mohr's salt, ammonium iron(II) sulfate hexahydrate
(ferrous ammonium sulfate)
➢ **Nährsalz** nutrient salt
➢ **Steinsalz (Halit)/Kochsalz/ Tafelsalz/Natrium chlorid (NaCl)**
rock salt (halite), common salt,
table salt, sodium chloride
Salzbrücke (Ionenpaar)
salt bridge (ion pair)
salzen salt
Salzgehalt/Salzigkeit
salinity, saltiness
salzig salty, saline
Salzigkeit saltiness
Salzperlen salt beads
Salzschmelze molten salt, salt melt
Salzwasser saltwater
Samenbank seed repository
Sammelbegriff/Sammelname
generic name

Sammelbehälter/Sammelgefäß
  storage container; sump
Sammelgel *electrophor* stacking gel
Sammelglas (Behälter) specimen jar
Sammellinse *micros*
  collecting lens, focusing lens
> **parallel-richtende Sammellinse**
  collimating lens
sammeln
  collect, put/come/bring together
Sammlung/Kollektion collection
Sandfang (Kläranlage) grit chamber
Sandstrahlgebläse
  sandblasting apparatus
Sandstrahlreinigung sandblasting
sanitäre Einrichtungen
  sanitary facilities,
  sanitary installations
Sanitärzubehör sanitary
  supplies/equipment,
  plumbing supplies/equipment
Sanitäter
  first-aid attendant, nurse
Sanitätsbedarf medical supplies
Sanitätsdienst medical service
Sanitätskasten first-aid kit
Sanitätspersonal medical personnel
Sanitätswagen/Rettungswagen
  ambulance
Saprobien (Organismen)
  saprobes, saprobionts
saprogen/fäulniserregend saprogenic
satt/gesättigt
  full, having eaten enough, saturated
sättigen (gesättigt) saturate (saturated)
Sättigung saturation
> **ungesättigter Zustand**
  unsaturation
Sättigungsbereich/Sättigungszone
  range of saturation,
  zone of saturation
Sättigungshybridisierung
  saturation hybridization
Sättigungskinetik
  saturation kinetics
Sättigungsverlust/Sättigungsdefizit
  saturation deficit
Satz/Garnitur set

Satzkultur/
  diskontinuierliche Kultur/
  Batch-Kultur batch culture
Satzverfahren batch process
säubern
  clean, cleanse, tidy up; mop up
Säuberungsaktion cleanup
sauer/azid acid, acidic
säuerlich acidic
Sauerstoff (O) oxygen
Sauerstoffbedarf oxygen demand
> **biologischer S. (BSB)**
  biological oxygen demand (BOD)
> **chemischer S. (CSB)**
  chemical oxygen demand (COD)
sauerstoffbedürftig/aerob aerobic
Sauerstoffpartialdruck
  oxygen partial pressure
Sauerstoffschuld/
  Sauerstoffverlust/Sauerstoffdefizit
  oxygen debt
Sauerstofftransferrate
  oxygen transfer rate (OTR)
Sauerstoffverlust/
  Sauerstoffschuld/Sauerstoffdefizit
  oxygen debt
Säuerung acidification
Saugball/Pipettierball/Pipettierbällchen
  pipet bulb, rubber bulb
saugen *allg* suck; (aufsaugen) absorb,
  take up, soak up
saugfähig absorbent
Saugfähigkeit absorbency
Saugfiltration suction filtration
Saugflasche/Filtrierflasche
  suction flask, filter flask,
  filtering flask, vacuum flask,
  aspirator bottle
Saugfüßchen suction-cup feet
Saugheber siphon
Saugkissen (zum Aufsaugen von
  verschütteten Chemikalien)
  spill containment pillow
Saugkolbenpipette
  piston-type pipet
Saugkraft suction force
Saugluftabzug forced-draft hood
Saugnapf/Saugscheibe suction disk

Saugpapier (Löschpapier) absorbent paper, bibulous paper (for blotting dry)
Saugpipette suction pipet (patch pipet)
Saugpumpe/Vakuumpumpe aspirator pump, vacuum pump
Saugspannung suction tennsion; (Boden) soil-moisture tension
Säule pillar, column; des Mikroskops) pillar
Säulenchromatographie column chromatography
➢ **Normaldruck-S.** gravity column chromatography
Säulenfüllung/Säulenpackung *chromat* column packing
Säulenreaktor/Turmreaktor column reactor
Säulenwirkungsgrad *chromat* column efficiency
Saum/Rand seam, border, edge, fringe
Säure acid
➢ **'aktivierte Essigsäure'/Acetyl-CoA** acetyl CoA, acetyl coenzyme A
➢ **Abietinsäure** abietic acid
➢ **Acetessigsäure (Acetacetat)/ 3-Oxobuttersäure** acetoacetic acid (acetoacetate), acetylacetic acid, diacetic acid
➢ **Aconitsäure (Aconitat)** aconitic acid (aconitate)
➢ **Adenylsäure (Adenylat)** adenylic acid (adenylate)
➢ **Adipinsäure (Adipat)** adipic acid (adipate)
➢ **Akkusäure/Akkumulatorsäure** accumulator acid, storage battery acid (electrolyte)
➢ **Alginsäure (Alginat)** alginic acid (alginate)
➢ **Allantoinsäure** allantoic acid
➢ **Ameisensäure (Format)** formic acid (formate)
➢ **Aminosäure** amino acid
➢ **Anthranilsäure/2-Aminobenzoesäure** anthranilic acid, 2-aminobenzoic acid
➢ **Äpfelsäure (Malat)** malic acid (malate)
➢ **Arachidonsäure** arachidonic acid, icosatetraenoic acid
➢ **Arachinsäure/Arachidinsäure/ Eicosansäure** arachic acid, arachidic acid, icosanic acid
➢ **Ascorbinsäure (Ascorbat)** ascorbic acid (ascorbate)
➢ **Asparaginsäure (Aspartat)** asparagic acid, aspartic acid (aspartate)
➢ **Azelainsäure/Nonandisäure** azelaic acid, nonanedioic acid
➢ **Behensäure/Docosansäure** behenic acid, docosanoic acid
➢ **Benzoesäure (Benzoat)** benzoic acid (benzoate)
➢ **Bernsteinsäure (Succinat)** succinic acid (succinate)
➢ **Blausäure/Cyanwasserstoff** hydrogen cyanide, hydrocyanic acid, prussic acid
➢ **Borsäure (Borat)** boric acid (borate)
➢ **Brenztraubensäure (Pyruvat)** pyruvic acid (pyruvate)
➢ **Buttersäure/Butansäure (Butyrat)** butyric acid, butanoic acid (butyrate)
➢ **Caprinsäure/Decansäure (Caprinat/Decanat)** capric acid, decanoic acid (caprate/decanoate)
➢ **Capronsäure/Hexansäure (Capronat/Hexanat)** caproic acid, capronic acid, hexanoic acid (caproate/hexanoate)
➢ **Caprylsäure/Octansäure (Caprylat/Octanat)** caprylic acid, octanoic acid (caprylate/octanoate)
➢ **Carbonsäuren/Karbonsäuren (Carbonate/Karbonate)** carboxylic acids (carbonates)
➢ **Cerotinsäure/Hexacosansäure** cerotic acid, hexacosanoic acid
➢ **Chinasäure** chinic acid, kinic acid, quinic acid (quinate)
➢ **Chinolsäure** chinolic acid
➢ **chlorige Säure** $HClO_2$ chlorous acid
➢ **Chlorogensäure** chlorogenic acid

Sä

- Chlorsäure HClO₃ chloric acid
- Cholsäure (Cholat) cholic acid (cholate)
- Chorisminsäure (Chorismat) chorismic acid (chorismate)
- Chromschwefelsäure chromic-sulfuric acid mixture for cleaning purposes
- Cinnamonsäure/Zimtsäure (Cinnamat) cinnamic acid
- Citronensäure/Zitronensäure (Citrat/Zitrat) citric acid (citrate)
- Crotonsäure/Transbutensäure crotonic acid, α-butenic acid
- Cysteinsäure cysteic acid
- einwertige/einprotonige Säure monoprotic acid
- Eisessig glacial acetic acid
- Ellagsäure ellagic acid, gallogen
- Erucasäure/Δ¹³-Docosensäure erucic acid, (Z)-13-docosenoic acid
- Essigsäure/Ethansäure (Acetat) acetic acid, ethanoic acid (acetate)
- Ferulasäure ferulic acid
- Fettsäure (*siehe auch dort*) fatty acid
- Flechtensäure lichen acid
- Fluoroschwefelsäure/Fluorsulfonsäure fluorosulfonic acid
- Fluorwasserstoffsäure/Flusssäure hydrofluoric acid, phthoric acid
- Flusssäure/Fluorwasserstoffsäure hydrofluoric acid, phthoric acid
- Folsäure (Folat)/Pteroylglutaminsäure folic acid (folate), pteroylglutamic acid
- Fumarsäure (Fumarat) fumaric acid (fumarate)
- Galakturonsäure galacturonic acid
- Gallussäure (Gallat) gallic acid (gallate)
- gamma-Aminobuttersäure gamma-aminobutyric acid
- Gelbbrennsäure/Scheidewasser (konz. Salpetersäure) aquafortis (nitric acid used in metal etching)
- Gentisinsäure gentisic acid
- Geraniumsäure geranic acid
- Gerbsäure (Tannat) tannic acid (tannate)
- Gibberellinsäure gibberellic acid
- Glucarsäure/Zuckersäure glucaric acid, saccharic acid
- Gluconsäure (Gluconat) gluconic acid (gluconate)
- Glucuronsäure (Glukuronat) glucuronic acid (glucuronate)
- Glutaminsäure (Glutamat)/2-Aminoglutarsäure glutamic acid (glutamate), 2-aminoglutaric acid
- Glutarsäure (Glutarat) glutaric acid (glutarate)
- Glycyrrhetinsäure glycyrrhetinic acid
- Glykolsäure (Glykolat) glycolic acid (glycolate)
- Glyoxalsäure (Glyoxalat) glyoxalic acid (glyoxalate)
- Glyoxylsäure (Glyoxylat) glyoxylic acid (glyoxylate)
- Goldsäure auric acid
- Guanylsäure (Guanylat) guanylic acid (guanylate)
- Gulonsäure (Gulonat) gulonic acid (gulonate)
- Harnsäure (Urat) uric acid (urate)
- Homogentisinsäure homogentisic acid
- Huminsäure humic acid
- Hyaluronsäure hyaluronic acid
- Ibotensäure ibotenic acid
- Iminosäure imino acid
- Indolessigsäure indolyl acetic acid, indoleacetic acid (IAA)
- Iodwasserstoffsäure hydroiodic acid, hydrogen iodide
- Isovaleriansäure isovaleric acid
- Jasmonsäure jasmonic acid
- Kaffeesäure caffeic acid
- Ketosäure keto acid
- Kieselsäure silicic acid
- Kohlensäure (Karbonat/Carbonat) carbonic acid (carbonate)
- Kojisäure kojic acid

- **Laktat (Milchsäure)**
  lactate (lactic acid)
- **Laurinsäure/Dodecansäure (Laurat/Dodecanat)**
  lauric acid, decylacetic acid, dodecanoic acid (laurate/dodecanate)
- **Lävulinsäure** levulinic acid
- **Lignocerinsäure/Tetracosansäure**
  lignoceric acid, tetracosanoic acid
- **Linolensäure** linolenic acid
- **Linolsäure**
  linolic acid, linoleic acid
- **Liponsäure/Thioctsäure (Liponat)**
  lipoic acid (lipoate), thioctic acid
- **Lipoteichonsäure** lipoteichoic acid
- **Litocholsäure** litocholic acid
- **Lötsäure** soldering acid
- **Lysergsäure** lysergic acid
- **Magensäure**
  stomach acid, gastric acid
- **Magische Säure**
  magic acid (HSO_3F/SbF_5)
- **Maleinsäure (Maleat)**
  maleic acid (maleate)
- **Malonsäure (Malonat)**
  malonic acid (malonate)
- **Mandelsäure/Phenylglykolsäure**
  mandelic acid,
  phenylglycolic acid, amygdalic acid
- **Mannuronsäure** mannuronic acid
- **Mevalonsäure (Mevalonat)**
  mevalonic acid (mevalonate)
- **Milchsäure (Laktat)**
  lactic acid (lactate)
- **Muraminsäure** muramic acid
- **Myristinsäure/**
  Tetradecansäure (Myristat)
  myristic acid, tetradecanoic acid
  (myristate/tetradecanate)
- **N-Acetylmuraminsäure**
  N-acetylmuramic acid
- **Nervonsäure/Δ^{15}-Tetracosensäure**
  nervonic acid,
  (Z)-15-tetracosenoic acid,
  selacholeic acid
- **Neuraminsäure** neuraminic acid
- **Nikotinsäure (Nikotinat)**
  nicotinic acid (nicotinate), niacin
- **Ölsäure/Δ^9-Octadecensäure (Oleat)**
  oleic acid (oleate),
  (Z)-9-octadecenoic acid
- **Orotsäure** orotic acid
- **Orsellinsäure**
  orsellic acid, orsellinic acid
- **Osmiumsäure** osmic acid
- **Oxalbernsteinsäure (Oxalsuccinat)**
  oxalosuccinic acid (oxalosuccinate)
- **Oxalsäure (Oxalat)**
  oxalic acid (oxalate)
- **Oxoglutarsäure (Oxoglutarat)**
  oxoglutaric acid (oxoglutarate)
- **Palmitinsäure/Hexadecansäure (Palmat/Hexadecanat)**
  palmitic acid, hexadecanoic acid
  (palmate/hexadecanate)
- **Palmitoleinsäure/**
  Δ^9-Hexadecensäure
  palmitoleic acid,
  (Z)-9-hexadecenoic acid
- **Pantoinsäure** pantoic acid
- **Pantothensäure (Pantothenat)**
  pantothenic acid (pantothenate)
- **Pektinsäure (Pektat)** pectic acid (pectate)
- **Penicillansäure** penicillanic acid
- **Perameisensäure** performic acid
- **Perchlorsäure** perchloric acid
- **Phosphatidsäure** phosphatidic acid
- **phosphorige Säure** phosphorous acid
- **Phosphorsäure (Phosphat)**
  phosphoric acid (phosphate)
- **Phthalsäure** phthalic acid
- **Phytansäure** phytanic acid
- **Phytinsäure** phytic acid
- **Pikrinsäure (Pikrat)**
  picric acid (picrate)
- **Pimelinsäure** pimelic acid
- **Plasmensäure** plasmenic acid
- **Prephensäure (Prephenat)**
  prephenic acid (prephenate)
- **Propionsäure (Propionat)**
  propionic acid (propionate)
- **Prostansäure** prostanoic acid
- **Pyrethrinsäure** pyrethric acid
- **rauchend** fuming
- **Retinsäure** retinic acid

- Salicylsäure (Salicylat) salicic acid (salicylate)
- Salpetersäure nitric acid
- Salpetrige Säure nitrous acid
- Salzsäure/Chlorwasserstoffsäure hydrochloric acid
- Schleimsäure/Mucinsäure mucic acid
- Schwefelsäure sulfuric acid
- Schweflige Säure/Schwefligsäure sulfurous acid
- Shikimisäure (Shikimat) shikimic acid (shikimate)
- Sialinsäure (Sialat) sialic acid (sialate)
- Sinapinsäure sinapic acid
- Sorbinsäure (Sorbat) sorbic acid (sorbate)
- Stearinsäure/Octadecansäure (Stearat/Octadecanat) stearic acid, octadecanoic acid (stearate/octadecanate)
- Suberinsäure/Korksäure/Octandisäure suberic acid, octanedioic acid
- Sulfanilsäure sulfanilic acid, p-aminobenzenesulfonic acid
- Supersäure superacid
- Teichonsäure teichoic acid
- Teichuronsäure teichuronic acid
- Uridylsäure uridylic acid
- Urocaninsäure (Urocaninat)/Imidazol-4-acrylsäure urocanic acid (urocaninate)
- Uronsäure (Urat) uronic acid (urate)
- Usninsäure usnic acid
- Valeriansäure/Pentansäure (Valeriat/Pentanat) valeric acid, pentanoic acid (valeriate/pentanoate)
- Vanillinsäure vanillic acid
- Weinsäure (Tartrat) tartaric acid (tartrate)
- Zimtsäure/Cinnamonsäure (Cinnamat) cinnamic acid
- Zitronensäure/Citronensäure (Zitrat/Citrat) citric acid (citrate)
- Zuckersäure/Aldarsäure (Glucarsäure) saccharic acid, aldaric acid (glucaric acid)
- zweiwertige/zweiprotonige Säure diprotic acid

Säureamid acid amide
Säure-Basen-Gleichgewicht acid-base balance
Säure-Basen-Titration/Neutralisationstitration acid-base titration
Säurebehandlung acid treatment
säurebeständig acid-proof, acid-fast
säurebildend/säurehaltig acidic
Säurebildung acidification
Säureester acid ester
säurefest acid-fast
Säurefestigkeit acid-fastness
Säuregrad/Säuregehalt/Azidität acidity
Säurenkappenflasche acid bottle (with pennyhead stopper)
saurer Regen/Niederschlag acid rain, acid deposition
Säureschrank acid storage cabinet
Säureschutzhandschuhe acid gloves, acid-resistant gloves
Säureverätzung acid burn
Scatchard-Diagramm Scatchard plot
schaben scrape
Schabepräparat *micros* scraping (mount)
Schaber scraper
Schablone *tech* template; (Zeichenschablone für Formeln etc.) stencil
Schachtel box
Schaden damage
schadhaft defective
Schädigung damage
Schädigungskurve *ecol* damage response curve
Schadinsekt pest insect
schädlich harmful, causing damage, damaging
- unschädlich harmless, not harmful; inactive
Schädling(e)/Ungeziefer pest(s)
Schädlingsbefall pest infestation

**Schädlingsbekämpfung/
Schädlingskontrolle** pest control
➢ **biologische Schädlingsbekämpfung**
biological pest control
➢ **integrierte Schädlingsbekämpfung/
integrierter Pflanzenschutz**
integrated pest management (IPM)
**Schädlingsbekämpfungsmittel/
Pestizid/Biozid**
pesticide, biocide
**Schädlingsbekämpfungsmittel-
resistenz/Pestizidresistenz**
pesticide resistance
Schadorganismus
harmful organism, harmful lifeform
Schadstoff pollutant,
harmful substance, contaminant
Schadstoffbelastung pollution level
Schaft (Griff) handle; (Welle) shaft
Schäkel shakle
Schale *allg* shell; husk, coat, cover;
bowl; (Flachbehälter) tray
Schall (Geräusch) sound;
(Widerhall) resonance, echo,
reverberation
Schallwellen sound waves
Schaltanlage switchboard
Schalter switch
Schalthebel
control lever; *electr* switch lever
Schaltkreis/Schaltsystem *neuro*
circuit (neural circuit)
Schalttafel control panel, switchboard
Schamotte fireclay
scharf *micro/photo* in focus, sharp
➢ **unscharf** *micro/photo* not in focus,
out of focus, blurred
**scharfe Gegenstände
(scharfkantige/spitze G.)** sharps
Schärfe *micro/photo* sharpness, focus
➢ **Sehschärfe** visual acuity
➢ **Unschärfe** *micro/photo*
blurredness, blur, obscurity,
unsharpness
Scharfeinstellung focussing
schärfen (Messer/Scheren) sharpen
Schärfentiefe/Tiefenschärfe
depth of focus, depth of field

Scharfstellung/Akkommodation
opt accommodation
Scharnier/Schloss/Schlossleiste
hinge
Schatten *allg* shade;
(eines bestimmten Gegenstandes)
shadow
schattieren shade
schattig shady
schätzen/annehmen estimate, assume
Schätzfehler *stat* error of estimation
Schätzung/Annahme
estimate, estimation, assumption
Schätzverfahren *stat*
method of estimation
Schätzwert estimate
Schaufel shovel; scoop
➢ **Kehrschaufel/Kehrblech** dustpan
➢ **Messschaufel** measuring scoop
➢ **Radschaufel** paddle, vane
➢ **Turbinenschaufel** blade, bucket
Schaufelmischer blade mixer
Schaufelrad/Laufrad
paddle wheel,
bucket wheel, blade wheel
Schaukasten/Vitrine showcase
Schaukelbewegung
see-saw motion
Schaukelvektor/bifunktionaler Vektor
shuttle vector, bifunctional vector
Schaum foam;
froth (fein: auf Flüssigkeit);
lather (Seifenschaum)
**Schaumbrecher-Aufsatz/
Spritzschutz-Aufsatz
(Rückschlagsicherung)** *dest*
antisplash adapter,
splash-head adapter
**Schaumdämpfer/
Schaumverhütungsmittel**
antifoaming agent,
defoamer, foam inhibitor;
(Gerät) antifoam controller
schäumen foam; lather
Schäumer/Schaumbildner
foamer, foaming agent
Schaumgummi
foam rubber, plastic foam, foam

Schaumhemmer *chem/lab*
 anti-foaming agent
Schaumlöscher
 foam fire extinguisher
Schaumstoff
 foamed plastic, plastic foam
Scheibe disk, disc (*Br*);
 (Platte) plate, saucer
➢ **Berstscheibe/Sprengscheibe/ Sprengring/Bruchplatte**
 bursting disk
➢ **Frontscheibe (Sicherheitswerkbank)**
 sash
➢ **Fensterscheibe/Glasscheibe**
 pane
➢ **Schutzscheibe/Schutzschirm**
 protective screen/shield, workshield
➢ **Sichtscheibe**
 viewing window
➢ **Unterlegscheibe** washer
➢ **Wählscheibe/Einstellscheibe**
 dial
scheibenförmig disk-shaped
Scheibenmühle plate mill
Scheibenversprüher disk atomizer
Scheide/Umhüllung sheath
scheiden/trennen/abtrennen separate
scheidenförmig sheathed
Scheidetrichter *lab* separatory funnel
Scheidewand/Septe/Septum
 dividing wall, cross-wall,
 partition, dissepiment, septum
Scheidewasser/Gelbbrennsäure
 (konz. Salpetersäure) aquafortis
 (nitric acid used in metal etching)
Scheidung/Trennung separation
Scheitelpunkt apex,
 peak (highest among other high points), vertex, summit
Scheitelwert/Höchstwert/Maximum
 peak value, maximum (value)
Schelle/Klemme
 clip, clamp, band clamp
Schenkel *chem/biochem/immun* arm
Schere scissors
➢ **Blechschere**
 sheet-metal shears, plate shears
➢ **chirurgische Schere** surgical scissors

➢ **Drahtschere**
 wire shears, wire cutter
➢ **Drahtseilschere/Kabelschere**
 wire cable shears, cable shears
➢ **Irisschere/Listerschere** iris scissors
➢ **Präparierschere** dissecting scissors
➢ **spitze Schere** sharp point scissors
➢ **stumpfe Schere** blunt point scissors
➢ **Verbandsschere** bandage scissors
scheren shear, cut, clip,
Scherfestigkeit/Schubfestigkeit (Holz)
 shear strength, shearing strength
Schergefälle/Schergradient
 shear gradient
Scherkraft shear force; shear stress
 (shear force per unit area)
Scherrate shear rate, rate of shear
Scherspannung shear stress
 (shear force per unit area)
Scheuerbürste (Schrubbbürste)
 scrubbing brush
Scheuermittel scouring agent, abrasive
scheuern scrub, scour; (reiben) rub
Schicht layer, story, stratum, sheet
Schichtenbildung stratification
 (act/process of stratifying)
Schichtung stratification
 (state of being stratified),
 layering
Schiebefenster/Frontschieber/ verschiebbare Sichtscheibe (Abzug/Werkbank)
 sash (> hood)
Schieberventil slide valve
schief oblique
Schiene *med* splint
Schießbaumwolle nitrocotton,
 guncotton (12.4–13% N);
 pyroxylin (11.2–12.4% N)
Schießofen
 Carius furnace, bomb furnace,
 bomb oven, tube furnace
Schießpulver gunpowder
Schießrohr/Bombenrohr/ Einschlussrohr
 bomb tube, Carius tube
Schießstoff/Schießmittel
 low explosive

Schild (Schutz~)
shield, screen (protective ~)
schillern opalesce
Schirm/Blende (Sicht~) visor
schlachten slaughter, butcher
Schlachthof slaughterhouse
Schlacke *tech/metall/geol*
cinders, slag, dross, scoria
schlaff (welk) limp
Schlagbohrer percussion drill
schlagen/hauen beat, hit, strike
Schlamm/Aufschlämmung
slurry
Schlangenkühler
coil condenser,
coil distillate condenser,
coiled-tube condenser,
spiral condenser
Schlauch
tube, tubing; hose
➤ **Gartenschlauch** garden hose
➤ **Hochdruckschlauch**
high-pressure tubing
(mit größerem Durchmesser: hose)
Schlauchklemme/Quetschhahn
tubing clamp, pinch clamp,
pinchcock clamp;
(Schlauchklemme: Installationen zur
Schlauchbefestigung) hose clamp,
hose connector clamp
Schlauchkupplung
tubing connection, tube coupling
Schlauchpumpe tubing pump
Schlauch-Rohr-Verbindungsstück
pipe-to-tubing adapter
Schlauchschelle
tube clip, hose clip;
(Abrutschsicherung/Befestigung an
Verbindungsstück)
hose/tubing bundle
Schlauchsperre
(tube) compressor clamp
Schlauchtülle
(z.B. am Gasreduzierventil)
tubing/hose attachment socket,
tubing/hose connection gland
Schlauchventil (Klemmventil)
(tubing) pinch valve

Schlauchverbinder/
Schlauchverbindung(en)
tubing connector (for connecting
tubes), tube coupling, fittings
Schlauchverschlussklemme *dial*
tubing closure
Schlaufe *tech/gen/biochem* loop
Schlaufenradreaktor
paddle wheel reactor
Schlaufenreaktor/Umlaufreaktor
loop reactor, circulating reactor,
recycle reactor
Schleifenkonformation/
Knäuelkonformation
loop conformation, coil conformation
Schleifer/Schleifmaschine
grinder, grinding machine
Schleifstein/Abziehstein
sharpening stone,
grindstone, honing stone
Schleim mucus, slime, ooze;
mucilage (speziell pflanzlich)
Schleimhautreizung
irritation of the mucosa
schleimig
slimy, mucilaginous, glutinous
Schleimsäure/Mucinsäure mucic acid
Schlempe dried distillers' solubles
Schleppdampfdestillation
distillation by steam entrainment
Schleppgas *chromat* carrier gas
Schleppmittel
(Gas/Flüssigkeit) *chromat* carrier;
dest entrainer, separating agent
Schleuder *tech* spinner;
(Zentrifuge) centrifuge
schleudern *tech* spin;
(zentrifugieren) centrifuge
Schleuse sluice
➤ **Luftschleuse** airlock
schleusen sluice, channel
Schliere streak, ream, striation
Schlierenbildung
streak formation, streaking, striation
schlierenfrei
free from streaks, free from reams
schlierig streaky, streaked
Schließfach locker

Schliff ground joint
> **festgebackener Schliff**
jammed joint, stuck joint,
caked joint, 'frozen' joint
> **Kegelschliff (N.S. = Normalschliff)**
ground-glass joint,
tapered ground joint
(S.T. = standard taper)
> **Kugelschliff**
spherical ground joint
> **Planschliff (glatte Enden)**
flat-flange ground joint,
flat-ground joint,
plane-ground joint
Schliff-Fett
lubricant for ground joints
Schliffgerät
ground-glass equipment
Schliffhülse
('Futteral'/Einsteckstutzen)
socket, ground socket,
ground-glass socket
(female: ground-glass joint)
Schliffkern (Steckerteil)
cone, ground cone, ground-glass cone
(male: ground-glass joint)
Schliffklammer/Schliffklemme
(Schliffsicherung)
joint clip, ground-joint clip,
ground-joint clamp
Schliffkolben ground-jointed flask
Schliffkugel ball (male: spherical joint)
Schliffpfanne
socket (female: spherical joint)
Schliffstopfen ground-glass stopper,
ground-in stopper, ground stopper
Schliffverbindung/
Glasschliffverbindung
ground joint, ground-glass joint;
(Kegelschliffverbindung)
tapered joint
> **Manschette** sleeve, joint sleeve
schlimmster anzunehmender Fall
worst-case scenario
Schlittenmikrotom
sliding microtome
Schlitzlochplatte slot blot
Schloss (Verschluss) lock

Schlosserhammer
fitter's hammer,
locksmith's hammer
Schlosserhammer mit Kugelfinne
ball pane hammer,
ball peen hammer,
ball pein hammer
Schloss-Schlüssel-Prinzip
lock-and-key principle
Schlucken *n* swallowing
schlucken *vb* swallow
Schlüssel key;
(Schrauben~/Schraubschlüssel)
wrench, spanner (*Br*)
> **Drehmomentschlüssel**
torque wrench
> **Gabelschlüssel/Maulschlüssel**
open-end wrench,
open-end spanner (*Br*)
> **Inbusschlüssel** Allen wrench
> **Kreuzschlüssel** spider wrench
> **Ringschlüssel**
ring spanner wrench,
box wrench, box/ring spanner (*Br*)
> **Sechskant-Steckschlüssel**
hex nutdriver
> **Sechskant-Stiftschlüssel**
hex socket wrench
> **Steckschlüssel**
nutdriver (wrench or screwdriver)
> **Stiftschlüssel**
socket wrench, box spanner
Schlüssel-Schloss-Prinzip/
Schloss-Schlüssel-Prinzip
lock-and-key principle
Schlussventil cutoff valve
Schmalz/Schweineschmalz/
Schweinefett lard
schmecken taste
schmelzbar fusible
Schmelzdraht *electr* fusible wire
Schmelze melt
Schmelzelektrolyse/
Schmelzflusselektrolyse
molten-salt electrolysis
schmelzen/aufschmelzen *chem/gen*
melt
schmelzflüssig fused, fusible, molten

Schmelzkurve *chem* melting curve
Schmelzling ingot (zone melting)
Schmelzmittel flux, fluxing agent
Schmelzofen
  melting furnace, smelting furnace
Schmelzorgan enamel organ
Schmelzplombe *electr* fusible plug
Schmelzpunkt *chem* melting point
Schmelztemperatur
  melting temperature
Schmelztiegel crucible
Schmelzwasser meltwater
Schmerz pain
Schmerzempfindlichkeit
  sensitivity to pain
schmerzen hurt, be painful
Schmerzgefühl pain sensation
schmerzhaft painful
schmieren lubricate, grease, oil
Schmierfett/Schmiere
  grease, lubricating grease
➢ **Apiezonfett** apiezon grease
➢ **Silikon-Schmierfett** silicone grease
Schmierinfektion smear infection
Schmiermittel/Schmierstoff/Schmiere
  lubricant
Schmieröl lubricating oil, lube oil
Schmierseife soft soap
Schmierung lubrication
Schmirgel emery
Schmirgelleinen emery cloth
Schmirgelpapier
  sandpaper, emery paper (*Br*)
Schmorpfanne/Kasserole stewpan
Schmutz/Dreck dirt, filth
schmutzig/dreckig dirty, filthy
Schmutzstoffe pollutants
Schnalle buckle
Schnappdeckel/Schnappverschluss
  snap cap, push-on cap
Schnappdeckelglas/
Schnappdeckelgläschen
  snap-cap bottle, snap-cap vial
Schnappriegel/Schnappschloss
  latch
Schnauze/Mundstück spout
Schneckenantriebspumpe
  progressing cavity pump

Schneckenbohrer/
Windenschneckenbohrer
  gimlet bit
Schneckengetriebe worm gear
Schneckengewinde worm thread
Schneebesen whisk
Schneide (Grat: Messer etc.)
  edge, cutting edge (of blade etc.)
Schneidmühle cutting-grinding mill,
  shearing machine
Schneidbrenner cutting torch
schneiden cut
Schneidwerkzeug cutting tool
Schnellfärbung *micros* quick-stain
Schnellgefrieren rapid freezing
Schnellkupplung
(z.B. Schlauchverbinder)
  quick-disconnect fitting
Schnellscan-Detektor
  fast-scanning detector (FSD),
  fast-scan analyzer
Schnellspannverschluss
  quick-release clamp (seal)
Schnellverbindung
(Rohr/Glas/Schläuche etc.)
  quick-fit connection
Schnellverdampfer (GC)
  flash vaporizer
schnellwachsend
  fast-growing, rapid-growing
Schnitt cut; section
➢ **Dünnschnitt** thin section
➢ **Gefrierschnitt** frozen section
➢ **Hirnschnitt/Querschnitt**
  transverse section, cross section
➢ **Querschnitt** cross section
➢ **Sagittalschnitt**
(parallel zur Mittelebene)
  sagittal section, median longisection
➢ **Schnellschnitt** quick section
➢ **Semidünnschnitt** semithin section
➢ **Serienschnitte** *micros/anat*
  serial sections
➢ **Ultradünnschnitt** ultrathin section
Schnittdicke thickness of section,
  section thickness
Schnittfläche/Schnittebene
  cutting face, cutting plane

schnittig/geschnitten/eingeschnitten
cut, incised
Schnittstelle *electr* interface
Schnittverletzung *med* cut, incision
Schnittwunde *med*
cut, incision; slash wound
Schnur string
Schockgefrieren shock freezing
Schockwelle/Stoßwelle shock wave
Schokoladenagar/Kochblutagar
chocolate agar
schonend gentle, mild; careful
Schöpfer/Schöpfgefäß/Schöpflöffel
dipper, scoop
Schöpfkelle ladle
Schorf (Wundschorf)/Grind *zool/med*
scab
schorfig/Schorf...
scurfy, scabby, furfuraceous
Schorfwunde
scab lesion (crustlike disease lesion)
Schornstein stack, smokestack
➤ **Abzugschornstein** exhaust stack
Schrägkultur (Schrägagar) *micb*
slant culture
Schrank cabinet, cupboard
Schraubdeckel/Schraubkappe
screw-cap, screwtop
Schraubdeckelgläschen
screw-cap vial
Schraube screw
➤ **Bügelmessschraube**
outside micrometer
➤ **Daumenschraube** thumbscrew
➤ **Einstellschraube**
adjustment screw; tuning screw
➤ **Feinjustierschraube/Feintrieb** *micros*
fine adjustment knob
➤ **Flügelschraube** thumbscrew
➤ **Grobjustierschraube/
Grobtrieb** *micros*
coarse adjustment knob
➤ **Inbusschraube**
socket screw, socket-head screw
➤ **Justierschraube/
Justierknopf/Triebknopf** *micros*
adjustment knob,
focus adjustment knob
➤ **Mikrometerschraube** *micros*
micrometer screw, fine-adjustment,
fine-adjustment knob
➤ **Rändelschraube**
knurled screw, knurled thumbscrew
➤ **Stellschraube**
adjusting screw, setting screw,
adjustment knob, fixing screw
Schraube/Spirale/Helix spiral, helix
Schraubenbolzen/Bolzen bolt
Schraubenschlüssel
wrench, screw wrench; spanner (*Br*)
➤ **Engländer/Rollgabelschlüssel**
adjustable wrench
➤ **Gabelschlüssel/Maulschlüssel**
open-end wrench,
open-end spanner (*Br*)
➤ **Kreuzschlüssel**
spider wrench, spider spanner (*Br*)
➤ **Ringschlüssel** ring spanner wrench,
box wrench, box/ring spanner (*Br*)
➤ **Sechskant-Steckschlüssel**
hex nutdriver (wrench)
➤ **Sechskant-Stiftschlüssel**
hex socket wrench
➤ **Stiftschlüssel**
socket wrench, box spanner
Schraubenzieher/Schraubendreher
screwdriver
➤ **Akkuschrauber** cordless screwdriver
➤ **Elektroschrauber** power screwdriver
➤ **Kreuzschraubenzieher/
Kreuzschlitzschraubenzieher**
Phillips®-head screwdriver;
Phillips® screwdriver
➤ **Schlitzschraubenzieher**
slotted screwdriver
➤ **Sechskantschraubenzieher**
hexagonal screwdriver,
hex screwdriver
➤ **Uhrmacherschraubenzieher**
watchmaker's screwdriver,
jeweler's screwdriver
➤ **Winkelschrauber** offset screwdriver
Schraubflasche screw-cap bottle
Schraubgewinde screw thread
Schraubgewindeverschluss
threaded top

Schraubgläschen
screw-cap vial, screw-cap jar
schraubig/spiralig/helical spiraled, helical, spirally twisted, contorted
Schraubkappe/
Schraubkappenverschluss
screw-cap, screw cap, screwtop
Schraubklemme
screw clam, pinch clamp
Schraubstock vise, vice (*Br*)
Schraubverschluss/Schraubdeckel
screwtop (threaded top)
Schraubzwinge screw clamp
Schreckstoff/Abschreckstoff deterrent, repellent
Schreckstoff/
Alarmstoff/Alarm-Pheromon
alarm substance, alarm pheromone
Schreiber (Gerät zur Aufzeichnung)
recorder; plotter
Schreibkraft secretarial help, secretarial assistant, typist
Schreibpapier bond paper
Schreibwaren stationery
Schreiner carpenter
Schrittmacher
pacemaker (*siehe:* Sinusknoten)
Schrumpffolie
shrink film, shrink wrap, shrink foil, shrinking foil
Schub *aer* thrust
Schubfestigkeit/Scherfestigkeit (Holz)
shear strength, shearing strength
Schubkraft/Vortriebkraft
thrust, forward thrust
Schublehre
slide caliper, caliper square
Schulung/Fortbildung training
Schüttdichte
bulk density (powder density)
Schüttelbad
shaking water bath, water bath shaker
Schüttelflasche/Schüttelkolben
shaker bottle, shake flask
Schüttelkultur *micb* shake culture
Schütteln shaking
schütteln *vb* shake

Schüttelwasserbad
shaking water bath, water bath shaker
schütten pour; (vollschütten) fill; (verschütten) spill; (ausschütten) pour out, empty out
Schüttgut bulk goods
Schüttgutbehälter bulk container
Schüttler shaker
➤ **Drehschüttler (rotierend)**
shaker with spinning/rotating motion
➤ **Federklammer (für Kolben)**
(four-prong) flask clamp
➤ **Inkubationsschüttler**
shaking incubator, incubating shaker, incubator shaker
➤ **Kreisschüttler/Rundschüttler**
circular shaker, orbital shaker, rotary shaker
➤ **Reziprokschüttler/**
Horizontalschüttler/
Hin- und Herschüttler (rütteln)
reciprocating shaker (side-to-side motion)
➤ **Rundschüttler/Kreisschüttler**
circular shaker, orbital shaker, rotary shaker
➤ **Rüttler (hin und her/rauf-runter)**
rocker, rocking shaker (side-to-side/up-down)
➤ **Taumelschüttler**
nutator, nutating mixer, 'belly dancer' (shaker with gyroscopic, i.e., threedimensional circular/orbital & rocking motion)
➤ **Überkopfmischer**
mixer/shaker with spinning/rotating motion (vertically rotating 360°)
➤ **Vortexmischer/**
Vortexschüttler/Vortexer
vortex shaker, vortex
➤ **Wippschüttler**
rocking shaker (see-saw motion)
Schüttung filling
Schüttvolumen bulk volume
Schutz
protection; cover; screen, shield

Schutzanzug (Ganzkörperanzug)
coverall (one-piece suit),
boilersuit, protective suit
Schutzbelag protective covering
Schutzbrille
(einfach) safety spectacles;
(ringsum geschlossen) goggles,
safety goggles
Schutzcreme (Gewebeschutzsalbe/ Arbeitsschutzsalbe)
barrier cream
schützen protect
Schutzgas
protective gas, shielding gas
(in welding)
Schutzglas/Sicherheitsglas safety glass
Schutzgruppe (*chem* Synthese)
protective group, protecting group
Schutzhandschuhe protective gloves
Schutzhaube protective hood
Schutzhelm safety helmet; hard hat
Schutzimpfung protective
immunization, vaccination
Schutzkittel/Schutzmantel
protective coat, protective gown
Schutzkleidung protective clothing
Schutzmaßnahme
protective/precautionary measure
Schutzring/Stoßschutz
(Prellschutz für Messzylinder)
bumper guard
Schutzsäule/Vorsäule
guard column, precolumn
Schutzscheibe/
Schutzschirm/Schutzschild
protective screen/shield, workshield
Schutzversuch/Schutzexperiment
protection assay,
protection experiment
Schutzvorhang protective curtain
Schutzvorrichtung
guard, protective device
Schwalbenschwanzbrenner/
Schlitzaufsatz für Brenner
wing-tip (for burner),
burner wing top
Schwalbenschwanzverbindung *micros*
dovetail connection

Schwammstopfen sponge stopper
Schwanenhals gooseneck
Schwanenhalskolben
swan-necked flask,
S-necked flask, gooseneck flask
schwanken (fluktuieren) fluctuate;
(variieren) variate
Schwankung (Fluktuation) fluctuation;
(Variation) variation
Schwanz (z.B. des Fettmoleküls) tail
Schwanzbildung/Signalnachlauf
chromat tailing
Schwebedichte/Schwimmdichte
buoyant density
schweben (schwebend)
float (floating), suspend (suspended)
Schwebeteilchen suspended particle
Schwebstoff(e)
suspended substance,
suspended matter
Schwefel (S) sulfur
Schwefelbakterien sulfur bacteria
Schwefelblüte flowers of sulfur
schwefelhaltig
sulfurous, sulfur-containing
Schwefelkies pyrite
Schwefelkreislauf sulfur cycle
schwefeln (z.B. Fässer)
sulfurize (e.g. vats)
Schwefeln/Schwefelung (z.B. Fässer)
sulfuring (e.g. vats)
Schwefelsäure H_2SO_4 sulfuric acid
Schwefelverbindung/
schwefelhaltige Verbindung
sulfur compound
Schwefelwasserstoff H_2S
hydrogen sulfide
schweflig sulfurous
schweflige Säure/
Schwefligsäure H_2SO_3
sulfurous acid
Schweiß sweat, perspiration
Schwelen/Schwelung
smoldering, smouldering
➤ **Verschwelung** carbonization
Schwelle (z.B. Reizschwelle/
Geschmacksschwelle etc.)
threshold

schwellen/anschwellen (turgeszent)
swell (turgescent)
Schwelleneffekt threshold effect
Schwellenkonzentration
threshold concentration
Schwellenmerkmal threshold trait
Schwellenwert threshold value
Schwellung
swelling; (Turgeszenz) turgescence
Schwellungsgrad turgidity
schwenken (Flüssigkeit in Kolben)
swirl
Schwenkrollen/Lenkrollen/
Schwenkrollfüße swivel casters
Schwerefeld gravitational field
Schwerelosigkeit weightlessness
Schweresinn gravitational sense
schwerflüchtig nonvolatile
➤ **flüchtig** volatile
schwergewicht *adj/adv* heavyweight
Schwerkraft gravity, gravitational force
schwerlöslich of low solubility
Schwermetall heavy metal
Schwermetallbelastung
heavy metal contamination
Schwermetallvergiftung
heavy metal poisoning
Schwimmdichte/Schwebedichte
buoyant density
Schwimmer (z.B. am
Flüssigkeitsstandregler)
float
Schwimmerschalter float switch
Schwimmständer/
Schwimmgestell/
Schwimmer (für Eiswanne)
floating rack
Schwingphase
swing phase, suspension phase
Schwingung oscillation, vibration
➤ **Deformationsschwingung (IR)**
deformation vibration,
bending vibration
➤ **Oberschwingung (IR)** overtone
➤ **Streckschwingung**
stretching vibration
➤ **Wippschwingung (IR)**
wagging vibration

Schwingungsbewegung
vibrational motion
Schwingungsspektrum
vibrational spectrum
Schwitzen *n*
sweating, perspiration, hidrosis
schwitzen *vb* sweat, perspire
Sechskant-Steckschlüssel
hex nutdriver
Sechskant-Stiftschlüssel
hex socket wrench
Sechskantstopfen
hex-head stopper, hexagonal stopper
Sediment sediment;
centrif (Pellet) pellet
Sedimentationsgeschwindigkeits-
analyse *biochem*
sedimentation analysis
Sedimentationskoeffizient
sedimentation coefficient
segmentieren segment
Segmentierung segmentation
Segregation/Aufspaltung segregation
segregieren/aufspalten segregate
Sehen *n* seeing, vision
sehen/anschauen/erblicken *vb*
see, view
Sehfeld/Blickfeld/Gesichtsfeld
field of view, scope of view,
field of vision, range of vision,
visual field
Sehfeldblende/Gesichtsfeldblende
field stop (a field diaphragm)
➤ **Gesichtsfeldblende des**
Okulars/Okularblende
ocular diaphragm, eyepiece
diaphragm, eyepiece field stop
Sehkraft/Sehvermögen eyesight
Sehschärfe visual acuity
Sehvermögen vision, sight; eyesight;
(Sehstärke) strength of vision
Sehweite
range of vision, visual distance
Seide silk (fibroin/sericin)
seiden/Seiden... silken
seidenartig/seidenhaarig/seidig
silky, sericeous, sericate
Seidenfaden silk suture

Seife soap
> **ein Stück Seife** a bar of soap
> **Flüssigseife**
liquid soap, liquid detergent
> **Kernseife (fest)**
curd soap (domestic soap)
> **Schmierseife** soft soap
Seifenschaum/Seifenwasser suds
Seifenspender (Flüssigseife)
soap dispenser (liquid soap)
Seiher/Abtropfsieb colander
Seil rope
Seilklampe rope cleat
Seilrolle pulley
Seilwinde
winch (for rope/cable/chains etc.)
Seitenachse lateral axis, lateral branch
Seitenarm/Tubus (Kolben etc.)
sidearm, tubulation
Seitenkette *chem* side chain
Seitenschneider
diagonal cutter, diagonal pliers,
diagonal cutting nippers
seitlich/lateral lateral
Sekret secretion
Sekretär(in) secretary
Sekretariat secretary's office, office
Sekretion secretion
sekretorisch secretory
Sekundärinfekt/Sekundärinfektion
secondary infection
selbstabgleichend self-balancing
Selbstassoziierung/
  Selbstzusammenbau/
  spontaner Zusammenbau
  (molekulare Epigenese)
self-assembly
selbstdichtend self-sealing
selbstentzündlich
spontaneously ignitable,
self-ignitable, autoignitable
Selbstentzündung spontaneous ignition,
self-ignition, autoignition
Selbstentzündungstemperatur
spontaneous ignition temperature
(SIT)
selbsthärtend (Harze/Polymere)
self-curing

selbstklebend self-adhesive,
self-adhering, gummed
Selbstmord-Substrat suicide substrate
Selbstorganisation self-organization
selbstregulierend/selbsteinstellend
self-regulating, self-adjusting
Selbstreinigung self-cleansing
Selbstschutz self-protection
Selbsttoleranz/Eigentoleranz
self-tolerance
selbstverschließend self-locking
Selbstzersetzung
spontaneous decomposition,
autodecomposition
selbstzündend self-igniting
Selbstzusammenbau/
  Spontanzusammenbau/
  Selbstassoziierung/
  spontaner Zusammenbau
  (molekulare Epigenese)
self-assembly
selektieren/auslesen select
Selektionsdruck selective pressure,
selection pressure
Selektionsnachteil
selective disadvantage
Selektionsvorteil selective advantage
Selektionswert/Selektionskoeffizient
selection coefficient,
coefficient of selection
selektiv selective
Selektivität selectivity
Selen (Se) selenium
selten/rar scarce, rare
Seltenheit/Rarität scarcity, rarity
Semidünnschnitt semithin section
semikonservative Replikation
semiconservative replication
Senföl mustard oil
sengen singe
Sengen *n* singeing
Senke/Verbrauchsort
  (von Assimilaten)
sink (importer of assimilates)
Senkgrube/Sickergrube sump, cesspit,
cesspool, soakaway (*Br*)
sensibilisieren sensitize
Sensibilisierung sensitization

Sensitivität/Empfindlichkeit sensitivity
sensorisch sensory
Sepsis/Septikämie/Blutvergiftung
 sepsis, septicemia, blood poisoning
Septum (*pl* **Septen**)
 septum (*pl* septa or septums)
sequentielle Reaktion/Kettenreaktion
 sequential reaction, chain reaction
Sequenz sequence
Sequenzierungsautomat *gen* sequencer
Serienschnitte serial sections
Serologie serology
serologisch serologic(al)
serös serous
Serum (*pl* **Seren**)
 serum (*pl* sera or serums)
Servierwagen
 service cart, service trolley (*Br*)
Sesselform (Cycloalkane) *chem* chair conformation
Seuche/Epidemie epidemic
sexuell übertragbare Krankheit/
 Geschlechtskrankheit/
 venerische Krankheit
 sexually transmitted disease (STD), venereal disease (VD)
sezernieren/abgeben (Flüssigkeit)
 secrete (excrete)
Sezierbesteck
 dissection equipment (dissecting set)
sezieren dissect
Seziernadel
 dissecting needle (teasing needle);
 (Stecknadel) dissecting pin
Sezierpinzette dissecting forceps
Sezierschere dissecting scissors
Sezierung dissection
Shikimisäure (Shikimat)
 shikimic acid (shikimate)
Sialinsäure (Sialat) sialic acid (sialate)
sicher *tech* safe;
 (*personal protection*) secure
sicherer Umgang safe handling
Sicherheit *tech* safety;
 (*personal protection*) security
 ➢ **erhöhte S.** increased safety
Sicherheitbestimmungen
 safety regulations

Sicherheitsbeauftragter
 safety officer
 ➢ **biologischer S./**
 Beauftragter für biol. Sicherheit
 biosafety officer
Sicherheitsbehälter
 (Abfallbox zur Entsorgung von Nadeln/Skalpellklingen/Glas etc.) sharps collector;
 (Sicherheitskanne) safety vessel, safety container, safety can
Sicherheitsdaten safety data
Sicherheitsdatenblatt safety data sheet;
 U.S.: Material Safety Data Sheet (MSDS)
Sicherheitsglas safety glass
Sicherheitsingenieur safety engineer
Sicherheitskennzeichnung
 safety labeling
Sicherheitsmaßnahmen/
Sicherheitsmaßregeln
 security measures,
 safety measures, containment
 ➢ **biologische S.**
 biological containment
 ➢ **physikalische/technische S.**
 physical containment
Sicherheitsmerkmal safety feature
Sicherheitspersonal security personnel, security
Sicherheitsraum/Sicherheitsbereich/
Sicherheitslabor (S1-S4)
 biohazard containment (laboratory)
 (classified into biosafety containment classes)
Sicherheitsrichtlinien safety guidelines
Sicherheitsrohr (Laborglas) guard tube
Sicherheitsschrank safety cabinet
Sicherheitsspielraum margin of safety
Sicherheitsstufe (Laborstandard)
 physical containment level;
 (Risikostufe) risk class,
 security level, safety level
 ➢ **Biologische S.** (Laborstandard)
 biological containment level;
 (Risikostufe) biosafety level
 ➢ **für Tierhaltungseinheit**
 animal containment level

Si

**Sicherheitsüberprüfung/
Sicherheitskontrolle**
safety check, safety inspection
Sicherheitsvektor containment vector
Sicherheitsventil
security valve, security relief valve
Sicherheitsverhaltensmaßregeln
safety policy
**Sicherheitsvorkehrungen/
Sicherheitsvorbeugemaßnahmen/
Absicherungen**
safety precautions,
safety measures,
safeguards
Sicherheitsvorrichtung safety device
Sicherheitsvorschriften
safety instructions,
safety protocol, safety policy
Sicherheitswerkbank
clean bench, safety cabinet
➢ **biologische S.**
biosafety cabinet
➢ **mikrobiologische S. (MSW)**
microbiological safety cabinet (MSC)
sichern/absichern secure
Sicherung securing, safeguarding;
safety device;
electr fuse, circuit breaker
➢ **rausfliegen/durchbrennen
(auslösen)** *electr* trip,
blow (fuse/circuit breaker)
Sicherungskasten *electr* fuse box,
fuse cabinet, cutout box
Sicherungsringzange
snap-ring pliers, circlip pliers
Sicht sight, view
sichtbar visible
➢ **unsichtbar** invisible
Sichtfenster/Sichtscheibe
viewing window
➢ **verschiebbare S./Schiebefenster/
Frontschieber (Abzug/Werkbank)**
sash (: hood)
Sichtgerät
visualizer, visual indicator,
viewing unit, display unit
Sichtschutz/Visier
visor, vizor (*Br*), face visor

Sieb sieve, sifter, strainer
➢ **Molekularsieb/Molekülsieb/Molsieb**
molecular sieve
➢ **Seiher/Abtropfsieb**
colander
Siebanalyse
sieve analysis, screen analysis
**Siebbodenkaskadenreaktor/
Lochbodenkaskadenreaktor**
sieve plate reactor
**Siebdurchgang/
Siebunterlauf/Unterkorn**
sievings, screenings, siftings;
undersize
sieben sieve, sift, screen
➢ **abseihen** strain
Siebgut
sieve material, sieving material,
material to be sieved
Siebmaschine (Schüttler) sieve shaker
Siebnummer mesh size, mesh
Siebplatte sieve plate, perforated plate
Siebrückstand/Sieböberlauf/Überkorn
sieve residue, screenings; oversize
Siebtuch straining cloth;
(Mull/Gaze) cheesecloth
Siebung screening, siftage, size
separation by screening
Siedebereich boiling range
Siedegefäß boiling flask
Siedekapillare *dest* capillary air bleed,
boiling capillary, air leak tube
Sieden/Aufwallen ebullition; boiling
sieden/kochen
boil; (leicht kochen) simmer
siedend simmering, ebullient;
(kochend) boiling
➢ **höhersiedend** less volatile
(boiling/evaporating at higher temp.)
Siedepunkt boiling point
Siedepunkterhöhung
boiling point elevation
Siedepunkterniedrigung
boiling point depression,
lowering of boiling point
Siederöhrchen ebullition tube
Siedestab bumping rod, bumping stick,
boiling rod, boiling stick

Siedestein/Siedesteinchen
boiling stone, boiling chip
Siedeverzug (durch Überhitzung)
defervescence, delay in boiling (due to superheating)
Siegel seal
Signalband/Warnband warning tape
Signal-Rausch-Verhältnis
signal-to-noise ratio (S/N ratio)
Signalstoff signal substance
Signalübertragung signal transduction
Signalwandler signal transducer
Signifikanzniveau/ Irrtumswahrscheinlichkeit
significance level, level of significance (error level)
Signifikanztest *stat* significance test, test of significance
Silber (Ag) silver
Silicium/Silizium (Si) silicon
Siliciumdioxid silica, silicon dioxide
Silicon/Silikon silicone (silicoketone)
Silicon-Schmierfett silicone grease
Siliconkautschuk silicone rubber
Sinapinalkohol sinapic alcohol
Sinapinsäure sinapic acid
Singulettzustand singulet condition
Sinterglas fritted glass
sintern sinter, sintering
Siphon siphon; siphon trap
SIP-Sterilisation (ohne Zerlegung/Öffnung der Bauteile)
sterilization in place (SIP)
Skala (*pl* **Skalen**) scale
Skalpell scalpel
Skalpellklinge scalpel blade
Smogverordnung smog ordinance
Sockelleiste baseboard, washboard
Sodaauszug soda extraction
Sodbrennen
heartburn, acid indigestion
Sofortmaßnahme
immediate measure (instant action)
Sog/Zug (Wasserleitung)
tension, suction, pull
Sol *chem* sol
Solarenergie/Sonnenenergie
solar energy

Solarzelle solar cell, photovoltaic cell
Sole brine (salt water)
Soll (Plan/Leistung/Produktion)
target, quota
Soll-Leistung
nominal output, rated output
Sollfrequenz nominal frequency
Sollwert nominal value, rated value, desired value, set point
➢ **Istwert** actual value, effective value
Sollwertgeber
set-point adjuster, setting device
Sollwertkorrektur set-point correction
Solubilisierung/Solubilisation
solubilization
Solvatation solvation
Solvathülle solvation shell
solvatisieren solvate
solvatisierter Stoff (Ion/Molekül)
solvate
Solvens/Lösungsmittel
solvent; dissolver
Sonde (Mikrosonde) probe, microprobe
➢ **mit Hilfe einer heterologen Sonde**
heterologous probing
➢ **Protonensonde** proton microprobe
Sondergenehmigung
special license, special permit
Sondermüll/Sonderabfall
hazardous waste
Sondermülldeponie
hazardous waste dump
Sondermüllentsorgung
hazardous waste disposal
Sondermüllentsorgungsanlage
hazardous waste treatment plant
Sondermüllverbrennungsanlage
hazardous waste incineration plant
Sonifikation/ Beschallung/ Ultraschallbehandlung
sonification, sonication
Sonnenbrand/Rindenbrand sunscald
Sonneneinstrahlung insolation
Sonnenenergie/Solarenergie
solar energy
Sonnenstich
sunstroke (heatstroke: Hitzschlag)

Sonnenstrahlung solar radiation
Sonogramm sonogram
Sonographie/Ultraschalldiagnose
 sonography, ultrasound,
 ultrasonography
Sorbens (*pl* **Sorbentien**) sorbent
Sorbinsäure (Sorbat)
 sorbic acid (sorbate)
Sorbit sorbitol
Sorte sort, type, kind, variety, cultivar
Sortenreinheit
 purity of variety, variety purity
sortieren sort
Sortimentkasten
 compartmentalized case
Spachtel trowel; (Schaber) scraper
Spachtelmesser/Kittmesser
 putty knife
spaltbar
 cleavable, crackable; *nucl* fissionable
Spaltbarkeit cleavage
Spalte crevice, crack
spalten cleave, break, open, crack,
 split, break down; *nucl* fission
Spaltfusion cleavage fusion
Spaltprodukt *chem* cleavage product,
 breakdown product;
 nucl fission product
Spaltung cleavage, breakage, opening,
 cracking, splitting, breakdown;
 (Furchung) cleavage;
 nucl fission(ing)
Spanne (Mess~) range
spannen stretch, tighten;
 (einspannen) clamp, fix into
Spannfutter (Bohrer)
 chuck, collet chuck
Spannkraft *physiol* tonicity
Spannschloss turnbuckle
Spannung/Potentialdifferenz
 potential difference, voltage
➢ **Hochspannung** high voltage
Spannungsklemme voltage clamp
Spannungsmessgerät voltmeter
Spannungsprüfer (Schraubenzieher)
 neon screwdriver (*Br*),
 neon tester (*Br*),
 voltage tester screwdriver

**Spannungsreihe (der
 Metalle)/Normalpotentiale**
 standard electrode potentials
 (tabular series),
 standard reduction potentials,
 electrochemical series (of metals)
Spannweite *stat* range
Spannzange (Kabelbinder)
 tensioning tool, tensioning gun
 (cable ties/wrap-it-ties)
Spanplatte flakeboard, chipboard
Sparflamme pilot flame, pilot light
Sparpackung economy pack
Sparpreis budget price
sparsam
 economical, thrifty; *adv* sparingly
Spatel spatula
➢ **Kolbenwischer/Gummiwischer
 (zum mechanischen Loslösen
 von Kolbenrückständen)**
 policeman, rubber policeman
 (rod with rubber or Teflon tip)
➢ **Löffelspatel**
 scoop, scoopula
➢ **Mundspatel/Zungenspatel**
 tongue depressor
➢ **Pulverspatel** powder spatula
➢ **Wägespatel**
 weighing spatula
spatelförmig spathulate, spatulate
Spatelplattenverfahren *micb*
 spread-plate method/technique
Spätfolgen *med* late sequelae
Spätschaden delayed damage
Spediteur carrier, shipper
Spedition freight company,
 shipping company, shipper
Speichel saliva
Speicher (Lager) storage;
 (Lagerhaus) storehouse, warehouse;
 (Reservoir) reservoir, storage basin;
 comp memory
speichern/anreichern/akkumulieren
 store, save, accumulate
Speichertank storage tank
Speicherung storage
speien spit
Spektralfarben spectral colors

Spektrometrie spectrometry
- **Elektronenstoß-Spektrometrie**
  electron-impact spectrometry (EIS)
- **Flugzeit-Massenspektrometrie (FMS)**
  time-of-flight mass spectrometry (TOF-MS)
- **Ionen-Fallen-Spektrometrie**
  ion trap spectrometry
- **Massenspektrometrie (MS)**
  mass spectrometry (MS)
- **Photoelektronenspektrometrie**
  photoelectron spectrometry (PES)

Spektroskopie
spectroscopy
- **Atom-Absorptionsspektroskopie (AAS)**
  atomic absorption spectroscopy (AAS)
- **Atom-Emissionsspektroskopie (AES)** atomic emission spectroscopy (AES)
- **Atom-Fluoreszenzspektroskopie (AFS)**
  atomic fluorescence spectroscopy (AFS)
- **Auger-Elektronenspektroskopie (AES)**
  Auger electron spectroscopy (AES)
- **Elektronen-Energieverlust-Spektroskopie**
  electron energy loss spectroscopy (EELS)
- **Elektronen-Spinresonanz-spektroskopie (ESR)/ elektronenparamagnetische Resonanz (EPR)**
  electron spin resonance spectroscopy (ESR), electron paramagnetic resonance (EPR)
- **Flammenemissionsspektroskopie (FES)**
  flame atomic emission spectroscopy (FES), flame photometry
- **Infrarot-Spektroskopie/ IR-Spektroskopie**
  infrared spectroscopy
- **Kernspinresonanz-Spektroskopie/ kernmagnetische Resonanzspektroskopie**
  nuclear magnetic resonance spectroscopy, NMR spectroscopy
- **Massenspektroskopie (MS)**
  mass spectroscopy (MS)
- **Mikrowellenspektroskopie**
  microwave spectroscopy
- **photoakustische Spektroskopie (PAS)/optoakustische S.**
  photoacoustic spectroscopy (PAS)
- **Röntgenabsorptionsspektroskopie**
  X-ray absorption spectroscopy (XAS)
- **Röntgenemissionsspektroskopie**
  X-ray emission spectroscopy (XES)
- **Röntgenfluoreszenzspektroskopie (RFS)**
  X-ray fluorescence spectroscopy (XFS)
- **UV-Spektroskopie**
  ultraviolet spectroscopy, UV spectroscopy

Spektrum (*pl* **Spektren**)
spectrum (*pl* spectra/spectrums)
- **Absorptionsspektrum**
  absorption spectrum, dark-line spectrum
- **Bandenspektrum/Molekülspektrum (Viellinienspektrum)**
  band spectrum, molecular spectrum
- **elektromagnetisches Spektrum**
  electromagnetic spectrum
- **Flammenspektrum** flame spectrum
- **Funkenspektrum** spark spectrum
- **Lichtbogenspektrum** arc spectrum
- **Linienspektrum/Atomspektrum**
  line spectrum
- **Rotationsspektrum**
  rotational spectrum
- **Schwingungsspektrum**
  vibrational spectrum
- **Umkehrspektrum** reversal spectrum
- **Wirtsspektrum** *ecol* host range

Spender (für Flüssigseife etc.)
dispenser (liquid detergent etc.); (Donor) donor

Sp

Sperrfilter cutoff filter;
  micros selective filter, barrier filter, stopping filter, selection filter
Sperrflüssigkeit barrier fluid
Sperrholz plywood
Sperrholzplatte plywood board
Sperrrelais *electr* interlocking relay
Spezialisierung specialization
speziell (zu einem bestimmten Zweck bestimmt) dedicated
spezifisch specific
➢ **unspezifisch** nonspecific
spezifische Wärme specific heat
spezifisches Gewicht
  specific gravity
Spezifität specificity
spezifizieren specify
Spind/Schließfach locker
Spindel/Zapfen/Stift/Achse
  pivot
Spindeldiagramm spindle diagram
Spinentkopplung (NMR)
  spin decoupling
Spinne/Eutervorlage/Verteilervorlage
  dest multi-limb vacuum receiver adapter, cow receiver adapter, 'pig' (receiving adapter for three/four receiving flasks)
Spinnerflasche/Mikroträger *micb*
  spinner flask
Spin-Spin-Aufspaltung (NMR)
  spin-spin splitting
Spinumkehr (NMR) flipping
Spirale/Helix spiral, helix
spiralig spiral, spiraled, twisted, helical
spiralig aufgewickelt spirally coiled
Spiralwindung
  spiral winding, coiling
Spiritus spirit
Spiritusbrenner/Spirituslampe
  alcohol burner
spitz acute, sharp, pointed, sharp-pointed
spitz zulaufen (spitz zulaufend)
  taper (tapering/tapered), attenuate
Spitze point, tip, spike;
  (Gipfel/Scheitelpunkt/Höhepunkt) apex, summit, peak

Spitzkolben
  pear-shaped flask (small, pointed)
Spitzpinzette sharp-point tweezers, sharp-pointed tweezers
spleißen *gen* splice
Splitter splinter;
  (Glassplitter) bits of broken glass
splitterfrei (Glas)
  shatterproof (safety glass)
Spontanzusammenbau/ Selbstzusammenbau
  self-assembly
sporadisch sporadic
Sporn (Immunodiffusion) spur
Sprechanlage
  intercom, intercom system
➢ **Wechselsprechanlage/ Gegensprechanlage**
  two-way intercom, two-way radio
Spreitung spreading
Sprengkraft
  explosive force, explosive power
Sprengstoff (Explosivstoff)
  explosive
➢ **brisanter S.** high explosive
➢ **hochbrisanter S.**
  high energy explosive (HEX)
➢ **Schießstoff/Schießmittel**
  low explosive
➢ **verpuffender S.** low explosive
sprießen sprout, grow, bud
springen jump, spring, bound, leap
Sprinkleranlage (Beregnungsanlage/ Berieselungsanlage: Feuerschutz)
  fire sprinkler system
Spritze syringe, hypodermic syringe;
  (Injektion) shot, injection;
  med hypodermic injection
➢ **Kanüle/Hohlnadel**
  needle, syringe needle
➢ **Luer T-Stück** Luer tee
➢ **Luerhülse** female Luer hub (lock)
➢ **Luerkern** male Luer hub (lock)
➢ **Luerlock/Luerverschluss** Luer lock
➢ **Luerspitze** Luer tip
➢ **Nadeladapter** syringe connector
➢ **Spritzenkolben/Stempel/Schieber**
  syringe piston, syringe plunger

spritzen (verspritzen/herumspritzen: auch versehentlich) splash, splatter; (injizieren) inject
Spritzenkolben/Stempel/Schieber syringe piston/plunger
Spritzennadel/Spritzenkanüle syringe needle, syringe cannula
Spritzenvorsatzfilter/Spritzenfilter syringe filter
Spritzer (verspritzte Chemikalie) splash (chemical)
spritzfest splash-proof
Spritzflasche wash bottle, squirt bottle
Spritzfleck splash
Spritzguss/Spritzgießen injection molding
Spritzschutzadapter/ Spritzschutzaufsatz/ Schaumbrecher-Aufsatz (Rückschlagsicherung: Reitmeyer-Aufsatz) *dest* splash protector, antisplash adapter, splash-head adapter
Sprossung/Knospung sprouting, budding; (Hefe) budding
sprudeln bubble
Sprühdose/Druckgasdose spray can, aerosol can
sprühen spray
Sprühflasche spray bottle
Sprühgerät/Zerstäuber atomizer
Sprühkolonne *dest* spray column
Sprung (Glas/Keramik etc.) crack
Spülbecken sink
Spülbürste dishwashing brush
Spüle sink; (Abtropfbrett) drainboard, dish board
Spule spool, coil
Spüleimer dishwashing bucket, dishpan
spülen/abspülen wash; clean
➢ **ausspülen** rinse
Spülgas purge gas
Spülicht (Rückstand vom Schmutz~/Spülwasser) slops
Spülküche washup room
Spüllappen dishwashing cloth, dishcloth, dishrag

Spülmaschine dishwasher, dishwashing machine
Spülmaschinenreiniger dishwasher detergent
Spülmittel detergent
➢ **Geschirrspülmittel** dishwashing detergent
Spülschwamm dishwashing pad; (Topfkratzer/Topfreiniger) scouring pad, pot cleaner
Spültisch sink, sink unit
Spülventil (Inertgas) T-purge (gas purge device)
Spülvorrichtung (z.b. Inertgas) purge assembly, purge device
Spülwanne dishwashing tub
Spülwasser/Abwaschwasser dishwater
Spundschlüssel (für Fässer) plug wrench (bung removal)
Spur/Überrest (meist *pl* **Überreste)** trace, remainder (meist *pl* remains)
Spurenanalyse trace analysis
Spurenelement/Mikroelement trace element, microelement, micronutrient
sputtern/besputtern (Vakuumzerstäubung) sputter
Sputtern/Besputtern/ Besputterung (Metallbedampfung) sputtering
staatlich kontrolliert/geprüft certified, registered (official)
staatlich subventioniert state/government-subsidized
staatliche Einrichtung governmental institution
staatliche Mittel public funds
Staatsbedienstete(r) public servant, civil servant; (staatl. Angestelleter) governmental employee, federal employee
Stäbchen rod
Stabdiagramm bar diagram, bar graph
stabil stable
➢ **instabil/nicht stabil** unstable (instable)

St

Stabilisator stabilizer
stabilisieren stabilize
Stabilisierung stabilization
Stab-Kugel-Modell/
  Kugel-Stab-Modell *chem*
  stick-and-ball model,
  ball-and-stick model
Stadium (*pl* **Stadien**) stage
Stahl steel
➤ **Edelstahl**
  high-grade steel, high-quality steel
➤ **rostfreier Stahl** stainless steel
Stahlbürste wire brush
Stahlflasche (Gasflasche)
  steel cylinder (gas cylinder)
Stamm
  stem; stock; *micb* strain
➤ **Bakterienstamm** bacterial strain
➤ **Inzuchtstamm** inbred strain
➤ **Referenzstamm** *micb*
  reference strain
Stammkultur/Impfkultur
  stem culture, stock culture
Stammlösung stock solution
Standard standard; (Typus) type
Standardabweichung *stat*
  standard deviation,
  root-mean-square deviation
Standardbedingung standard condition
Standardfehler/mittlerer Fehler *stat*
  standard error
  (standard error of the means)
standardisieren/vereinheitlichen
  standardize
Standardisierung/Vereinheitlichung
  standardization
Standardlösung
  standard solution
Standardpotential/Normalpotential
  standard potential,
  standard electrode potential
Standardtisch *micros* plain stage
Standardverfahren standard procedure
Ständer stand, rack
Standflasche/Laborstandflasche
  lab bottle, laboratory bottle
Standort site, location;
  biol habitat, place of growth

Stange pole
Stanniol (Aluminiumfolie/Alufolie)
  tinfoil (aluminum foil)
Stapel stack
Stapelkräfte stacking forces
stapeln stack
Stärke starch
Starkionendifferenz
  strong ion difference (SID)
Starterkultur (Anzuchtmedium)
  starter culture (growth medium)
stationäre Phase
  stationary phase,
  stabilization phase
stationärer Zustand/
  gleichbleibender Zustand
  steady state
Statistik statistics
statistische Abweichung
  statistical deviation
statistische Auswertung
  statistical evaluation
statistische Verteilung
  statistical distribution
statistischer Fehler statistical error
Stativ/Bunsenstativ
  support stand, ring stand,
  retort stand, stand
Stativklemme support clamp
Stativplatte support base
Stativring
  ring (for support stand/ring stand)
Stativstab support rod
Staub dust
➤ **Feinstaub** mist, fine dust, fines
➤ **Grobstaub** dust (coarse)
➤ **Inertstaub** inert dust
staubdicht dustproof
Staubexplosion dust explosion
staubig dusty
Staublunge/
  Staublungenerkrankung/
  Pneumokoniose
  pneumoconiosis
Staubkorn dust particle
staubsaugen vacuum-clean
Staubsauger vacuum cleaner, vacuum
Staubschutz dust cover

Staubschutzmaske (Partikelfilter)
dust mask, particulate respirator
(U.S. safety levels N/R/P according to regulation 42 CFR 84)
Staubwischen dusting
stauchen compress
Stauchung compression
stauen congest; stop;
accumulate, pile up, build up
Stauraum storage, stowage
Stearinsäure/Octadecansäure (Stearat/Octadecanat)
stearic acid, octadecanoic acid (stearate/octadecanate)
stechen sting, pierce, puncture
stechend/beizend/ätzend (Geruch)
pungent
Stechheber thief, thief tube, sampling tube (pipet); plunging siphon
Steckdose outlet, socket, wall (socket); receptacle; jack (mains electricity supply *Br*)
➢ **Mehrfachsteckdose**
outlet strip
➢ **Stecker in Steckdose stecken**
plug in (plug into the wall)
➢ **Wandsteckdose**
wall outlet
Stecker *electr/tech* plug (male/female), jack (female), connector, coupler
➢ **Bananenstecker** *electr* banana plug
➢ **Flachstecker** flat plug
➢ **Mehrfachstecker/ Vielfachstecker (~steckdose)**
oulet strip
➢ **Netzstecker** power plug
➢ **S. einstecken/reinstecken**
plug in, connect
➢ **S. herausziehen**
unplug, disconnect
➢ **Zwischenstecker/Adapter**
adapter
Steckschlüssel
socket wrench, box spanner
Steckschlüsseleinsatz/ Stecknuss/Nuss
socket, chuck, nut

Steckverbindung/Steckvorrichtung
tech/electr coupler, fitting; plug connection
➢ **Hochdruck-Steckverbindung**
compression fitting
➢ **Gleitverbindung**
slip-joint connection
Stehhilfe support
Stehkolben/Siedegefäß
Florence boiling flask, Florence flask (boiling flask with flat bottom)
Stehleiter/Treppenleiter stepladder
Steigrohr riser tube, riser pipe. riser, chimney; dip tube
Steinkohle bituminous coal, soft coal (*siehe unter:* Kohle)
Steinsalz (Halit)/ Kochsalz/Natrium chloride
rock salt (halite), table salt, sodium chloride
Steinwolle rock wool
Stellantrieb/Stellmotor
actuator
Stellglied
controlling element, adjuster, actuator
Stellgröße adjustable variable
Stellschraube
adjusting screw, setting screw, adjustment knob, fixing screw
Stempel-Methode *micb*
replica plating
Stengel stalk; *bot* stipe
sterben *vb* die
Stereoisomer stereoisomer
stereoselektiv stereoselective
Stereospezifität stereospecificity
steril (desinfiziert) sterile, disinfected; (unfruchtbar) sterile, infertile
sterile Werkbank sterile bench
Sterilfilter sterile filter
Sterilfiltration
sterile filtration
Sterilisation/Sterilisierung
sterilization, sterilizing
sterilisierbar sterilizable
Sterilisierbarkeit sterilizability

st

sterilisieren
(keimfrei machen) sterilize, sanitize; (sterilisieren/unfruchtbar machen) sterilize
Sterilität/Unfruchtbarkeit
sterility, infertility
Sterin/Sterol sterol
sterisch/räumlich
steric, sterical, spacial
sterische Hinderung/sterische Behinderung steric hindrance
Sternriss (im Glas) star-crack
Stetigkeit constancy, presence degree
Steuergerät
control unit, control gear, controller
steuern (in eine Richtung lenken) steer, steering; (regulieren) regulate, control
Steuerung control
Steuerungsmechanismus
regulatory mechanism
Steuerungstechnik control engineering
Stichflamme
explosive flame, sudden flame
Stichkultur/Einstichkultur (Stichagar)
stab culture
Stichprobe
sample, spot sample, aliquot
➤ **Teilstichprobe** subsample
➤ **Zufallsstichprobe**
random sample, sample taken at random
Stichprobenerhebung sampling
Stichprobenfunktion *stat*
sample function, sample statistic
Stichprobenumfang *stat*
sample size
Stichverletzung (Nadel etc.) *med*
stick injury (needle)
stickig stifling, stuffy
Stickstoff nitrogen
➤ **Flüssigstickstoff**
liquid nitrogen
stickstoffhaltig/stickstoffenthaltend/Stickstoff...
nitrogen-containing, nitrogenous
Stickstoffmangel nitrogen deficiency
Stickstoffverbindung
nitrogenous compound, nitrogen-containing compound
Stift
(Metall~) tack; (Nadel) pin; (Nagel) nail; *electr* (Stecker/Anschluss) pin, (Kontakt) lead
Stilett stylet, stiletto
Stöchiometrie stoichiometry
stöchiometrisch stoichiometric(al)
Stoff(e) substance, matter; material; (Gewebe) fabric, textile; cloth;
➤ **Wirkstoff** agent
Stoffaustausch
mass exchange, substance exchange
Stofffluss material flow, chemical flow
Stoffhandschuhe fabric gloves
Stoffkreislauf *ecol*
nutrient cycle
➤ **Mineralstoffkreislauf** mineral cycle
➤ **Phosphorkreislauf** phosphorus cycle
➤ **Sauerstoffkreislauf** oxygen cycle
➤ **Schwefelkreislauf** sulfur cycle
➤ **Stickstoffkreislauf** nitrogen cycle
➤ **Wasserkreislauf**
water cycle, hydrologic cycle
Stoffmenge
amount of substance (quantity)
Stoffmengenanteil/Molenbruch
mole fraction
Stoffübergang/Massenübergang/Stofftransport/Massentransport/Massentransfer
mass transfer
Stoffübergangszahl/Stofftransportkoeffizient/Massentransferkoeffizient
mass transfer coefficient
Stoffwechsel/Metabolismus
metabolism
Stoffwechselprodukt/Metabolit
metabolite
Stoffwechselstörung
metabolic derangement, metabolic disturbance
Stopfbuchse (Rührer: Wellendurchführung)
stuffing gland, packing box seal

Stopfen/Korken/Stöpsel
  stopper, cork
  ➤ **Achtkantstopfen**
    octa-head stopper, octagonal stopper
  ➤ **Gummistopfen/Gummistöpsel**
    rubber stopper, rubber bung (*Br*)
  ➤ **Sechskantstopfen**
    hex-head stopper, hexagonal stopper
Stopfenschlüssel
  plug wrench, bung wrench
Stöpsel/Stopfen stopper, bung (*Br*)
Storchschnabelzange
  needle-nose pliers,
  snipe-nose pliers,
  snipe-nosed pliers
  ➤ **gebogene S.** dip needle-nose pliers
Störfall incident, accident; breakdown
Störfallverordnung
  industrial accident directive,
  statutory order on hazardous
  incidents
Störgröße disturbance value,
  interference factor
Störung disturbance,
  interference, disruption;
  (Perturbation) perturbation
Stoßaktivierung collision activation
Stößel/Pistill (und Mörser)
  pestle (and mortar)
Stoßen
  (Sieden/Überhitzung/Siedeverzug)
  bumping
stoßen/umstoßen
  (umkippen/umwerfen) tip over;
  (dranstoßen) bump (into),
  knock (into);
  (mit dem Fuß/Bein/Körper)
  kick over (knock over)
Stoßverbindung butt joint
Strafe penalty
Strahl ray; beam; jet
  ➤ **Lichtstrahl** beam of light
  ➤ **Röntgenstrahl** X-ray
  ➤ **Sonnenstrahl**
    ray (of sunshine), sunbeam
  ➤ **Wasserstrahl** jet of water
strahlen shine; radiate
Strahlendiagramm *opt* ray diagram

Strahlenschäden
  radiation hazards, radiation injury
Strahlenschutz radiation control,
  radiation protection,
  protection from radiation
Strahlenschutzplakette film badge
Strahlentherapie
  radiation therapy, radiotherapy
Strahler (Licht) light, illuminator,
  beamer; (Wärme) radiator, heater;
  ➤ **Punktstrahler/Spot**
    spotlight, spot
Strahlreaktor *biot* jet reactor
Strahlung radiation
  ➤ **Ausstrahlung/Emission/Ausstoss**
    emission
  ➤ **Bestrahlung** irradiation
  ➤ **elektromagnetische Strahlung**
    electromagnetic radiation
  ➤ **Globalstrahlung** global radiation
  ➤ **Hintergrundsstrahlung**
    background radiation
  ➤ **ionisierende Strahlung**
    ionizing radiation
  ➤ **Kernstrahlung** nuclear radiation
  ➤ **photosynthetisch aktive Strahlung**
    photosynthetically active radiation
    (PAR)
  ➤ **radioaktive Strahlung**
    radioactive radiation
  ➤ **Sonneneinstrahlung** insolation
  ➤ **Sonnenstrahlung** solar radiation
  ➤ **Streustrahlung**
    scattered radiation, diffuse radiation
  ➤ **Teilchenstrahlung**
    corpuscular radiation
  ➤ **Wärmestrahlung**
    thermal radiation
  ➤ **zulässige Strahlung**
    permissible radiation
Strahlungsenergie radiant energy
Strahlungsintensität radiation intensity
Strahlungsvermögen/
  Emissionsvermögen
  (Wärmeabstrahlvermögen)
  emissivity
Strahlungswärme radiant heat
Strang (*pl* **Stränge**) cord; *gen* strand

stranggepresst *polym*
extruded, extrusion-molded
strangpressen/extrudieren *polym*
extrude
Strangpressen/Extrudieren/Extrusion
polym extrusion (extrusion molding)
strecken (in die Länge ziehen)
elongate, extend
Streckschwingung
stretching vibration
Streckspannung/Fließspannung
yield stress
Streckung/Verlängerung
elongation, extension
Stress/Belastungszustand//Spannung
phys stress
stressen/belasten stress
Stretchfolie stretch film/foil
Streu (für Tierkäfige etc.) litter
Streudiagramm scatter diagram
(scattergram/scattergraph/scatterplot)
streuen/
verstreuen/ausstreuen/verteilen
scatter, spread, distribute; sprinkle
Streuer shaker; dredger
Streulicht
scattered light, stray light
Streulichtmessung/
Nephelometrie nephelometry
Streulichtschirm *photo* diffusing screen
Streustrahlung
scattered radiation, diffuse radiation
Streuung (Lichtstreuung)
optical diffusion, dispersion,
dissipation, scattering (light);
(Ausbreitung) dispersal,
dissemination;
(Verstreuen/Verteilung) scattering,
spreading, distribution
Streuungsverhalten *stat*
scedasticity,
heterogeneity of variances
Streuverlust *electr* leakage
Strichdiagramm line diagram
Strichliste tally chart
stringente Bedingungen/
strenge Bedingungen
stringent conditions

Stringenz
(von Reaktionsbedingungen)
stringency (of reaction conditions)
Stroboskop
stroboscope, strobe, strobe light
Strom (Flüssigkeit) stream, flow;
(Volumen pro Zeit) flow rate
➤ **Elektrizität** *colloquial/general*
electricity, power, juice;
(Ladung/Zeit) current
➤ **Stromstärke** current,
electric current, amperage, amps
stromaufwärts upstream
Stromausfall
electricity failure, power failure
Strombrecher
(z.B. an Rührer von Bioreaktoren)
baffle
strömen stream, flow
Stromgerät power supply
Stromkabel power cord, electric cord,
electrical cord, power cable,
electric cable
Stromkontakt power lead
Stromkreis
electric circuit, electrical circuit
Stromleiter current carrier; conductor
Stromleitung/Hauptstromleitung
mains (*Br*)
Strommessgerät/Amperemeter
(Stromstärke) ammeter
Stromquelle/Stromzufuhr *electr*
power supply
Stromschlüssel (Salzbrücke) *electrolyt*
salt bridge
Strömung (Flüssigkeit) current, flow;
electr (Strömung) flux
➤ **Konvektionsströmung/**
Konvektionsstrom
convection current
➤ **Konzentrationsströmung**
density current
➤ **Schichtströmung/**
laminare Strömung
laminar flow
➤ **turbulente Strömung** turbulent flow
➤ **Wirbelstrom (Vortex-Bewegung)**
eddy current

Strömungsmesser
current meter, flowmeter
Strömungsmuster flow pattern
Strömungswiderstand
flow resistance, resistance to flow
Stromversorgung
electric power supply,
power supply, mains (*Br*)
Stromzähler electric meter
Strontium (Sr) strontium
Strudel eddy, swirl
strudeln whirl, swirl, eddy
Struktur structure; (Textur/Faser/
Fibrillenanordnung: Holz) grain
Strukturanalyse structural analysis
Strukturaufklärung
structure elucidation
Strukturformel structural formula
Stufe (einer Treppe) step, stair;
(Leiter) rung; (Niveau) level;
(Rang) rank, position;
chem stage, tray;
math degree, order, rank
**Stufenfolge/Rangordnung/
Rangfolge/Hierarchie**
order of rank, ranking, hierarchy
Stufengradient step gradient
Stufenleiter stepladder
**stufenlos
(regulierbar/regelbar/einstellbar etc.)**
variable (variably adjustable)
stufenlos regelbar
continually variable,
infinitely variable
stufenlos regulierbar
continuously adjustable,
variably adjustable
Stufenschalter step switch
Stufenwiderstand step resistance
Stufung zonation; grading, staggering
Stuhl chair; (Hocker) stool;
(Laborhocker) lab stool;
(Fäzes/Kot: Mensch) stool, feces
Stuhlprobe stool sample
stumme Infektion/stille Feiung
silent infection
stumpf obtuse, blunt
Stütze support, prop

stützen support, prop up
**Stutzen
(Anschlussstutzen/Rohrstutzen)**
nozzle, socket;
connecting piece, connector
➤ **Ansatzstutzen**
(Kolben) side tubulation, side arm;
(Schlauch) hose connection
➤ **Ausgussstutzen (Kanister)**
nozzle (attachable/detachable)
➤ **Beschickungsstutzen (Kolben)**
delivery tube (flask)
➤ **Gewindestutzen**
threaded socket (connector/nozzle)
➤ **Hülse/Schliffhülse
('Futteral'/Einsteckstutzen)**
socket (female: ground-glass joint)
➤ **Olive (meist geriffelter
Ansatzstutzen: Schlauch/Kolben)**
barbed hose connection
(flask: side tubulation/side arm)
Styrol styrene
Suberinsäure/Korksäure/Octandisäure
suberic acid, octanedioic acid
Subklonierung subcloning
**Subkultur/Subkultivierung/
Passage (einer Zellkultur)**
subculture, passage (of cell culture)
subletal sublethal
Sublimation sublimation
sublimieren sublimate
Submerskultur submerged culture
Subsistenz subsistence
Substanzgemisch substance mixture
substituieren substitute
Substitution substitution
Substitutionsvektor
replacement vector
Substrat substrate
➤ **Folgesubstrat**
following substrate
➤ **Leitsubstrat**
leading substrate
Substraterkennung
substrate recognition
**Substrathemmung/
Substratüberschusshemmung**
substrate inhibition

Substratkonstante (K_S)
substrate constant
Substratsättigung substrate saturation
Substratspezifität
substrate specificity
Subtypisierung subtyping
Suchtest *gen/med*
screening, screening test
Suchtmittel/Droge drug
Sulfanilsäure sulfanilic acid,
p-aminobenzenesulfonic acid
Sulfat sulfate
Sulfierkolben sulfonation flask
Sulfurikanten sulfuricants
Summe sum, total
**Summenformel/
Elementarformel/
empirische F./Verhältnisformel**
empirical formula
**Summenhäufigkeit/
kumulative Häufigkeit** *stat*
cumulative frequency
Summenpotential
gross potential
Summenregel sum rule
Sumpf (Rückstand in Dest.-Blase)
dest bottoms
Superinfektion/Überinfektion
superinfection
Supersäure superacid
**superspiralisiert/superhelikal/
überspiralisiert** supercoiled
superstark/verstärkt/Hochleistungs...
heavy-duty,
superior performance
Suppression/Unterdrückung
suppression
**supprimieren/unterdrücken/
zurückdrängen** suppress

**suspendieren
(schwebende Teilchen in
Flüssigkeit)** suspend
Suspension suspension
**Suspensionstechnik
(IR-Spektroskopie)**
mull technique
süß sweet
Süße sweetness
Süßstoff sweetener
Symbiose *allg* symbiosis;
(gemeinnützige) mutualistic
symbiosis, mutualism
Symmetrie symmetry
Synchronkultur synchronous culture
Syndrom/Symptomenkomplex
syndrome, complex of symptoms
Synthese synthesis
➢ **Biosynthese** biosynthesis
➢ **Chemosynthese**
chemosynthesis
➢ **DNA-Synthese** DNA synthesis
➢ **Halbsynthese** semisynthesis
➢ **Neusynthese/*de-novo* Synthese**
de-novo-synthesis
➢ **Photosynthese**
photosynthesis
synthetisieren synthesize
Systemanalyse systems analysis
Systematik systematics
systematisch systematic
systemisch systemic
Szintillationsgläschen
scintillation vial
Szintillationszähler ('Blitz'zähler)
scintillation counter, scintillometer
**szintillieren/funkeln/
Funken sprühen/glänzen**
scintillate

Tablet tray
Tageslichtprojektor/ Overhead-Projektor
overhead projector
Takt cycle time, stroke, time
Taktung cycle timing
Taktgeber clock, clock generator, timing generator; pulse generator; synchronizer
Taktrate clock frequency
Talg *med* sebaceous matter, sebum; *zool* tallow (extracted from animals), suet (from abdominal cavity of ruminants)
Talgdrüse sebaceous gland
talgig/Talg... sebaceous, tallowy
tamponieren tampon, plug, pack
Tangentialschnitt tangential section
Tank/Kessel tank, vessel
Tannat (Gerbsäure)
tannate (tannic acid)
Tannin (Gerbstoff)
tannin (tanning agent)
Tara (Gewicht des Behälters/ der Verpackung)
tare (weight of container/packaging)
tarieren tare (determine weight of container/packaging in order to substract from gross weight)
Tarnung camouflage
Tasche pocket; (Vertiefung: Elektrophorese-Gel) well, depression (at top of gel)
Taschenlampe flashlight, torch (*Br*)
Tastatur
(groß) keyboard, (klein) keypad
Taste
key, button, knob, push-button
tasten feel, touch, palpate
Tastkopf *micros* probe, probing head
taub
(gefühllos) numb; (gehörlos) deaf
Taubheit (Gehörlosigkeit) deafness; (Gefühllosigkeit) numbness
Tauchbad immersion bath
tauchfähig (Pumpe) submersible
Tauchflächenreaktor
immersing surface reactor

Tauchkanalreaktor
immersed slot reactor
Tauchpumpen-Wasserbad/ Einhängethermostat
immersion circulator
Tauchsieder
immersion heater, 'red rod' (*Br*)
Tauchstrahlreaktor
plunging jet reactor, deep jet reactor, immersing jet reactor
Tauchtank dip tank
Taumelbewegung, dreidimensionale
nutation, gyroscopic motion (threedimensional circular/orbital & rocking motion)
taumeln tumble, sway, stagger; nutate (gyroscopic motion); (Bakterien) tumble
Taumelschüttler
nutator, nutating mixer, 'belly dancer' (shaker with gyroscopic, i.e., threedimensional circular/orbital & rocking motion)
Täuschung
deception, delusion; illusion
tautomere Umlagerung tautomeric shift
Tautropfen dewdrop
Technik (einzelnes Verfahren/Arbeitsweise)
technique, technic
➢ **Technologie (Wissenschaft)**
technology
➢ **Umweltverfahrenstechnik**
environmental process engineering
Technikfolgenabschätzung
technology assessment
➢ **US-Büro für Technikfolgenabschätzung** OTA (Office of Technology Assessment)
technisch technic(al); (Laborchemikalie) lab grade
Technische Anweisung Lärm (TALärm)
Technical Instructions on Noise Reduction
Technische Anweisung Luft (TALuft)
Technical Instructions on Air, Clean Air Act

technischer Assistent
(**technische Assistentin**)/
Laborassistent (Laborassistentin)/
Laborant (Laborantin)
laboratory technician, lab technician,
technical lab assistant
**Technischer Überwachungsverein
(TÜV)**
technical inspection agency/
authority,
technical supervisory association
Technologie technology
technologisch technologic(al)
Teichonsäure teichoic acid
Teichuronsäure teichuronic acid
Teil (des Ganzen) moiety, part, section;
(Anteil/Hälfte) moiety
Teilchen/Partikel particle
Teilchengröße (Bodenpartikel)
particle size, soil texture
teilen divide, fission, separate
Teilerhebung *stat* partial survey
Teilkorrelationskoeffizient *stat*
partial correlation coefficient
Teilmenge/Portion/Fraktion
portion, fraction
Teilmengenauswahl *stat*
subset selection
Teilreaktion
partial reaction;
(electrode potentials:) half-reaction
Teilstichprobe *stat* subsample
Teilung division, fission, separation
Teilungsphase division phase
Telefonzentrale switchboard
Tellermühle disk mill
Tellur (Te) tellurium
Temperatur temperature
➢ **Arbeitstemperatur**
operating temperature
➢ **Körpertemperatur**
body temperature
➢ **Phasenübergangstemperatur**
phase transition temperature
➢ **Raumtemperatur**
room temperature,
ambient temperature
➢ **Schmelzpunkt** melting point
➢ **Schmelztemperatur**
melting temperature
➢ **Siedepunkt** boiling point
➢ **Umgebungstemperatur**
ambient temperature
➢ **Vorzugstemperatur**
cardinal temperature
➢ **Zündpunkt/Zündtemperatur**
ignition point, kindling temperature,
flame temperature, flame point,
spontaneous-ignition temperature
(SIT)
temperaturabhängig
temperature-dependent
Temperaturempfindlichkeit
sensitivity to temperature
Temperaturfühler temperature sensor
Temperaturgradient
temperature gradient
Temperaturorgel *ecol*
temperature-gradient apparatus
Temperaturregler
temperature controller
Temperaturschwankung
fluctuation of temperature
Tempereisen/Temperguss
malleable iron,
malleable cast iron, wrought iron
temperenter Phage temperate phage
Tempergluhofen
malleable annealing furnace
Temperierbecher cooling beaker,
chilling beaker, tempering beaker
(jacketed beaker)
temperieren
bring to a moderate temperature;
to have an agreeable temperature
tempern temper; *polym* anneal
**teratogen/
Missbildungen verursachend**
teratogenic
Teratogenese/Missbildungsentstehung
teratogenesis, teratogeny
Teratologie (Lehre von Missbildungen)
teratology
Teratom teratoma
Terminus/Ende (Molekülende)
terminus

Terpentinharz
pitch (resin from conifers)
Test (Prüfung/Bestimmungsmethode)
test, examination, assay;
(Untersuchung) investigation
Testkreuzung testcross
Testmedium/Prüfmedium
(zur Diagnose) test medium
Testpartner *gen* tester
Testverfahren
test procedure, testing procedure
**Tetrachlorkohlenstoff/
Tetrachlormethan**
carbon tetrachloride,
tetrachloromethane
tetraedrisch tetrahedral
Textilfaser textile fiber
Textilveredlung textile finishing
Thein/Koffein theine, caffeine
theoretisch theoretic, theoretical
Theorie theory
Thermoanalyse/thermische Analyse
thermal analysis
Thermodynamik thermodynamics
➢ **1./2. Hauptsatz (der Thermodynamik)**
first/second law of thermodynamics
Thermoelement thermocouple
**Thermoelement-Schutzrohr/
Thermohülse**
thermowell (for thermocouples)
Thermoelementsonde
thermocouple probe
**Thermogravimetrie (TG)
(=thermogravimetrische Analyse)**
thermogravimetry (TG)
(=thermogravimetric analysis)
Thermometer thermometer
➢ **Bimetallthermometer**
bimetallic thermometer
➢ **Dampfdruckthermometer** vapor
pressure thermometer
➢ **Gasthermometer** gas thermometer
➢ **Quarzthermometer**
quartz thermometer
➢ **Quecksilberthermometer**
mercury-in-glass thermometer
➢ **Pyrometer/Hitzemessgerät**
pyrometer
➢ **Rauschthermometer**
noise thermometer
Thermoregulation thermoregulation
Thermoskanne/Thermosflasche
thermos
Thermospray thermospray
Thermostat thermostat
Tiefenätzung deep etching
Tiefenschärfe/Schärfentiefe *opt*
depth of focus, depth of field
Tieffeldverschiebung low-field shift
Tiefkühlfach (des Kühlschranks)
deep-freeze compartment
Tiefkühltruhe
deep-freeze, deep freezer (chest)
Tiefkühlung deep freeze
Tiegel/Schmelztiegel crucible
➢ **Filtertiegel** filter crucible
➢ **Gooch-Tiegel** Gooch crucible
Tiegeldreieck crucible triangle
Tiegelofen crucible furnace
Tiegelzange crucible tongs
Tierarzt/Veterinär veterinarian, vet
Tierhaltungseinheit (DIN) animal unit
tierisch (von tierischer Herkunft)
animal
Tierkäfig animal cage
➢ **kleiner Tierkäfig/Verschlag
(z.B. Geflügelstall)** hutch, coop
Tierlabor
animal laboratory, animal lab,
animal research lab
**Tiermedizin/Tierheilkunde/
Veterinärmedizin**
veterinary medicine,
veterinary science
Tiermodell animal model
Tierpfleger/Tierwärter
(animal) keeper, warden
Tierseuche/Viehseuche
epizooic disease, pest;
livestock epidemic
Tierversuch animal experiment
Tierzwinger (staatl. Verwahrung
verwaister Tiere) pound;
(z.B. in Zoos) cage, enclosure
Tinktur tincture
Tinte ink

Ti

Tisch table
- Arbeitstisch worktable
- Drehtisch *micros* rotating stage
- höhenverstellbare Plattform (Labor) laboratory jack
- Kreuztisch *micros* mechanical stage
- Laborarbeitstisch lab bench
- Labortisch/Labor-Werkbank laboratory/lab table, laboratory/lab bench, laboratory/lab workbench
- Objekttisch *micros* stage, microscope stage
- Spültisch sink, sink unit
- Standardtisch *micros* plain stage
- Wägetisch weighing table

Tischzentrifuge tabletop/benchtop centrifuge
Titan (Ti) titanium
Titer titer
Titration titration
- amperometrische T./Amperometrie amperometric titration
- coulometrische T./Coulometrie coulometric titration
- Endpunktverdünnungsmethode (Virustitration) end-point dilution technique
- Fällungstitration precipitation titration
- Fließinjektions-T. flow-injection titration
- Leitfähigkeitstitration/ konduktometrische Titration/ Konduktometrie conductometric titration
- Oszillometrie/ oszillometrische Titration/ Hochfrequenztitration oscillometry, high-frequency titration
- photometrische T. photometric titration
- Redoxtitration redox titration
- Rücktitration back titration
- Säure-Basen-Titration/ Neutralisationstitration acid-base titration

Titrationskurve titration curve
Titrationsmittel/Titrant titrant
titrieren titrate
Tochterion daughter ion
Tod *n* death
Todesursache cause of death
tödlich/letal deadly, lethal
Toleranzbereich tolerance range
Toleranzgrenze tolerance limit
Tomographie tomography
Ton *acust* tone, sound; *geol* clay
Tondreieck/Drahtdreieck clay triangle, pipe clay triangle
Tönung/Schattierung (Farbton) hue
Topferde potting soil (potting mixture: soil & peat a.o.)
Topfkratzer/Topfreiniger scouring pad, pot cleaner
Topflappen potholder
Topfpflanze potted plant
Topfzeit/Verarbeitungsdauer pot life
Torsion/Drehung torsion
tot dead
tot geboren stillborn
Totenkopf (Giftzeichen) skull and crossbones
Totenstarre/Leichenstarre rigor mortis
Totgeburt stillbirth
Totraum deadspace, headspace
totstellen feign death, play dead
Totvolumen (Spritze/GC) dead volume, deadspace volume, holdup (volume)
Totzeit/Durchflusszeit (GC) holdup time
Toxikologie toxicology
Toxin/Gift toxin
toxisch/giftig toxic, poisonous
- cytotoxisch/zellschädigend cytotoxic
- embryotoxisch embryotoxic
- fetotoxisch fetotoxic
- hepatotoxisch/leberschädigend hepatotoxic
- hochgiftig highly toxic
- mindergiftig moderately toxic
- neurotoxisch neurotoxic
- phytotoxisch/pflanzenschädlich phytotoxic
- sehr giftig extremely toxic (T+)

Toxizität/Giftigkeit
toxicity, poisonousness
trächtig/schwanger gravid, pregnant
Trächtigkeit/
 Schwangerschaft/Gravidität
pregnancy, gravidity
träg/träge *chem* inert
Trage/Krankentrage/Krankenbahre
stretcher
Träger carrier (auch: *chromat*)
Trägerarm *micros* arm
Trägerelektrophorese
carrier electrophoresis
Trägergas/Schleppgas (GC)
carrier gas (an inert gas)
Trägermolekül carrier molecule
Trägheit inertia
Trägheitskraft inertial force
Traglast carrying capacity
Tran/Fischöl train oil, fish oil
(also from whales)
➢ **Lebertran** cod-liver oil
Träne tear
tränen tear
tränken/einweichen (durchfeuchten)
soak, drench, steep
Transferöse transfer loop
transformieren transform
Transplantat transplant, graft
transplantieren transplant
Transport transport, transportation
➢ **Gefahrguttransport**
transport of dangerous goods,
transport of hazardous materials
➢ **Krankentransport** ambulance service
Transportfahrzeug transport vehicle
transportieren transport
Transportkiste tote box
Transportwagen/Transportkarren
bogie
Traubenzucker/
 Glukose/Glucose/Dextrose
grape sugar, glucose, dextrose
Treber/Biertreber brewers' grains
Treibgas (z.B. für Sprühflaschen)
propellant
Treibhaus greenhouse, hothouse
Treibhauseffekt greenhouse effect

Treibmittel (z.B. in Druckflaschen)
propellant (pressure can);
(Gärmittel/Gärstoff) leavening,
raising agent
Treibstoff/Kraftstoff fuel
Treibstoffalkohol/Gasohol gasohol
trennen separate; divide;
(lösen/entkuppeln/auskuppeln)
separate, disconnect
Trennfaktor/Separationsfaktor *analyt*
separation factor
Trenngel separating gel (running gel)
Trenngrenze/Ausschlussgrenze
 (Teilchentrennung) cutoff
Trennkammer *chromat* **DC**
developing chamber,
developing tank (TLC)
Trennleistung separation efficiency
(column efficiency)
Trennmethode separation method
Trennsäule separating column,
fractionating column
Trennschärfe *chromat*
resolution, separation accuracy
Trennstufe *chromat* **HPLC**
plate
Trennung separation
Trennungsgang *chem/analyt*
analytical (separation) procedure
Trennverfahren/Trennmethode
separation technique,
separation procedure,
separation method
Trennwand (Gebäude) partition wall
Trennwirkungsgrad separation
efficiency
Treppe stairs
➢ **die Treppe hoch (oben)** upstairs
➢ **die Treppe runter (unten)** downstairs
Treppenhaus stairs, staircase, stairway
Treppenhauseingang/~ausgang
stairway entry/exit
Treppenschacht stairwell
Trester/Treber (*siehe auch dort*)
(Fruchtpressrückstand/
Traubenpressrückstand) marc;
(Malzrückstand) draff
Treteimer (Mülleimer) step-on pail

Tr

Trichter funnel
> **Analysentrichter** analytical funnel
> **Einfülltrichter** addition funnel
> **Filternutsche/Nutsche (Büchner-Trichter)**
  nutsch, nutsch filter, filter funnel, suction funnel, suction filter, vacuum filter (Buechner funnel)
> **Fülltrichter** filling funnel
> **Glastrichter** glass funnel
> **Heißwassertrichter**
  hot-water funnel (double-wall funnel)
> **Hirsch-Trichter** Hirsch funnel
> **Kurzhalstrichter/Kurzstieltrichter**
  short-stem funnel, short-stemmed funnel
> **Pulvertrichter** powder funnel
> **Scheidetrichter**
  separatory funnel
> **Tropftrichter**
  dropping funnel
> **Zulauftrichter**
  addition funnel
Trichterrohr funnel tube
Triebkraft *phys/mech*
  (> Antrieb) ropulsive force
Trimmblock *micros* trimming block
Trimmschere trimming shears
Trinkbrunnen/Trinkfontäne
  fountain (for drinking water)
Trinkwasser drinking water
Trinokularaufsatz/Tritubus *micros*
  trinocular head
Tripelpunkt/Dreiphasenpunkt
  triple point
Triplettbindungsversuch
  triplet binding assay
Trittleiter stepladder, steps
Trittschall impact sound
Trittschalldämmung
  impact sound insulation
trittschallgedämpft
  impact sound-reduced
Trittschemel
  step-stool
> **'Elefantenfuß'/Rollhocker**
  rolling step-stool

trocken dry, arid
Trockenbatterie *electr*
  dry cell, dry cell battery
Trockenblotten dry blotting
Trockendestillation dry distillation
Trockeneis (CO_2) dry ice
Trockenextrakt dry extract
Trockengestell drying rack
Trockengewicht
  (*sensu stricto*: **Trockenmasse**)
  dry weight (*sensu stricto*: dry mass)
Trockengut dry product, dry substance
Trockenlauf/Probelauf test run
trockenlaufen *chromat* run dry
Trockenlegung drainage
Trockenmasse/Trockensubstanz
  dry mass, dry matter
Trockenmittel/Sikkativ
  siccative, desiccant, drying agent, dehydrating agent
Trockenofen drying oven
Trockenpistole/Röhrentrockner
  drying pistol
trockenresistent
  drought resistant, xerophytic
Trockenrohr/Trockenröhrchen
  drying tube
Trockenschrank
  drying cabinet, drying oven
Trockensubstanz
  dry matter
Trockenturm/Trockensäule
  drying tower, drying column
trocknen dry
> **austrocknen** desiccate
Trocknungsverlust
  loss on drying
Trog/Wanne trough
Trommel/Zylinder drum, barrel
Trommelmischer
  barrel mixer, drum mixer
Trommelmühle
  drum mill, tube mill, barrel mill
Trommelzentrifuge
  basket centrifuge, bowl centrifuge
Tröpfcheninfektion droplet infection
Tropfen *n* drop
tropfen *vb* drip

Tropfenfänger drip catcher, drip catch; splash trap, antisplash adapter (distillation apparatus); (Reitmeyer-Aufsatz: Rückschlagschutz: Kühler/Rotationsverdampfer etc.) splash adapter, antisplash adapter, splash-head adapter
tropfenweise
dropwise, drop by drop
Tropfflasche
drop bottle, dropping bottle
Tropfglas/Tropfpipette dropper
Tropfkörper
(Tropfkörperreaktor/ Rieselfilmreaktor)
trickling filter
Tropfpunkt
dropping point, drop point
Tropftrichter
dropping funnel (addition funnel)
trüb (Flüssigkeit) cloudy, turbid
Trübheit/Trübung (Flüssigkeit)
cloudiness, turbidity
tuberkulös tuberculous
Tuberkulose tuberculosis
tuberös tuberous, tuberal
tubulär tubular

Tubus *micros* tube, body tube; (Steckhülse für Okular) draw tube
Tülle (ausgießen) nozzle, spout; (Fassung) socket; (Schlauchtülle) hose connection gland
Tullgren-Apparat *ecol* Tullgren funnel
Tumor/Wucherung/Geschwulst
tumor
Tunnelmikroskopie
tunneling microscopy
Tüpfelplatte spot plate
Tüpfelprobe spot test
tupfen/abtupfen dab, swab
Tupfer pad, gauze pad; (Abstrichtupfer) swab; (Wattebausch) cotton pledget
Tupferklemme
sponge forceps
Turbidimetrie/Trübungsmessung
turbidimetry
turbulente Strömung turbulent flow
Turbulenzdiffusion/Wirbeldiffusion
eddy diffusion
Turgor/hydrostatischer Druck
turgor, hydrostatic pressure
Turgordruck turgor pressure
Turmreaktor column reactor

Üb

Übelkeit/Übelsein
nausea, sickness, illness
übelriechend/stinkend
fetid, smelly, smelling bad, malodorous, stinking
Überbleibsel relic
Überbrückungskabel/Starterkabel
jumper cable, coupling cable
Überdauerung persistance, survival
Überdosis overdose
Überdrehung overwinding
Überdruck positive pressure
Überdruckventil pressure valve, pressure relief valve; safety valve
Überdüngung overfertilization
Überempfindlichkeit hypersensitivity
Überfluss excess
überführen transfer
Überführung transfer
Überfunktion overactivity, hyperactivity
Übergabe handing over, delivery
Übergang/Entwicklungsübergang
transition, developmental transition
Übergangsmetall transition metal
Übergangsphase transition phase
Übergangsstück (Laborglas)
adapter, connector, transition piece
➢ **Expansionsstück**
expansion adapter, enlarging adapter
➢ **Reduzierstück**
reducing adapter, reduction adapter
Übergangszustand (Enzymkinetik)
transition state
übergreifen *med* spread
(e.g. disease/epidemic)
Überhitzen/Überhitzung
overheating, superheating
Überinfektion/Superinfektion
superinfection
Überkorn (Siebrückstand) oversize
überkritisch (Gas/Flüssigkeit)
supercritical (gas/fluid)
überlasten *tech/electr* overload
Überlastung *tech/electr* overload
Überlauf overflow, overrun; (Abflusskanal) spillway
Überleben *n* survival
überleben *vb* survive

überlegen/vorherrschend/dominant
superior, dominant
Überlegenheit/Dominanz
superiority, dominance
überprüfen check, examine, confirm, inspect, review; verify, control
Überprüfung
check-up, examination, inspection, reviewal; verification, control
überragen
protrude, project, stand/stick out, rise over
übersättigt supersaturated
überschreiten exceed
Überschuhe/
Überziehschuhe (Einweg~)
shoe covers,
shoe protectors (disposable)
Überschuss (Menge) excess
überschüssig in excess (of)
Überschussproduktion
surplus production
Überschwingen (aufheizen)
overswing
Übersender/Konsignant consignor
Überspannung overpotential
Überspannungsfilter/
Überspannungsschutz
surge suppressor
überspiralisiert/
superspiralisiert/superhelikal
supercoiled
Überspiralisierung supercoiling
Überstand *chem* supernatant
übersteuern overshoot
Übertrag carryover, carry forward
übertragbar
transmissible, communicable
übertragbare Krankheit
transmissible disease, communicable disease
Übertragbarkeit transferability
übertragen *med/tech/electr*
transmit (e.g. a disease)
Überträger/Überträgerstoff/Transmitter
transmitter; (Vektor) vector
Übertragung *phys/tech*
transmission, transfer

Übertragungsrate (im Datentransfer)
throughput rate, transfer rate
Übervölkerung overpopulation
überwachen
monitor, survey, supervise
Überwachung
monitoring, surveillance, supervision, surveyance
Überwachungskamera
monitoring camera
überwuchert overgrown
Überwurfmutter/ Überwurfschraubkappe (z.B. am Rotationsverdampfer)
swivel nut, coupling nut, mounting nut, cap nut, sleeve nut
Überzug coating
ubiqitär/weitverbreitet/ überall verbreitet
ubiquitous, widespread, existing everywhere
Uhrglas/Uhrenglas
watch glass, clock glass
Uhrmacherpinzette
watchmaker forceps, jeweler's forceps
Uhrmacherschraubenzieher
watchmaker's screwdriver, jeweler's screwdriver
Ultradünnschnitt *micros*
ultrathin section
Ultrafiltration ultrafiltration
Ultrakryomikrotom/ Ultragefriermikrotom
ultracryomicrotome
Ultramikrotom ultramicrotome
Ultraschall ultrasound, ultrasonics
Ultraschall betreffend/Ultraschall...
ultrasonic
Ultraschalldiagnose/Sonographie
ultrasound, ultrasonography, sonography
Ultrastruktur ultrastructure
Ultrazentrifugation ultracentrifugation
Ultrazentrifuge ultracentrifuge
Umdrehungen pro Minute (UpM)
revolutions per minute (rpm)
Umfang girth

umfüllen (Chemikalie)
transfer (a chemical); decant (in case of a liquid)
Umgang (Verhalten) handling
Umgebung surroundings, environs, environment, vicinity
Umgebungsdruck
ambient pressure
Umgebungstemperatur
ambient temperature
Umkehrosmose/Reversosmose
reverse osmosis
Umkehrphase/Reversphase
reversed phase, reverse phase
Umkehrphasenchromatographie
reversed phase chromatography, reverse-phase chromatography (RPC)
Umkehrpinzette/Klemmpinzette
reverse-action tweezers (self-locking tweezers)
Umkehrpotential reversal potential
umkippen (Gewässer)
turn over, become oxygen-deficient, turn anaerobic
Umkleide change area; (mit Spinden) locker room
Umkleidekabine
changing room, changing cubicle; (cleanroom/containment level: gowning room)
umkleiden change, change cloths
Umkristallisation recrystallization; (fraktionierte K.) fractional crystallization
umkristallisieren recrystallize
umlagern/umordnen *chem* rearrange
Umlagerung/Umordnung *chem*
rearrangement
➤ **tautomere Umlagerung**
tautomeric shift
Umlaufdüsen-Reaktor/ Düsenumlaufreaktor
nozzle loop reactor, circulating nozzle reactor
Umlaufreaktor/Umwälzreaktor/ Schlaufenreaktor
loop reactor, circulating reactor, recycle reactor

Umluft
  forced air, recirculating air;
  air circulation
Umluftofen forced-air oven
ummanteln (ummantelt)
  jacket (jacketed)
umpflanzen/versetzen
  transplant, replant
umrechnen convert
Umrechnungstabelle
  conversion table
Umriss contour, outline
Umsatz turnover
Umsatzgeschwindigkeit/Umsatzrate
  turnover rate, rate of turnover
Umsatzzeit turnover period
umsetzen turn, convert, transfer,
  process; *metabol* metabolize;
  (verkaufen) sell, turn over
Umsetzung, chemische
  chemical reaction, transformation,
  chemical change
umtopfen repot
Umverpackung overpacking
Umwälzkühler/
  Kältethermostat/Kühlthermostat
  refrigerated circulating bath
Umwälzpumpe circulation pump
Umwälzthermostat/Badthermostat
  circulating bath
umwandeln convert;
  (transformieren) transform
Umwandlung conversion;
  (Transformation) transformation
Umwelt environment
Umweltanalyse
  environmental analysis
Umweltanalytik
  environmental analytics
Umweltansprüche
  environmental requirements
Umweltaudit/Öko-Audit
  environmental audit
Umweltbedingungen
  environmental conditions
Umweltbelastung
  environmental burden,
  environmental load

Umweltchemie
  environmental chemistry
Umweltfaktor
  environmental factor
umweltgerecht
  environmentally compatible
Umweltkapazität/
  Grenze der ökologischen
  Belastbarkeit
  carrying capacity
Umweltkriminalität
  environmental crime
Umweltmedizin
  environmental medicine
Umweltmesstechnik
  environmental monitoring technology
Umweltpolitik environmental politics
Umweltrecht environmental law
Umweltschutz
  environmental protection;
  pollution control
Umweltschützer environmentalist
Umweltschutzpapier recycled paper
Umweltsünder person who litters or
  commits an environmental crime
Umweltvarianz
  environmental variance
Umweltverfahrenstechnik
  environmental process engineering
Umweltverhältnisse
  environmental conditions
Umweltverschmutzer polluter
Umweltverschmutzung
  environmental pollution
umweltverträglich
  environmentally compatible,
  environmentally friendly
Umweltverträglichkeit
  environmental compatibility
Umweltverträglichkeitsprüfung (UVP)
  environmental impact assessment
  (EIA)
Umweltwiderstand
  environmental resistance
Umweltwissenschaft
  environmental science
Umweltzerstörung
  environmental degradation

unbedenklich
  safe, without risk, unrisky
unbefruchtet unfertilized
unbelebt inanimate, lifeless, nonliving
unbeweglich/bewegungslos/fixiert
  nonmotile, immotile, immobile, motionless, fixed
undicht/leck leaking, leaky
undicht sein
  leak (doesn't close tightly)
Undichtigkeit leak, leakiness
undurchlässig/impermeabel
  impervious, impenetrable; impermeable
Undurchlässigkeit/Impermeabilität
  imperviousness, impermeability
unempfindlich insensitive
unersättlich insatiable
Unersättlichkeit insatiability
Unfall accident
  ➢ **Arbeitsunfall** occupational accident
  ➢ **Betriebsunfall**
    industrial accident, accident at work
  ➢ **größter anzunehmender U.**
    worst-case accident
  ➢ **Zwischenfall** incident
Unfallgefahr danger of accident
unfallträchtig hazardous
Unfallverhütung
  prevention of accidents
Unfallversicherung accident insurance
unfruchtbar/steril infertile, sterile
Unfruchtbarkeit/Sterilität
  infertility, sterility
ungefährlich (sicher) not dangerous, harmless (safe); (nicht gesundheitsgefährdend) nonhazardous
ungelöst undissolved
ungesättigt unsaturated
  ➢ **einfach u.** monounsaturated
  ➢ **mehrfach u.** polyunsaturated
Ungeziefer pest
ungleich/nicht identisch/anders
  unequal, different
Ungleichgewicht imbalance, disequilibrium
ungleichmäßig irregular, nonuniform

Unkrautbekämpfung/
Unkrautvernichtung weed control
Unkrautbekämpfungsmittel/
Unkrautvernichtungsmittel/
Herbizid herbicide, weed killer
unlöslich insoluble
Unlöslichkeit insolubility
unnatürlich unnatural
unpolar apolar
unregelmäßig/irregulär/anomal
  irregular, anomalous
Unregelmäßigkeit/Anomalie
  irregularity, anomaly
unreif unripe, immature
Unreife immaturity, immatureness
unscharf *micro/photo*
  not in focus, out of focus, blurred
Unschärfe *micro/photo*
  blurredness, blur, obscurity, unsharpness
unsicher/gefährlich unsafe
unspezifisch nonspecific
unteilbarer Faktor unit factor
unterbrechen interrupt
Unterbrecher/Trennschalter *electr*
  circuit breaker
Unterbrechung interruption
Unterdruck negative pressure
unterdrückbar suppressible
unterdrücken suppress
Unterdrückung suppression
Untereinheit subunit
Unterfunktion/Insuffizienz
  hypofunction, insufficiency
  ➢ **Überfunktion**
    hyperfunction, hyperactivity
untergärig bottom fermenting
  ➢ **obergärig** top fermenting
untergetaucht/submers
  submerged, submersed
untergliedern (untergliedert)
  subdivide(d)
Untergliederung subdivision
Unterkorn (Siebdurchgang) undersize
unterkühlen undercool, supercool
unterkühlte Flüssigkeit
  undercooled liquid, supercooled liquid

Unterkühlung
undercooling, supercooling
unterlegen inferior, put underneath
Unterlegenheit inferiority; defeat
Unterlegscheibe washer
unterordnen subordinate, submit
Unterphase (flüssig-flüssig)
lower phase
Unterscheidungsmerkmal
differentiating characteristic
Unterseite
underside, undersurface
untersuchen/
prüfen/testen/analysieren
investigate, examine, test,
assay, analyze; probe
Untersuchung/
Prüfung/Test/Probe/Analyse
investigation, examination (exam),
study, search, test, trial,
assay, analysis
➢ **Fruchtwasseruntersuchung**
analysis of amniotic fluid
(for prenatal diagnosis)
➢ **medizinische/ärztliche**
Untersuchung medical examination,
medical exam, medical checkup,
physical examination, physical
➢ **Wasseruntersuchung** water analysis
Untersuchungsgerät
testing equipment/apparatus
Untersuchungsliege
examination couch
Untersuchungslösung
test solution, solution to be analyzed
Untersuchungsmedium/
Prüfmedium/Testmedium
assay medium
unterteilt/kompartimentiert
divided, subdivided,
compartmentalized
Unterteilung subdivision

unterweisen instruct, train, teach
Unterweisung
instruction(s), training, teaching;
briefing
unvermischbar immiscible
Unvermischbarkeit immiscibility
unverschmiert/schmutzfrei
smudge-free
unverschmutzt uncontaminated
unverträglich/inkompatibel
incompatible
Unverträglichkeit/Inkompatibilität
incompatibility
Unverträglichkeitreaktion/
Inkompatibilitätreaktion
incompatibility reaction
unverzerrt/unverfälscht *math/stat*
unbiased
unverzweigt (Kette) *chem*
unbranched (chain)
unvorsichtig
careless, incautious, unwary
Unwucht unbalanced state
unwuchtig unbalanced
unzerbrechlich unbreakable
Uridylsäure uridylic acid
Urin/Harn urine
urinieren/harnlassen/harnen urinate
Urocaninsäure (Urocaninat)/
Imidazol-4-acrylsäure
urocanic acid (urocaninate)
Uronsäure (Urat) uronic acid (urate)
Ursprung origin
ursprünglich
(originär) original, basic, simple,
primitive; (urtümlich) pristine
ursprungsgleich/homolog
homologous
Usninsäure usnic acid
UV-Spektroskopie
ultraviolet spectroscopy,
UV spectroscopy

Vakuum vacuum
Vakuumdestillation
vacuum distillation,
reduced-pressure distillation
Vakuumdrehfilter/
Vakuumtrommeldrehfilter *micb*
rotary vacuum filter
Vakuumfalle vacuum trap
vakuumfest vacuum-proof
Vakuumfiltration
vacuum filtration,
suction filtration
Vakuumofen vacuum furnace
Vakuumpumpe vacuum pump
Vakuumverteiler (mit Hähnen)
vacuum manifold
Vakzination/Vakzinierung/Impfung
vaccination
Vakzine/Impfstoff vaccine
Valenz valence, valency
Valeriansäure/
Baldriansäure/Pentansäure
(Valeriat/Pentanat)
valeric acid, pentanoic acid
(valeriate/pentanoate)
Validierung validation
Vanillinsäure vanillic acid
Variabilität/
Veränderlichkeit/Wandelbarkeit
(*auch*: **Verschiedenartigkeit**)
variability
Variabilitätsrückgang
decay of variability
Varianz/
mittlere quadratische Abweichung/
mittleres Abweichungsquadrat *stat*
variance, mean square deviation
➢ **additive genetische Varianz**
additive genetic variance
➢ **Dominanzvarianz**
dominance variance
➢ **Umweltvarianz**
environmental variance
Varianzanalyse
analysis of variance (ANOVA)
Varianzheterogenität/
Heteroskedastizität *stat*
heteroscedasticity

Varianzhomogenität/
Varianzgleichheit/Homoskedastizität
stat homoscedasticity
Variationsbreite *stat*
range of variation,
range of distribution
Variationskoeffizient *stat*
coefficient of variation
Vaterschaftsbestimmung/
Vaterschaftstest paternity test
Vegetation vegetation, plant life
Vektor vector
venerische Übertragung
venereal transmission
Venerologie venereology
Ventil valve, vent
➢ **Abschaltventil/Absperrventil**
shut-off valve
➢ **Abzweigventil** *chromat* split valve
➢ **Ausatemventil (an Atemschutzgerät)**
exhalation valve
➢ **Ausgleichsventil** relief valve
(pressure-maintaining valve)
➢ **Auslaufventil** plug valve
➢ **Begrenzungsventil** limit valve
➢ **Dosierventil** metering valve
➢ **Drosselventil** throttle valve
➢ **Druckluftventil** pneumatic valve
➢ **Druckminderventil/**
Druckminderungsventil/
Druckreduzierventil
pressure-relief valve
(gas regulator, gas cylinder pressure
regulator)
➢ **Druckregelventil**
pressure control valve
➢ **Einspritzventil**
injection valve, syringe port
➢ **Entlüftungsventil** purge valve,
pressure-compensation valve
➢ **Fassventil (Entlüftung)** drum vent
➢ **Flügelhahnventil**
butterfly valve
➢ **hydraulisch vorgesteuert**
pilot-operated
➢ **Kegelventil**
cone valve, mushroom valve,
pocketed valve

Ve

- **Kugelventil** ball valve
- **Lufteinlassventil**
  air inlet valve, air bleed
- **Magnetventil (Zylinderspule)**
  solenoid valve
- **Membranventil** diaphragm valve
- **Nadelventil** needle valve
- **Quetschventil** pinch valve
- **Reduzierventil/
  Druckminderventil/
  Druckminderungsventil/
  Druckreduzierventil
  (für Gasflaschen)**
  pressure-relief valve (gas regulator, gas cylinder pressure regulator)
- **Regelventil** control valve
- **Rückflusssperre/Rücklaufsperre/
  Rückstauventil**
  backflow prevention, backstop (valve)
- **Rückschlagventil**
  check valve, backstop valve
- **Schieberventil** slide valve
- **Schlauchventil (Klemmventil)**
  (tubing) pinch valve
- **Schlussventil** cutoff valve
- **Sicherheitsventil**
  security valve, security relief valve
- **Sperrventil/Kontrollventil**
  check valve, control valve
- **Spülventil (Inertgas)**
  T-purge (gas purge device)
- **Überdruckventil**
  pressure valve, pressure relief valve
- **Verdrängerventil**
  positive-displacement valve
- **Zulaufventil/Beschickungsventil**
  delivery valve

Ventilation ventilation
Ventilator fan
**ventilieren/belüften/entlüften/
  durchlüften/Rauch abziehen lassen**
  ventilate, vent
**Veränderlichkeit/
  Wandelbarkeit/Variabilität** variability
verändern change, modify, vary
Veränderung
  change, modification, variation

verankern (befestigen)
  anchor (fasten/attach)
Verankerung anchorage
Verantwortliches Handeln (Rio '92)
  'Responsible Care'
verarbeiten process, processing, treat
Verarbeitung processing, treatment
Verarbeitungsdauer/Topfzeit
  pot life
veraschen incinerate, reduce to ashes
Veraschung ashing
verätzen (Chemikalien/Alkali/Säure)
  (chemicals/alkali/acid) burn;
  med cauterize
Verätzung (Chemikalien/Alkali/Säure)
  (chemicals/alkali/acid) burn,
  caustic burn; *med* cauterization
Verätzungsgefahr caustic hazard
**verbacken/festgebacken/
  festgesteckt (Schliff/Hahn)**
  jammed, seized-up, stuck,
  'frozen', caked
Verband (Vereinigung) association,
  union, federation, society;
  med dressing, bandage
Verbandskasten
  first-aid box, first-aid kit
Verbandsschrank first-aid cabinet
Verbandstisch instrument table
Verbandszeug
  bandaging/dressing material
verbinden connect, bond, link; *med*
  (e. Verband anlegen) dress, bandage
Verbinder (*siehe auch:* **Kupplung)**
  (Adapter) fitting(s), adapter;
  (Kupplung) coupling, coupler,
  connector
- **Flachsteckverbinder**
  flat-plug connector
- **Kabelverbinder** cable connector
- **Rohrverbinder/Rohrverbindung(en)**
  pipe fitting(s), fittings
- **Schlauchverbinder/
  Schlauchverbindung(en)**
  tubing connector (for connecting tubes), tube coupling, fittings
- **Schnellverbinder**
  quick-disconnect fitting

verbindlich obligatory, binding, mandatory, compulsory
Verbindung *allg* connection, bond, linkage; *chem* compound
➢ **chemische Verbindung** (chemical) compound
➢ **energiereiche Verbindung** high energy compound
Verbindungsmuffe (Kupplung: Rohr/Schlauch etc.) fittings, couplings, couplers
Verbindungsschnur *electr* connecting cord
Verbindungsstück (von Bauteilen) coupling, coupler
verblassen fade
Verbot prohibition, ban
➢ **Rauchverbot!** No Smoking!
verboten forbidden, prohibited
➢ **strengstens verboten** strictly forbidden, strictly prohibited
➢ **Zutritt verboten!/ Betreten verboten!** off-limits!, Do Not Enter!, No Entrance!, No Trespassing!
Verbrauch consumption, use, usage
Verbraucher/Konsument consumer
➢ **Großverbraucher** bulk consumer
Verbraucherschutz consumer protection
Verbrauchsmaterial consumable goods, consumables
verbrennen combust, incinerate, burn
Verbrennung combustion, incineration; *med* burn
➢ **chemische V.** chemical burn
Verbrennungsofen combustion furnace
Verbrennungsrohr (Glas) incinerating tube
Verbrennungswärme combustion heat, heat of combustion
Verbrühung/Verbrühungsverletzung scald, scalding
Verbund *tech* composite; composite construction; compound

Verbundfolie composite foil/film
Verbundglas laminated glass
Verbundwerkstoff composite material
verchromen (verchromt) chrome-plate(d)
verchromtes Messing chrome-plated brass
Verdacht (auf eine Erkrankung) suspicion (of a disease)
Verdachtsstoff *med* suspected toxin
verdampfen evaporate, vaporize
Verdampferkolben evaporating flask
Verdampfungs-Lichtstreudetektor evaporative light scattering detector (ELSD)
Verdampfungswärme heat of vaporization
Verdau (enzymatischer) digest (enzymatic)
verdauen digest
verdaulich digestible
Verdaulichkeit/Bekömmlichkeit digestibility
verderblich perishable; (Früchte: leicht verderblich) highly perishable
verdichten compress, condense; compact; concentrate; thicken
Verdichtung compression, condensation
verdicken thicken
Verdickung thickening; *med* swelling
Verdopplungszeit (Generationszeit) doubling time (generation time)
verdrahten wire
Verdrahtung wiring
verdrängen displace
Verdrängerventil positive-displacement valve
Verdrängung displacement
Verdrängungspumpe/ Kolbenpumpe (HPLC) displacement pump
Verdrängungsreaktion *biochem* displacement reaction
verdünnen dilute, thin down
Verdünner/Verdünnungsmittel/ Diluent/Diluens diluent
Verdünnung dilution, thinning down

Verdünnungsausstrich
  dilution streak, dilution streaking
Verdünnungs-Schüttelkultur
  dilution shake culture
verdunsten evaporate, vaporize
Verdunstung evaporation, vaporization
Verdunstungsbrenner
  evaporation burner
Verdunstungskälte/
  Verdunstungsabkühlung
  evaporative cooling
Verdunstungswärme
  heat of vaporization
veredeln
  refine, improve, process, finish
Veredlung refinement, improvement, processing, finishing
Veredlungsprozess refinement process
Vereinbarung agreement
verengen/einschnüren constrict
Verengung/Enge/Einschnürung
  constriction
verestern esterify
Veresterung esterification
Verfahren procedure, technique
Verfahrenstechnik
  process engineering
  ➢ **biologische V./Bioingenieurwesen/**
    Biotechnik bioengineering
Verfallsdatum expiration date
verfaulen (verfault)/
  zersetzen (zersetzt)
  foul, rot (rotten),
  decompose(d), decay(ed)
verflochten
  interwoven, intertwined, entangled
verflüchtigen volatilize
Verflüchtigung volatilization
verflüssigen liquefy, liquify
Verflüssiger liquefier
Verflüssigung liquefaction
Verformung/
  Formänderung/Deformation
  deformation
Verfügbarkeit availability
vergällen/denaturieren (z.B. Alkohol)
  denature
  ➢ **unvergällt** pure (not denatured)

vergären/fermentieren ferment
Vergärung/Fermentation fermentation
vergiften poison, intoxicate;
  (durch Tiergift) envenom
Vergiftung
  (Intoxikation) poisoning,
  intoxication;
  (durch Tiergift) envenomation,
  envenomization
Vergiftungszentrale/
  Entgiftungszentrale
  poison control center
Vergleich comparison; reference
Vergleichsgas (GC) reference gas
Vergleichspräzision
  reproducibility
Vergleichssubstanz
  comparative substance
vergolden gilding
Vergossenes/Übergelaufenes
  spillage, spill
vergrößern magnify, enlarge
Vergrößerung
  magnification, enlargement
  ➢ **x-fache Vergrößerung**
    magnification at x diameters
Vergrößerungsglas
  magnifying glass, magnifier, lens
Verhalten
  behavior, behaviour (*Br*), conduct
Verhaltensregeln rules of conduct
Verhältnis (Quotient/Proportion) ratio, quotient, proportion;
  (Beziehung) relationship
Verhältnisskala/Ratioskala *stat*
  ratio scale
Verhütung
  (Verhinderung: Unfälle/Vorsorge)
  prevention (provision);
  (Kontrazeption) contraception
verjüngen/regenerieren
  rejuvenate, regenerate
Verjüngung/Regeneration
  rejuvenation, regeneration
Verkabelung *electr*
  wiring, electrical wiring;
  (Leitungen) circuitry
verkalken (verkalkt) calcify (calcified)

Verkalkung/Kalkeinlagerung/
Kalzifizierung/Calcifikation
calcification
Verkäufer (Firma/Lieferant) vendor
Verkernung medullation
verketten concatenate
Verkettung concatenation
Verkleinerung *photo* (size) reduction
verkohlen char, carbonize
Verkohlung/Verkohlen
charring, carbonization
Verlängerung
elongation; (Ausdehnung) extension
Verlängerungskabel *electr*
extension cable
Verlängerungsklemme
extension clamp
Verlängerungsschnur *electr*
extension cord (power cord)
Verlauf course;
(Verlauf: z.B. einer Krankheit)
course (of a disease),
progress, development, trend;
(einer Kurve) path, course, trend
verleimen glue together;
(verkleben) stick together
verletzen injure
verletzlich vulnerable
Verletzung injury
vermehren/fortpflanzen/reproduzieren
propagate, reproduce
Vermehrung
(Vervielfältigung/Multiplikation)
mulplication;
(Fortpflanzung/Reproduktion)
propagation, reproduction;
(Amplifikation/Vervielfältigung)
amplification
Vermeidung avoidance
vermischbar miscible
➢ **unvermischbar** immiscible
Vermischbarkeit miscibility
➢ **Unvermischbarkeit**
immiscibility
vermischen mix
Vermischung mix, mixing
vermitteln mediate; arrange between
Vermittler/Mediator mediator

vermodern/modern
rot, decay, decompose, putrefy
vermuten/annehmen
hunch, guess, assume
Vermutung/Annahme
hunch, guess, assumption
vernachlässigbar negligible
Vernachlässigung negligence
Vernässung waterlogging
Vernebler nebulizer
vernetzen interconnect, network
vernetzt
netted, interconnected,
meshy, reticulate
Vernetzung interconnection, mesh,
network, networking, webbing
vernichten destroy, eliminate
Vernichtung destruction, elimination
veröden *med* obliterate;
(Landschaft) become desolate,
become deserted, obliterate
verödet *med* obliterate(d);
(Landschaft) desolate(d),
deserted, obliterate(d)
Verödung obliteration; desolation
Verordnung ordinance, decree
➢ **Arbeitsstättenverordnung**
 (ArbStättV)
Workplace Safety Ordinance,
Working Site Ordinance
➢ **Arbeitsstoffverordnung (AStoffV)**
Ordinance on
Occupational Substances
➢ **Gefahrgutverordnung (GefahrgutV)**
Hazardous Materials
Transportation Ordinance
➢ **Gefahrstoffverordnung (GefStoffV)**
Ordinance on Hazardous Substances
➢ **Gentechnik-Sicherheitsverordnung**
 (GenTSV)
Genetic Engineering
Safety Ordinance
➢ **Smogverordnung**
German smog ordinance
➢ **Störfallverordnung (StörfallV)**
Statutory Order
on Hazardous Incidents,
Industrial Accidents Directive

> **Strahlenschutzverordnung (StSV)**
  Radiation Protection Ordinance
> **Trinkwasserverordnung (TrinkwV)**
  Drinking Water Ordinance,
  Safe Drinking Water Ordinance
Verpackung packaging;
  (mit Folie/Papier) wrapping
> **in vitro-Verpackung**
  in vitro packaging
Verpackungsflasche packaging bottle
Verpackungsgläser packaging glasses
Verpackungsklebeband
  packaging tape
Verpackungsmittel packaging material
verpflanzen *med* (transplantieren)
  transplant; *bot* (umpflanzen/
  umsetzen/versetzen) replant
Verpflanzung/Transplantation
  transplantation
verpuffen deflagrate
Verpuffung deflagration
Verputz (innen/außen) finish
verriegeln
  bolt, bar, interlock
Verriegelung
  bolt(ing), barring, interlock
Versalzung (Boden) salinization
Versand shipment, dispatch
> **Auslieferung** delivery
versandfertig
  ready for dispatch/shipment/delivery
Versandkosten
  shipment costs, shipping charges,
  carriage charges
Versandpapiere
  shipping documents
Versauerung acidification
verschicken send, ship
Verschiebung shift
> **chemische V.**
  spectros chemical shift
Verschleiß
  (Abnutzung) wear, attrition, erosion;
  (Abrieb) abrasion
verschleißen/abnutzen wear (out)
verschleißfest
  resistant to wear
Verschleißteile expendable parts

Verschleppung displacement;
  (*zeitlich*) protraction, delay,
  procrastination
> **Kreuzkontamination** *chromat*
  carry-over, cross-contamination;
> **Übertragung** *med*
  transmission, spreading;
  protraction (through neglect)
verschließbar sealable
verschließen lock; (mit Deckel) cap;
  (mit Stopfen) stopper;
  (zustopfen) plug
Verschluss lock; closure;
  (Deckel) cap, lid, cover;
  seal (air-tight)
Verschlusskappe seal, cap, closure
**Verschlussklammer/Verschlussclip
  (Dialysierschlauch)**
  clamping closure (dialysis tubing)
Verschluss-Scheibe *electrophor*
  gate (gel-casting)
verschmelzen/fusionieren fuse
Verschmelzung/Fusion fusion
verschmutzen
  pollute, contaminate
verschmutzt
  polluted, contaminated
> **beschmutzt/fleckig** *allg*
  dirty, stained (fleckig)
> **unverschmutzt**
  unpolluted, uncontaminated
Verschmutzung
  pollution, contamination
> **Lärmverschmutzung**
  noise pollution
> **Luftverschmutzung**
  air pollution
> **Umweltverschmutzung**
  environmental pollution
> **Wasserverschmutzung**
  water pollution
Verschmutzungsgrad
  amount of pollution,
  degree of contamination
verschütten
  spill; (ausschütten) pour out,
  empty out; (überlaufen) overflow,
  run over

Verschütten
spill; (Ausschütten) pouring out, emptying out; (Überlaufen) overflow, run over
verseifen saponify
Verseifung saponification
versetzen/umpflanzen
transplant, replant
verseucht
contaminated, poisoned, polluted; (mit Mikroorganismen/Ungeziefer etc.) infested
Verseuchung contamination, pollution; (mit Mikroorganismen/Ungeziefer etc.) infestation
Verseuchungsgefahr risk of contamination
Versorgung *tech/mech/electr* supply
Versorgungsanschluss (Zubehörteil/Armatur) fixture
Versorgungseinrichtungen utilities
Versorgungsleitung
supply line, utility line, service line
verspritzen splash, squirt, spatter
Verständigung/Kommunikation
communication
verstärken *tech* amplify; *metabol* enhance; (fest/solide) reinforce, amplify (stimulus)
Verstärker *tech* amplifier; *metabol* (Substanz) enhancer
Verstärkerfolie (Autoradiographie)
intensifying screen (autoradiography)
Verstärkung
reinforcement, amplification
Verstärkungsstoff booster (substance)
versteifen (versteift) stiffen(ed)
verstellbar (einstellbar)
adjustable; variable
verstellen adjust, regulate, move, shift; (falsch einstellen) set the wrong way; (herumdrehen an) tamper with
verstopft clogged, blocked
verstrahlt radioactively contaminated
verstreuen (ausstreuen) spread, scatter, disseminate; (verstreut liegen) intersperse, disperse

Versuch experiment, test, trial; (Ansatz) attempt; (Bemühung) endeavor
➤ **Doppelblindversuch**
double blind assay, double-blind study
➤ **Feldversuch/ Freilanduntersuchung/ Freilandversuch**
field study, field investigation, field trial
➤ **Isotopenversuch** isotope assay
➤ **Schutzversuch/ Schutzexperiment**
protection assay, protection experiment
➤ **Tierversuch**
animal experiment
➤ **Triplettbindungsversuch**
triplet binding assay
➤ **Vorversuch**
pretrial, preliminary experiment
versuchen
try, attempt; (bemühen) endeavor
Versuchsanlage/Pilotanlage
pilot plant
Versuchsanordnung/ Versuchsaufbau
experiment setup, experimental arrangement
Versuchsbedingungen
experimental conditions
Versuchsdurchführung
performing an experiment, performance of an experiment
Versuchsreihe
experimental series, trial series
Versuchstier
laboratory animal, experimental animal
Versuchsverfahren
experimental procedure/protocol, experimental method
Verteidigung defense
verteilen distribute
Verteiler distributor; diffuser; manifold
➤ **Gasverteiler/Luftverteiler (Düse in Reaktor)** *biot* sparger

Verteilung *chem/stat* distribution;
(Zerstreuung) dispersion, spreading
> **Affinitätsverteilung**
  affinity partitioning
> **Altersverteilung**
  age distribution
> **bimodale Verteilung**
  bimodal distribution
> **Binomialverteilung**
  binomial distribution
> **freie/unabhängige Verteilung** *gen*
  independent assortment
> **F-Verteilung/Fisher-Verteilung/
  Varianzquotientenverteilung**
  F-distribution, Fisher distribution,
  variance ratio distribution
> **Gauß-Verteilung/Normalverteilung/
  Gauß'sche Normalverteilung**
  Gaussian distribution (Gaussian
  curve/normal probability curve)
> **Gegenstromverteilung**
  countercurrent distribution
> **Häufigkeitsverteilung**
  frequency distribution (FD)
> **Lognormalverteilung/
  logarithmische Normalverteilung**
  lognormal distribution,
  logarithmic normal distribution
> **nicht-zufallsgemäße Verteilung**
  nonrandom disjunction
> **Normalverteilung**
  normal distribution
> **Poissonsche Verteilung/
  Poisson Verteilung**
  Poisson distribution
> **Randverteilung**
  marginal distribution
> **statistische Verteilung**
  statistical distribution
> **Varianzquotientenverteilung/
  F-Verteilung/Fisher-Verteilung**
  variance ratio distribution,
  F-distribution, Fisher distribution
Verteilungsfunktion *stat*
  distribution function
Verteilungskoeffizient *chromat*
  partition coefficient,
  distribution constant

Verteilungsmuster
  distribution pattern
**vertikale Luftführung
(Vertikalflow-Biobench)**
  vertical air flow (clean bench with
  vertical air curtain)
Vertikalrotor *centrif* vertical rotor
verträglich/kompatibel/tolerant
  compatible, tolerant
> **unverträglich/
  inkompatibel/intolerant**
  incompatible, intolerant
Verträglichkeit/Kompatibilität/Toleranz
  compatibility, tolerance
> **Unverträglichkeit/
  Inkompatibilität/Intoleranz**
  incompatibility, intolerance
Vertrauensintervall/Konfidenzintervall
  stat confidence interval
Vertreter representative, rep;
  (Verkauf) sales representative
verunreinigen contaminate, pollute
verunreinigt/schmutzig/unsauber
  impure
Verunreinigung/Kontamination
  impurity, contamination
**Vervielfältigung/
Vermehrung/Amplifikation**
  amplification
verwachsen/angewachsen *allg*
  fused, coalescent
Verwachsung *allg*
  fusion; coalescence, symphysis
Verwaltung administration
> **öffentliche V.**
  civil service, public service
Verwaltungsangestellte(r)
  administrative employee
verwandt akin, related;
  gen (zugehörig) cognate
**Verweilzeit/Verweildauer/
Aufenthaltszeit/Verweildauer**
  residence time;
  (Retentionszeit) retention time
verwelken
  wither, wilt, fade (shrivel up)
Verwendbarkeitsdauer/Nutzungsdauer
  working life

Verwendung use, usage
➢ **Weiterverwendung**
continued use/usage
➢ **Wiederverwendung** reuse
verwerfen *chem* discard, dispose of
verwerten *metabol/ecol* utilize
Verwertung *metabol/ecol* utilization
verwesen/zersetzen
putrefy, rot, decompose
Verwesung/Zersetzung
putrefaction, rotting, decomposition
verwittern *geol* weather; *bot* waste
Verwitterung
geol weathering; *bot* wastage
Verwitterungsbeständigkeit durability
verzehren/verschlingen/
herunterschlingen
devour, gulp down
verzerrt/verfälscht *math/stat* biased
verzögern delay, retard
Verzögerung delay, retardation
verzuckern saccharify
Verzuckerung saccharification
verzweigen, sich branch out, ramify
verzweigt branched, ramified
verzweigtkettig *chem* branched-chained
Verzweigung *chem* branching
Verzweigungsstelle branch site
Vesikel *nt*/**Bläschen** vesicle
vesikulär/bläschenartig
vesicular, bladderlike
Veste, kugelsichere bulletproof vest
Vibrationsbewegung vibrating motion
Vibrationsenergie
vibrational energy, vibration energy
vibrieren vibrate
Viehfutter animal feed
Vielfachmessgerät/
Universalmessgerät/Multimeter
electr multimeter
Vielfachschale/Multischale *micb*
multiwell plate
Vielfachzucker/Polysaccharid
multiple sugar, polysaccharide
Vielfalt/Vielfältigkeit/
Vielgestaltigkeit/Mannigfaltigkeit
diversity
Vielkanalgerät multichannel instrument

vielschichtig/mehrschichtig
multilayered
Vierfuß (für Brenner) quadrupod
Vierkantflasche square bottle
Viertelswert/Quartil *stat* quartile
vierwertig *chem* tetravalent
Vigreux-Kolonne Vigreux column
Virologie virology
Virose/Viruserkrankung virosis
Virostatikum virostatic
virtuelles Bild *micros* virtual image
virulent virulent
Virulenz/Infektionskraft
virulence (disease-evoking
power/ability of cause disease)
Virus (*pl* **Viren)** virus
Viruserkrankung/Virose
viral infection, virosis
viruzid virucidal, viricidal
viskos/viskös/
zähflüssig/dickflüssig
viscous, viscid
(glutinous consistency)
Viskosität/
Dickflüssigkeit/Zähflüssigkeit
viscosity, viscousness
Viskositätskoeffizient
coefficient of viscosity
Vitalfarbstoff vital dye, vital stain
Vitalfärbung/Lebendfärbung
vital staining
Vitalität/Lebenskraft vitality
Vitalkapazität vital capacity
Vitamin(e) vitamin(s)
Vitrine/Schaukasten showcase
Vlies(stoff) fleece
voll aufdrehen (Wasserhahn etc.)
full blast
vollgesogen (mit Wasser)
waterlogged
vollgestellt/zugestellt
(Schränke/Abzug etc.)
cluttered
Vollmedium complete medium
Vollpipette/volumetrische Pipette
transfer pipet, volumetric pipet
Vollzeitbeschäftigte(r)
full-time employee (worker)

Voltammetrie voltammetry
➢ **lineare Voltammetrie**
  linear scan voltammetry,
  linear sweep voltammetry
➢ **Stripping-Analyse/ Inversvoltammetrie**
  stripping analysis,
  stripping voltammetry
➢ **cyclische Voltammetrie/ Cyclovoltammetrie**
  cyclic voltammetry
Volumenanteil (Volumenbruch) volume fraction
Vorarbeiten preparatory work
Vorauflaufbehandlung *agr* pre-emergence treatment
Voraussage prediction
Voraussagemodell predictive model
voraussagend predictive
Vorbehandlung pretreatment
Vorbereitung preparation
➢ **Probenvorbereitung** sample preparation
Vorderseite (Gerät etc.) front side, front, face
vorderseitig (bauchseitig) front side, ventral
Vordruck/Eingangsdruck (Hochdruck: Gasflasche) initial pressure, initial compression, high pressure
Vorfilter prefilter
Vorfluter recipient; discharge; (Gewässer: Abwassergraben etc.) drainage ditch, outfall ditch, receiving water
vorgereinigt precleaned
Vorhängeschloss (für Laborspind etc.) padlock
vorherrschen predominate
Vorhersage/Prognose prognosis
Vorkehrung precaution
Vorkehrungen treffen take precautions (precautionary measures)
vorkeimen pregerminate

Vorkeimung pregermination
Vorkommen occurrence, presence
Vorkultur preculture
Vorlage *dest* distillation receiver adapter, receiving flask adapter
Vorlagekolben recovery flask, receiving flask, receiver flask (collection vessel)
Vorlauf *dest* forerun; forshot (alcohol)
Vorläufer/Präkursor precursor
Vorprobe preliminary test, crude test
Vorrat stock, store, supply (*meist pl* supplies), provisions, reserve
Vorratshaltung hoarding of food
Vorratskammer storage chamber
Vorratsschädling storage pest
Vorratsschrank storage cabinet; (Schränkchen) cupboard
vorreinigen prepurify
Vorreinigung precleaning
Vorrichtung device
Vorsäule (HPLC) guard column
Vorschaltdrossel *electr* ballast, choke
Vorschaltgerät *electr* ballast unit; (Starter: Leuchtstoffröhren) starter
Vorschlaghammer sledge hammer
Vorschrift(en) (Anweisungen) instructions, specifications, directions, prescription; (Regeln) policy, rule
Vorschub *micros* advance
Vorsicht caution, cautiousness, care, carefulness, precaution; (Vorsicht!) caution! (careful!)
vorsichtig cautious, careful
Vorsichtsmaßnahme/ Vorsichtsmaßregel precaution, precautionary measure, safety warning
Vorsorge provision
Vorsorgemaßnahme provisional measure, precautionary measure
Vorsorgeuntersuchung preventive medical checkup

Vorstoß *lab/chem*
(*siehe auch unter:* **Adapter**) adapter
- **Destilliervorstoß** receiver adapter
- **Filtervorstoß** adapter for filter funnel
- **Vakuumvorstoß** vacuum adapter
- **Vakuumfiltrationsvorstoß** vacuum-filtration adapter

Vortex/Mixer/Mixette/Küchenmaschine vortex, mixer

Vortex-Bewegung (Schüttler: kreisförmig-vibrierende Bewegung) vortex motion, whirlpool motion

Vortexmischer/ Vortexschüttler/Vortexer (für Reagenzgläser etc.) vortex shaker, vortex

Vortrieb/Anschub thrust

Vorverstärker preamplifier

Vorversuch pretrial, preliminary experiment

Vorwärmer preheater

Vorzugstemperatur cardinal temperature

Vulkanasche volcanic ash

Wa

Waage scale (weight), balance (mass)
- **Analysenwaage** analytical balance
- **Balkenwaage** beam balance
- **Federzugwaage/Federwaage** spring balance, spring scales
- **Feinwaage/Präzisionswaage** precision balance
- **Kontrollwaage** checkweighing scales
- **Laborwaage** laboratory balance
- **Tafelwaage** pan balance
- **Tischwaage** bench scales
- **Wasserwaage** level

Waagschale scalepan, weigh tray, weighing tray, weighing dish
Wachs wax
- **Plastilin** plasticine

wachsartig waxy, wax-like, ceraceous
wachsen grow; thrive
Wachsfüßchen (Plastilinfüßchen an Deckgläschen) *micros* wax feet, plasticine supports on edges of coverslip
Wachspapier wax paper
Wachstum growth
wachstumsfördernd growth-stimulating
Wachstumsgeschwindigkeit/Wachstumsrate/Zuwachsrate growth rate
wachstumshemmend growth-retarding, growth-inhibiting
Wachstumshemmer/Wuchshemmer/Wuchshemmstoff growth inhibitor
Wachstumskurve growth curve
Wachstumsphase growth phase
- **Absterbephase** decline phase, phase of decline, death phase
- **Adaptationsphase/Anlaufphase/Latenzphase/Inkubationsphase/lag-Phase** lag phase, latent phase, incubation phase, establishment phase
- **Beschleunigungsphase/Anfahrphase** acceleration phase
- **Eingewöhnungsphase** establishment phase
- **exponentielle Wachstumsphase/exponentielle Entwicklungsphase** exponential growth phase
- **lag-Phase/Adaptationsphase/Anlaufphase/Latenzphase/Inkubationsphase** lag phase, incubation phase, latent phase, establishment phase
- **logarithmische Phase** logarithmic phase (log-phase)
- **Ruhephase/Ruheperiode** dormancy period
- **stationäre Phase** stationary phase, stabilization phase
- **Teilungsphase** division phase
- **Verlangsamungsphase/Bremsphase/Verzögerungsphase** deceleration phase, retardation phase

Wächter/Wachmann guard, security guard
Wägebürette weight buret, weighing buret
Wägeglas weighing bottle
Wägelöffel weighing spoon
wägen/wiegen weigh
Wägepapier weighing paper
Wägeschiffchen weighing boat, weighing scoop
Wägespatel weighing spatula
Wägetisch weighing table
Wägung weighing
Wählscheibe/Einstellscheibe dial
wahrnehmen/empfinden (Reiz) perceive
Wahrnehmung/Empfindung/Perzeption (Reiz) perception
Wahrscheinlichkeit probability, likelihood
Walrat spermaceti
Walratöl spermaceti oil, sperm oil
Walze/Rolle/Zylinder barrel, cylinder
Wanderung/Migration *chromat/electrophor* migration

Wanderungsgeschwindigkeit/ Migrationsgeschwindigkeit *chromat/electrophor*
migration speed (velocity)
Wandler/Umwandler
transducer, converter
Wandschrank
wall cabinet, cupboard
Wandtafel wall chart
Wangenabstrich *med* buccal swab
Wanne tub; *electrophor*
reservoir, tray
Wannenform *chem*
(Cycloalkane) boat conformation
Wannen-Stapel *micb* multi-tray
Ware(n) ware, articles, products, goods
Warenkontrolle
inspection, checking of goods
Warenlager
stockroom, repository, warehouse
Warenprobe sample, specimen
Warensendung
consignment of goods,
shipment of goods
Warenzeichen/Markenbezeichnung
brand name, trade name;
(eingetragenes Warenzeichen)
registered trademark
Wärme/Hitze warmth, heat
➢ **Abwärme** waste heat
➢ **Bildungswärme** heat of formation
➢ **Erwärmung** warming
➢ **globale Erwärmung** global warming
➢ **Lösungswärme** heat of solution
➢ **Mischungswärme** heat of mixing
➢ **Reaktionswärme/Wärmetönung**
heat of reaction
➢ **spezifische Wärme** specific heat
➢ **Strahlungswärme** radiant heat
➢ **Umwandlungswärme/latente Wärme**
heat of transition, latent heat
➢ **Verbrennungswärme**
heat of combustion
➢ **Verdunstungswärme**
heat of evaporation,
heat of vaporization
Wärmeabstrahlung heat dissipation
Wärmebehandlung heat treatment

Wärmedurchgangszahl (C)
thermal conductance
Wärmekapazität
heat capacity, thermal capacity
Wärmeleitfähigkeit
heat conductivity,
thermal conductivity
Wärmeleitfähigkeitsdetektor/ Wärmeleitfähigkeitsmesszelle (WLD)
thermal conductivity detector (TCD)
Wärmeofen heating oven,
heating furnace (more intense)
Wärmepumpe heat pump
Wärmeregler thermoregulator
Wärmeschrank incubator
Wärmeschrumpfen heat-shrinking
Wärmestrahlung thermal radiation
wärmesuchend/thermophil
thermophilic
Wärmetauscher heat exchanger
Wärmetönung heat tone, heat tonality;
heat of reaction, heat effect
Wärmetransport heat transport
Wärmeübergang heat transfer
Wärmeübertragung heat transmission
Wärmeverlust heat loss
Wärmezufuhr
heat supply, addition of heat
Warnband warning tape
warnen warn
warnend warning, precautionary
Warnetikett warning label
Warnruf/Alarm alarm
Warnschild danger sign, warning sign
Warntafel warning sign
Warnung warning, caution
Warnzeichen/Warnhinweis
warning, warning sign,
precaution sign
Wartezeit waiting time/period;
(Verzögerung) delay
Wartung/Instandhaltung
maintenance
Wartungsdienst maintenance service
wartungsfrei maintenance-free
Wartungsmonteur
maintenance worker,
maintenance man

Wa

Wartungspersonal
  maintenance personnel
Wartungsvertrag
  maintenance contract
Waschbecken wash basin
➢ **verstopftes W.** blocked drain
Wäsche washing; clothes, linen;
  (schmutzige Kleider) laundry
Wascheinrichtung
  washing facilities
Wäschekorb
  laundry hutch, laundry hamper
Wäscherei laundry;
  (Schnellwäscherei) laundrette
Waschmittel detergent
Waschraum/Toilette
  washroom, lavatory
Wasser water
➢ **Abwasser** wastewater
➢ **Bidest** double distilled water
➢ **Brauchwasser/Betriebswasser**
  (nicht trinkbares Wasser)
  process water, service water;
  (Industrie-B.) industrial water
  (nondrinkable water)
➢ **Brunnenwasser** well water
➢ **destilliertes Wasser** distilled water
➢ **entionisiertes Wasser**
  deionized water
➢ **gereinigtes Wasser/**
  aufgereinigtes Wasser/
  aufbereitetes Wasser
  purified water
➢ **Grundwasser** ground water
➢ **Haftwasser**
  film water, retained water
➢ **hartes Wasser** hard water
➢ **Kristallisationswasser**
  water of crystallization
➢ **Kristallwasser**
  crystal water,
  water of crystallization
➢ **Leitungswasser** tap water
➢ **Meerwasser** seawater, saltwater
➢ **Mineralwasser** mineral water
➢ **Peptonwasser** peptone water
➢ **Quellwasser** springwater
➢ **salziges Wasser** saline water

➢ **Salzwasser** saltwater
➢ **schweres Wasser D_2O** heavy water
➢ **Selterswasser/Sprudel** soda water
➢ **Süßwasser** freshwater
➢ **trinkbares Wasser** potable water
➢ **Trinkwasser**
  drinking water, potable water
➢ **Warmwasser** hot water
➢ **weiches Wasser** soft water
Wasserabscheider
  water separator, water trap
wasserabstoßend/wasserabweisend
  water-repellent, water-resistant
Wasseraktivität/Hydratur
  water activity
wasseranziehend/hygroskopisch
  (Feuchtigkeit aufnehmend)
  hygroscopic
Wasseraufbereitung
  water purification
Wasseraufbereitungsanlage
  water purification plant/facility,
  water treatment plant/facility
Wasseraufnahme water uptake
Wasserbad water bath
Wasserdampf water vapor
Wasserdestillierapparat
  water still
wasserdicht/wasserundurchlässig
  watertight, waterproof
Wassereinlagerung/
  Wasseranlagerung/Hydratation
  hydration
Wasserenthärter water softener
Wasserenthärtung water softening
wasserentziehend/dehydrierend
  dehydrating
Wasserentzug dehydration
wasserfest waterproof
wasserfrei free from water;
  moisture-free; anhydrous
Wassergefahrenklasse (WGK)
  water hazard class
Wassergehalt water content
Wasserglas $M_2O \times (SiO_2)_x$
  water glass, soluble glass
Wassergüte/Wasserqualität
  water quality

Wasserhahn faucet
Wasserhärte water hardness
➢ **bleibende Härte/permanente Härte**
  permanent hardness
➢ **Gesamthärte** total hardness
➢ **Karbonathärte/Carbonathärte/
  vorübergehende Härte/
  temporäre Härte**
  carbonate hardness,
  temporary hardness
Wasserhülle/Hydrationsschale *chem*
  hydration shell
Wasserkapazität moisture capacity,
  water-holding capacity of soil
wasserleitend water-conducting
wasserlöslich water-soluble
Wasserlöslichkeit water solubility
Wassermantel (Kühler) water jacket
Wasserpotential/Hydratur/Saugkraft
  water potential
Wasserprobe water sample
Wasserpumpe water pump
Wasserpumpenzange/Pumpenzange
  water pump pliers,
  slip-joint adjustable water pump
  pliers (adjustable-joint pliers)
wasserreaktiv water reactive
Wassersättigung water saturation
Wassersättigungsdefizit
  water saturation deficit (WSD)
Wassersäule
  water column, column of water
Wasserschieber/Wasserabzieher
  squeegee (for floors)
Wassersog
  water tension, water suction
wasserspaltend/hydrolytisch
  hydrolytic
Wasserspaltung/Hydrolyse
  hydrolysis
Wasserstoff (H) hydrogen
**Wasserstoffbrücke/
Wasserstoffbrückenbindung**
  hydrogen bond
Wasserstoffelektrode
  hydrogen electrode
Wasserstoffion (Proton)
  hydrogen ion (proton)

Wasserstoffperoxid hydrogen peroxide
Wasserstrahl jet of water
Wasserstrahlpumpe
  water pump, filter pump,
  vacuum filter pump
Wasserstress water stress
Wasserströmung water flow
wasserundurchlässig
  watertight, waterproof
Wasserundurchlässigkeit
  watertightness, waterproofness
wasserunlöslich insoluble in water
Wasserunlöslichkeit
  water-insolubility
Wasseruntersuchung/Wasseranalyse
  water analysis
Wasserverbrauch
  water consumption, water usage
Wasserverlust water loss
Wasserverschmutzung water pollution
Wasserversorgung water supply
Wasserwaage level
Wasserzufuhr water supply
**Wasserzulauf/Wasserzapfstelle
(Wasserhahn)** water outlet
wässrig aqueous
Watte absorbent cotton;
  (Baumwolle) cotton
Wattebausch/Baumwoll-Tupfer
  cotton ball, cotton pad
Wattestopfen cotton stopper
Wechselbeziehung
  interrelation, interrelationship
**Wechselfeld-Gelelektrophorese/
Puls-Feld-Gelelektrophorese**
  pulsed field gel electrophoresis
  (PFGE)
**Wechselsprechanlage/
Gegensprechanlage**
  two-way intercom, two-way radio
Wechselstrom alternating current (AC)
Wechselvorlage *chromat*
  fraction cutter; ('Spinne'/Euter-
  vorlage/Verteilervorlage) *dest* multi-
  limb vacuum receiver adapter, cow
  receiver adapter, 'pig' (receiving
  adapter for three/four receiving
  flasks)

Wechselwirkung interaction
Wechselzahl
  k_{cat} **(katalytische Aktivität)**
  turnover number
wegführend/ausführend/ableitend
  efferent
Wegwerf.../Einweg.../Einmal...
  disposable
Weichmacher/Plastifikator
  softener (esp. in foods),
  plasticizer (in plastics a.o.)
Weingeist spirit of wine
  (rectified spirit: alcohol)
Weinsäure/Weinsteinsäure (Tartrat)
  tartaric acid (tartrate)
Weinstein/Tartarus
  (Kaliumsalz der Weinsäure)
  tartar
Weiterbildung continuing education
weiterleiten forward; refer; fedirect;
  (fortleiten) pass on, propagate
weiterverarbeiten/prozessieren
  process, finish
Weiterverarbeitung/Prozessierung
  processing, finishing
Weithals ...
  wide-mouthed, widemouthed,
  wide-neck, widenecked
Weithalsfass
  wide-mouth vat, wide-neck vat
Weithalsflasche
  wide-mouth flask,
  wide-neck bottle
weitverbreitet/ubiquitär
  (überall verbreitet)
  widespread, ubiquitous
  (existing everywhere)
Weitwinkel *micros* widefield
welk/schlaff wilted, withered, faded,
  limp, flaccid
welken wilt, wither, fade
welkend wilting, withering, fading,
  flaccid, deficient in turgor
Welkepunkt wilting point
Welkungsgrad, permanenter
  permanent wilting percentage
Welle shaft, spindle
Wellendichtung (Rotor) shaft seal

Wellenlänge wavelength
Wellenzahl (IR) wavenumber
Wellpappe corrugated board
Werk/Fabrik
  factory, plant, manufacturing plant
Werkbank (Labor-Werkbank)
  bench, workbench (lab bench)
  ➤ **Fallstrombank** vertical flow
    workstation/hood/unit
  ➤ **Handschuhkasten/**
    Handschuhschutzkammer
    glove box
  ➤ **Labor-Werkbank**
    laboratory bench, lab bench
  ➤ **Querstrombank**
    laminar flow workstation,
    laminar flow hood,
    laminar flow unit
  ➤ **Reinraumwerkbank**
    clean-room bench
  ➤ **Sicherheitswerkbank**
    clean bench
  ➤ **sterile Werkbank** sterile bench
Werkstatt workshop, 'shop'
Werkstoff material
Werkstück workpiece
Werkstückkasten/Teilekasten
  tote tray
Werkzeug tools
Werkzeugkasten tool box
Wertigkeit valency
  ➤ **einwertig** univalent
  ➤ **zweiwertig** bivalent, divalent
  ➤ **dreiwertig** trivalent
  ➤ **vierwertig** tetravalent
  ➤ **fünfwertig** pentavalent
wetterbeständig weatherproof
Widerstand resistance
  ➤ **spezifischer W.** resistivity
widerstandsfähig
  resistive, resistant, hardy
Widerstandsfähigkeit
  resistance, resistivity, hardiness
Widerstandsheizung resistive heating
Widerstandsthermometer
  resistance thermometer
wiederaufladbar rechargeable
Wiederaufnahme *physiol* re-uptake

Wiederbefall reinfestation
Wiederbelebung/Reanimation
  resuscitation
  ➢ **kardiopulmonale Reanimation**
    cardiopulmonary resuscitation (CPR)
  ➢ **Mund-zu-Mund Beatmung**
    mouth-to-mouth resuscitation/
    respiration
Wiederbelebungsversuch
  attempt at resuscitation
wiedergewinnen/
  rückgewinnen/aufbereiten
  retrieve, recover
Wiedergewinnung
  retrieval, recovery
Wiederholbarkeit repeatability
Wiederholung
  repeat, repetition; recurrence
Wiederholungsrisiko
  recurrence risk
wiederverwenden reuse
Wiederverwendung reuse
wiederverwerten recycle
Wiederverwertung recycling
wiegen weigh
  ➢ **abwiegen (eine Teilmenge)**
    weigh out
  ➢ **auswiegen (genau wiegen)**
    weigh out precisely
  ➢ **einwiegen (nach Tara)**
    weigh in (after setting tare)
willkürlich
  generell arbitrary, random;
  med/psych voluntary
Winde/Kurbel winch
winden wind, twist, coil
Windkessel
  air chamber, air receiver,
  air vessel, surge chamber
Windmesser/Anemometer
  air meter, anemometer
Windung
  (Spirale) twist, coil, spiral
  (a series of loops);
  Krümmung/Biegung) winding,
  contortion, turn, bend; (Bewegung)
  spiral movement, spiral coiling
Winkel angle

Winkelrohr/
  Winkelstück/Krümmer
  (Glas/Metall etc. zur Verbindung)
  bend, elbow, elbow fitting, ell,
  bent tube, angle connector
Winkelrotor *centrif*
  angle rotor, angle head rotor
winterfest/winterhart hardy
Wippbewegung
  see-saw motion,
  rocking motion
Wippe/Schwinge/Rüttler rocker
Wippschwingung (IR)
  wagging vibration
Wirbel
  whirl, swirl, spin; eddy, vortex
Wirbelschichtreaktor/Wirbelbettreaktor
  fluidized bed reactor
Wirbelstrom eddy current
wirken act, work, be effective,
  causing an effect, take effect
Wirkschwelle
  no adverse effect level (NOAEL)
Wirkstoff/Wirksubstanz
  active ingredient, active principle,
  active component
Wirkung effect, action
Wirkungsgrad efficiency
Wirkungsspezifität specificity of action
Wirkungsweise/Mechanismus
  mode of action, mechanism
Wirrwarr/Durcheinander/Unordnung
  clutter
Wirtsspektrum host range
Wirtsspezifität host specificity
Wirtswechsel alternation of hosts
wischen wipe
Wischer wipe, wiper
  ➢ **Fensterwischer/Fensterabzieher**
    squeegee (for windows)
  ➢ **Wasserschieber/Wasserabzieher**
    squeegee (for floors)
Wischtuch/Wischlappen
  cloth, wiping cloth, rag;
  (Wischtücher) wipes
Wölbung/Koeffizient der Wölbung *stat*
  kurtosis
Wolfram (W) tungsten

Wo

Wolle wool
Wollfettdrüse wool fat gland
Woulff'sche Flasche
    Woulff bottle
Wuchs growth, habit
Wuchsform/Habitus
    growth form, appearance, habit
Wuchshemmer/
    Wachstumshemmer/
    Wuchshemmstoff
    growth inhibitor
Wuchskraft growth vigor

Wuchsstoff (Pflanzenwuchsstoff)/
    Phytohormon growth regulator,
    phytohormone, growth substance
Wunde wound
➢ **offene Wunde** open wound
➢ **Schnittwunde** cut
Wundgewebe/Wundcallus/Wundholz
    wound tissue, callus
Wundsalbe/Wundheilsalbe
    healing ointment,
    wound healing ointment
Wundheilung wound healing

Xanthangummi xanthan gum
Xanthogensäure
xanthogenic acid,
xanthic acid, xanthonic acid,
ethoxydithiocarbonic acid
Xenobiotikum (*pl* **Xenobiotika**)
xenobiotic (*pl* xenobiotics)

Xenotransplantat/
  Fremdtransplantat xenograft
  (xenogeneic graft: from other
  species)
Xylit xylitol/xylite
Xylose xylose
Xylulose xylulose

zä

zäh tough, rigid
zähflüssig/dickflüssig/viskos/viskös
  viscous, viscid
Zähflüssigkeit/
  Dickflüssigkeit/Viskosität
  viscosity, viscousness
Zählkammer counting chamber
Zählplatte counting plate
Zahlung (einer Rechnung) payment
Zählung count
Zahlungsbedingungen
  conditions of payment,
  terms of payment
Zahlungsbeleg/Quittung
  confirmation of payment, receipt
Zahlungsfrist term of payment
Zahlungstermin date of payment
Zahlungsweise mode of payment
Zahnstocher toothpick
Zange plier, pliers;
  (Labor: Haltezangen) tongs
➤ **Becherglaszange**
  beaker clamp
➤ **Beißzange/Kneifzange**
  pliers, nippers (*Br*)
➤ **Crimpzange/**
  Aderendhülsenzange
  (Quetschzange)
  crimping pliers, crimper
➤ **Eckrohrzange**
  rib joint pliers, rib-lock pliers
➤ **Extraktionszange** *dent*
  extraction forceps
➤ **Flachzange** flat-nosed pliers
➤ **Greifzange** grippers
➤ **Gripzange** Vise-Grip® pliers
➤ **Kneifzange** cutting pliers
➤ **Knochenzange**
  bone-cutting forceps,
  bone-cutting shears
➤ **Kolbenzange** flask tongs
➤ **Kombizange**
  combination pliers, linesman pliers;
  (verstellbar) slip-joint pliers
➤ **Lochzange** punch pliers
➤ **Mehrzweckzange** utility pliers
➤ **Monierzange/Rabitzzange**
  end nippers

➤ **Pumpenzange/Wasserpumpenzange**
  water pump pliers,
  slip-joint adjustable
  water pump pliers
➤ **Revolverlochzange**
  revolving punch pliers
➤ **Rohrzange** pipe wrench,
  griplock pliers (US),
  channellock pliers (US)
➤ **Seitenschneider** diagonal pliers
➤ **Sicherungsringzange**
  snap-ring pliers, circlip pliers
➤ **Spitzzange** longnose pliers,
  long-nose pliers
➤ **Spitzzange, gebogen**
  bent longnose pliers,
  bent long-nose pliers
➤ **Storchschnabelzange**
  needle-nose pliers,
  snipe-nose pliers,
  snipe-nosed pliers
➤ **Storchschnabelzange, gebogen**
  dip needle-nose pliers
➤ **Telefonzange/Kabelzange**
  linesman pliers
➤ **Tiegelzange** crucible tongs
Zapfen (Fass~) faucet
Zapfhahn/Fasshahn spigot
Zeiger pointer; indicator
Zeigerokular *micros*
  pointer eyepiece
Zeigerwerte indicator value
zeitaufgelöst time-resolved
Zeitgeber Zeitgeber, synchronizer
Zeitschaltuhr/Zeitschalter timer
Zellaufschluss
  (Öffnen der Zellmembran) cell lysis;
  (Zellfraktionierung) cell
  fractionation; (Zellhomogenisierung)
  cell homogenization
Zellaufschlussgerät cell disrupter
Zellextrakt cell extract
Zellfraktionierung cell fractionation
zellfreier Extrakt cell-free extract
Zellfusion/Zellverschmelzung
  cell fusion
Zellgift/Zytotoxin/Cytotoxin
  cytotoxin

Zellhomogenisation/
Zellhomogenisierung
  cell homogenization
zellig cellular
➢ **nicht zellig/azellulär**
  acellular, noncellular
Zellkultur cell culture
Zelllinie cell lineage, cell line, celline
Zellobiose/Cellobiose cellobiose
Zellschaber cell scraper
zellschädigend/zytopathisch/
  cytopathisch (zytotoxisch)
  cytopathic (cytotoxic)
Zellsorter/Zellsortierer/
  Zellsortiergerät (Zellfraktionator)
  cell sorter
Zellsortierung cell sorting
Zellstoff wood pulp
Zellstoffwatte wood wool
Zelltod cell death
➢ **programmierter Zelltod (Apoptose)**
  programmed cell death (apoptosis)
zelltötend/zytozid cytocidal
Zellulose/Cellulose cellulose
Zentil/Perzentil/Prozentil *stat*
  centile, percentile
Zentrierbohrer center drill
zentrieren center
Zentrifugalkraft
  centrifugal force
Zentrifugation centrifugation
➢ **analytische Zentrifugation**
  analytical centrifugation
➢ **Dichtegradientenzentrifugation**
  density gradient centrifugation
➢ **Differentialzentrifugation/**
  differentielle Zentrifugation
  differential centrifugation
  ('pelleting')
➢ **isopyknische Zentrifugation**
  isopycnic centrifugation,
  isodensity centrifugation
➢ **präparative Zentrifugation**
  preparative centrifugation
➢ **Ultrazentrifugation**
  ultracentrifugation
➢ **Zonenzentrifugation**
  zonal centrifugation

Zentrifuge centrifuge
➢ **Hochgeschwindigkeitszentrifuge**
  high-speed centrifuge,
  high-performance centrifuge
➢ **Kammerzentrifuge**
  multichamber centrifuge,
  multicompartment centrifuge
➢ **Kühlzentrifuge**
  refrigerated centrifuge
➢ **Mikrozentrifuge**
  microfuge
➢ **Röhrenzentrifuge**
  tubular bowl centrifuge
➢ **Schälschleuder**
  knife-discharge centrifuge,
  scraper centrifuge
➢ **Siebkorbzentrifuge**
  screen basket centrifuge
➢ **Siebschleuder**
  screen centrifuge
➢ **Tischzentrifuge**
  tabletop centrifuge,
  benchtop centrifuge
  (multipurpose c.)
➢ **Trommelzentrifuge**
  basket centrifuge, bowl centrifuge
➢ **Ultrazentrifuge**
  ultracentrifuge
➢ **Vollmantelzentrifuge/**
  Vollwandzentrifuge
  solid-bowl centrifuge
Zentrifugenröhrchen centrifuge tube
Zentrifugenröhrchenständer
  centrifuge tube rack
zentrifugieren centrifuge, spin
zerbrechen break, shatter; collapse
zerbrechlich fragile;
  (Vorsicht, zerbrechlich!) Fragile!
  Handle with care!
Zerfall (Abbau/Zusammenbruch)
  breakdown; (Zersetzung/Verrottung/
  Verfaulen) decay, disintegration,
  decomposition
➢ **radioaktiver Zerfall**
  radioactive decay,
  radioactive disintegration
zerfallen decay, disintegrate,
  decompose, fall apart

Zerfließen/Zerschmelzen/Zergehen
 deliquescence
zerfließend/zerfließlich/ zerschmelzend/zergehend
 deliquescent
zerkleinern reduce (to small pieces); break up
zermahlen
 (grob) grind, (fein) pulverize; (im Mörser) triturate
Zermahlen (grob) grinding; (fein: Pulverisierung) pulverization; (im Mörser) trituration
zermalmen
 crush; (zermahlen) grind
zerreiben rub, grind; (im Mörser) triturate
Zerreißfestigkeit/Reißfestigkeit/ Zugfestigkeit (Holz)
 tensile strength
zersetzen disintegrate, decay, decompose, degrade
Zersetzer/Destruent/Reduzent
 decomposer
Zersetzung disintegration, decay, decomposition, degradation
Zersetzungsprodukt
 degradation product
Zersetzungstemperatur *chem*
 decomposition temperature, disintegration temperature
zerstäuben atomize; spray
Zerstäuber/Sprühgerät (z.B. für DC)
 atomizer, sprayer; (Wasserzerstäuber) humidifier, mist blower
Zerstäuberdüse spray nozzle
zerstoßen crush
zerstreuen/dispergieren
 scatter, disperse
Zerstreuung/Dispergierung
 scattering, dispersion
Zeuge witness; testimony
zeugen/fortpflanzen
 procreate, reproduce, propagate
Zeugung/Fortpflanzung
 procreation, reproduction, propagation

Ziehklinge draw blade, (cabinet) scraper, (Rakel) drawing knife; spokeshave
 ➢ **Schabhobel** scraper
Zifferblatt dial, face
Zimtaldehyd
 cinnamic aldehyde, cinnamaldehyde
Zimtalkohol
 cinnamic alcohol, cinnamyl alcohol
Zimtsäure/Cinnamonsäure (Cinnamat)
 cinnamic acid
Zink (Zn) zinc
Zinkblende zinc blende, blackjack
Zinkfinger *gen* zinc finger
Zinn (Sn) tin
Zippverschluss zip-lip seal, zipper-top
Zippverschlussbeutel
 zip-lip bag, zipper-top bag
Zirconium (Zr) zirconium
Zirconiumdioxid ZrO_2 zirconia (zirconium oxide/zirconium dioxide)
Zirkel compass; divider
Zirkon $ZrSiO_4$ zircon
zirkular/zirkulär/kreisförmig/rund
 circular, round
Zirkularchromatographie
 circular chromatography
Zirkulardichroismus: Circulardichroismus
 circular dichroism
Zirkularisierung/Ringschluss
 circularization
zirkulieren circulate
zirkulierend/Zirkulations...
 circulating, circulatory
Zitronensäure/Citronensäure (Zitrat/Citrat) citric acid (citrate)
Zivildienst civilian social service (in place of military service for conscientious objectors)
Zivildienstleistender
 civilian social servant (conscientious objector)
Zonenschmelze(n) zone refining
Zonensedimentation
 zone sedimentation, zonal sedimentation
Zonierung zonation

Zubehör accessories, supplies; (Kleinteile an Geräten etc.) fittings, fixing
Zubehörlager supplies storage, supplies 'shop', 'supplies'
Zubehörlieferant supplier, accessories supplier, vendor
Zubehörteile (Kleinteile/Passteile) fittings
Zuber tub
Zubereitung/Herstellung preparation
züchten/kultivieren/aufziehen *bot/micb* breed, cultivate, grow; *zool* raise, rear
➤ **anzüchten** (*einer Kultur*) establish, start (a culture)
Züchtung/Kultivierung breed, breeding, cultivation, growing; raising, rearing
Züchtungsexperiment breeding experiment
Zucker sugar
➤ **Aminozucker** amino sugar
➤ **Blutzucker** blood sugar
➤ **Doppelzucker/Disaccharid** double sugar, disaccharide
➤ **Einfachzucker/ einfacher Zucker/ Monosaccharid** single sugar, monosaccharide
➤ **Fruchtzucker/Fruktose** fruit sugar, fructose
➤ **Holzzucker/Xylose** wood sugar, xylose
➤ **Invertzucker** invert sugar
➤ **Isomeratzucker/Isomerose** high fructose corn syrup
➤ **Malzzucker/Maltose** malt sugar, maltose
➤ **Milchzucker/Laktose** milk sugar, lactose
➤ **Rohrzucker/Rübenzucker/ Saccharose/Sukrose/Sucrose** cane sugar, beet sugar, table sugar, sucrose
➤ **Rohzucker** raw sugar, crude sugar (unrefined sugar)
➤ **Traubenzucker/ Glukose/Glucose/Dextrose** grape sugar, glucose, dextrose
➤ **Verzuckerung** saccharification
➤ **Vielfachzucker/Polysaccharid** multiple sugar, polysaccharide
zuckerbildend sacchariferous, saccharogenic
zuckerhaltig sugar-containing
Zuckerkrankheit/Diabetes mellitus diabetes mellitus
Zuckersäure/Aldarsäure saccharic acid, aldaric acid
zuckerspaltend saccharolytic
Zufall chance; accident; coincidence
zufällig by chance, at random; (aus Versehen) accidentally
Zufallsabweichung *stat* random deviation
Zufallsauslese random screening
Zufallsereignis random event
Zufallsfehler *stat* random error
Zufallsstichprobe/Zufallsprobe *stat* random sample, sample taken at random
Zufallsvariable *stat* random variable
Zufallsverteilung *stat* random distribution
Zufallszahl *stat* random number
Zufluss influx, inflow; supply; inlet; tributary, affluent
Zufuhr supply; influx
Zufuhröffnung inlet opening
Zug/Sog tension, suction, pull
Zugang access, admission, admittance, entry
Zugfestigkeit/Zerreißfestigkeit/ Reißfestigkeit (Holz) tensile strength
Zugluft draft
Zugspannung (Wasserkohäsion) water tension
Zulage compensation, bonus, extra pay
➤ **Gefahrenzulage** hazard bonus
Zulassung/Lizenz/Erlaubnis admission, licence, permit; registration

Zulauf inlet, feed, feed inlet; intake, supply; (process) inflow; (eintretende Flüssigkeit) feed (incoming fluid); (Eintrittsstelle einer Flüssigkeit) inlet
Zulaufkultur/Fedbatch-Kultur (semi-diskontinuierlich) fed-batch culture
Zulaufschlauch feed tube
Zulauftrichter addition funnel
Zulaufventil/Beschickungsventil delivery valve
Zulaufverfahren/Fedbatch-Verfahren (semi-diskontinuierlich) fed-batch process, fed-batch procedure
Zuleitung feed, inlet
Zuleitungsrohr inlet pipe
Zulieferung supply, shipment
Zuluft input air
Zunahme gain, increase, increment
Zündbarkeit ignitability
zünden ignite, fire, spark; start
Zünder igniter, primer; fuse
Zündflamme pilot flame, pilot light (from a pilot burner)
Zündfunke (ignition) spark
Zündpunkt/Zündtemperatur/ Entzündungstemperatur ignition point, kindling temperature, flame temperature, flame point, spontaneous-ignition temperature (SIT)
Zündröhrchen/Glühröhrchen ignition tube
Zündschnur fuse
Zündstein/Feuerstein/Flintstein/Flint flint, flint stone
Zündstoff/Brandstoff incendiary
Zündung ignition
zunehmen gain, increase; gain weight
Zurrgurt lashing strap
zurücksetzen reset
Zusammenarbeit collaboration, cooperation
Zusammenbau/Assemblierung *chem/gen* assembly

Zusammenbruch/Abbau/Zerfall breakdown; *ecol* (population) crash
Zusammenhang/Verhältnis/ Verbindung relation, correlation, interrelationship, connection
Zusatz/Zusatzstoff/Additiv additive
➢ **Lebensmittelzusatzstoff** food additive
Zusatzbezeichnung/Epitheton epithet
Zuschlag *metall* addition
zusetzen/hinzufügen add
zuspitzen (zugespitzt) taper(ed)
Zustand state, condition
➢ **gleichbleibender/ stationärer Zustand** steady state
Zuständigkeit responsibility; competence, jurisdiction
zustöpseln stopper
Zutritt/Zugang access, admission, admittance, entry
➢ **für Unbefugte verboten!** off-limits to unauthorized personnel
➢ **nur für Befugte** authorized personnel only
zutrittsberechtigt have admission, have access, having permitted access
Zutrittsbeschränkung restricted access, access control
Zutrittsverweigerung denial of access
zuverlässig reliable
Zuverlässigkeit reliability
Zuwachs increase, increment
zuwachsen/überwachsen overgrow
Zweifelsfall (im) in case of doubt
Zweihalskolben two-neck flask
Zweistoffgemisch binary mixture
Zweisubstratreaktion/ Bisubstratreaktion bisubstrate reaction
zweiteilig dimeric
zweiwertig/bivalent/divalent *chem* bivalent, divalent
Zweiwertigkeit *chem* bivalence, divalence

zweizählig/dimer dimerous
Zwinge clamp, vise, vice (*Br*)
Zwirn/Garn twine
Zwischenbild *micros*
 intermediate image
Zwischenfall
 incident (Unfall: accident)
zwischengeschlechtlich
 intersexual
Zwischenlager
 interim storage,
 temporary storage
Zwischenprodukt/Zwischenform
 biochem intermediate (product),
 intermediate form
➢ **doppelköpfiges/janusköpfiges Z.**
 double-headed intermediate
➢ **tetraedrisches Z.**
 tetrahedral intermediate
Zwischenstadium/Zwischenstufe
 intermediate state/stage
Zwischenstecker/Adapter adapter
Zwischenstufe/Übergangsform
 intergrade, intermediary form,
 transitory form, transient
Zwitterion zwitterion
Zwitterkontakt *electr*
 hermaphroditic contact
zyklisch/
 ringförmig (siehe: cyclisch) cyclic
Zyklisierung/Ringschluss
 (siehe: Cyclisierung) *chem*
 cyclization
Zyklus (*siehe*: Cyclus) cycle
Zylinder cylinder;
 (Hahn) barrel (stopcock barrel)
➢ **Messzylinder**
 graduated cylinder
➢ **Mischschzylinder**
 volumetric flask
➢ **Zylinderglas/Becherglas** beaker
zylindrisch/cylindrisch/walzenförmig
 cylindric, cylindrical
Zyto... (*siehe*: Cyto...) cyto...

English – German

aberration
Aberration, Abweichung, Anomalie; Abbildungsfehler, Bildfehler
ability test Eignungstest
ablate entfernen, abtragen; amputieren
abortive abortiv, verkümmert, unfertig, unvollständig entwickelt, rudimentär, rückgebildet; vorzeitig, verfrüht
abrasion Abrieb, Abreiben, Verschleiß; Abschürfen, Abschürfung, Abschaben
absorb (take up/soak up) absorbieren, saugen, aufsaugen; aufnehmen
absorbance/absorbancy (extinction: optical density)
Absorbanz (Extinktion)
absorbance index/absorptivity
Absorptionsindex
absorbency
Absorptionsvermögen, Absorptionsfähigkeit, Aufnahmefähigkeit, Saugfähigkeit
absorbent (absorbant) *adj/adv*
absorbierend, absorptionsfähig, saugfähig
absorbent (absorbant) *n*
Absorptionsmittel, Aufsaugmittel
absorbent paper/bibulous paper (for blotting dry)
Saugpapier ('Löschpapier')
absorption Absorption
absorption coefficient
Absorptionskoeffizient
absorption spectrum/dark-line spectrum Absorptionsspektrum
absorptive absorbierend, aufsaugend
accelerate beschleunigen
accelerating voltage *micros*
Beschleunigungsspannung (EM)
acceleration of gravity
Erdbeschleunigung
acceleration phase
Beschleunigungsphase, Anfahrphase
access (admission/admittance/entry)
Zutritt, Zugang
➤ **have access/**
 having permitted access/
 have admission zutrittsberechtigt

accessories/supplies/fittings
Zubehör, Ausrüstung
accident Unfall
➤ **danger of accident** Unfallgefahr
➤ **industrial accident/accident at work**
Industrieunfall, Betriebsunfall
➤ **occupational accident** Arbeitsunfall
➤ **prevention of accidents**
Unfallverhütung
➤ **worst-case accident**
größter anzunehmender Unfall
accident insurance
Unfallversicherung
accidental release
störungsbedingter Austritt (unerwartetes Entweichen von Prozessstoffen)
accidentally/
 by chance/at random
zufällig, versehentlich, aus Versehen
acclimation/acclimatization
Eingewöhnung
accommodation *opt*
Akkommodation, Scharfstellung
accumulation Anhäufung, Kumulation
accumulator acid/storage battery acid (electrolyte)
Akkusäure, Akkumulatorsäure
acellular/noncellular
azellulär, nicht zellig
acetaldehyde/acetic aldehyde/ethanal
Acetaldehyd, Ethanal
acetic acid/ethanoic acid (acetate)
Essigsäure, Ethansäure (Acetat/Azetat)
acetic anhydride/
 ethanoic anhydride/
 acetic acid anhydride
Essigsäureanhydrid
acetoacetic acid (acetoacetate)/
 acetylacetic acid/diacetic acid
Acetessigsäure (Acetacetat), 3-Oxobuttersäure
acetone/dimethyl ketone/2-propanone
Aceton (Azeton), Propan-2-on, 2-Propanon, Dimethylketon
acetyl CoA/acetyl coenzyme A
Acetyl-CoA, 'aktivierte Essigsäure'

ac

achromatic condenser/ achromatic substage *micros* achromatischer Kondensor
achromatic objective *micros* achromatisches Objektiv
acid/acidic *adj/adv* azid, acid, sauer
acid *n* Säure
➢ **abietic acid (7,13-abietadien-18-oic acid)** Abietinsäure
➢ **accumulator acid/ storage battery acid (electrolyte)** Akkusäure, Akkumulatorsäure
➢ **acetic acid/ethanoic acid (acetate)** Essigsäure, Ethansäure (Acetat)
➢ **acetoacetic acid (acetoacetate)/ acetylacetic acid/diacetic acid** Acetessigsäure (Acetacetat), 3-Oxobuttersäure
➢ **acetyl CoA/acetyl coenzyme A** Acetyl-CoA, 'aktivierte Essigsäure'
➢ **aconitic acid (aconitate)** Aconitsäure (Aconitat)
➢ **adenylic acid (adenylate)** Adenylsäure (Adenylat)
➢ **adipic acid (adipate)** Adipinsäure (Adipat)
➢ **alginic acid (alginate)** Alginsäure (Alginat)
➢ **allantoic acid** Allantoinsäure
➢ **amino acid** Aminosäure
➢ **anthranilic acid/2-aminobenzoic acid** Anthranilsäure, 2-Aminobenzoesäure
➢ **aqua regia** Königswasser
➢ **arachic acid/arachidic acid/ icosanic acid** Arachinsäure, Arachidinsäure, Eicosansäure
➢ **arachidonic acid/ icosatetraenoic acid** Arachidonsäure
➢ **ascorbic acid (ascorbate)** Ascorbinsäure (Ascorbat)
➢ **asparagic acid/ aspartic acid (aspartate)** Asparaginsäure (Aspartat)
➢ **auric acid** Goldsäure
➢ **azelaic acid/nonanedioic acid** Azelainsäure, Nonandisäure

➢ **behenic acid/docosanoic acid** Behensäure, Docosansäure
➢ **benzoic acid (benzoate)** Benzoesäure (Benzoat)
➢ **boric acid (borate)** Borsäure (Borat)
➢ **butyric acid/butanoic acid (butyrate)** Buttersäure, Butansäure (Butyrat)
➢ **caffeic acid** Kaffeesäure
➢ **capric acid/decanoic acid (capratedecanoate)** Caprinsäure, Decansäure (Caprinat, Decanat)
➢ **caproic acid/capronic acid/ hexanoic acid (caproatehexanoate)** Capronsäure, Hexansäure (CapronatHexanat)
➢ **caprylic acid/octanoic acid (caprylateoctanoate)** Caprylsäure, Octansäure (Caprylat, Octanat)
➢ **carbonic acid (carbonate)** Kohlensäure (Karbonat/Carbonat)
➢ **carboxylic acids (carbonates)** Carbonsäuren, Karbonsäuren (Carbonate, Karbonate)
➢ **cerotic acid/hexacosanoic acid** Cerotinsäure, Hexacosansäure
➢ **chinic acid/kinic acid/ quinic acid (quinate)** Chinasäure
➢ **chinolic acid** Chinolsäure
➢ **chloric acid** $HClO_3$ Chlorsäure
➢ **chlorogenic acid** Chlorogensäure
➢ **chlorous acid** $HClO_2$ chlorige Säure
➢ **cholic acid (cholate)** Cholsäure (Cholat)
➢ **chorismic acid (chorismate)** Chorisminsäure (Chorismat)
➢ **chromic(VI) acid** H_2CrO_4 Chromsäure
➢ **chromic-sulfuric acid mixture for cleaning purposes** Chromschwefelsäure
➢ **cinnamic acid/ 3-phenyl-2-propenoic acid** Cinnamonsäure, Zimtsäure (Cinnamat), 3-Phenylprop-2-ensäure
➢ **citric acid (citrate)** Citronensäure, Zitronensäure (Citrat, Zitrat)

- **crotonic acid/α-butenic acid**
  Crotonsäure, Transbutensäure
- **cysteic acid** Cysteinsäure
- **diprotic acid**
  zweiwertige, zweiprotonige Säure
- **ellagic acid/gallogen** Ellagsäure
- **erucic acid/**
  (Z)-13-docosenoic acid
  Erucasäure, Δ^{13}-Docosensäure
- **fatty acid** Fettsäure
- **ferulic acid** Ferulasäure
- **fluorosulfonic acid**
  Fluoroschwefelsäure, Fluorsulfonsäure
- **folic acid (folate)/**
  pteroylglutamic acid
  Folsäure (Folat), Pteroylglutaminsäure
- **formic acid (formate)**
  Ameisensäure (Format)
- **fumaric acid (fumarate)**
  Fumarsäure (Fumarat)
- **galacturonic acid** Galakturonsäure
- **gallic acid (gallate)**
  Gallussäure (Gallat)
- **gamma-aminobutyric acid (GABA)**
  Aminobuttersäure, γ-Aminobuttersäure (GABA)
- **gentisic acid** Gentisinsäure
- **geranic acid** Geraniumsäure
- **gibberellic acid** Gibberellinsäure
- **glacial acetic acid** Eisessig
- **glucaric acid/saccharic acid**
  Glucarsäure, Zuckersäure
- **gluconic acid (gluconate)/**
  dextronic acid
  Gluconsäure (Gluconat)
- **glucuronic acid (glucuronate)**
  Glucuronsäure (Glukuronat)
- **glutamic acid (glutamate)/**
  2-aminoglutaric acid
  Glutaminsäure (Glutamat), 2-Aminoglutarsäure
- **glutaric acid (glutarate)**
  Glutarsäure (Glutarat)
- **glycolic acid (glycolate)**
  Glykolsäure (Glykolat)
- **glycyrrhetinic acid** Glycyrrhetinsäure
- **glyoxalic acid (glyoxalate)**
  Glyoxalsäure (Glyoxalat)
- **glyoxylic acid (glyoxylate)**
  Glyoxylsäure (Glyoxylat)
- **guanylic acid (guanylate)**
  Guanylsäure (Guanylat)
- **gulonic acid (gulonate)**
  Gulonsäure (Gulonat)
- **homogentisic acid**
  Homogentisinsäure
- **humic acid** Huminsäure
- **hyaluronic acid** Hyaluronsäure
- **hydrochloric acid**
  Salzsäure, Chlorwasserstoffsäure
- **hydrofluoric acid/phthoric acid**
  Flusssäure, Fluorwasserstoffsäure
- **hydrogen cyanide/**
  hydrocyanic acid/prussic acid
  Blausäure, Cyanwasserstoff
- **hydroiodic acid/hydrogen iodide**
  Iodwasserstoffsäure
- **ibotenic acid** Ibotensäure
- **imino acid** Iminosäure
- **indolyl acetic acid/**
  indoleacetic acid (IAA)
  Indolessigsäure
- **isovaleric acid** Isovaleriansäure
- **jasmonic acid** Jasmonsäure
- **keto acid** Ketosäure
- **kojic acid** Kojisäure
- **lactate (lactic acid)**
  Laktat (Milchsäure)
- **lactic acid (lactate)**
  Milchsäure (Laktat)
- **lauric acid/decylacetic acid/**
  dodecanoic acid
  (laurate/dodecanate)
  Laurinsäure, Dodecansäure (Laurat/Dodecanat)
- **levulinic acid** Lävulinsäure
- **lichen acid** Flechtensäure
- **lignoceric acid/tetracosanoic acid**
  Lignocerinsäure, Tetracosansäure
- **linolenic acid** Linolensäure
- **linolic acid/linoleic acid** Linolsäure
- **lipoic acid (lipoate)/thioctic acid**
  Liponsäure, Dithiooctansäure, Thioctsäure, Thioctansäure (Liponat)

- lipoteichoic acid Lipoteichonsäure
- litocholic acid Litocholsäure
- lysergic acid Lysergsäure
- magic acid (HSO_3F/SbF_5) Magische Säure
- maleic acid (maleate) Maleinsäure (Maleat)
- malic acid (malate) Äpfelsäure (Malat)
- malonic acid (malonate) Malonsäure (Malonat)
- mandelic acid/ phenylglycolic acid/ amygdalic acid Mandelsäure, Phenylglykolsäure
- mannuronic acid Mannuronsäure
- mevalonic acid (mevalonate) Mevalonsäure (Mevalonat)
- monoprotic acid einwertige/einprotonige Säure
- mucic acid Schleimsäure, Mucinsäure
- muramic acid Muraminsäure
- myristic acid/tetradecanoic acid (myristate/tetradecanate) Myristinsäure (Myristat), Tetradecansäure
- N-acetylmuramic acid N-Acetylmuraminsäure
- nervonic acid/ (Z)-15-tetracosenoic acid/ selacholeic acid Nervonsäure, Δ^{15}-Tetracosensäure
- neuraminic acid Neuraminsäure
- nicotinic acid (nicotinate)/niacin Nikotinsäure, Nicotinsäure (Nikotinat)
- nitric acid Salpetersäure
- nitrous acid salpetrige Säure
- oleic acid/(Z)-9-octadecenoic acid (oleate) Ölsäure (Oleat), Δ^9-Octadecensäure
- orotic acid Orotsäure
- orsellic acid/orsellinic acid Orsellinsäure
- osmic acid Osmiumsäure
- oxalic acid (oxalate) Oxalsäure (Oxalat)
- oxalosuccinic acid (oxalosuccinate) Oxalbernsteinsäure (Oxalsuccinat)
- oxoacid Oxosäure
- oxoglutaric acid (oxoglutarate) Oxoglutarsäure (Oxoglutarat)
- palmitic acid/hexadecanoic acid (palmate/hexadecanate) Palmitinsäure, Hexadecansäure (Palmat, Hexadecanat)
- palmitoleic acid/ (Z)-9-hexadecenoic acid Palmitoleinsäure, Δ^9-Hexadecensäure
- pantoic acid Pantoinsäure
- pantothenic acid (pantothenate) (vitamin B_3) Pantothensäure (Pantothenat)
- pectic acid (pectate) Pektinsäure (Pektat)
- penicillanic acid Penicillansäure
- perchloic acid Perchlorsäure
- performic acid Perameisensäure
- phosphatidic acid Phosphatidsäure
- phosphoric acid (phosphate) Phosphorsäure (Phosphat)
- phosphorous acid phosphorige Säure
- phthalic acid Phthalsäure
- phytanic acid Phytansäure
- phytic acid Phytinsäure
- picric acid (picrate) Pikrinsäure (Pikrat)
- pimelic acid Pimelinsäure
- plasmenic acid Plasmensäure
- prephenic acid (prephenate) Prephensäure (Prephenat)
- propionic acid (propionate) Propionsäure (Propionat)
- prostanoic acid Prostansäure
- pyrethric acid Pyrethrinsäure
- pyruvic acid (pyruvate) Brenztraubensäure (Pyruvat)
- retinic acid Retinsäure
- saccharic acid/aldaric acid (glucaric acid) Zuckersäure, Aldarsäure (Glucarsäure)
- salicic acid (salicylate) Salicylsäure (Salicylat)

- shikimic acid (shikimate)
  Shikimisäure (Shikimat)
- sialic acid (sialate)
  Sialinsäure (Sialat)
- silicic acid Kieselsäure
- sinapic acid Sinapinsäure
- soldering acid Lötsäure
- sorbic acid (sorbate)
  Sorbinsäure (Sorbat)
- stearic acid/
  octadecanoic acid
  (stearate/octadecanate)
  Stearinsäure, Octadecansäure
  (Stearat/Octadecanat)
- stomach acid Magensäure
- stomach acid/gastric acid
  Magensäure
- suberic acid/octanedioic acid
  Suberinsäure, Korksäure,
  Octandisäure
- succinic acid (succinate)
  Bernsteinsäure (Succinat)
- sulfanilic acid/
  p-aminobenzenesulfonic acid
  Sulfanilsäure
- sulfuric acid Schwefelsäure
- sulfurous acid
  schweflige Säure, Schwefligsäure
- superacid Supersäure
- tannic acid (tannate)
  Gerbsäure (Tannat)
- tartaric acid (tartrate)
  Weinsäure,
  Weinsteinsäure (Tartrat)
- teichoic acid Teichonsäure
- teichuronic acid Teichuronsäure
- uric acid (urate) Harnsäure (Urat)
- uridylic acid Uridylsäure
- urocanic acid (urocaninate)
  Urocaninsäure (Urocaninat),
  Imidazol-4-acrylsäure
- uronic acid (urate) Uronsäure (Urat)
- usnic acid Usninsäure
- valeric acid/pentanoic acid
  (valeriate/pentanoate)
  Valeriansäure, Pentansäure
  (Valeriat/Pentanat)
- vanillic acid Vanillinsäure

acid amide Säureamid
acid bottle (with pennyhead stopper)
   Säurenkappenflasche
acid burn Säureverätzung
acid ester Säureester
acid gloves/acid-resistant gloves
   Säureschutzhandschuhe
acid rain/acid deposition
   saurer Regen, Niederschlag
acid storage cabinet Säureschrank
acid treatment Säurebehandlung
acid-base balance
   Säure-Basen-Gleichgewicht
acid-base titration
   Säure-Basen-Titration,
   Neutralisationstitration
acid-fast säurefest
acid-fastness Säurefestigkeit
acidic sauer, säuerlich;
   säurebildend, säurehaltig
acidification Säuerung;
   Säurebildung; Versauerung
acidify ansäuern
acidifying agent Säuerungsmittel
acidity Acidität, Azidität,
   Säuregrad, Säuregehalt
acidosis Azidose, Acidose
acid-proof/acid-fast säurebeständig
aconitic acid (aconitate)
   Aconitsäure (Aconitat)
acorn nut Hutmutter
acoustical panel/tile Schalldämmplatte
acquire erwerben;
   (record) erfassen, aufnehmen
acquired characteristic
   erworbenes Merkmal
acquired immune deficiency syndrome (AIDS)
   erworbenes Immunschwächesyndrom
acquired immunity/adaptive immunity (active/passive)
   erworbene Immunität
   (aktive/passive)
acquirement Erwerbung, Erlangung;
   Erfassung, Aufnahme (von Daten)
acquisition Anschaffung,
   Erwerb, Erwerbung, Ankauf
acquisition cost Anschaffungskosten

ac

acquisition time *vir* Aufnahmezeit
acridine dye Acridinfarbstoff
acrylic glass
  Acrylglas; (plexiglass) Plexiglas
act (work/be effective/causing an effect/take effect) wirken; (effect/contact/attack/interact) einwirken
activated carbon Aktivkohle
activated sludge
  Belebtschlamm, Rücklaufschlamm
active ingredient/
  active principle/
  active component
  Wirkstoff, Wirksubstanz
active metabolic rate
  Arbeitsumsatz, Leistungsumsatz
active metabolism
  Leistungsstoffwechsel, Arbeitsstoffwechsel
active transport/uphill transport
  aktiver Transport
actual value/effective value
  Istwert
actuator Stellantrieb, Stellmotor
acute/sharp/pointed/sharp-pointed
  spitz
adapter Adapter, Zwischenstecker, Manschette; (connector: glass) *lab/chem* Vorstoß, Übergangsstück; (for filter funnel) Filtervorstoß
➢ bellows Balg
➢ receiver adapter Destilliervorstoß
➢ cow receiver adapter/'pig'/multi-limb vacuum receiver adapter (receiving adapter for three/four receiving flasks) *dist* Eutervorlage, Verteilervorlage, 'Spinne'
➢ adapter for filter funnel
  Filtervorstoß
➢ anticlimb adapter *dist*
  Kriechschutzadapter
➢ antisplash adapter/
  splash-head adapter *dist*
  Schaumbrecher-Aufsatz, Spritzschutz-Aufsatz (Rückschlagsicherung)
➢ bent adapter/bend Krümmer

➢ cone/screwthread adapter
  Kern-/Gewindeadapter, Gewinde-mit-Kern Adapter
➢ distillation receiver adapter/
  receiving flask adapter Vorlage
➢ drip catcher/drip catch
  Tropfenfänger
➢ expansion adapter/enlarging adapter
  Expansionsstück
➢ filter adapter Filterstopfen; (Guko) Filtermanschette, Guko
➢ offset adapter
  Übergangsstück
  mit seitlichem Versatz
➢ pipe-to-tubing adapter
  Schlauch-Rohr-Verbindungsstück
➢ reducing adapter/reduction adapter/reducer Reduzierstück
➢ septum-inlet adapter
  Septum-Adapter
➢ splash adapter/
  splash-head adapter/
  antisplash adapter
  (distillation apparatus)
  Tropfenfänger; (Reitmeyer-Aufsatz: Rückschlagschutz: Kühler/ Rotationsverdampfer etc.)
➢ syringe connector Nadeladapter
➢ transition piece Übergangsstück
➢ tubing adapter Schlauchadapter
➢ two-neck (multiple) adapter
  Zweihalsaufsatz
➢ vacuum adapter Vakuumvorstoß
➢ vacuum-filtration adapter
  Vakuumfiltrationsvorstoß
Additionsverbindung
  addition compound (of two compounds), additive compound (saturation of multiple bonds)
add zusetzen, hinzufügen, hinzugeben, ergänzen; beimengen
addition Zusatz, Hinzufügen, Ergänzung; Beimengung
addition compound
  (union of two compounds)
  Additionsverbindung
addition funnel
  Einfülltrichter, Zulauftrichter

additive *n* Zusatzstoff, Zusatz, Additiv
**additive compound
 (saturation of double/triple bonds)**
 Additionsverbindung
adhere (stick/cling)
 haften (kleben)
adhesion/adhesive power
 Adhäsion, Haftung, Anheftung
adhesive *adj/adv* haftend, klebend
adhesive *n*
 (glue/gum) Kleber, Klebstoff, Leim;
 (cement) Kitt, Kittsubstanz
➤ **multicomponent adhesive/
 multiple-component adhesive**
 Mehrkomponentenkleber
**adhesive strip/band-aid/
 sticking plaster/patch** *med*
 Pflaster, Heftpflaster (Streifen)
adhesive tape
 Klebeband, Klebestreifen
adjust einstellen, regulieren, justieren;
 (focus: *fine/coarse*) justieren,
 fokussieren (Scharfeinstellung des
 Mikroskops: *fein/grob*); (equalize)
 abgleichen
adjustable
 einstellbar, verstellbar,
 regulierbar, justierbar
adjustable variable Stellgröße
adjustable wrench
 Rollgabelschlüssel, 'Engländer'
**adjusting screw/
 setting screw/
 adjustment knob/fixing screw**
 Stellschraube
adjustment Anpassung, Angleichung;
 Einstellung, Regulierung;
 (focus adjustment/focus: *fine/coarse*)
 Justierung, Fokussierung
 (Scharfeinstellung des Mikroskops:
 fein/grob)
adjustment knob
 Einstellknopf
➤ **coarse-adjustment knob** *micros*
 Grobjustierschraube, Grobtrieb
➤ **condenser adjustment knob/
 substage adjustment knob** *micros*
 Kondensortrieb

➤ **fine-adjustment/
 fine-adjustment knob/
 micrometer screw** *micros*
 Mikrometerschraube,
 Feinjustierschraube,
 Feintrieb
➤ **focus adjustment knob** *micros*
 Justierschraube, Justierknopf,
 Triebknopf
adjustment screw/tuning screw
 Einstellschraube
administration Verwaltung
admission
 Zulassung; Eintritt, Zutritt;
 Aufnahme
adsorb adsorbieren, haften, anhaften
adsorbate Adsorbat
adsorbent
 Adsorptionsmittel, Adsorbens,
 adsorbierende Substanz
adsorption Adsorption; Haftung
advance *micros* Vorschub
advantage Vorteil, Nutzen
aerate belüften, durchlüften
aeration Belüftung, Durchlüftung
aeration tank/aerator
 Belebtschlammbecken,
 Belebungsbecken,
 Belüftungsbecken (Kläranlage)
aerobic aerob, sauerstoffbedürftig
aerobic respiration aerobe Atmung
afferent/rising aufsteigend
affinity chromatography
 Affinitätschromatographie
affinity labeling Affinitätsmarkierung
affinity partitioning Affinitätsverteilung
affix/attach
 fixieren (befestigen, fest machen)
afterburning Nachverbrennung
afterdischarge *n neuro* Nachfeuerung,
 Nachentladung; *vb* nachfeuern
after-ripening Nachreifen
aftertreatment Nachbehandlung
agar diffusion test Agardiffusionstest
agar medium Agarnährboden
agar plate Agarplatte
agate mortar Achatmörser
agency/department Amt, Behörde

ag

agent Agens, Agenz (*pl* Agentien)
- **antifeeding agent/
  antifeeding compound/
  feeding deterrent**
  Fraßhemmer,
  fraßverhinderndes Mittel
- **cleaning agent/cleanser**
  Reinigungsmittel
- **crosslinking agent/cross linker**
  quervernetzendes Agens
- **etiological agent**
  Krankheitsverursacher
  (Wirkstoff, Agens, Mittel)
- **fire-extinguishing agent**
  Löschmittel, Feuerlöschmittel
- **fluxing agent/fusion reagent**
  Flussmittel, Schmelzmittel, Zuschlag
- **intercalation agent/
  intercalating agent**
  interkalierendes Agens
- **oxidizing agent (oxidant/oxidizer)**
  Oxidationsmittel
- **reducing agent** Reduktionsmittel
- **scouring agent/abrasive**
  Scheuermittel
- **toxic agent** Giftstoff
- **wetting agent
  (wetter/surfactant/spreader)**
  Benetzungsmittel;
  Entspannungsmittel
  (oberflächenaktive Substanz)
- **workplace agent** Arbeitsstoff

agitator vessel
Rührkessel, Rührbehälter
agreement Einwilligung, Zustimmung;
(consent) Vereinbarung,
Einverständniserklärung
agriculture/farming Landwirtschaft
air *n* Luft; Brise, Wind, Luftzug
air *vb* **(ventilate/aerate)**
lüften, belüften
air bath Luftbad
air bubble Luftblase
air capacity Luftkapazität
air capillary Luftkapillare
**air chamber/air receiver/
air vessel/surge chamber**
Windkessel
air circulation
Luftumwälzung, Luftzirkulation
air condenser Luftkühler
air conditioner Klimaanlage
air conditioning Klimatisierung;
Klimaanlage; Klimatechnik
**air current/
airflow/current of air/air stream**
Luftstrom, Luftströmung
air curtain/air barrier
Luftvorhang, Luftschranke
(z.b. an Vertikalflow-Biobench)
air-cushion foil Luftpolster-Folie
air dryer Luftentfeuchter
air duct/air conduit/airway
Lüftungskanal, Luftkanal
air humidity (absolute/realtive)
Luftfeuchtigkeit (absolute/relative)
air inlet valve/air bleed
Lufteinlassventil
air jet Luftstrahl
air meter/anemometer
Windmesser, Anemometer
air pollutant Luftschadstoff
air pollution
Luftverschmutzung,
Luftverunreinigung
air pressure Luftdruck
air reactive luftreaktiv
air recirculation Luftrückführung
air-sensitive luftempfindlich
air shaft/air duct
Lüftungsschacht, Luftschacht
air speed
Luftströmung, Luftgeschwindigkeit
(Sicherheitswerkbank)
air supply Luftzufuhr
**air threshold value/
atmospheric threshold value**
Luftgrenzwert
airborne luftgetragen
airborne dust Flugstaub
airfoil Profil (hervorstehende Teile
eines Gerätes, z.B.
Sicherheitswerkbank); Tragfläche
airlift reactor/pneumatic reactor
Airliftreaktor, pneumatischer
Reaktor (Mammutpumpenreaktor)

airlock Luftschleuse
airtight/airproof luftdicht
aisle/corridor Gang, Flur, Korridor
alarm *vb* alamieren, warnen
alarm (alert) *n* Alarm
➤ **false alarm** falscher Alarm
➤ **fire alarm** Feueralarm
➤ **drill/emergency drill**
 Probealarm, Probe-Notalarm
alarm signal Alarmsignal
alarm siren/
 air-raid siren Alarmsirene
alarm substance/alarm pheromone
 Schreckstoff, Alarmstoff,
 Alarm-Pheromon
alarm system Alarmanlage
albedo Albedo, Rückstrahlvermögen
alcohol Alkohol
➤ **absolute alcohol**
 absoluter Alkohol
➤ **amyl alcohols/**
 pentyl alcohols/pentanols
 Amylalkohole, Pentanole
➤ **butyl alcohols/butanols**
 Butylalkohole, Butanole
➤ **cinnamyl alcohol/**
 3-phenyl-2-propen-1-ol
 Zimtalkohol, 3-Phenylprop-2-enol
➤ **denatured alcohol**
 vergälltes Alkohol
➤ **ethyl alcohol/ethanol**
 (grain alcohol/spirit of wine)
 Ethylalkohol, Ethanol,
 Äthanol (Weingeist)
➤ **isopropyl alcohol/**
 isopropanol/
 1-methyl ethanol (rubbing alcohol)
 Isopropylalkohol, Propan-2-ol
➤ **methanol/methyl alcohol**
 (wood alcohol)
 Methylalkohol, Methanol
 (Holzalkohol)
➤ ***n*-propyl alcohol/propanol**
 Propylalkohol, Propan-1-ol
➤ **rubbing alcohol** 70% v/v Ethanol
➤ **sinapic alcohol** Sinapinalkohol
➤ **wax alcohols** Wachsalkohole
➤ **wool alcohols** Wollwachsalkohole

alcohol burner
 Spiritusbrenner, Spirituslampe
aldehyde Aldehyd
➤ **acetaldehyde/**
 acetic aldehyde/ethanal
 Acetaldehyd, Ethanal
➤ **anisic aldehyde/anisaldehyde**
 Anisaldehyd
➤ **cinnamic aldehyde/cinnamaldehyde**
 Zimtaldehyd, 3-Phenylprop-2-enal
➤ **formaldehyde/methanal**
 Formaldehyd, Methanal
➤ **glutaraldehyde/1,5-pentanedione**
 Glutaraldehyd, Glutardialdehyd,
 Pentandial
alert *n* Alarm, Alarmbereitschaft
align (tune) abgleichen
alignment (tuning) Abgleich
aliquot Aliquote, aliquoter Teil
 (Stoffportion als Bruchteil einer
 Gesamtmenge)
alive lebendig; (living) lebend
alkali blotting Alkali-Blotting
alkali burn
 Alkaliverätzung, Basenverätzung
alkaline/basic alkalisch, basisch
alkaliproof
 alkalibeständig, laugenbeständig
alkaloid(s) Alkaloid(e)
all-clear! Entwarnung!
all-purpose/general-purpose/utility ...
 Allzweck..., Allgemeinzweck...,
 Mehrzweck..., Universal...
all-purpose cleaner ...
 Allzweckreiniger, Universalreiniger
all-purpose glue/adhesive ...
 Alleskleber
Allen wrench Inbusschlüssel
allergen (sensitizer) Allergen
allergic (sensitizing) allergisch
alligator clip/alligator connector clip
 Krokodilklemme
Allihn condenser/bulb condenser
 Kugelkühler
alloy Legierung
alternating current (AC) Wechselstrom
alum Alaun, Aluminiumsulfat
aluminum (Al) Aluminium

aluminum foil
  Aluminiumfolie, Alufolie
amber Bernstein
amber glass Braunglas
ambient pressure Umgebungsdruck
ambient temperature
  Umgebungstemperatur
ambulance
  Ambulanz;
  Rettungswagen, Sanitätswagen
American Chemical Society (ACS)
  Amerikanische Chemische
  Gesellschaft
Ames test Ames-Test
amidation Amidierung
amide Amid
amination Aminierung
amine Amin
amino acid Aminosäure
  ➤ **essential amino acids**
  essentielle Aminosäure
  ➤ **nonessential amino acids**
  nicht-essentielle Aminosäure
amino sugar Aminozucker
aminoacylation Aminoacylierung
ammeter Strommessgerät,
  Amperemeter (Stromstärke)
ammonia Ammoniak
ammonium chloride
  (sal ammoniac/salmiac)
  Salmiak, Ammoniumchlorid
ammonium hydroxide/
  ammonia water (ammonia solution)
  Salmiakgeist, Ammoniumhydroxid
  (Ammoniaklösung)
amplification
  Amplifikation, Vermehrung,
  Vervielfältigung, Verstärkung
amplifier Verstärker
amplify verstärken, vermehren,
  vervielfältigen
ampule/ampoule Ampulle
  (Glasfläschchen)
  ➤ **prescored ampule/ampoule**
  vorgeritzte Spießampulle
analeptic amine Weckamin
analog/analogue
  Analogon (*pl* Analoga)

analogize analogisieren
analogous analog, funktionsgleich
analog-to-digital converter (ADC)
  Analog-Digital-Wandler
analogy Analogie
analysis (*pl* analyses)
  Analyse; (evaluation) Auswertung
analyte Analyt, zu analysierender Stoff,
  analysierte Substanz
analytic(al) analytisch
analytical balance Analysenwaage
analytical separation procedure
  Trennungsgang
analyze analysieren
analyzer Analysator
anchor *vb* **(fasten/attach)**
  verankern (befestigen)
anchorage Verankerung
ancillary unit of equipment
  Hilfseinrichtung (Apparat das nicht
  direkt mit dem Produkt in Berührung
  kommt)
anesthesia Narkose
  ➤ **general anesthesia**
  Vollnarkose, Vollanästhesie
  ➤ **local anesthesia** Lokalanästhesie
anesthetic Anästhetikum,
  Narkosemittel, Betäubungsmittel
angle rotor/angle head rotor
  Winkelrotor
angular aperture *micros*
  Öffnungswinkel
animal cage Tierkäfig
animal containment level
  Sicherheitsstufe für
  Tierhaltungseinheit
animal feed Viehfutter
animal laboratory/animal lab Tierlabor
animal model Tiermodell
animal unit Tierhaltungseinheit (DIN)
animate(d) beleben (belebt)
  ➤ **inanimate/lifeless/nonliving**
  unbelebt
anion (negatively charged ion)
  Anion (negatives Ion/negativ
  geladenes Ion)
anion exchange resin
  Anionenaustauscherharz

ap

anion exchanger Anionenaustauscher
anisic aldehyde/anisaldehyde
　Anisaldehyd
anneal (Glas) ausglühen, kühlen;
　(Keramik) einbrennen; härten;
　polym tempern
annealing Glühen (spannungsfrei
　Stabilglühen); Entspannen;
　polym Tempern
annealing cup (clay crucible)
　Schmelztiegel
　(unglasiert mit hohem Rand)
annealing furnace Glühofen
annular/cyclic ringförmig, cyclisch
anodal current Anodenstrom
anode rays
　Anodenstrahlen
anodization/
　anodic oxidation Eloxierung
anodize/
　anodically oxidize/
　oxidize by anodization
　(electrolytic oxodation) eloxieren
ANOVA (analysis of variance)
　Varianzanalyse
Anschütz head *dist* Anschütz-Aufsatz
answer/respond antworten
antacid Antazidum,
　säureneutralisierendes Mittel
antagonism Antagonismus
antenatal diagnosis/
　prenatal diagnosis
　pränatale Diagnose
anthracite/hard coal
　Anthrazit, Kohlenblende
antibiotic(s)
　Antibiotikum (*pl* Antibiotika)
➢ **broad-spectrum antibiotic**
　Breitspektrumantibiotikum
antibody
　Antikörper
➢ **monoclonal antibody**
　monoklonaler Antikörper
anticlimb adapter *dist*
　Kriechschutzadapter
antidote/antitoxin/antivenin
　Antidot, Gegengift,
　Gegenmittel (tierische Gifte)

antifeeding agent/
　antifeeding compound/
　feeding deterrent
　Fraßhemmer,
　fraßverhinderndes Mittel
antifoam/
　antifoaming agent/
　defoamer/defrother
　Entschäumer, Antischaummittel
antifoam controller
　Schaumdämpfer,
　Schaumverhütungsmittel
antimony (Sb) Antimon, Stibium
antisplash adapter/
　splash-head adapter
　Schaumbrecher-Aufsatz,
　Spritzschutz-Aufsatz
　(Rückschlagsicherung)
aperture (opening/orifice) Öffnung,
　Mündung; *micros* Apertur (Blende)
aperture protection factor
　(open bench)
　Schutzfaktor für die Arbeitsöffnung
　(Werkbank)
apex/summit/peak
　(highest among other high points)/
　vertex
　Gipfel, Scheitelpunkt, Höhepunkt
apiezon grease Apiezonfett
apolar unpolar
appearance Erscheinung,
　Erscheinungsbild, Erscheinungsform
appearance energy (MS)
　Auftrittsenergie
appliance Gerät; Anwendung
➢ **electric appliance**
　Elektrogerät
application Antrag, Bewerbung;
　Anwendung, Nutzen; *chromat*
　Applikation, Auftrag, Auftragung
application rod Auftragestab,
　Applikator
applied chemistry Angewandte Chemie
apply anwenden;
　chromat applizieren, auftragen
apportioning/proportioning
　Dosierung, Dosieren
　(im Verhältnis, anteilig)

apprentice/trainee (on-the-job)
Lehrling
apprenticeship (training period)
Lehre, Lehrjahre, Lehrzeit
approach *n* **(method)**
Ansatz (Methode)
approach *vb* **(e.g., a value)** *math/stat*
erreichen, sich annähern, näherkommen,
annähern (z.b. einen Wert)
approval
Genehmigung, Erlaubnis, Zusage
➢ **subject to approval (requiring permission/authorization)**
genehmigungspflichtig
approve genehmigen, erlauben, zusagen, gut heißen
approximate value
Richtwert, Näherungszahl
approximation *math* Näherung
aqua regia (3 pts. HCl/1 pt. HNO$_3$)
Königswasser
aquarium/fishtank Aquarium
aqueous wässrig
aqueous solution wässrige Lösung
arachic acid/arachidic acid/ icosanic acid Arachinsäure, Arachidinsäure, Eicosansäure
arachidonic acid/icosatetraenoic acid
Arachidonsäure
arbitrary/random willkürlich
arc flame Bogenflamme
arc furnace Lichtbogenofen
arc lamp Bogenlampe
arc spectrum Lichtbogenspektrum
area/region/zone/territory
Raum, Gebiet, Gegend, Region, Zone
areometer Aräometer (Densimeter/Senkwaage)
arithmetic growth
arithmetisches Wachstum
arithmetic mean *stat*
arithmetisches Mittel
arm *chem/biochem/immun*
Schenkel; (microscope) Trägerarm
aroma/fragrance/pleasant odor
Aroma, Wohlgeruch

aromatic aromatisch
arousal/excitement
Erregung, Aufregung
arrangement Regelung, Vereinbarung; (set-up: of an experiment) Ansatz (Versuchsansatz/Versuchsaufbau)
arrest/stop/lock
arretieren, feststellen; (fixate) fixieren
arsenic (As) Arsen
arsine
Arsenwasserstoff, Arsan, Monoarsan
artery forceps/artery clamp
Arterienklemme
artifact/artefact Artefakt
artificial colors/artificial coloring
künstliche Farbstoffe
artificial flavor/artificial flavoring
künstlicher Geschmackstoff
artificial light(ing)
künstliche Beleuchtung
artificial resin/synthetic resin
Kunstharz
artificial respiration
künstliche Beatmung
asbestos Asbest
➢ **blue asbestos/ crocidolite**
Blauasbest, Krokydolith
➢ **white asbestos/ chrysotile/Canadian asbestos**
Weißasbest, Chrysotil
asbestos board Asbestplatte
asbestosis
Asbestose, Asbeststaublunge, Bergflachslunge
ascending *chromat* **(TLC)** aufsteigend
ascorbic acid (ascorbate)
Ascorbinsäure (Ascorbat)
ash Asche
ashing Veraschung
ashless (quantitative filter)
aschefrei (quantitativer Filter)
asparagic acid/aspartic acid (aspartate)
Asparaginsäure (Aspartat)
asparagine/aspartamic acid
Asparagin

asphyxiant erstickend (chem. Gefahrenbezeichnung)
aspirator pump/vacuum pump Saugpumpe, Vakuumpumpe
assay (test/trial/examination/ exam/investigation) Probe, Versuch, Untersuchung, Test, Prüfung
➢ **isotope assay** Isotopenversuch
➢ **protection assay/ protection experiment** Schutzversuch, Schutzexperiment
➢ **triplet binding assay** Triplettbindungsversuch
assay material/ test material/examination material Probe, Probensubstanz, Untersuchungsmaterial
assay medium Untersuchungsmedium, Prüfmedium, Testmedium
assembly Assemblierung, Zusammenbau
assess erfassen, bewerten
assessment Erfassung, Bewertung
assimilate ***n*** Assimilat
assimilate ***vb*** assimilieren
atmosphere Atmosphäre
atmospheric atmosphärisch, Luft...
atmospheric moisture Luftfeuchtigkeit
atmospheric nitrogen Luftstickstoff
atmospheric oxygen Luftsauerstoff
atmospheric pressure atmosphärischer Luftdruck
atomic atomar, Atom...
atomic absorption spectroscopy (AAS) Atom-Absorptionsspektroskopie (AAS)
atomic bond Atombindung
atomic emission detector (AED) Atomemissionsdetektor (AED)
atomic emission spectroscopy (AES) Atom-Emissionsspektroskopie (AES)
atomic fluorescence spectroscopy (AFS) Atom-Fluoreszenzspektroskopie (AFS)
atomic force microscopy (AFM) Rasterkraftmikroskopie
atomic number Atomzahl
atomic weight (actually: atomic mass) Atomgewicht (*sensu strictu:* Atommasse)
atomize/spray zerstäuben, sprühen
atomizer/sprayer Atomisator, Zerstäuber, Sprühgerät
ATP (adenosine triphosphate) ATP (Adenosintriphosphat)
attachment Befestigung, Anheftung; (extension piece) Ansatzstück (Glas); (fixture) Aufsatz (auf ein Gerät)
attempt ***n*** Versuch, Bemühung; Ansatz
attenuate attenuieren, abschwächen (die Virulenz vermindern, mit herabgesetzter Virulenz)
attenuated vaccine attenuierte(r)/abgeschwächte(r) Impfstoff/Vakzine
attenuation Attenuation, Attenuierung, Abschwächung
attractant Attraktans (*pl* Attraktantien), Lockmittel, Lockstoff
audibility Hörbarkeit
audit Audit, Prüfung (Sachverständigenprüfung)
auditing Audit, Prüfung (z.B. Rechnungsprüfung)
audition Hörvermögen, Gehör
Auger electron spectroscopy (AES) Auger-Elektronenspektroskopie (AES)
auric Gold(III)...
auric acid $HAuO_2$ Goldsäure
aurous Gold(I)...
aurous sulfide Au_2S Goldsulfid
authorization procedure Genehmigungsverfahren
authorized personnel only Zutritt/Zugang nur für Befugte
auto-shutoff automatische Abschaltung (elektron. Gerät)
autocatalysis Autokatalyse

autoclavable autoklavierbar
autoclave *vb* autoklavieren
autoclave *n* Autoklav
➢ **cooling time/
cool-down period**
Abkühlzeit, Fallzeit
➢ **heat-up time**
Aufheizphase
➢ **preheating time/rise time**
Anheizzeit, Steigzeit
➢ **setting time**
Ausgleichszeit,
thermisches Nachhinken
autoclave bag Autoklavierbeutel
**autoclave tape/
autoclave indicator tape**
Autoklavier-Indikatorband
**autodecomposition/
spontaneous decomposition**
Selbstzersetzung
autologous autolog
autolysis Autolyse

**autoradiography/
radioautography**
Autoradiographie
auxiliary drug/adjuvant
Hilfsstoff, Adjuvans
availability Verfügbarkeit
average/mean
Durchschnitt (Mittelmaß)
average yield Durchschnittsertrag
averaging Mittelwertbildung
avoidance Vermeidung
awareness Bewusstheit
awl/pricker Ahle (reamer: Reibahle)
axe Axt
➢ **fire axe** Brandaxt
azelaic acid/nonanedioic acid
Azelainsäure, Nonandisäure
azeotropic azeotrop
azeotropic distillation
Azeotropdestillation
azeotropic mixture
azeotropes Gemisch

back extraction/stripping
Rückextraktion, Strippen
back titration Rücktitration
backdraft preventer/protection
Rückstauschutz
backflow preventer/backflow protection/backstop
Rückströmsperre
backflow prevention/backstop (valve)
Rückflusssperre, Rücklaufsperre, Rückstauventil
backflush/backflushing *chromat*
Rückspülen, Rückspülung (der Säule)
backlit hintergrundbeleuchtet
backorder ausstehende Lieferung
➤ **on backorder**
ausstehende Lieferung wird nachgeliefert (sobald wieder auf Lager)
backscatter
Rückstreuung, Reflexion
backstop Rücklaufsperre
backstop valve/check valve
Rückschlagventil
bacteria (*sg* bacterium) Bakterien (*sg* Bakterie/Bakterium)
bacterial bakteriell
bacterial culture Bakterienkultur
bacterial flora Bakterienflora
bacterial infection
bakterielle Infektion
bacterial lawn Bakterienrasen
bacterial strain Bakterienstamm
bacteriocidal/bactericidal
bakterizid, keimtötend
bacteriologic/bacteriological
bakteriologisch
bacteriology Bakteriologie
bacteriophage/phage/bacterial virus
Bakteriophage, Phage
bacteriosis Bakteriose
bacterium (*pl* bacteria)
Bakterie, Bakterium (*pl* Bakterien)
badge Kennzeichen, Abzeichen, Marke, Banderole
➤ **film badge** Strahlenschutzplakette
badging Bezettelung

baffle Prall..., Ablenk...;
biot (impeller: bioreactor)
Strombrecher (z.b. an Rührer von Bioreaktoren)
baffle plate Prallblech, Prallplatte, Leitblech, Ablenkplatte (Strombrecher z.b. an Rührer von Bioreaktoren)
baffle screen Prallschirm
bagging
Eintüten (Tüten, Säcke einfüllen)
baghouse (fabric filter dust collector)
Sackkammer, Sackraum (Staubabscheider)
bail (out) *vb* ausschöpfen (Wasser)
bailer (small bucket)
Schöpfer, kleiner Schöpfeimer
baker's yeast Backhefe, Bäckerhefe
baking soda/ sodium hydrogencarbonate
Natron, Natriumhydrogencarbonat, Natriumbicarbonat
balance *vb* **(balance out)**
ausbalancieren
balance *n* Bilanz (Energiebilanz/Stoffwechselbilanz); (equilibrium) Gleichgewicht; (scales) Waage
➤ **analytical balance**
Analysenwaage
➤ **beam balance** Balkenwaage
➤ **ecological balance/ ecological equilibrium**
ökologisches Gleichgewicht
➤ **natural balance**
natürliches Gleichgewicht (Naturhaushalt)
➤ **pan balance** Tafelwaage
➤ **precision balance**
Feinwaage, Präzisionswaage
➤ **spring balance/spring scales**
Federzugwaage, Federwaage
balanced equation
'eingerichtete' Gleichung
balanced lethal balanciert letal
ball (male: spherical joint) Schliffkugel
ball-and-socket joint/spheroid joint
Kugelgelenk

ball-and-stick model/
  stick-and-ball model *chem*
  Kugel-Stab-Modell,
  Stab-Kugel-Modell
ball bearing(s) Kugellager
  (Achsenlager: Rührer etc.)
ball mill/bead mill Kugelmühle
ball pane hammer/
  ball peen hammer/
  ball pein hammer
  Schlosserhammer mit Kugelfinne
ball valve Kugelventil
ballast (choke) *electr* Vorschaltdrossel
ballast group Ballastgruppe (*chem* Synthese)
banana plug Bananenstecker
band *chromat/electrophor* Bande
band-aid (adhesive strip)/
  sticking plaster/patch *med*
  Heftpflaster (Streifen)
band broadening *chromat*
  Bandenverbreiterung
band spectrum/
  molecular spectrum
  Bandenspektrum, Molekülspektrum
  (Viellinienspektrum)
bandage scissors *med*
  Verbandsschere
bandaging material/
  dressing material *med* Verbandszeug
banded/fasciate
  gebändert, breit gestreift
banding pattern (of chromosomes)
  Bänderungsmuster, Bandenmuster
banding technique Bänderungstechnik
bandwidth Bandbreite
bar diagram/bar graph Stabdiagramm
bar magnet/stir bar/
  stirrer bar/stirring bar/'flea'
  Magnetstab, Magnetstäbchen,
  Magnetrührstab, 'Fisch', Rühr'fisch'
barbed/fluted/serrated
  (e.g. tubing adapters)
  geriffelt (z.B. Schlauchadapter)
barophilic/barophilous barophil
barrel (drum/vat/tub/keg/tun) Fass;
  (cylinder) Walze, Rolle, Zylinder
▷ **stopcock barrel** Zylinder (Hahn)

barrel mixer/drum mixer
  Trommelmischer
barrel opener Fassöffner
barrel pump/
  drum pump Fasspumpe
barricade
  Absperrung, Sperre, Barrikade
barricade tape
  Absperrband, Markierband
barrier cream Gewebeschutzsalbe,
  Arbeitsschutzsalbe, Schutzcreme
barrier fluid Sperrflüssigkeit
barrier layer Grenzschicht,
  Randschicht, Sperrschicht
basal medium
  Basisnährboden, Basisnährmedium
base
  Basis; (foundation) Grundlage,
  Unterlage; *chem* Base
▷ **nitrogenous base**
  stickstoffhaltige Base, 'Base'
  (Purine/Pyrimidine)
base material (starting material/raw
  material) Grundstoff, Rohstoff;
  (ground substance/matrix)
  Grundsubstanz, Grundgerüst, Matrix
base peak (MS) Basispeak
baseboard/washboard Sockelleiste
basic/alkaline basisch, alkalisch
basic building block Grundbaustein
basic research Grundlagenforschung
basicity Basizität, Baseität
basket Korb
▷ **collapsible basket**
  zusammenlegbarer Korb, Faltkorb
basket centrifuge/bowl centrifuge
  Trommelzentrifuge
baster Fettgießer (große 'Pipette'),
  Ölabscheidepipette
batch Charge (in einem Arbeitsgang
  erzeugt), Partie, Posten, Füllung,
  Ladung, Los, Menge;
  kleine Stückzahl
batch culture
  diskontinuierliche Kultur,
  Batch-Kultur, Satzkultur
batch extraction
  diskontinuierliche Extraktion

batch number
Chargen-Bezeichnung, Chargen-B.
batch operation/batch process
diskontinuierliche Arbeitsweise/ Verfahren
batch process
Satzverfahren
bath Bad
➢ **circulating bath**
Umwälzthermostat, Badthermostat
➢ **refrigerated circulating bath**
Kältethermostat, Kühlthermostat, Umwälzkühler
bead *n* Kugel, Kügelchen, Perle; (beaded rim/flange) Bördelrand
bead *vb* **(flange/seam/edge)** bördeln
bead mill (shaking motion)
Schwing-Kugelmühle
bead-bed reactor Kugelbettreaktor
beaded rim Bördelrand, Rollrand (Glas: Ampullen etc.)
beaded rim bottle Rollrandflasche
beaker Becherglas, Zylinderglas (ohne Griff)
➢ **tempering beaker (jacketed beaker)/ cooling beaker/chilling beaker**
Temperierbecher
beaker brush Becherglasbürste
beaker clamp Becherglaszange
beaker tongs Becherglaszange
beam *n* Strahl; Balken
beam balance Balkenwaage
beam of light Lichtstrahl, Lichtbündel
beamer Strahler
bearing(s)
Lager (Achsen~, Rührer etc.)
➢ **ball bearing(s)**
Kugellager (Achsenlager: Rührer etc.)
➢ **bearing housing**
(Kugel)Lagergehäuse
beat/hit/strike schlagen, hauen
beeswax Bienenwachs
beet sugar/cane sugar/ table sugar/sucrose
Rübenzucker, Rohrzucker, Sukrose, Sucrose

behenic acid/docosanoic acid
Behensäure, Docosansäure
bell-shaped curve (Gaussian curve)
Glockenkurve (Gauß'sche Kurve)
bellows Balg
bellows pump Balgpumpe
'belly dancer'/nutator/nutating mixer (shaker with gyroscopic, i.e., threedimensional circular/orbital & rocking motion)
Taumelschüttler
ben oil/benne oil
Behenöl
bench (workbench/lab bench)
Werkbank (Labor-Werkbank/Laborbank)
➢ **clean bench/safety cabinet**
Sicherheitswerkbank
➢ **clean-room bench**
Reinraumwerkbank
➢ **lab bench/laboratory bench**
Labor-Werkbank, Laborbank, Laborarbeitstisch
➢ **sterile bench**
sterile Werkbank
bench grinder Doppelschleifer
bench row Laborzeile
bench-scale/lab-scale *adj*
im Labormaßstab
bench scales Tischwaage
benchtop
Arbeitsfläche (auf der Laborbank)
benchtop procedure
Laborverfahren (im Kleinmaßstab/auf der Laborbank)
bending vibration/ deformation vibration
Deformationsschwingung
beneficial/useful nützlich
beneficial species/beneficient species
Nützling, Nutzart
benefit *n* **(positive/favorable use)**
Nutzen
benefit *vb* nützen
benign benigne, gutartig
benignity/benign nature
Benignität, Gutartigkeit

**bent longnose pliers/
  bent long-nose pliers**
  gebogene Spitzzange
benzene Benzol
benzoic acid (benzoate)
  Benzoesäure (Benzoat)
berl saddle (column packing) *dist*
  Berlsattel (Füllkörper)
Berlese funnel *ecol* Berlese-Apparat
bevel (metal/glass/cannulas etc.)
  abkanten, abschrägen
beveled/bevelled
  abgeschrägt, abgekantet
  (Kanülenspitze/Pinzette etc.)
biased *math/stat* verzerrt, verfälscht
bibulous paper (for blotting dry)
  Löschpapier
bile Galle, Gallflüssigkeit
bile salts Gallensalze
bill (invoice)
  Rechnung (Waren~), Faktura
bill of lading Frachtbrief
bimetallic thermometer
  Bimetallthermometer
**bimodal distribution/
  two-mode distribution**
  bimodale Verteilung
bin (container) Behälter, Kasten;
  Tonne; Container
binary mixture Zweistoffgemisch
**binder/binding agent/
  absorbent/absorbing agent**
  Bindemittel, Saugmaterial
  (saugfähiger Stoff)
binder (office/paper)
  Aktenhefter, Umschlag;
  Ringbuch, Ordner
binder clip Halteklammer (Büro)
binding Bindung
➢ **cooperative binding**
  kooperative Bindung
binding curve Bindungskurve
binding energy/bond energy
  Bindungsenergie
binoculars Binokular
binomial distribution
  Binomialverteilung
binomial formula binomische Formel

bioassay/biological assay
  biologischer Test
bioavailability Bioverfügbarkeit
**biochemical oxygen demand/
  biological oxygen demand (BOD)**
  biochemischer Sauerstoffbedarf,
  biologischer Sauerstoffbedarf (BSB)
biochemistry Biochemie
biocide Biozid
biodegradability biologische
  Abbaubarkeit
biodegradable biologisch abbaubar
biodegradation
  Biodegradation, biologischer Abbau
bioenergetics Bioenergetik
bioengineering
  biologische Verfahrenstechnik,
  Biotechnik, Bioingenieurwesen
bioequivalence Bioäquivalenz
bioethics Bioethik
biogenic biogen
biohazard Biogefährdung;
  biologische Gefahrenquelle;
  biologische Gefahr,
  biologisches Risiko
**biohazard containment (laboratory)
  (classified into biosafety
  containment classes)**
  Sicherheitsraum, Sicherheitslabor
  (S1-S4)
biohazardous substance
  biologischer Gefahrstoff
biohazardous waste
  Abfall mit biologischem
  Gefährdungspotential
bioindicator/indicator species
  Bioindikator, Indikatorart, Zeigerart,
  Indikatororganismus
bioinorganic bioanorganisch
**biolistics/microprojectile
  bombardment** Biolistik
biologic(al)/biotic biologisch, biotisch
biological containment
  biologische Sicherheit(smaßnahmen)
biological containment level
  biologische Sicherheitsstufe
biological equilibrium
  biologisches Gleichgewicht

**biological oxygen demand/
biochemical oxygen demand (BOD)**
biologischer Sauerstoffbedarf,
biochemischer Sauerstoffbedarf
(BSB)
biological pest control
biologische Schädlingsbekämpfung
biological warfare agent
biologischer Kampfstoff
biologist/bioscientist/life scientist
Biologe, Biologin
biology/bioscience/life sciences
Biologie, Biowissenschaften
**biology lab technician/
biological lab assistant**
BTA
(biologisch-technischer Assistent)
bioluminescence Biolumineszenz
biomass Biomasse
biomedicine Biomedizin
bionics Bionik
biophysics Biophysik
bioreactor (see also: reactors)
Bioreaktor
biosafety
biologische Sicherheit
biosafety cabinet
biologische Sicherheitswerkbank
biosafety level
biologische Sicherheitsstufe/
Risikostufe
biosafety officer
biologischer Sicherheitsbeauftragter,
Beauftragter für biol. Sicherheit
**bioscience (meist *pl* biosciences)/
life science (meist *pl* life sciences)**
Biowissenschaft
biostatics Biostatik
biostatistics Biostatistik
biosynthesis Biosynthese
biosynthesize biosynthetisieren
biosynthetic(al) biosynthetisch
**biosynthetic reaction
(anabolic reaction)**
Biosynthesereaktion
biotechnology Biotechnologie
biotransformation/bioconversion
Biotransformation, Biokonversion

birefringence/double refraction
Doppelbrechung
birefringent/double-refracting
doppelbrechend
bismuth (Bi) Bismut (Wismut)
bisubstrate reaction
Zweisubstratreaktion,
Bisubstratreaktion
bit (drill bit/drill) Bohrer, Bohrspitze,
Bohraufsatz; (of a wrench) Maul
(Öffnung am Schraubenschlüssel);
(of a key) Bart (eines Schlüssels)
➢ **gimlet bit** Schneckenbohrer,
Windenschneckenbohrer
➢ **nail bit** Nagelbohrer
bitter bitter
bitterness Bitterkeit
bitters Bitterstoffe
bituminous coal/soft coal
Steinkohle, bituminöse Kohle
bivalence/divalence *chem*
Zweiwertigkeit
bivalent/divalent *chem*
zweiwertig, bivalent, divalent
blackout (short) Ohnmacht, 'Aussetzer'
bladder Blase
bladderlike/bladdery/vesicular
blasenartig, blasenförmig
blade Klinge; (cutting edge of a knife
blade) Schneide (Messer etc.)
blade mixer Schaufelmischer
blank *n math/stat* Blindwert
blast furnace Hochofen
bleach *n* Bleiche, Bleichmittel
bleach *vb* ausbleichen, bleichen
(*activ*: weiss machen, aufhellen)
bleaching Ausbleichen, Bleichen
bleed/bleeding Bluten
bleed (spotting) *vb chromat* (TLC)
durchschlagen, bluten
blender (vortex) Mixette,
Küchenmaschine (Vortex)
blind *adj/adv*
blind; verdeckt, matt, nicht poliert
blind *n* Rolladen, Rollo, Rouleau;
Markise; Blende
blind rivet Blindniete
blind rivet nut Blindnietmutter

bl

blindness Blindheit
bloat blähen
bloating/gas Blähungen, Flatulenz
block holder *micros* Blockhalter
block synthesis Blockverfahren
blocked/clogged/choked (drain) verstopft (Abfluss)
blocking reagent Blockierungsreagens
blood Blut
➢ fresh blood Frischblut
➢ stored blood/banked blood Blutkonserve
➢ whole blood Vollblut
blood agar Blutagar
blood bank Blutbank
blood cell/blood corpuscle/ blood corpuscule Blutkörperchen
blood clot Blutgerinnsel, Blutkoagulum
blood clotting Blutgerinnung
blood count Blutzellzahlbestimmung, Blutkörperchenzählung; (hematogram) Blutbild, Blutstatus, Hämatogramm
blood culture Blutkultur
blood donation Blutspende
blood group Blutgruppe
blood group incompatibility Blutgruppenunverträglichkeit
blood plasma Blutplasma
blood poisoning Blutvergiftung, Sepsis
blood pressure Blutdruck
blood smear Blutausstrich
blood substitute Blut-Ersatz
blood sugar Blutzucker
blood sugar level (elevated/reduced) Blutzuckerspiegel (erhöhter/erniedrigter)
blood-typing Blutgruppenbestimmung
blot *vb* blotten (klecksen, Flecken machen, beflecken)
blot hybridization Blothybridisierung
blotting (blot transfer) Blotten, Blotting
➢ affinity blotting Affinitäts-Blotting
➢ alkali blotting Alkali-Blotting
➢ capillary blotting Diffusionsblotting
➢ dry blotting Trockenblotten
➢ genomic blotting genomisches Blotting
➢ ligand blotting Liganden-Blotting
➢ Western blot/immunoblot Western-Blot, Immunoblot
➢ wet blotting Nassblotten
blower/fan Gebläse (Föhn)
blowpipe assay/test Lötrohrprobe
blowtorch Gebläselampe
blunt end/flush end *gen* glattes Ende, bündiges Ende
blurredness/blur/ obscurity/unsharpness Unschärfe
boat conformation (cycloalkanes) Wannenform
bobbin *electr* Spule, Induktionsrolle
body (soma) Körper; (dead body/corpse) Leichnam; (female fitting/female) *tech/mech* weibliche Kupplung, Körper
body fluid Körperflüssigkeit
body temperature Körpertemperatur
bogie Transportwagen, Transportkarren
boil *vb* sieden, kochen
➢ simmer leicht kochen
boilerstone/boiler scale (incrustation) Kesselstein (Ablagerung)
boiling flask Siedegefäß
boiling point Siedepunkt
boiling point depression/ lowering of boiling point Siedepunkterniedrigung
boiling point elevation Siedepunkterhöhung
boiling range Siedebereich
boiling rod/boiling stick/ bumping rod/bumping stick Siedestab
boiling stone/boiling chip Siedestein, Siedesteinchen
bolt *n* Schraubenbolzen, Bolzen; (bolting/barring/interlock) Verriegelung
bolt *vb* (bar/interlock) verriegeln
bolt cutter Bolzenschneider

bomb calorimeter
  Kalorimeterbombe,
  Bombenkalorimeter,
  Verbrennungsbombe
bomb tube/Carius tube/sealing tube
  Bombenrohr, Schießrohr,
  Einschlussrohr
bond (link) *vb* *chem* binden
bond *n* **(linkage)** *chem* Bindung
  ➢ **atomic bond** Atombindung
  ➢ **carbon bond** Kohlenstoffbindung
  ➢ **chemical bond** chemische Bindung
  ➢ **conjugated bond** *chem*
    konjugierte Bindung
  ➢ **double bond** Doppelbindung
  ➢ **glycosidic bond/glycosidic linkage**
    glykosidische Bindung
  ➢ **heteropolar bond**
    heteropolare Bindung
  ➢ **high energy bond**
    energiereiche Bindung
  ➢ **homopolar bond/nonpolar bond**
    homopolare Bindung
  ➢ **hydrophilic bond**
    hydrophile Bindung
  ➢ **hydrophobic bond**
    hydrophobe Bindung
  ➢ **ionic bond** Ionenbindung
  ➢ **multiple bond** Mehrfachbindung
  ➢ **nonpolar bond** unpolare Bindung
  ➢ **peptide bond/peptide linkage**
    Peptidbindung
  ➢ **triple bond** Dreifachbindung
bond angle Bindungswinkel
bond paper/stationery Schreibpapier
bonded phase gebundene Phase
bonded-phase chromatography
  Festphasenchromatographie
bonding strength Bindefähigkeit
bone Knochen
bone-cutting forceps/
  bone-cutting shears
    Knochenzange
bone meal Knochenmehl
bone saw Knochensäge
bony knöchern, Knochen...
book cart Bücherwagen
booster (substance) Verstärkungsstoff

borax/sodium tetraborate Borax,
  Natriumtetraborat decahydrat
bore *n* Bohrung
boric acid (borate) Borsäure (Borat)
boron (B) Bor
borosilicate glass Borosilikatglas
bottle Flasche
  ➢ **beaded rim bottle** Rollrandflasche
  ➢ **drop bottle/dropping bottle**
    Tropfflasche
  ➢ **dropping bottle/dropper vial**
    Pipettenflasche; Tropfflasche
  ➢ **gas bottle/**
    gas cylinder/
    compressed-gas cylinder
      Gasflasche
  ➢ **lab bottle/laboratory bottle**
    Laborstandflasche, Standflasche
  ➢ **narrow-mouthed bottle**
    Enghalsflasche
  ➢ **packaging bottle**
    Verpackungsflasche
  ➢ **reagent bottle** Reagentienflasche
  ➢ **roller bottle** Rollerflasche
  ➢ **screw-cap bottle** Schraubflasche
  ➢ **spray bottle** Sprühflasche
  ➢ **square bottle** Vierkantflasche
  ➢ **wash bottle/squirt bottle**
    Spritzflasche
  ➢ **weighing bottle** Wägeglas
  ➢ **wide-mouthed bottle** Weithalsflasche
  ➢ **Woulff bottle** Woulff'sche Flasche
bottle brush
  (beaker/jar/cylinder brush)
    Flaschenbürste
bottle cart (barrow)/
  bottle pushcart/cylinder trolley (*Br*)
    Flaschenwagen
bottle shelf/bottle rack Flaschenregal
bottleneck Engpass; Flaschenhals
bottom fermenting untergärig
bottom yeast
  niedrigvergärende Hefe ('Bruchhefe')
bottoms/deposit
  (sediment/precipitate/settlings)
    Bodenkörper;
    dist Sumpf (Rückstand in Blase)
bouffant cap Haarschutzhaube

bo

boundary layer Grenzschicht
box (crate) Schachtel, Kasten, Kiste
box wrench/
   box spanner/ring spanner (*Br*)
   Ringschlüssel
brackish water (somewhat salty)
   Brackwasser
bran Kleie
branch out/ramify sich verzweigen
branch site Verzweigungsstelle
branched/ramified verzweigt
branched-chained *chem*
   verzweigtkettig
branching *chem* Verzweigung
brand Marke (Ware/Handel)
brand name
   Markenbezeichnung, Warenzeichen
brass Messing
➢ chrome-plated brass
   verchromtes Messing
breach of duty Pflichtverletzung
break (glass: shatter)
   zerbrechen, zerspringen
➢ break up zerkleinern
breakable zerbrechlich
➢ nonbreakable/unbreakable/
   crashproof
   bruchsicher
breakage Bruch
breakdown
   Abbau, Zusammenbruch, Zerfall;
   (incident/accident) Störfall
breaking pressure (valve)
   Öffnungsdruck (Ventil)
breakpoint Bruchstelle
breath *n* Atem
breathe *vb* (respire) atmen
➢ breathe in/inhale einatmen
➢ breathe out/exhale ausatmen
breathing (respiration) Atmung
breathing apparatus/respirator
   Atemschutzgerät, Atemgerät
breathing protection/
   respiratory protection Atemschutz
breed/breeding/cultivation/growing
   Züchtung, Kultivierung
breed *vb* (cultivate/grow)
   züchten, kultivieren, aufziehen

breeding period/incubation period
   Brutdauer, Inkubationszeit
brewers' grains Treber, Biertreber
brewers' yeast Bierhefe, Brauhefe
bright (luminous) lichtstark
bright field *micros* Hellfeld
brightener/brightening agent/
   clearant/clearing agent
   (optical brightener)
   Aufheller,
   Aufhellungsmittel
   (optischer Aufheller)
brightfield microscopy
   Hellfeld-Mikroskopie
bristle Borste
British Standard Pipe (BSP)
   thread/fittings
   Britisches Standard Gewinde
broad-spectrum antibiotic
   Breitspektrumantibiotikum
brochure/pamphlet
   Broschüre,
   Informationsschrift
bromine (Br) Brom
brood/breed/incubate
   bebrüten, brüten, inkubieren
broom Besen, Kehrbesen
brown paper/kraft Packpapier
bruise/hematoma
   Bluterguss, Hämatom
brush *n* Bürste;
   electr (motor brush) Schleifkontakt
➢ beaker brush Becherglasbürste
➢ bottle brush Flaschenbürste
➢ dishwashing brush
   Spülbürste, Geschirrspülbürste
➢ flask brush Kolbenbürste
➢ funnel brush Trichterbürste
➢ laboratory brush Laborbürste
➢ paint brush Malpinsel
➢ pipe cleaner
   Pfeifenreiniger, Pfeifenputzer
➢ pipet brush Pipettenbürste
➢ scrubbing brush/scrub brush
   Scheuerbürste,
   Schrubbbürste
➢ test tube brush Reagenzglasbürste
➢ wire brush Drahtbürste, Stahlbürste

bubble *n* Blase
(Gasblase/Luftblase/Seifenblase);
(small bubble) Bläschen;
(vesicle) Vesikel
bubble *vb* sprudeln, brodeln, 'blubbern'
bubble column reactor
Blasensäulen-Reaktor
bubble counter/bubbler/gas bubbler
Blasenzähler
bubble linker PCR *gen*
Blasen-Linker-PCR
bubble-shaped/bulliform
bläschenförmig
bubble tube
 (slightly bowed glass tube/
 vial in spirit level)
Libelle (Glasröhrchen der
Wasserwaage)
bubbler/bubble counter/gas bubbler
Blasenzähler
buccal swab Wangenabstrich
bucket (plastic)/pail (metal)
Eimer
bucking circuit *electr*
Kompensationsspule
buckle *n* Schnalle
buckle strap Schnallenriemen
Buechner funnel
Büchner-Trichter
(Schlitzsiebnutsche)
buff *vb tech* polieren
buffer *n* Puffer
buffer *vb* puffern
buffer solution Pufferlösung
buffer zone Pufferzone
buffering Pufferung
buffering capacity Pufferkapazität
building (construction)
Gebäude; Bau...
building block Baustein, Bauelement
➢ **basic building block**
Grundbaustein
building cleaners
Gebäudereinigungspersonal
building code
Bauvorschriften
building evacuation plan
Gebäudeevakuierungsplan

bulb Kugel, Kolben, Ballon
bulb-to-bulb distillation
Kugelrohrdestillation
bulk Masse, Fülle; Mengen...
bulk cargo Bulkladung (Transport)
bulk consumer Großverbraucher
bulk container
Schüttgutbehälter
bulk delivery/bulk shipment
Großlieferung
bulk density (powder density)
Schüttdichte
bulk package Großpackung
bulk shipment/bulk delivery
Großlieferung
bulk volume Schüttvolumen
bulking sludge Blähschlamm
bull horn Megaphon, Megafon
bulletproof glass Panzerglas
bulletproof vest
kugelsichere Veste
bump (into)/knock (into)
dranstoßen
bump tube Rückschlagschutz
bumper guard
Schutzring, Stoßschutz
(Prellschutz für Meßzylinder)
bumping Stoßen
(Sieden/Überhitzung/Siedeverzug)
bumping rod/bumping stick/
 boiling rod/boiling stick
Siedestab
bundle/bunch/
 lashing/packaging
 (larger quantities of items fastened
 together) Gebinde
Bunsen burner/flame burner
Bunsenbrenner
buoyancy
Auftrieb (hydrostatisch);
Schwimmkraft, Tragkraft
buoyant schwimmend, tragend,
schweben, federnd
buoyant density
Schwebedichte, Schwimmdichte
bur (bit on a dental drill)
Bohrer, Bohrspitze, Bohraufsatz
burden Belastung

bu

buret (burette *Br***)** Bürette
- **weight buret/weighing buret**
Wägebürette

buret clamp Bürettenklemme

burn *n* *med* Verbrennung;
(caustic burn: chemical/alkali/acid)
Verätzung (Chemikalie/Alkali/Säure)
- **respiratory tract burn/**
(alkali/acid)
caustic burn of the respiratory tract
Atemwegsverätzung

burn *vb* brennen, verbrennen
- **burn through/out**
durchbrennen

burner (flame: oven)
Brenner (Flamme: Ofen)
- **alcohol burner**
Spiritusbrenner, Spirituslampe
- **Bunsen burner/flame burner**
Bunsenbrenner
- **cartridge burner**
Kartuschenbrenner
- **evaporation burner**
Verdunstungsbrenner
- **gas burner** Gasbrenner, Gaskocher

burner wing top/wing-tip (for burner)
Schwalbenschwanzbrenner,
Schlitzaufsatz für Brenner

burst *n* Aufbrechen, Ausbruch;
Bersten, Sprengen; Salve

bursting disk
Berstscheibe, Sprengscheibe,
Sprengring, Bruchplatte

bush hammer Scharrierhammer

bushing/guide bushing
Buchse, Gleitlager, Spannhülse,
Führungsbuchse (Rührwelle etc.)

business/company/firm/enterprise
Betrieb, Unternehmen

butcher Schlachter, Fleischer, Metzger

butt joint Stoßverbindung

butterfly valve Flügelhahnventil

button Knopf; (control) Regler

butyl rubber Butylkautschuk
(Isobutylen-Isopren-Kautschuk)

butyric acid/butanoic acid (butyrate)
Buttersäure, Butansäure (Butyrat)

by-product/residual product/
side product Nebenprodukt

cabinet (cupboard) Schrank
> **wall cabinet/cupboard**
  Wandschrank
cable Kabel
cable connector Kabelverbinder
cable drum Kabeltrommel
cable lug Kabelöse, Kabelschuh;
  Ringkabelschuh
cable stripping knife Kabelmesser
cable tie(s)/
  wrap-it tie(s)/wrap-it tie cable
  Kabelbinder, Spannband
> **tensioning tool/tensioning gun**
  Spannzange (Kabelbinder)
cadaver/carcass/corpse
  Kadaver, Tierleiche
cadaverous smell Leichengeruch
caffeic acid Kaffeesäure
caffeine/theine Koffein, Thein
cage/enclosure (for animals)
  Käfig, Tierzwinger
cake *n* Klumpen, Kruste
  (fest verbackender Niederschlag)
cake *vb* klumpen,
  zusammenbacken (Präzipitat)
calcification Verkalkung,
  Kalkeinlagerung, Kalzifizierung,
  Calcifikation
calcify (calcified) verkalken (verkalkt)
calcination Kalzinierung
calcine kalzinieren
calcite Kalkspat
calcium (Ca) Kalzium, Calcium
calculate berechnen
calculation Berechnung
calibrate
  (adjust/standardize/gage/gauge)
  kalibrieren, eichen; standardisieren
  (Maße/Gewichte)
calibrated kalibriert, geeicht
calibrating instrument/calibrator
  Eichgerät
calibrating mark Eichmarke
calibrating standard/
  standard (measure) Eichmaß
calibration/adjustment/
  adjusting/standardization
  Kalibrierung, Eichung

caliper Tastzirkel, Taster, Lehre
> **inside caliper**
  Innentaster, Lochtaster
> **micrometer caliper**
  Messschraube,
  Mikrometerschraube, 'Mikrometer'
> **odd-leg caliper** einseitiger Tastzirkel
> **outside caliper** Außentaster
> **slide caliper/caliper square**
  Schublehre, Schieblehre
> **vernier caliper**
  Schublehre mit Nonius
caliper gage (gauge *Br***)**
  Messschieber
caliper rule
  (one fixed/one adjustable jaw)
  Schieblehre
callus Kallus, Callus;
  (wound tissue) Wundkallus,
  Wundcallus, Wundgewebe,
  Wundholz
callus culture
  Callus-Kultur, Kallus-Kultur
caloric value Brennwert
calorie Kalorie
calorimeter
  Kalorimeter,
  Brennwertbestimmungsgerät
> **bomb calorimeter**
  Kalorimeterbombe,
  Bombenkalorimeter,
  Verbrennungsbombe
> **continuous-flow calorimeter**
  Durchflusskalorimeter
calorimetry
  Kalorimetrie,
  Brennwertbestimmung
> **differential scanning calorimetry**
  (DSC)
  Differentialkalorimetrie
> **power-compensated DSC**
  Leistungskompensations-
  Differentialkalorimetrie
> **scanning calorimetry**
  Raster-Kalorimetrie
camouflage Tarnung
canal/duct/tube Kanal (zum
  Weiterleiten von Flüssigkeiten)

cancer
 (malignant neoplasm/carcinoma)
 Krebs (malignes Karzinom)
cancer causing/
 oncogenic/oncogenous
 krebserzeugend, onkogen, oncogen
cancer risk Krebsrisiko
cancer suspect agent/
 suspected carcinogen
 krebsverdächtiger Stoff/Substanz
cancerous krebsartig
cane sugar/beet sugar/
 table sugar/sucrose
 Rohrzucker, Rübenzucker,
 Saccharose, Sukrose, Sucrose
cannula Kanüle
canola oil (rapeseed oil)
 Speise-Rapsöl, Rüböl
caoutchouc/rubber/india rubber
 Kautschuk
cap *n* **(lid/cover)**
 Deckel, Verschluss
cap *vb* mit Deckel verschließen
capacitance (C)
 elektrische Kapazität
capacitative current
 kapazitiver Strom
capacitor Kondensator
capacity Kapazität; Fassungsvermögen;
 (volume) Rauminhalt (Volumen)
capacity factor
 Kapazitätsfaktor,
 Verteilungsverhältnis
capillary Kapillare, Haargefäß
capillary air bleed/
 boiling capillary/
 air leak tube *dist*
 Siedekapillare
capillary blotting
 Diffusionsblotting
capillary chromatography (CC)
 Kapillarchromatographie
capillary column *chromat*
 Kapillarsäule
capillary electrophoresis (CE)
 Kapillarelektrophorese
capillary pipet/capillary pipette
 Kapillarpipette
capillary tube/tubing Kapillarrohr
capillary zone electrophoresis (CZE)
 Kapillar-Zonenelektrophorese
capnophilic
 kohlendioxidliebend, kapnophil
capric acid/decanoic acid
 (caprate/decanoate)
 Caprinsäure, Decansäure
 (Caprinat/Decanat)
caproic acid/capronic acid/
 hexanoic acid (caproate/hexanoate)
 Capronsäure, Hexansäure
 (Capronat/Hexanat)
caprylic acid/octanoic acid
 (caprylate/octanoate)
 Caprylsäure, Octansäure
 (Caprylat/Octanat)
carbohydrate Kohlenhydrat
carbon Kohlenstoff
 ➢ **activated carbon** Aktivkohle
carbon bond
 Kohlenstoffbindung
carbon brush *tech*
 Kohlebürste (Motor)
carbon compound
 Kohlenstoffverbindung
carbon dioxide CO_2 Kohlendioxid
carbon monoxide CO Kohlenmonoxid
carbon source
 Kohlenstoffquelle
carbon tetrachloride/
 tetrachloromethane
 Tetrachlorkohlenstoff,
 Tetrachlormethan
carbonate hardness/
 temporary hardness
 Karbonathärte, Carbonathärte,
 vorübergehende Härte,
 temporäre Härte
carbonic acid (carbonate)
 Kohlensäure (Karbonat/Carbonat)
carbonization
 Karbonisation; Verschwelung;
 paleo/geol (coalification) Inkohlung
carboxylic acids (carbonates)
 Carbonsäuren, Karbonsäuren
 (Carbonate, Karbonate)
carboy Ballonflasche

carcinogen Karzinogen
carcinogenic (Xn)
  krebserzeugend, karzinogen, kanzerogen, carcinogen, krebserregend
carcinoma Karzinom
cardboard (paperboard/fiberboard)
  Karton, Kartonpapier; (pasteboard) Pappe, Pappdeckel
cardboard box Pappkarton
cardinal temperature
  Vorzugstemperatur
cardiopulmonary resuscitation (CPR)
  kardiopulmonale Reanimation
careless/incautious/unwary
  nachlässig, unachtsam, fahrlässig, leichtsinnig, unvorsichtig
carelessness Nachlässigkeit, Unachtsamkeit, Fahrlässigkeit, Leichtsinn, Unvorsichtigkeit
caretaker/janitor/custodian
  Hausmeister, Hausverwalter
**caretaker's office/
  custodian's office**
  Büro des Hausmeisters/Hausverwalters
cargo Frachtgut, Ladung
cargo tank Frachtkessel
**Carius furnace/bomb furnace/
  bomb oven/tube furnace**
  Schießofen
Carius tube/bomb tube/sealing tube
  Bombenrohr, Schießrohr, Einschlussrohr
carob gum/locust bean gum
  Karobgummi, Johannisbrotkernmehl
carpenter Schreiner
carrageenan/carrageenin
  (*Irish moss* **extract**)
  Carrageen, Carrageenan
carriage
  Schlitten, Fahrgestell; Beförderung
carrier Träger; *chromat*
  Trägersubstanz; Schleppmittel; (shipper) Spediteur
carrier electrophoresis
  Trägerelektrophorese, Elektropherografie
carrier gas (an inert gas) *chromat* **(GC)**
  Trägergas, Schleppgas
carrier molecule Trägermolekül
carry off (drain/discharge) ableiten
carrying capacity
  Belastungsfähigkeit, Grenze der ökologischen Belastbarkeit, Kapazitätsgrenze, Umweltkapazität; Traglast
carryover (carry forward) Übertrag; *chromat* (cross-contamination) Verschleppung, Kreuzkontamination
cartilage Knorpel
cartilage forceps Knorpelpinzette
cartilage knife Knorpelmesser
cartilaginous knorpelig
cartridge Kartusche, Patrone; Filterkerze; (cassette) Kassette
cartridge burner Kartuschenbrenner
cascade Kaskade, Kascade
cascade system (enzymes)
  Kaskadensystem
case Etui; *med* Fall
case report *med/jur* Fallbericht
case report form (CRF)/case record form Prüfprotokoll
cask Holzfass
cast aluminum Aluminiumguss
cast iron Gusseisen
casters/castors
  Rollfüße, Laufrollen, Rollen (zum Schieben, für Laborwagen etc.)
➢ **swivel casters**
  Lenkrollen, Schwenkrollen, Schwenklaufrollen
castor oil/ricinus oil Rizinusöl
casualty
  Verunglückter, Opfer, Verwundeter
catabolic katabolisch/abbauend; (degradative reactions) katabol, catabol
catabolite Katabolit, Stoffwechselabbauprodukt
catalysis Katalyse
catalyst Katalysator
catalytic/catalytical katalytisch
catalytic antibody
  katalytischer Antikörper

ca

catalytical unit/
unit of enzyme activity (*katal*)
katalytische Einheit,
Einheit der Enzymaktivität (*katal*)
catalyze katalysieren
categorization
Einstufung, Kategorisierung
catenation
Catenation, Ringbildung
cation Kation
cation exchanger
(*strong:* **SCX**/*weak:* **WCX**)
Kationenaustauscher
(starker/schwacher)
caustic/corrosive/mordant
ätzend, beizend, korrosiv *chem*
caustic agent Ätzmittel (Beizmittel)
caustic hazard Verätzungsgefahr
caustic lime CaO Branntkalk
caustic potash/
potassium hydroxide KOH
Ätzkali, Kaliumhydroxid
caustic soda/
sodium hydroxide NaOH
Ätznatron, Natriumhydroxid
cauterization *med*
Ätzen, Ätzung, Verätzung;
Ätzverfahren
cauterize *vb med* ätzen
caution/cautiousness/
care/carefulness/
precaution
Vorsicht
caution! (careful!) Vorsicht!
caution, danger!
Vorsicht, Lebensgefahr!
cautious/careful vorsichtig
cavity/chamber/ventricle
Höhle, Kammer, Ventrikel
(kleine Körperhöhle); (lumen)
Hohlraum, Höhlung, Lumen
CBA-paper
(cyanogen bromide activated paper)
CBA-Papier
CDC
(Centers for Disease Control)
U.S. Gesundheitsbehörde
(entspricht in etwa: BGA)

ceiling level (CL)
(with reference to TLV)
Maximalwert,
maximale Konzentration
ceiling temperature *polym*
Ceiling-Temperatur (meist nicht
übersetzt), Gipfeltemperatur
cell Zelle; Elektrolysezelle,
Elektrolysierzelle; Küvette,
Probenbehälter, Messzelle;
elektrochemisches Element, Zelle
cell count/germ count Keimzahl
(Anzahl von Mikroorganismen)
cell culture Zellkultur
cell death Zelltod
cell disrupter Zellaufschlussgerät
cell extract Zellextrakt
cell fractionation Zellfraktionierung
cell-free extract zellfreier Extrakt
cell fusion Zellfusion,
Zellverschmelzung
cell holder *analyt* Küvettenhalter
cell homogenization
Zellhomogenisation,
Zellhomogenisierung
cell hybridization Zellhybridisierung
cell lineage/cell line/celline
Zelllinie
cell lysis Zellaufschluss
(Öffnen der Zellmembran)
cell scraper Zellschaber
cell sorter Zellsorter, Zellsortierer,
Zellsortiergerät (Zellfraktionator)
cell sorting Zellsortierung
cellobiose Zellobiose, Cellobiose
cellular zellulär, zellig
cellulose Zellulose, Cellulose
cement (adhesive)
Kitt, Kittsubstanz (Kleber/Klebstoff)
➤ **multicomponent cement/**
multiple-component cement
Mehrkomponentenkleber
center *vb* zentrieren
center drill (tool) Zentrierbohrer
center punch (tool)
Körner (Werkzeug)
centile/percentile *stat*
Zentil, Perzentil, Prozentil

centrifugal zentrifugal
centrifugal extractor
Zentrifugalextraktor
centrifugal force Zentrifugalkraft
centrifugal grinding mill
Rotormühle,
Zentrifugalmühle,
Fliehkraftmühle
centrifugal pump Kreiselpumpe
centrifugation Zentrifugation
➢ **analytical centrifugation**
analytische Zentrifugation
➢ **density gradient centrifugation**
Dichtegradientenzentrifugation
➢ **differential centrifugation**
('pelleting')
Differentialzentrifugation,
differentielle Zentrifugation
➢ **equilibrium centrifugation/**
equilibrium centrifuging
Gleichgewichtszentrifugation
➢ **isopycnic centrifugation/**
isodensity centrifugation
isopyknische Zentrifugation
➢ **preparative centrifugation**
präparative Zentrifugation
➢ **refrigerated centrifuge**
Kühlzentrifuge
➢ **ultracentrifugation**
Ultrazentrifugation
➢ **zonal centrifugation**
Zonenzentrifugation
centrifuge (spin) *vb* zentrifugieren
centrifuge *n* Zentrifuge
➢ **basket centrifuge/**
bowl centrifuge
Trommelzentrifuge
➢ **screen basket centrifuge**
Siebkorbzentrifuge
➢ **high-speed centrifuge/**
high-performance centrifuge
Hochgeschwindigkeitszentrifuge
➢ **knife-discharge centrifuge/**
scraper centrifuge
Schälschleuder
➢ **microfuge** Mikrozentrifuge
➢ **screen centrifuge**
Siebschleuder
➢ **tabletop centrifuge/**
benchtop centrifuge
(multipurpose c.)
Tischzentrifuge
➢ **tubular bowl centrifuge**
Röhrenzentrifuge
➢ **ultracentrifuge** Ultrazentrifuge
centrifuge tube
Zentrifugenröhrchen
centrifuge tube rack
Zentrifugenröhrchenständer
ceramic filter Tonfilter
cermet Cermet, Kermet,
Metallkeramik
cerotic acid/hexacosanoic acid
Cerotinsäure, Hexacosansäure
certificate Zertifikat, Zeugnis,
Urkunde, Bescheinigung,
Beglaubigung
certificate of performance
Qualitätszertifikat
certification
Zertifizierung, Bescheinigung,
Beglaubigung (amtliche)
certified zertifiziert, bestätigt,
bescheinigt, beglaubigt
cesium (Cs) Cäsium
cesium chloride gradient
Cäsiumchloridgradient
CFCs (chlorofluorocarbons/
chlorofluorinated hydrocarbons)
FCKW
(Fluorchlorkohlenwasserstoffe)
chain *n* **(branched/unbranched)**
Kette (verzweigte/unverzweigte)
chain (to) *vb*
(e.g., gas bottles/cylinders)
anketten (Gasflaschen etc.)
chain clamp Kettenklammer
chain form/
open-chain form *chem*
Kettenform
chain formula/open-chain formula
Kettenformel
chain length Kettenlänge
chain reaction Kettenreaktion
chain-terminating technique *gen*
Kettenabbruchverfahren

chair Stuhl
➤ **swivel chair** Drehstuhl
chair conformation (cycloalkanes) Sesselform
chalcocite Cu_2S Kupferglanz
chamber/tank *electrophor* Kammer
chance Zufall
change area Umkleide; (locker room) Umkleide(raum) mit Spinden
changing room/changing cubicle; (cleanroom/containment level: gowning room) Umkleidekabine
chaotropic agent chaotrope Substanz
chaotropic series chaotrope Reihe
char *vb* verkohlen, ankohlen; verschwelen
char residue Verkohlungsrückstand
characteristic value Kennwert
charcoal Holzkohle
charge *n* **(electrical charge)** Ladung (elektrische Ladung)
charge *vb* beladen, aufladen; (feed) *micb* beschicken
charge separation *electr* Ladungstrennung
charger *electr* Ladegerät
charring Verkohlung, Verschwelung
chart Karte, Tafel, Schaubild, Tabelle
chart paper Registrierpapier, Aufzeichnungspapier; Tabellenpapier
check *vb* **(examine/confirm/inspect/review)** überprüfen, bestätigen, inspizieren
check sum Prüfsumme
checkup/ examination/inspection/reviewal Überprüfung, Inspektion; (physical examination/physical) ärztliche/medizinische Untersuchung
checkweighing scales Kontrollwaage
cheesecloth (gauze) Mull (Gaze)
chelate *n* Chelat, Komplex
chelate *vb* komplexieren
chelating agent/chelator Chelatbildner, Komplexbildner
chelation/chelate formation Chelatbildung, Komplexbildung

chemical bond chemische Bindung
chemical burn chemische Verbrennung
chemical compound chemische Verbindung
chemical equation chemische Gleichung, Reaktionsgleichung
chemical oxygen demand (COD) chemischer Sauerstoffbedarf (CSB)
chemical reaction/ transformation/ chemical change chemische Reaktion, chemische Umsetzung
chemical shift *spectr* chemische Verschiebung
chemical stockroom counter Chemikalienausgabe
chemical warfare agent chemischer Kampfstoff
chemical waste Chemieabfälle
chemically pure (CP) chemisch rein
chemicals Chemikalien
➤ **existing chemicals/ existing substances** Altstoffe
➤ **new chemicals/ new substances** Neustoffe
chemiosmosis Chemiosmose
chemiosmotic hypothesis/theory chemiosmotische Hypothese/Theorie
chemisorption Chemisorption, chemische Adsorption
chemistry Chemie
➤ **analytical chemistry** Analytische Chemie
➤ **biochemistry** Biochemie
➤ **inorganic chemistry** Anorganische Chemie
➤ **organic chemistry** Organische Chemie
➤ **physical chemistry** Physikalische Chemie
chemoaffinity hypothesis Chemoaffinitäts-Hypothese
chemostat Chemostat

chemosynthesis Chemosynthese
chemotherapy Chemotherapie
chew/masticate
 kauen, zerkauen
chicken embryo culture
 Eikultur (Hühnerei)
chief association
 Hauptassoziation
chiller
 Kühler, Kühlgerät, Kühlaggregat
 ➢ **refrigerated chiller**
  with immersion probe
  Eintauchkühler (mit Kühlsonde)
chilling damage/injury
 Kälteschaden, Kälteschädigung
chimney/flue Rauchfang
chinic acid/kinic acid/
 quinic acid (quinate)
 Chinasäure
chinolic acid Chinolsäure
chip/chipping (e.g., glass)
 anschlagen, Ecke abschlagen
chiral chiral
chiral chromatography
 enantioselektive Chromatographie
chirality Chiralität
chisel
 Meißel; Beitel, Stechbeitel
chloric acid $HClO_3$
 Chlorsäure
chlorinate chlorieren
chlorination Chlorierung
chlorine (Cl) Chlor
chlorine bleach Chlorbleiche
chlorobenzene Chlorbenzol
chlorofluorocarbons/
 chlorofluorinated hydrocarbons
 (CFCs)
 Fluorchlorkohlenwasserstoffe
 (FCKW)
chloroform/trichloromethane
 Chloroform, Trichlormethan
chlorogenic acid Chlorogensäure
chlorous acid chlorige Säure
chocolate agar
 Schokoladenagar, Kochblutagar
choke/throttle/slow down/dampen
 drosseln, herunterfahren, dämpfen

cholesterol Cholesterin, Cholesterol
cholic acid (cholate)
 Cholsäure (Cholat)
chorismic acid (chorismate)
 Chorisminsäure (Chorismat)
chromaffin/chromaffine/chromaffinic
 chromaffin
chromatid conversion
 Chromatidenkonversion
chromatogram Chromatogramm
chromatograph Chromatograph
chromatography
 Chromatographie
 ➢ **affinity chromatography**
  Affinitätschromatographie
 ➢ **bonded-phase chromatography**
  Festphasenchromatographie
 ➢ **capillary chromatography (CC)**
  Kapillarchromatographie
 ➢ **chiral chromatography**
  enantioselektive Chromatographie
 ➢ **circular chromatography/**
  circular paper chromatography
  Zirkularchromatographie,
  Rundfilterchromatographie
 ➢ **column chromatography**
  Säulenchromatographie
 ➢ **electrochromatography (EC)**
  Elektrochromatographie (EC)
 ➢ **flash chromatography (FC)**
  Blitzchromatographie,
  Flash-Chromatographie
 ➢ **gas chromatography**
  Gaschromatographie
 ➢ **gas-liquid chromatography**
  Gas-Flüssig-Chromatographie
 ➢ **gel filtration/**
  molecular sieving chromatography/
  gel permeation chromatography
  (GPC) Gelfiltration,
  Molekularsiebchromatographie,
  Gelpermeations-Chromatographie
 ➢ **gel permeation chromatography/**
  molecular sieving chromatography
  Gelpermeationschromatographie
  (GPC), Molekularsiebchromatographie
 ➢ **gravity column chromatography**
  Normaldruck-Säulenchromatographie

ch

- high-pressure liquid chromatography/
  high performance liquid chromatography (HPLC)
  Hochdruckflüssigkeits-chromatographie,
  Hochleistungsflüssigkeits-chromatographie
- immunoaffinity chromatography
  Immunaffinitätschromatographie
- immunofluorescence chromatography
  Immunfluoreszenz-chromatographie
- ion-exchange chromatography (IEX)
  Ionenaustauschchromatographie
- ion-pair chromatography (IPC)
  Ionenpaarchromatographie (IPC)
- liquid chromatography (LC)
  Flüssigkeitschromatographie
- medium-pressure liquid chromatography (MPLC)
  Mitteldruckflüssigkeits-chromatographie
- membrane chromatography (MC)
  Membranchromatographie (MC)
- paper chromatography
  Papierchromatographie
- partition chromatography/
  liquid-liquid chromatography (LLC)
  Verteilungschromatographie,
  Flüssig-flüssig-Chromatographie
- preparative chromatography
  präparative Chromatographie
- recognition site affinity chromatography
  Erkennungssequenz-Affinitätschromatographie
- reversed phase chromatography/
  reverse-phase chromatography (RPC)
  Umkehrphasenchromatographie
- salting-out chromatography
  Aussalzchromatographie
- size exclusion chromatography (SEC)
  Größenausschlusschromatographie,
  Ausschlusschromatographie
- supercritical fluid chromatography (SFC)
  überkritische Fluidchromatographie,
  superkritische Fluidchromatographie,
  Chromatographie mit überkritischen Phasen (SFC)
- thin-layer chromatography (TLC)
  Dünnschichtchromatographie (DC)

chrome-plated verchromt
chrome-plated brass
  verchromtes Messing
chromic(VI) acid H_2CrO_4
  Chromsäure
chromic-sulfuric acid mixture for cleaning purposes
  Chromschwefelsäure
chromium (Cr) Chrom
chromium mordant Chrombeize
chronic/chronical chronisch
chuck/collet chuck (drill)
  Spannfutter (Bohrer)
cinders/slag/dross/scoria
  Schlacke
cinnamic acid Zimtsäure,
  Cinnamonsäure (Cinnamat)
cinnamic alcohol/cinnamyl alcohol
  Zimtalkohol
cinnamic aldehyde/cinnamaldehyde
  Zimtaldehyd
'circles'/filter paper disk/round filter
  Rundfilter
circuit *electr* (also: neural circuit)
  Schaltkreis, Schaltsystem
circuit board Schaltungsplatte
- **printed circuit board**
  Platine, Leiterplatte, Board
circuit breaker *electr*
  Unterbrecher, Trennschalter
circuit diagram Schaltbild, Schaltplan
circuitry Leitungen, Schaltungen;
  Schaltungsbauteile
circular (round)
  zirkular, zirkulär (kreisförmig/rund)
**circular chromatography/
  circular paper chromatography**
  Circularchromatographie,
  Zirkularchromatographie,
  Rundfilterchromatographie

circular dichroism
Circulardichroismus, (Zirkulardichroismus)
circular shaker/
orbital shaker/rotary shaker
Rundschüttler, Kreisschüttler
circularization
Zirkularisierung; *chem* Ringschluss
circulate zirkulieren; umwälzen
circulating/circulatory
zirkulierend, Zirkulations...
circulating bath
Umwälzthermostat, Badthermostat
circulation Zirkulation, Zirkulieren, Umwälzung; (blood supply/blood circulation) Durchblutung
circulation pump Umwälzpumpe
circulator Umwälzer; Umlenker
➢ **immersion circulator**
Einhängethermostat, Tauchpumpen-Wasserbad
citric acid (citrate)
Citronensäure, Zitronensäure (Citrat)
Civil Code Bürgerliches Gesetzbuch
civil service/public service
öffentliche Verwaltung
civilian social servant
(conscientious objector)
Zivildienstleistender
civilian social service
(in place of military service for conscientious objectors)
Zivildienst
clamp *vb* **(fix/attach/mount)**
einspannen, festklemmen, befestigen
clamp *n* **(vise/vice** *Br***)** Zwinge; (clip) Klemme, Klammer
➢ **beaker clamp** Becherglaszange
➢ **buret clamp** Bürettenklemme
➢ **chain clamp** Kettenklammer
➢ **extension clamp**
Verlängerungsklemme
➢ **flask clamp**
Kolbenklemme, Kolbenklammer
➢ **four-prong flask clamp**
Federklammer (für Kolben)
➢ **hook clamp** Hakenklemme (Stativ)
➢ **pinch clamp** Schraubklemme
➢ **pinchcock clamp** Schlauchklemme
➢ **round jaw clamp**
Klemme mit runden Backen
➢ **three-finger clamp**
Dreifinger-Klemme
➢ **tubing clamp/**
pinch clamp/pinchcock clamp
Schlauchklemme, Quetschhahn
➢ **voltage clamp** Spannungsklemme
clamp holder/'boss'/clamp 'boss'
(rod clamp holder)
Doppelmuffe, Kreuzklemme
clamping closure (dialysis tubing)
Verschlussklammer, Verschlussclip (Dialysierschlauch)
clarification/purification
Klärung (z.B. absetzen/entfernen von Schwebstoffen aus einer Flüssigkeit)
clarifying filtration Klärfiltration
class frequency/cell frequency *stat*
Klassenhäufigkeit, Besetzungszahl, absolute Häufigkeit
class switch/class-switching
(isotype switching) *immun*
Klassenwechsel, Klassensprung
classification/classifying
Klassifikation, Klassifizierung, Gliederung, Einteilung, Gruppeneinteilung
classify klassifizieren, gliedern
claw hammer
Klauenhammer, Splitthammer
clay Ton
➢ **modeling clay**
Modellierknete
clay triangle/pipe clay triangle
Tondreieck, Drahtdreieck
clean *adj/adv* **(pure)** sauber (rein)
clean *vb* säubern, sauber machen; (cleanse/tidy up) putzen, säubern; (purify) reinigen, aufbereiten
➢ **clean up/tidy up**
sauber machen, aufräumen
Clean Air Act (US)
Gesetz zur Reinhaltung der Luft (entspricht in etwa: Technische Anweisung Luft TALuft)

clean bench/safety cabinet
Sicherheitswerkbank
clean room
Reinraum (*auch:* Reinstraum)
clean-room bench
Reinraumwerkbank
cleanability Reinigbarkeit;
Reinigungsmöglichkeit(en)
cleaner(s)/cleaning personnel
Reinigungskraft (Reinigungskräfte),
Reinigungspersonal
cleaning/cleansing
Reinigung, Saubermachen
➤ **cleaning in place (CIP)**
Reinigung ohne Zerlegung
von Bauteilen
cleaning pad/scrubber/sponge
Putzschwamm
cleaning squad(ron)
Putzkolonne,
Reinigungstrupp
cleanliness Reinlichkeit
cleanse/clean up/tidy (up)
reinigen, säubern
cleanser/cleaning agent
Reinigungsmittel;
(detergent) Detergens
cleansing tissue
Reinigungstuch (Papier)
cleanup Säuberungsaktion
clear (clarify/purify)
klären (z.B. absetzen/entfernen von
Schwebstoffen aus einer Flüssigkeit)
clear glass Klarglas
clearance Clearance, Klärung;
Abstand (Geräte/Möbel etc.);
Räumung, Ausverkauf
cleared geklärt
**cleavage (breakage/opening/
cracking/splitting/breakdown)**
Spaltung (Furchung);
Spaltbarkeit
cleavage fusion Spaltfusion
cleavage product Spaltprodukt
**cleave (break/open/crack/split/
break down)** spalten
clevis bracket
Bügel, U-Klammer, Gabelkopf

**cline/phenotypic character/
phenotypic gradient**
Merkmalsgefälle, Merkmalsgradient,
Cline, Kline, Klin
cling wrap/cling foil Frischhaltefolie
clinic Klinik, Krankenhaus
clinical waste Klinikmüll
clinically tested
klinisch getestet, klinisch geprüft
clip (clamp) Schelle, Klemme;
(for ground joint) Klemme
(Kegelschliffsicherung)
➤ **alligator clip/alligator connector clip**
Krokodilklemme
➤ **joint clip/ground-joint clip/
ground-joint clamp**
Schliffklammer, Schliffklemme
(Schliffsicherung)
➤ **paper clip** Büroklammer
➤ **stage clip** *micros* Objekttisch-
Klammer
clipping Scheren, Stutzen, Beschneiden
clogged/blocked verstopft
clone Klon
cloning Klonierung
clot *n* **(e.g., blood clot)**
Gerinnsel (z.B. Blut)
clot *vb* gerinnen, koagulieren
cloth Tuch, Gewebe, Lappen
➤ **wiping cloth/rag (wipes)**
Wischtuch, Wischlappen
cloth tape
Gewebeband, Textilband (einfach)
clothing/apparel Bekleidung, Kleidung
clotting Gerinnung
clotting factor
Gerinnungsfaktor
➤ **blood clotting factor**
Blutgerinnungsfaktor
cloud chamber Nebelkammer
cloudiness/turbidity
Trübheit, Trübung (Flüssigkeit)
cloudy/turbid trüb (Flüssigkeit)
club hammer Fäustel
cluster Gruppe
clutch *n* Griff (klammernd);
(coupling/coupler/attachment)
tech/mech Kupplung

clutter *n* Wirrwarr, Durcheinander, Unordnung
cluttered vollgestellt, zugestellt (Schränke/Abzug etc.)
coacervate Koazervat
coal Kohle
> **anthracite/hard coal** Anthrazit, Kohlenblende
> **bituminous coal/soft coal** Steinkohle, bituminöse Kohle
> **hard coal/anthracite** Glanzkohle, Anthrazit
> **subbituminous coal** Glanzbraunkohle, subbituminöse Kohle

coarse adjustment/
coarse focus adjustment *micros* Grobjustierung, Grobeinstellung (Grobtrieb)
coarse adjustment knob *micros* Grobjustierschraube, Grobtrieb
coarse-grained grobfaserig
coat *vb* überziehen; einhüllen; ummanteln, schützen
coat *n* Überzug; Überhang; (gown) Mantel, Kittel
> **protective coat/protective gown** Schutzkittel, Schutzmantel

coat hanger Kleiderbügel
coated (covered) beschichtet, überzogen
coating Überzug
> **protective coating** Schutzschicht, Schutzüberzug

cobalt (Co) Kobalt, Cobalt
coccus (*pl* cocci) Kokkus (*pl* Kokken), Kugelbakterium
cock/draincock Ablasshahn, Ablaufhahn
> **gas cock/gas tap** Gashahn
> **glass stopcock** Glashahn
> **pinchcock** Quetschhahn
> **single-way cock** Einweghahn
> **three-way cock/**
> **T-cock/three-way tap** Dreiweghahn, Dreiwegehahn
> **two-way cock** Zweiweghahn, Zweiwegehahn

coefficient of coincidence Coinzidenzfaktor, Koinzidenzfaktor
coefficient of variation *stat* Variationskoeffizient
coefficient of viscosity Viskositätskoeffizient
coffee mill/coffee grinder Kaffeemühle
coil condenser/
coil distillate condenser/
coiled-tube condenser/
spiral condenser Schlangenkühler; (Dimroth type) Dimroth-Kühler
coil conformation/
loop conformation *gen* Knäuelkonformation, Schleifenkonformation
coil distillate condenser/
coil condenser/
coiled-tube condenser Schlangenkühler
coiling Aufwinden
colander Seiher, Abtropfsieb
cold *n* (viral infection) Erkältung (viraler Infekt)
cold finger (finger-type condenser)/
suspended condenser Kühlfinger, Einhängekühler
cold frame (for plant forcing) Frühbeet, Anzuchtkasten (unbeheizt)
cold hardiness Kältetoleranz
cold house (greenhouse) Kalthaus, Frigidarium (kühles Gewächshaus)
cold resistance Kälteresistenz
cold room ('walk-in refrigerator')/
cold-storage room/
cold store/'freezer' Kühlraum, Gefrierraum
cold-sensitive kälteempfindlich, kältesensitiv
cold shock Kälteschock
cold spray Kälte-Spray
cold storage/deep freeze Kühlraum, Kühlkammer, Kühlhaus
cold-storage room/cold store/'freezer' Kühlraum, Gefrierraum
cold store Kühlhaus
cold trap/cryogenic trap Kühlfalle

colinearity
Colinearität, Kolinearität
collaboration/cooperation
Zusammenarbeit
**colleague/co-worker/
fellow-worker/collaborator**
Mitarbeiter, Kollege, Arbeitskollege
collect (put/come/bring together)
sammeln
collecting lens/focusing lens *micros*
Sammellinse
collection Sammlung, Kollektion
collector lens/collecting lens
Kollektorlinse
collimating lens *micros*
parallel-richtende Sammellinse
collimating slit *micros*
Kollimationsblende, Spaltblende
collimator Kollimator
collision activation Stoßaktivierung
collodion cotton Kollodiumwolle
colonial/colony-forming
kolonial, koloniebildend
colonization Kolonisation,
Kolonisierung, Besiedlung
colony bank Koloniebank
colony-forming/colonial
koloniebildend, kolonial
color *n* **(shade/tint/tone/
pigmentation/coloration)**
Färbung, Farbton, Pigmentation
color change Farbumschlag,
Farbänderung
color vision Farbensehen
color-matching Farbanpassung
column *dist/biot/chromat*
Säule, Kolonne; Turm (Bioreaktor)
> **capillary column** Kapillarsäule
> **separating column/
fractionating column**
Trennsäule
> **spinning band column**
Drehbandkolonne
> **spray column** Sprühkolonne
> **stripping column**
Abtriebsäule, Abtreibkolonne
> **Vigreux column**
Vigreux-Kolonne

column chromatography
Säulenchromatographie
column efficiency *chromat/dist*
Säulenwirkungsgrad
column packing *chromat*
Säulenfüllung, Säulenpackung;
Füllkörper (für Destillierkolonnen)
> **helice** Wendel
> **Raschig ring**
Raschig-Ring (Glasring)
> **saddle (berl saddles)**
Sattelkörper (Berlsättel)
> **spiral** Spirale
column reactor
Säulenreaktor, Turmreaktor
combination pliers/linesman pliers
Kombizange
combination vaccine/mixed vaccine
Kombinationsimpfstoff,
Mischimpfstoff, Mischvakzine
combust/incinerate/burn
verbrennen
combustibility/flammability
Brennbarkeit
combustible/flammable
brennbar
> **noncombustible/nonflammable**
nicht brennbar
combustion (incineration)
Verbrennung (Veraschung)
combustion furnace Verbrennungsofen
combustion gases Brandgase
**combustion heat/
heat of combustion**
Verbrennungswärme
combustion tube Glühröhrchen
comestible (eatable/edible)
genießbar (essbar)
commercial scale
kommerzieller Maßstab
commercial vendor Händler
commissioning
Begehung, Inspektion (zur
Abnahme); (certification: of a lab
upon completion) Abnahme
(eines Labors nach Fertigstellung)
communal installation
Gemeinschaftseinrichtung

**communicable disease/
transmissible disease**
übertragbare Krankheit
communication
Kommunikation, Verständigung
company doctor Betriebsarzt
comparative substance
Vergleichssubstanz
compare vergleichen
comparison Vergleich
**compartmenta(liza)tion/
sectionalization/division**
Fächerung, Kompartimentierung,
Unterteilung
compartmentalized case
Sortimentkasten
**compass saw (with open handle)/
pad saw** Stichsäge
compatibility (tolerance)
Verträglichkeit, Kompatibilität
(Toleranz)
> **incompatibility (intolerance)**
Inkompatibilität,
Unverträglichkeit (Intoleranz)
compatible (tolerant)
verträglich, kompatibel (tolerant)
> **incompatible (intolerant)**
inkompatibel,
unverträglich (intolerant)
compensation
Kompensation, Ausgleich;
(bonus/extra pay) Zulage
compensation point
Kompensationspunkt
compete
konkurrieren, in Wettstreit stehen
competence/jurisdiction Zuständigkeit
competition Kompetition,
Wettbewerb, Konkurrenz
competitive/competing
kompetitiv, konkurrierend
competitive inhibition
kompetitive Hemmung,
Konkurrenzhemmung
competitor Konkurrent, Mitbewerber
complete medium/rich medium
Vollmedium, Komplettmedium
complex medium komplexes Medium

complex salt Komplexsalz
**complexing
(chelation/chelate formation)**
Komplexbildung, Chelatbildung
**complexing agent/
chelating agent/chelator**
Komplexbildner, Chelatbildner
complexity Komplexität
compliance (observance)
Einhaltung (Vorschrift)
component Bestandteil
composite Mischung,
Zusammensetzung
composite foil/film
Verbundfolie
composite material
Kompositwerkstoff,
Verbundwerkstoff
compound *vb* verbinden, verdichten,
zusammenballen, mischen
compound *n*
Verbindung; Präparat;
Mischung, Masse
> **addition compound/
additive compound
(saturation of double/triple bonds)**
Additionsverbindung
> **carbon compound**
Kohlenstoffverbindung
> **chemical compound**
chemische Verbindung
> **high energy compound**
energiereiche Verbindung
> **inclusion compound**
Einschlussverbindung
> **nitrogenous compound/
nitrogen-containing compound**
Stickstoffverbindung
> **organic compound**
organische Verbindung
> **parent compound/
parent molecule (backbone)**
Grundkörper (Strukturformel)
> **saturated compound**
gesättigte Verbindung
> **sealing compound/
sealing material/sealant**
Dichtungsmasse, Dichtungsmittel

➤ **sulfur compound**
Schwefelverbindung, schwefelhaltige Verbindung
➤ **unsaturated compound**
ungesättigte Verbindung
compound microscope
zusammengesetztes Mikroskop
compress zusammenpressen, zusammendrücken, stauchen
compressed/contracted
gestaucht, zusammengezogen
compressed air (pressurized air)
Druckluft (Pressluft)
compressed air breathing apparatus
Pressluftatmer
compressed gas/pressurized gas
Druckgas
compression
Kompression, Verdichtung, Stauchung
compression fitting
Hochdruck-Steckverbindung
compression seal Druckverschluss
compressor clamp Schlauchsperre
computed tomography (CT)
Computertomographie
concatenate verketten
concatenation Verkettung
concave mirror Hohlspiegel
concentrate
konzentrieren; (enrich) anreichern
concentration Konzentration; (enrichment) Anreicherung
➤ **inhibitory concentration**
Hemmkonzentration
➤ **limiting concentration**
Grenzkonzentration
➤ **median lethal concentration (LC_{50})**
mittlere letale Konzentration (LC_{50})
➤ **minimal inhibitory concentration/ minimum inhibitory concentration (MIC)**
minimale Hemmkonzentration (MHK)
concentration gradient
Konzentrationsgefälle, Konzentrationsgradient
concrete drill (bit) Betonbohrer

condensate Kondensat; Kondenswasser
condensation Kondensation; Eindickung, Konzentration
condensation point/condensing point
Kondensationspunkt
condensation reaction/ dehydration reaction
Kondensationsreaktion, Dehydrierungsreaktion
condense kondensieren; eindicken, konzentrieren
condenser Kühler; *opt* Kondensator; *micros* Kondensor
➤ **air condenser** Luftkühler
➤ **Allihn condenser** Kugelkühler
➤ **coil condenser (Dimroth type)**
Dimroth-Kühler
➤ **coil distillate condenser/ coil condenser/ coiled-tube condenser**
Schlangenkühler
➤ **cold finger (finger-type condenser)**
Kühlfinger
➤ **jacketed coil condenser**
Intensivkühler
➤ **Liebig condenser** Liebigkühler
➤ **reflux condenser** Rückflusskühler
➤ **suspended condenser/cold finger**
Einhängekühler, Kühlfinger
condenser adjustment knob/ substage adjustment knob *micros*
Kondensortrieb
condenser diaphragm (iris diaphragm)
micros Kondensorblende, Aperturblende (Irisblende)
condenser jacket *dist* Kühlmantel
condensing point/condensation point
Kondensationspunkt
condition *n* Kondition, Zustand
condition *vb med/chromat*
konditionieren
conditional lethal
bedingt letal, konditional letal
conditioned medium
konditioniertes Medium
conditioning *med/chromat*
Konditionierung, Konditionieren

conduct *vb* (transport/translocate/lead) leiten (Elektrizität/Flüssigkeiten)
conductance Leitfähigkeit
conduction/conductance/ transport/translocation Leitung
conductive leitfähig
conductivity Leitfähigkeit
conductivity meter Leitfähigkeitsmessgerät
conductometric titration konduktometrische Titration, Konduktometrie
conductor *electr* Phase, Leiter
conduit Leitung, Rohr, Röhre, Kanal; Rohrkabel, Isolierrohr (für Kabel)
conduit box *electr* Abzweigdose
conduit pipe Leitungsrohr
cone/ground cone/ground-glass cone (male of a ground-glass joint) Schliffkern (Steckerteil)
confidence interval *stat* Konfidenzintervall, Vertrauensintervall, Vertrauensbereich
confidence level *stat* Konfidenzniveau, Konfidenzwahrscheinlichkeit
confidence limit *stat* Konfidenzgrenze, Vertrauensgrenze, Mutungsgrenze
confidential employer-employee relationship/ confidential working relationship Dienst- und Treueverhältnis
confirmatory data analysis konfirmatorische Datenanalyse
confocal laser scanning microscopy konfokale Laser-Scanning Mikroskopie
conformation Konformation
➢ **boat conformation** Wannenform
➢ **chair conformation** Sesselform
➢ **coil conformation/ loop conformation** *gen* Knäuelkonformation, Schleifenkonformation
➢ **repulsion conformation** Repulsionskonformation
➢ **ring conformation/ring form** Ringform

congenial kongenial, verwandt, gleichartig
conical socket (of ground-glass joint) Kegelhülse
conjugated bond *chem* konjugierte Bindung
connect (to/with)/bond/link verbinden, anschließen
connecting cord *electr* Verbindungsschnur
connecting piece/connector Stutzen (Anschlussstutzen/Rohrstutzen)
connection/bond/linkage Verbindung, Anschluss
➢ **quick-fit connection** Schnellverbindung (Rohr/Glas/Schläuche etc.)
➢ **three-way connection** Dreiwegverbindung
conscious bewusst
➢ **unconscious/unknowing(ly)** unbewusst
consciousness Bewusstsein
consent Einverständnis
➢ **informed consent** Einverständniserklärung nach ausführlicher Aufklärung
conservation Konservierung, Haltbarmachung, Erhalt, Erhaltung, Bewahrung; Schutz
conserve (store/keep) konservieren, haltbar machen, erhalten, bewahren
consign übergeben, anvertrauen; zusenden, übersenden, adressieren; einliefern
consignee Empfänger, Adressat, Konsignatar
consignment of goods/ shipment of goods Warensendung
consignor Übersender, Konsignant
consist konsistieren, beschaffen sein
consistant (consisting of) bestehend aus
consistency Konsistenz, Beschaffenheit

constrict verengen, einschnüren
constriction Verengung, Enge, Einschnürung
construction paper Bastelpapier
consumer Konsument, Verbraucher
➢ **bulk consumer** Großverbraucher
consumer protection Verbraucherschutz
consumption/use/usage Verbrauch
contact *n* **(exposure)** Berührung, Kontakt (z.B. mit Chemikalien)
contact adhesive Haftkleber
contact allergen Kontaktallergen
contact hazard Kontaktrisiko (Gefahr bei Berühren)
contact infection Kontaktinfektion
contact insecticide Kontaktinsektizid
contact pesticide Kontaktpestizid
contact time Einwirkzeit
contagion/infection Ansteckung, Infektion
contagious/infectious ansteckend, ansteckungsfähig, infektiös
contagious disease/infectious disease ansteckende Krankheit, infektiöse Krankheit
contagiousness Kontagiosität, Ansteckungsfähigkeit
contain *chem/micb* **(security)** eindämmen
container (large)/receptacle (small) Behälter, Behältnis
➢ **intermediate bulk container (IBC)** Großpackmittel
containment Eindämmung
➢ **biohazard containment (laboratory) (classified into biosafety containment classes)** Sicherheitslabor, Sicherheitsraum, Sicherheitsbereich (S1-S4)
➢ **biological containment** biologische Sicherheit(smaßnahmen)

containment level Einschlussgrad (phys./biol. Sicherheit)
containment vector Sicherheitsvektor
contaminate kontaminieren, verunreinigen, belasten (belastet/verschmutzt); (poisoned/polluted) vergiftet, verseucht
➢ **radioactively contaminated** radioaktiv verstrahlt
contamination Kontamination, Verunreinigung; (pollution) Belastung, Verschmutzung; Verseuchung
continually variable/infinitely variable stufenlos regelbar
continued use/usage Weiterverwendung
continuing education Weiterbildung
continuity tester *electr* Durchgangsprüfer
continuous culture/ maintenance culture kontinuierliche Kultur
continuous run (continuous operation/duty) Dauerbetrieb, Dauerleistung, Non-Stop-Betrieb
continuous use Dauernutzung
continuous wave (CW) technique (IR) CW-Technik
continuously adjustable/ variably adjustable stufenlos regulierbar
contort drehen, verdrehen
contour/outline Umriss
contract *vb* kontrahieren, zusammenziehen, verengen; *med* (eine Krankheit) zuziehen
contract *n* Kontrakt, Vertrag
contract of employment Arbeitsvertrag, Dienstvertrag
contract research Auftragsforschung
contractor Lieferant, Vertragslieferant
contrast *vb* kontrastieren, entgegensetzen, gegenüberstellen
contrast medium *med* Kontrastmittel

contrast staining/
differential staining *micros*
Kontrastfärbung, Differentialfärbung
control (regulation) Kontrolle,
Regelung, Regulierung; Steuerung
control element/control unit Regelglied
control engineering Steuerungstechnik
control panel/switchboard
Bedienfeld, Schalttafel
control technology
Regeltechnik, Regelungstechnik
control unit/control gear/controller
Regelgerät, Steuergerät
control valve Regelventil
controlled area
Kontrollbereich,
kontrollierter Bereich
controlled variable/
controlled condition Regelgröße
controlling element/adjuster/actuator
Stellglied
controlling instrument/
control instrument/
monitoring instrument
Kontrollgerät
convection current
Konvektionsströmung,
Konvektionsstrom
convection oven Konvektionsofen
conversion Umwandlung; Umrechnung
conversion table
Umrechnungstabelle
convert
umwandeln; umrechnen; wandeln
converter Konverter, Wandler
convey befördern, transportieren;
zuführen, fördern; übertragen; leiten
conveyor belt Förderband
cook/boil kochen
cooked-meat broth
Kochfleischbouillon, Fleischbrühe
cool/chill/refrigerate
kühlen; (let cool) erkalten lassen
➢ **supercool** unterkühlen
cool down/get cooler abkühlen
coolant *allg* Kältemittel,
Kühlflüssigkeit, Kühlmittel;
Kühlschmierstoff, Kühlschmiermittel

cool-down period/
cooling time (autoclave)
Abkühlzeit, Abkühlphase,
Fallzeit (Autoklav)
cooler Kühlbox
cooling beaker/chilling beaker
Temperierbecher
cooling coil/condensing coil
Kühlschlange
cooling pack/cooling unit
Kühlakku, Kälteakku
cooling time/
cool-down period (autoclave)
Abkühlzeit, Fallzeit
cooling water (coolant) Kühlwasser
cooperate/collaborate
kooperieren, zusammenarbeiten
cooperation/collaboration
Kooperation, Zusammenarbeit
cooperative binding
kooperative Bindung
cooperativity Kooperativität
coordinate koordinieren
coordination Koordination
coping saw Bogensäge
copper (Cu) Kupfer
copper filings
Kupferspäne, Kupferfeilspäne
copper grid *micros* Kupfernetz
copper grid mesh Kupferdrahtnetz
copper sulfate/
copper vitriol/cupric sulfate
Kupfersulfat, Kupfervitriol
coprecipitation Mitfällung
copy number Kopienzahl
copy machine/copying machine
Kopiergerät
cord/strand Schnur; Strang (*pl* Stränge)
core *n* **(center)**
Kern, Zentrum (Mark/Core);
vir Viruskern, Zentrum
(zentrale Virionstruktur)
core *vb* entkernen
cork ring Korkring
cork-borer Korkbohrer
corn oil Maisöl
corned gepökelt, eingesalzen;
(corned beef: gepökeltes Rindfleisch)

corner frequency Grenzfrequenz
cornsteep liquor Maisquellwasser
corpse/carcass/cadaver
 Gebeine, sterbliche Hülle;
 Leiche, Kadaver (Tierleiche)
correctness/exactness/accuracy *stat*
 Richtigkeit, Genauigkeit
corrosion Korrosion, Ätzen, Ätzung
corrosionproof
 korrosionsbeständig
corrosive *adj/adv*
 korrosiv, korrodierend,
 zerfressend, angreifend;
 (C) ätzend
corrosive *n*
 Korrosionsmittel, Ätzmittel
corrugated board Wellpappe
cost-benefit analysis
 Kosten-Nutzen-Analyse
cost-efficient production
 preisgünstige Produktion
cotton Baumwolle;
 (absorbent cotton) Watte
cotton ball/cotton pad
 Wattebausch,
 Baumwoll-Tupfer
cotton oil Baumwollsaatöl
cotton pledget Wattebausch
cotton stopper Wattestopfen
coulometric titration
 coulometrische Titration,
 Coulometrie
Coulter counter/cell counter
 Coulter-Zellzählgerät
counteract *vb*
 entgegenwirken; bekämpfen
counterbalance *n* (counterpoise)
 Gegengewicht; Ausgleich
counterbalance *vb* (counterpoise)
 ausbalancieren; ausgleichen
countercurrent Gegenstrom
countercurrent distribution
 Gegenstromverteilung
countercurrent electrophoresis
 Gegenstromelektrophorese,
 Überwanderungselektrophorese
countercurrent extraction
 Gegenstromextraktion
countercurrent immunoelectrophoresis/counterelectrophoresis
 Überwanderungsimmunelektrophorese,
 Überwanderungselektrophorese
counterion Gegenion
counterpart Gegenstück, Pendant;
 Kopie, Duplikat
counterpoise Gegengewicht
counterpressure Gegendruck
counterreaction Gegenreaktion
counterselection
 Gegenselektion, Gegenauslese
countershading Gegenschattierung
counterstain/counterstaining *n micros*
 Gegenfärbung
counterstain *vb micros* gegenfärben
countertop/benchtop Tischoberfläche
 (Labortisch), Arbeitsplatte,
 Arbeitsfläche (Labor-, Werkbank)
counting chamber Zählkammer
counting plate Zählplatte
couple *vb* koppeln, aneinander
 festmachen, verbinden
coupled reaction gekoppelte Reaktion
coupler/fitting
 Steckverbindung, Steckvorrichtung
coupling (coupler) Kupplung,
 Verbinder, Verbindungsstück (von
 Bauteilen); (linkage) Kopplung
➤ **fixed coupling** starre Kupplung
coupling reaction *chem*
 Kupplungsreaktion
course (of a disease) Verlauf
covalent bond kovalente Bindung
cover value Deckungswert
coverage percentage/coverage level
 Deckungsgrad
coverall/boilersuit/protective suit
 Arbeitsschutzanzug, Schutzanzug
coverglass/coverslip *micros* Deckglas
coverglass forceps Deckglaspinzette
coverslip/coverglass *micros* Deckglas
cow receiver adapter/'pig'
 (receiving adapter for three, four
 receiving flasks)
 'Spinne', Eutervorlage,
 Verteilervorlage

crack Sprung (Glas/Keramik etc.)
**craftsman
 (practicing a handicraft)/
 workman**
 Handwerker
**crashproof
 (nonbreakable/unbreakable)**
 bruchsicher
cream Creme, Kreme, Salbe; Sahne
➢ **barrier cream**
 Gewebeschutzsalbe,
 Arbeitsschutzsalbe
crevice/crack Spalte
crimp seal (for crimp-seal vials)
 Bördelkappe
➢ **cap crimper**
 Verschließzange für Bördelkappen
crimp-seal vial
 Rollrandgläschen, Rollrandflasche
crimping pliers/crimper
 Bördelzange; Aderendhülsenzange,
 Crimpzange (Quetschzange),
 Verschlusszange
➢ **decapper** Öffnungsschneider
➢ **decrimper** Öffnungszange
critical point kritischer Punkt
critical point drying (CPD)
 Kritisch-Punkt-Trocknung
crop (plant crop) Feldfrucht,
 Pflanzenkultur;
 (crop yield/harvest) Ernteertrag
crop plant/cultivated plant
 Kulturpflanze
cross *vb* **(crossbreed/breed/interbreed)**
 kreuzen, züchten
cross *n* **(breed)**
 Kreuzung, Kreuzungsprodukt
cross-contamination
 Kreuzkontamination
cross-flow filtration
 Kreuzstromfiltration,
 Querstromfiltration
cross-link quervernetzen
cross-linkage/cross-linking
 Vernetzung, Quervernetzung
cross-linker/crosslinking agent
 Vernetzer, quervernetzendes Agens
cross-matching *immun* Kreuzprobe

cross out
 auskreuzen, herauskreuzen *gen*
cross protection
 Kreuzimmunität,
 übergreifender Schutz
cross section Querschnitt
crosshairs Fadenkreuz
**crossing/cross/crossbre(e)d/
 breed/crossbreeding/
 interbreeding**
 Kreuzung, Züchtung
crotonic acid/α-butenic acid
 Crotonsäure, Transbutensäure
crowbar/jimmy Brecheisen
crown cap Kronenkorken
crucible
 Tiegel, Schmelztiegel
➢ **filter crucible** Filtertiegel
➢ **Gooch crucible** Gooch-Tiegel
crucible furnace Tiegelofen
crucible tongs Tiegelzange
crucible triangle Tiegeldreieck
crude *chem* roh (crudum/crd.)
crude death rate Bruttosterberate
crude extract Rohextrakt
crude oil/petroleum Rohöl
crude product
 Rohprodukt (unaufgereinigt)
crush zermalmen, zerstoßen
crusher Mühle (*grob*)
cryoprotectant
 Frostschutzmittel,
 Gefrierschutzmittel
cryoprotection
 Frostschutz, Gefrierschutz
cryosection/frozen section *micros*
 Gefrierschnitt
cryostat Kryostat
cryostat section *micros*
 Kryostatschnitt
cryoultramicrotomy
 Kryoultramikrotomie
crypt/cavity/cave Höhlung
crystal structure/crystalline structure
 Kristallstruktur
**crystal water/
 water of crystallization**
 Kristallwasser

crystallization Kristallisation
- **fractional crystallization** fraktionierte K.
- **recrystallization** Umkristallisation

crystallization nucleus Kristallisationskern, Kristallisationskeim

crystallize kristallisieren

crystallography Kristallographie

cullet/glass cullet Bruchglas

cultivate kultivieren, züchten, in Kultur züchten

cultivatible/arable kultivierbar

culture *vb* (**cultivate**) kultivieren, züchten, in Kultur züchten

culture *n* Kultur, Zucht, Züchtung; Anbau
- **batch culture** Satzkultur, Batch-Kultur, diskontinuierliche Kultur
- **blood culture** Blutkultur
- **cell culture** Zellkultur
- **chicken embryo culture** Eikultur (Hühnerei)
- **dilution shake culture** Verdünnungs-Schüttelkultur
- **enrichment culture** Anreicherungskultur
- **fed-batch culture** Zulaufkultur, Fedbatch-Kultur (semi-diskontinuierlich)
- **long-term culture** Dauerkultur
- **maintenance culture** Erhaltungskultur
- **mixed culture** Mischkultur
- **perfusion culture** Perfusionskultur
- **pure culture/axenic culture** Reinkultur
- **roller tube culture** Rollerflaschenkultur
- **shake culture** Schüttelkultur
- **slant culture/slope culture** Schrägkultur (Schrägagar)
- **stab culture** Stichkultur, Einstichkultur (Stichagar)
- **static culture** statische Kultur
- **stem culture/stock culture** Stammkultur, Impfkultur
- **streak culture/smear culture** Ausstrichkultur, Abstrichkultur
- **submerged culture** Submerskultur, Eintauchkultur
- **surface culture** Oberflächenkultur
- **synchronous culture** Synchronkultur
- **tissue culture** Gewebekultur

culture bottle Kulturflasche

culture dish Kulturschale

culture flask Kulturkolben

culture media flask Nährbodenflasche

culture tube Kulturröhrchen

cumulative effect Anreicherungseffekt, Gesamtwirkung

cumulative frequency *stat* Summenhäufigkeit, kumulative Häufigkeit

cumulative poison Summationsgift, kumulatives Gift

cup/bucket *centrif* Becher

cupboard Schränkchen

cupric ... Kupfer(II) ...

cupric oxide/black copper Kupfer(II)-oxid

cuprous ... Kupfer(I) ...

cuprous oxide Kupfer(I)-oxid

curd geronnene Milch

curd soap (domestic soap) Kernseife (feste Natronseife)

curdle/coagulate gerinnen, koagulieren

cure *n* (healing *med*) Heilung; *polym* Härten, Aushärten, Vulkanisieren

cure *vb* (heal *med*) heilen; (vulcanize *chem/polym*) härten, aushärten, vulkanisieren; (meat) pökeln (Fleisch)

curing *polym* Härten, Aushärten; (meat) Pökeln (Fleisch)

curing agent *polym* Härter, Aushärtungskatalysator

curing period *polym* Härtezeit, Aushärtungszeit, Abbindezeit

current (electric current/amperage/ amps) Strom, Stromstärke; (flow) Strömung (Flüssigkeit)
➢ **air current/ airflow/current of air/air stream** Luftstrom, Luftströmung
➢ **alternating current (AC)** Wechselstrom
➢ **capacitative current** kapazitiver Strom
➢ **convection current** Konvektionsströmung, Konvektionsstrom
➢ **countercurrent** Gegenstrom
➢ **density current** Konzentrationsströmung
➢ **direct current (DC)** Gleichstrom
➢ **eddy current** Wirbelstrom (Vortex-Bewegung)
➢ **ground fault current (leakage current)** Erdschlussstrom, Fehlerstrom
➢ **ionic current/ion current** Ionenstrom
➢ **leakage current/creepage** Kriechstrom
➢ **threshold current** Schwellenstrom
current density Stromdichte
current divider Stromteiler
current meter/flowmeter Strömungsmesser
current rectifier Stromgleichrichter
cushioned gepolstert, ausgepolstert, gefedert, gepuffert, abgepuffert, geschützt
custodial personnel/ security personnel Wachpersonal, Aufsichtspersonal
custodian Aufseher, Wächter; Hausmeister
customer service Kundendienst
cut *n* med Schnitt, Schnittwunde
cut *vb* schneiden, einschneiden
➢ **incised** schnittig, geschnitten, eingeschnitten
cutaneous respiration/breathing (integumentary respiration) Hautatmung
cutaway drawing Ausschnittszeichnung
cutis (skin) Cutis, Haut, eigentliche Haut
cutis vera/true skin/corium/dermis Lederhaut, Korium, Corium, Dermis
cutoff abrupte Beendigung; Abbruch, Ausschaltung; Trenngrenze, Ausschlussgrenze (Teilchentrennung)
cutoff filter Sperrfilter, Kantenfilter
cutoff valve Schlussventil
cutting edge (of a knife blade) Schneide (eines Messers)
cutting face/cutting plane Schnittfläche, Schnittebene
cutting-grinding mill/ shearing machine Schneidmühle
cuvette/spectrophotometer tube Küvette (für Spektrometer)
cyanogen bromide activated paper (CBA-paper) Bromcyan-aktiviertes Papier (CBA-Papier)
cycle *n* Cyclus (*med* auch: Zyklus)
cycle *vb* einen Kreislauf durchmachen, periodisch wiederholen, einen Prozess durchlaufen (lassen)
cycle time/running time Laufzeit (Gerät: für eine 'Runde')
cyclic cyclisch, zyklisch, ringförmig, kreisläufig; (periodic) periodisch
cyclization *chem* Cyclisierung, Zyklisierung, Ringschluss
cylinder Zylinder; (gas bottle) Druckflasche
➢ **gas cylinder** Druckgasflasche
➢ **graduated cylinder** Messzylinder
cylinder pressure gauge Flaschendruckmanometer
cylindric/cylindrical zylindrisch, walzenförmig

cyst Zyste
cysteic acid Cysteinsäure
cyto... cyto (*med* auch zyto...)
cytochemistry
    Cytochemie, Zellchemie
cytochrome Cytochrom
cytocidal
    zelltötend,
    cytocid (*med* auch: zytozid)
cytogenetics Cytogenetik
cytology/cell biology
    Cytologie, Zellenlehre, Zellbiologie
cytolytic cytolytisch

cytometry Cytometrie
cytopathic (cytotoxic)
    cytopathisch, zellschädigend
    (cytotoxisch)
cytoplasm Zellplasma, Cytoplasma
cytoplasmic cytoplasmatisch
cytoskeleton Cytoskelett
cytostatic agent/cytostatic
    Cytostatikum (meist *pl* Cytostatika)
cytotoxic cytotoxisch, zellschädigend
cytotoxicity
    Cytotoxizität, Zellschädigung
cytotoxin Zellgift, Cytotoxin

dab/swab *vb* tupfen, abtupfen
dairy product
  Milchprodukt, Molkereiprodukt
daisy-chain *vb* mehrere Gegenstände aneinander ketten
damage *n* Schaden; Schädigung
damage response curve
  Schädigungskurve
damp (humid) feucht
dampen/damp dämpfen, abschwächen
damper Befeuchter
damping Dämpfung (Waage)
danger (hazard/risk/chance)
  Gefahr, Gefährdung, Risiko
➤ **extreme danger** höchste Gefahr
➤ **immediate danger/imminent danger**
  akute Gefahr
➤ **imminent danger** drohende Gefahr
➤ **out of danger/safe/secure**
  außer Gefahr
➤ **source of danger (troublespot)**
  Gefahrenherd
danger allowance/hazard bonus
  Gefahrenzulage
danger area Gefahrenbereich
danger class/
  category of risk/class of risk
  Gefahrenklasse
danger of accident Unfallgefahr
danger of life/life threat Lebensgefahr
danger sign/warning sign Warnschild
danger zone Gefahrenzone
dangerous (hazardous/risky)
  gefährlich, riskant
dangerous for the environment
  (N = nuisant)
  umweltgefährlich
dangerous goods/
  hazardous materials
  Gefahrgut, Gefahrgüter
  (gefährliche Frachtgüter)
dangerous substance/
  hazardous substance/
  hazardous material
  Gefahrstoff
dark field *micros* Dunkelfeld
darkfield microscopy
  Dunkelfeld-Mikroskopie

darkroom Dunkelkammer
dashboard/dash
  Armaturenbrett (im Fahrzeug)
data *pl* **(used as** *sg* **&** *pl;* **often** *attrib***)**
  Daten; (fact) Tatsache, Angabe
data acquisition
  Datenerfassung, Datenermittlung
data analysis Datenanalyse
data processing Datenverarbeitung
data sheet
  Datenblatt, Merkblatt
  (für Chemikalien etc.)
➤ **safety data sheet**
  Sicherheitsdatenblatt
datalogger
  Datenerfassungsgerät,
  Messwertschreiber,
  Messwerterfasser, Registriergerät,
  Datensammler, Datenlogger
date of issue
  Ausstellungsdatum,
  Ausgabezeitpunkt
dating *n* Datierung
daughter ion Tochterion
de-novo **synthesis**
  Neusynthese, *de-novo*-Synthese
dead *adj/adv* tot
dead-end filtration Kuchenfiltration
dead volume
  deadspace volume/
  holdup volume/
  holdup
  Totvolumen (Spritze/GC)
deadline
  Abgabetermin, Ablieferungstermin
deadly/lethal tödlich, letal
deadspace/headspace Totraum
deaf taub, gehörlos
deafness Taubheit, Gehörlosigkeit
dealer Händler
➤ **retail dealer/retailer/retail vendor**
  Einzelhändler
➤ **wholesale dealer/wholesaler**
  Großhändler
deamidation/deamidization/
  desamidization Desamidierung
deamination/desamination
  Desaminierung

death Tod
> **cause of death** Todesursache
deburr abgraten, bördeln
deburred edge/beaded rim Bördelrand
decalcification
  Entkalkung, Dekalzifizierung
decant dekantieren, umfüllen,
  umgießen, vorsichtig abgießen
decantation Dekantieren, Umfüllen,
  vorsichtiges Abgießen
decanter Dekanter,
  Abklärflasche, Dekantiergefäß
decay *n* (disintegration) Zersetzung,
  Verrottung, Verfaulen;
  (rot/putrefaction) Fäulnis
decay *vb*
  (disintegrate/decompose/fall apart)
  zersetzen, verrotten, verfaulen;
  zerfallen
deceleration phase/retardation phase
  Verlangsamungsphase, Bremsphase,
  Verzögerungsphase
deception/delusion Täuschung
decerating agent
  (for removing paraffin) *micros*
  Entparaffinierungsmittel
decline phase/
  phase of decline/death phase
  Absterbephase
decoct abkochen, absieden;
  (digest: by heat/solvents) digerieren
decoction Abkochung, Absud, Dekokt
decoloration/bleaching
  Entfärbung, Bleichen
decompose zersetzen, zerfallen,
  abbauen; verrotten, verfaulen
decomposer
  Zersetzer, Destruent, Reduzent
decomposition (breakdown)
  Zersetzung, Zerfall,
  Abbau (Zusammenbruch)
decompression Dekompression
decontaminate dekontaminieren,
  reinigen, entseuchen
decontamination
  Dekontamination,
  Dekontaminierung,
  Reinigung, Entseuchung

decouple/uncouple/release
  entkoppeln
decoupling/uncoupling/release
  Entkopplung
decrimper/decrimping pliers
  Öffnungszange
dedicated gewidmet,
  zu einem bestimmten Zweck
  bestimmt, speziell
dedifferentiation
  Dedifferenzierung,
  Entdifferenzierung
deep etching Tiefenätzung
deep-freeze *vb* tiefkühlen, tiefgefrieren
deep-freeze compartment
  Tiefkühlfach (des Kühlschranks)
deep-freeze gloves
  Tiefkühlhandschuhe,
  Kryo-Handschuhe
deep freezer/deep-freeze/'cryo'
  Tiefkühltruhe, Gefriertruhe,
  Tiefkühlschrank
deep freezing Tiefkühlung
defecation/egestion
  Defäkation, Darmentleerung,
  Stuhlgang, Klärung, Koten
defect *n* Defekt, Schaden, Fehler
defect in material/
  flaw in material
  Materialfehler
defective defekt, schadhaft, fehlerhaft
defense Verteidigung, Schutz
defensive medicine Defensivmedizin
defervescence/delay in boiling
  (due to superheating)
  Siedeverzug (durch Überhitzung)
deficiency Defizienz, Mangel
deficiency medium Mangelmedium
deficiency symptom
  Defizienzerscheinung,
  Mangelerscheinung, Mangelsymptom
deficient/lacking
  defizient, mangelnd, Mangel...
defined medium synthetisches Medium
  (chem. definiertes Medium)
deflagrate
  verpuffen; rasch abbrennen (lassen)
deflagration Verpuffung

deflate Luft/Gas ablassen, Luft/Gas herauslassen
deflation Ablassen, Herauslassen, Entleeren, Entleerung (Luft/Gas)
deflect ablenken, umleiten
deflection Ablenkung, Umleitung
deflection voltage Ablenkungsspannung
deformation Deformation, Formänderung, Verformung
deformation vibration/bending vibration Deformationsschwingung
defrost abtauen (Kühl-/Gefrierschrank)
degas/degasify/outgas ausgasen, entgasen
degasing/gasing-out Ausgasung, Ausgasen, Entgasung, Entgasen
degeneracy Degeneration, Entartung
degenerate *adv* (IR) entartet, degeneriert
degenerate *vb* degenerieren, entarten; (regress) rückbilden
degeneration (degeneracy) Entartung; (regression) Rückbildung
degradability/decomposability Abbaubarkeit
degradation (decomposition/breakdown) Abbau, Zersetzung, Zerfall, Zusammenbruch
degradation product Abbauprodukt, Zersetzungsprodukt
degradative metabolism/catabolism Stoffwechsel-Abbau
degrade (decompose/break down) abbauen, zersetzen, zerfallen
degrease entfetten
degree of freedom (df) *stat* Freiheitsgrad
dehumidifier Entfeuchter
dehumidify entfeuchten
dehydrate dehydratisieren, entwässern
dehydrating dehydrierend, wasserentziehend, entwässernd
dehydration Dehydratation, Entwässerung
dehydrogenate dehydrieren
dehydrogenation Dehydrierung

deionize entionisieren
deionized water entionisiertes Wasser
deionizing/deionization Entionisierung
delay *n* (retardation) Verzögerung; (postponement) Verschiebung, Aufschiebung; Verspätung
delay *vb* (retard) verzögern; (postpone) verschieben, aufschieben; verspäten
delayed damage Spätschaden
delayed effect Verzögerungseffekt
deliberate release absichtliche/willkürliche Freisetzung
deliberate release experiment (environmental release experiment) Freisetzungsexperiment
deliquescence Zerfließen, Zerschmelzen, Zergehen
deliquescent zerfließend, zerfließlich, zerschmelzend, zergehend
deliver liefern, abliefern, ausliefern; beschicken; abgeben, einreichen
delivery Lieferung, Auslieferung; *chem* Beschickung; (handing in/dropoff) Abgabe, Einreichung (Ergebnisse etc.); (handing over) Übergabe
  ➤ **bulk delivery/bulk shipment** Großlieferung
  ➤ **cost of delivery/shipment costs** Lieferkosten
  ➤ **ready for delivery/dispatch/shipment** versandfertig
  ➤ **terms of delivery (terms and conditions of sale)** Lieferbedingungen
  ➤ **time of delivery** Lieferfrist
  ➤ **turn-key delivery** schlüsselfertige Lieferung
delivery tube (flask) Beschickungsstutzen (Kolben)
delivery valve Zulaufventil, Beschickungsventil
demanding (having high requirements or demands) anspruchsvoll
demethylation Demethylierung, Desmethylierung
demister Entfeuchter

demount (disassemble/dismantle/strip/take apart) demontieren
denaturation/denaturing Denaturierung
denature denaturieren (z.b. Eiweiß); vergällen (z.b. Alkohol)
denaturing gel denaturierendes Gel
dense (mass/vol) dicht
density (mass/vol) Dichte
density current Konzentrationsströmung
density gradient Dichtegradient
density gradient centrifugation Dichtegradientenzentrifugation
deodorant Deodorans, Desodorans, Deodorant, Desodorierungsmittel
deoxyribonucleic acid (DNA) Desoxyribonucleinsäure, Desoxyribonukleinsäure (DNS/DNA)
dephlegmation/fractional distillation Dephlegmation, fraktionierte Destillation, fraktionierte Kondensation
dephosphorylate dephosphorylieren
dephosphorylation Dephosphorylierung
deplete leeren, entleeren, erschöpfen
depletion Entleerung, Erschöpfung (Substanz: leer werden/'zu Ende gehen'); (stripping/downgrading) Abreicherung
depolarization Depolarisation
depolarize depolarisieren
deposit/sediment/precipitate Präzipitat, Niederschlag, Sediment, Fällung
depression/basin Mulde, Kuhle, Vertiefung
depth of focus/depth of field Tiefenschärfe, Schärfentiefe
deregister/sign out (schriftlich 'austragen') abmelden

derivative *chem* Derivat
derivatization Derivatisation
derivatize *chem* derivatisieren
derive herleiten, ableiten von; erhalten, gewinnen; herkommen, abstammen
derived characteristic abgeleitetes Merkmal
dermal/dermic/dermatic dermal, Haut...
desalinate entsalzen
desalination Entsalzung
descale entkalken (ein Gerät ~), Kesselstein entfernen
descending *chromat* (TLC) absteigend
description Beschreibung
deshielding (NMR) Entschirmung
desiccant Trockenmittel
desiccate (dry up/dry out) austrocknen, entwässern
desiccation Austrocknung, Entwässerung
desiccation avoidance Austrocknungsvermeidung
desiccator Exsikkator
design *n* Entwurf, Plan, Design
desorption Desorption
destroy (eliminate) vernichten, zerstören (entfernen)
destruction (elimination) Vernichtung, Zerstörung (Entfernung)
destructive distillation Zersetzungsdestillation
desulfurization/desulfuration Entschwefelung
desulfurize/desulfur entschwefeln
detect (prove) nachweisen
detection (proof) Nachweis
detection limit/limit of detection (LOD) Bestimmungsgrenze, Nachweisgrenze
detection method Bestimmungsmethode, Nachweismethode

detector Detektor, Nachweisgerät, Suchgerät, Prüfgerät; (sensor) Melder, Messfühler (Sensor)
- **atomic emission detector (AED)** Atomemissionsdetektor (AED)
- **electron capture detector (ECD)** Elektroneneinfangdetektor
- **evaporative light scattering detector (ELSD)** Verdampfungs-Lichtstreudetektor
- **fast-scanning detector (FSD)/ fast-scan analyzer** Schnellscan-Detektor
- **flame-ionization detector (FID)** Flammenionisationsdetektor (FID)
- **infrared absorbance detector (IAD)** Infrarot-Absorptionsdetektor
- **infrared detector (ID)** Infrarotdetektor
- **ion trap detector (ITD)** Ioneneinfangdetektor
- **mass-selective detector** massenselektiver Detektor
- **photo-ionization detector (PID)** Photoionisations-Detektor (PID)
- **resistance temperature detector (RTD)** Widerstands-Temperatur-Detektor
- **thermal conductivity detector (TCD)** Wärmeleitfähigkeitsdetektor, Wärmeleitfähigkeitsmesszelle

detergent Detergens, Reinigungsmittel, Waschmittel
- **dishwashing detergent** Geschirrspülmittel
- **liquid detergent/liquid soap** Flüssigseife

determination (Identifikation) Determinierung, Determination, Bestimmung
determine (identify) *chem* bestimmen
deterrent/repellent Schreckstoff, Abschreckstoff
detoxification Entgiftung
detoxify entgiften
develop entwickeln; (emerge/unfold) hervorgehen, entstehen
developer *photo* Entwickler
developing chamber/ developing tank *chromat* **(TLC)** Trennkammer (DC)
deviate from ... abweichen von ...
deviation Abweichung
- **statistical deviation** statistische Abweichung

device (piece of equipment) Vorrichtung; Gerät
devour/gulp down verzehren, verschlingen, herunterschlingen
Dewar vessel/Dewar flask Dewargefäß
dewdrop Tautropfen
diabetes mellitus Zuckerkrankheit, Diabetes mellitus
diagnosis Diagnose
- **differential diagnosis** Differentialdiagnose

diagnostic diagnostisch
diagnostic kit Diagnostikpackung (DIN)
diagnostics Diagnostik
diagonal cutter/ diagonal pliers/ diagonal cutting nippers Seitenschneider
diagram (plot/graph) Diagramm (*auch* Kurve)
- **bar diagram/bar graph** Stabdiagramm
- **dot diagram** Punktdiagramm
- **histogram/strip diagram** Histogramm, Streifendiagramm
- **phase diagram** Phasendiagramm
- **scatter diagram (scattergram/scattergraph/ scatterplot)** Streudiagramm
- **spindle diagram** Spindeldiagramm

dial/face *n* Wählscheibe, Einstellscheibe, Zifferblatt
dialysis Dialyse
dialyze dialysieren

diaphragm Diaphragma, Blende;
   med Zwerchfell
diaphragm aperture Blendenöffnung
diaphragm pressure regulator
   Membrandruckminderer
diaphragm pump Membranpumpe
diaphragm valve Membranventil
diarrhea Diarrhö
diatomaceous earth Kieselerde
dichlorodiphenyltrichloroethane (DDT)
   Dichlordiphenyltrichlorethan (DDT)
**dichlorodiphenyltrichloroethylene
   (DDE)** Dichlordiphenyldichlorethylen (DDE)
die *n* Würfel; Pressform; Düse
die *vb* sterben
die casting Druckgießen
die down ausschwingen, abklingen;
   zu Ende gehen
dielectric constant
   Dielektrizitätskonstante
diet (food/feed/nutrition)
   Kost, Essen, Speise, Nahrung; Diät
dietary Diät..., diät, die Diät betreffend
dietary fiber Ballaststoffe (diätätisch)
dietetic diätetisch
dietetics Diätetik
differential centrifugation ('pelleting')
   Differentialzentrifugation,
   differentielle Zentrifugation
differential diagnosis
   Differentialdiagnose
differential display (form of RT-PCR)
   differentieller Display
   (Form der RT-PCR)
differential equation
   Differentialgleichung
differential interference
   Differential-Interferenz (Nomarski)
differential medium
   Differenzierungsmedium
differential scanning calorimetry (DSC)
   Differentialkalorimetrie
differential staining/contrast staining
   Differentialfärbung, Kontrastfärbung
differential thermal analysis (DTA)
   Differentialthermoanalyse,
   Differenzthermoanalyse (DTA)

differentiating characteristic
   Unterscheidungsmerkmal
diffraction Diffraktion, Beugung
diffraction pattern Beugungsmuster
diffuse *adj/adv* diffus, zerstreut,
   ohne klare Abgrenzung
diffuse *vb* diffundieren;
   zerstreuen, vermischen
diffuse flux diffuser Fluss
diffuser Diffusor, Verteiler
diffusing screen *photo* Steulichtschirm
diffusion coefficient
   Diffusionskoeffizient
digest (enzymatic) *n*
   Verdau (enzymatischer)
digest *vb* verdauen;
   faulen (im Faulturm der Kläranlage)
**digester/digestor/
   sludge digester/sludge digestor**
   Faulturm
digestibility Verdaulichkeit,
   Bekömmlichkeit
digestible verdaulich, bekömmlich
digestive enzyme Verdauungsenzym
digitizer Digitalisiergerät
diluent Verdünner, Verdünnungsmittel,
   Diluent, Diluens
dilute (thin down) verdünnen; (water
   down) verwässern
dilute solution verdünnte Lösung
dilution (thinning down) Verdünnung
dilution rate
   Durchflussrate, Verdünnungsrate
dilution rule Mischregel
   (Mischungskreuz)
dilution shake culture Verdünnungs-
   Schüttelkultur
dilution streak/dilution streaking *micb*
   Verdünnungsausstrich
**dimensionless group/
   quantity/number** *math*
   Kenngröße
dimensions (height/width/depth)
   Abmessungen (Höhe/Breite/Tiefe)
dimeric zweiteilig
dimerization Dimerisierung
dimerize dimerisieren
dimerous dimer, zweizählig

diode array detection (DAD) Diodenarray-Nachweis, Diodenmatrixnachweis
diopter (D) Dioptrie
dioptric dioptrisch
dip needle-nose pliers gebogene Storchschnabelzange
dip switch DIP-Schalter, Mäuseklavier
dip tank Tauchtank
dip tube Steigrohr
diphasic diphasisch
dipole moment Dipolmoment
dipper/scoop Schöpfer, Schöpfgefäß, Schöpflöffel
diprotic acid zweiwertige/zweiprotonige Säure
dipstick *tech* Messstab (Öl etc.), Tauchmessstab
direct blotting electrophoresis/ direct transfer electrophoresis Blotting-Elektrophorese, Direkttransfer-Elektrophorese
direct current (DC) *electr* Gleichstrom
directive *n* Direktive, Weisung, Verhaltensmaßregel, Anweisung, Vorschrift; (EU *jur*) Richtlinie
dirt/filth Dreck, Schmutz
dirty (filthy) dreckig, schmutzig; (stained) verschmutzt, beschmutzt, fleckig
disadvantage Nachteil
disassemble (take equipment apart) abbauen (Apparatur/Experimentiergerät)
disassembly/dismantling/ dismantlement/takedown (of equipment) Abbau (einer Apparatur); (stripping) Demontage
discharge *n electr* Entladung; (drainage/outlet) Ableitung (von Flüssigkeiten); (secretion/flux) *med* Ausfluss
discharge *vb* **(drain/lead out/ lead away/carry away)** ausführen, wegführen, ableiten (Flüssigkeit)
discharge head (pump) Förderhöhe
discharge stroke (pump) Abgabepuls

disciplinary offense Dienstvergehen
disconnect/disassemble auseinandernehmen (Glas~/Versuchsaufbau)
disease/illness Krankheit
disease-causing (pathogenic) krankheitserregend (pathogen)
disease-causing agent/pathogen Krankheitserreger
dish (Servier)Platte, Schlüssel
dish towel Geschirrhandtuch
dishboard Geschirrablage (Spüle/Spültisch)
dishes Geschirr
➢ **rinse the dishes** Geschirr abspülen
➢ **wash the dishes** Geschirr spülen
dishwasher (dishwashing machine) Spülmaschine
dishwasher detergent Spülmaschinenreiniger
dishwashing brush Spülbürste, Geschirrspülbürste
dishwashing bucket/dishpan Spüleimer
dishwashing cloth/ dishcloth/dishrag Spüllappen
dishwashing detergent Geschirrspülmittel
dishwashing pad Spülschwamm
dishwashing tub Spülwanne
dishwater Spülwasser, Abwaschwasser
disinfect (disinfected) desinfizieren (desinfiziert), keimfrei machen
disinfectant Desinfektionsmittel, Desinfiziens
disinfection Desinfizierung, Desinfektion, Entseuchung, Entkeimung
disinhibition Disinhibition, Enthemmung
disintegrate/decay/ decompose/degrade zersetzen, zerfallen; (decompose/break up) aufschließen
disintegration/decay/ decomposition/degradation Zersetzung, Zerfall

disintegration temperature/ decomposition temperature
Zersetzungstemperatur
disjunction
Verteilung, Trennung, Disjunktion
disk (disc *Br*)
Scheibe, Platte; Ring, Teller
disk assay (for antibiotics) *micb*
Plattenverfahren (Kultur)
disk atomizer Scheibenversprüher
disk diaphragm (annular aperture) *micros*
Ringblende
disk electrophoresis
Diskelektrophorese, diskontinuierliche Elektrophorese
disk mill Tellermühle
disk-shaped scheibenförmig
disk turbine impeller
Scheibenturbinenrührer
dislodge losmachen, entfernen, lockern, befreien, ablösen
dismantling/dismantlement/takedown (disassembly of equipment)
Abbau (einer Apparatur); (stripping) Demontage
dispatch abschicken, absenden, versenden, abfertigen
➢ **ready for dispatch/shipment/delivery**
versandfertig
dispenser Ausgießer, Dosierspender, Probengeber; (e.g., for liquid detergent etc.) Spender (Flüssigseife etc.)
dispenser pump/dispensing pump
Dispenserpumpe
dispersal/dissemination
Streuung, Ausbreitung
disperse *chem/phys*
dispergieren, zerstreuen, fein verteilen
dispersion
Dispergierung, Dispersion; (colloid) Kolloid; (spreading) Verteilung, Zerstreuung
displacement
Verdrängung, Verlagerung, Verschiebung; Verschleppung

displacement pump
Verdrängungspumpe, Kolbenpumpe (HPLC)
displacement reaction
Verdrängungsreaktion
display *n* **(dial/scale/reading)**
Anzeige (an einem Gerät); (monitor) Bildschirm, Monitor
display *vb* **(show/read)** anzeigen
disposable
Einweg..., Einmal..., Wegwerf...
disposable gloves/ single-use gloves
Einweg~, Einmalhandschuhe
disposable syringe Einwegspritze
disposal Entsorgung
➢ **improper disposal**
unsachgemäße Entsorgung
disposal firm Entsorgungsfirma, Entsorgungsunternehmen
dispose of/remove entsorgen
disposition
Disposition, Veranlagung, Anfälligkeit
dissect präparieren; sezieren; zerlegen, zergliedern; analysieren
dissecting dish/dissecting pan
Präparierschale
dissecting forceps Seziorpinzette
dissecting instruments (dissecting set) Präparierbesteck
dissecting microscope
Präpariermikroskop
dissecting needle/probe (teasing needle)
Präpariernadel
dissecting pin (Stecknadel)
Seziernadel
dissecting scissors
Präparierschere, Sezierschere
dissection Präparation, Sezierung; Zerlegung, Zergliederung
dissection equipment (dissecting set)
Sezierbesteck
dissection tweezers/ dissecting forceps
Sezierpinzette, anatomische Pinzette

disseminate/disperse/ spread/release
ausstreuen
dissemination/dispersal/ spreading/releasing
Ausstreuung
dissepiment/septum/partition (dividing wall/cross-wall)
Scheidewand, Septe, Septum
dissociate dissoziieren
dissociation constant (K_i)
Dissoziationskonstante
dissociation rate
Dissoziationsgeschwindigkeit
dissolution Auflösung; (disintegration/decomposition/ digestion) Aufschluss
dissolve (dissolved) lösen (gelöst), auflösen (aufgelöst: in einem Lösungsmittel); aufschließen
➢ **undissolved** ungelöst
distil/distill/still destillieren
distilland/material to be distilled
Destillationsgut
distillate Destillat
distillation Destillation
➢ **azeotropic distillation**
Azeotropdestillation
➢ **bulb-to-bulb distillation**
Kugelrohrdestillation
➢ **dephlegmation/fractional distillation**
Dephlegmation, fraktionierte D., fraktionierte Kondensation
➢ **destructive distillation**
Zersetzungsdestillation
➢ **equilibrium distillation**
Gleichgewichtsdestillation
➢ **extractive distillation**
Extraktivdestillation, extrahierende Destillation
➢ **flash distillation**
Entspannungs-Destillation, Flash-Destillation
➢ **reaction distillation**
Reaktionsdestillation
➢ **repeated distillation/cohobation**
Redestillation, mehrfache Destillation

➢ **short-path distillation/ flash distillation**
Kurzwegdestillation, Molekulardestillation
➢ **simple distillation**
Gleichstromdestillation
➢ **spinning band distillation**
Drehband-Destillation
➢ **steam distillation**
Trägerdampfdestillation
➢ **straight-end distillation**
einfache, direkte Destillation
➢ **vacuum distillation/ reduced-pressure distillation**
Vakuumdestillation
distillation boiler flask/ reboiler/still pot/boiler
Blase, Destillierblase, Destillierrundkolben
distillation by steam entrainment
Schleppdampfdestillation
distillation head/stillhead
Destillieraufsatz, Destillierbrücke
distillation receiver adapter/ receiving flask adapter
Vorlage (Destillation)
distillation residue Destillierrückstand
distilled water destilliertes Wasser
distillers' solubles (dried) Schlempe
distiller's yeast Brennereihefe
distilling apparatus/still
Destilliergerät, Destillationsapparatur
distilling column Destillierkolonne
distilling flask/destallation flask/'pot'
Destillierkolben, Destillationskolben
distribute verteilen
distribution *chem/stat* Verteilung
➢ **binomial distribution**
Binomialverteilung
➢ **countercurrent distribution**
Gegenstromverteilung
➢ **frequency distribution (FD)**
Häufigkeitsverteilung
➢ **lognormal distribution/ logarithmic normal distribution**
Lognormalverteilung, logarithmische Normalverteilung

➢ **normal distribution**
Normalverteilung
➢ **statistical distribution**
statistische Verteilung
distribution function
Verteilungsfunktion
distribution pattern Verteilungsmuster
distributor Verteiler; Lieferant;
(wholesaler) Großhändler
disturbance
(interference/disruption)
Störung
disturbance value/interference factor
Störgröße
disulfide bond/
disulfide bridge/
disulfhydryl bridge
Disulfidbindung, Disulfidbrücke
diuresis
Diurese, Harnfluss,
Harnausscheidung
diverge
divergieren, abweichen; ablenken
diversity Diversität, Vielfalt,
Vielfältigkeit, Vielgestaltigkeit,
Mannigfaltigkeit
divide teilen, gliedern, einteilen;
trennen; (fission/separate) teilen
divided gegliedert, unterteilt;
(compartmentalized) unterteilt,
kompartimentiert;
(parted/partite/divided into parts)
geteilt
division Gliederung, Einteilung;
(fission/separation) Teilung
division phase Teilungsphase
DNA (deoxyribonucleic acid)
DNA/DNS
(Desoxyribonucleinsäure/
Desoxyribonukleinsäure)
DNA footprint
DNA-Fußabdruck, DNA-Footprint
DNA profiling/DNA fingerprinting
genetischer Fingerabdruck,
DNA-Fingerprinting
DNA sequencer
DNA-Sequenzierungsautomat
DNA synthesis DNA-Synthese

dolly Fahrgestell
(Kistenroller/Fassroller etc.);
Plattformwagen, -karren
donor Donor, Spender
DOP-PCR (degenerate oligonucleotide primer PCR)
DOP-PCR
(PCR mit degeneriertem
Oligonucleotidprimer)
dormancy period
Ruhephase, Ruheperiode
dorsal rückseitig, dorsal
dosage (dose) Dosis
dosage compensation
Dosiskompensation
dosage effect Dosiseffekt
dose *vb* **(give a dose)/**
measure out (proportion)
dosieren
dose *n* **(dosage)** Dosis
➢ **lethal dose**
letale Dosis,
Letaldosis, tödliche Dosis
➢ **maximum tolerated dose (MTD)**
maximal verträgliche Dosis
➢ **median effective dose (ED$_{50}$)**
mittlere effektive Dosis (ED$_{50}$),
mittlere wirksame Dosis
➢ **median lethal dose (LD50)**
mittlere letale Dosis (LD$_{50}$)
➢ **overdose** Überdosis
➢ **single dose** Einzeldosis
dose equivalent *rad* Dosisäquivalent
dose-response curve
Dosis-Wirkungskurve
dosing pump/
proportioning pump/
metering pump
Dosierpumpe
dot blot/spot blot Rundlochplatte
dot diagram Punktdiagramm
double-acting doppeltwirkend
double-acting pump
Druckpumpe, Saugpumpe,
doppeltwirkende Pumpe
double blind assay/double-blind study
Doppelblindversuch
double bond Doppelbindung

double-burner hot plate
Doppelkochplatte (Heizplatte)
double cross Doppelkreuzung
double digest *gen/biochem*
Doppelverdau
double distilled water Bidest
double-headed intermediate
doppelköpfiges,
janusköpfiges Zwischenprodukt
double layer/bilayer Doppelschicht
double salt Doppelsalz
double strand *gen* Doppelstrang
double sugar/disaccharide
Doppelzucker, Disaccharid
doubling time (generation time)
Verdopplungszeit (Generationszeit)
dovetail connection *micros*
Schwalbenschwanzverbindung
down regulation Herabregulation
downpipe Fallrohr
downstairs die Treppe runter; 'unten'
downstroke (pump)
Leerhub, Abwärtshub
downtime
Auszeit, Abschaltzeit, Ausfallzeit,
Standzeit, Stillstandzeit, Verlustzeit
draff Trester, Treber (Malzrückstand)
draft (*Br* **draught)** Zugluft, Luftzug
drain *vb* ablaufen lassen; entwässern,
drainieren, Flüssigkeit ablassen
drain *n* Abfluss, Ablauf; (of the sink)
Abguss (an der Spüle); (drainage)
Entleeren, Entleerung (Flüssigkeit)
➢ **blocked drain**
verstopftes Waschbecken (~abfluss)
➢ **pour s.th. down the drain**
etwas in den Abguss schütten
drainage/draining Dränung, Drainage;
Trockenlegung, Entwässerung
drainboard/dish board Ablaufbrett,
Abtropfbrett (Platte an der Spüle)
draincock Ablasshahn, Ablaufhahn
draining rack Abtropfgestell
**draw blade/drawing knife (>Rakel)/
(cabinet) scraper** Ziehklinge
**draw off/suction off/
siphon off/evacuate**
absaugen (Flüssigkeit)

dress *vb med* verbinden, einen
Verband anlegen; (coat/treat with
fungicides/pesticides) beizen
(Saatgut)
dress code Kleiderordnung
dressing *med*
Verband; Verband anlegen
dressing agent (pesticides/fungicides)
Saatgutbeizmittel
drift tube (TOF-MS) Driftröhre
drill *vb* bohren
drill *n* Bohrmaschine; Bohrer;
(bit) Bohraufsatz; (drilling/bore)
Bohrung (Prozess/Vorgang);
(exercise) Übung
➢ **concrete drill (bit)** Betonbohrer
➢ **emergency drill**
Probealarm, Probe-Notalarm
➢ **fire drill** Feueralarmübung,
Feuerwehrübung
➢ **metal drill (bit)** Metalbohrer
➢ **percussion drill** Schlagbohrer
➢ **rock drill (bit)** Steinbohrer
➢ **wood drill (bit)** Holzbohrer
drill chuck Bohrfutter
drill core ('core') *geol/paleo*
Bohrkern, Kern
drinking fountain (for drinking water)
Trinkbrunnen, Trinkfontäne
drinking water/potable water
Trinkwasser
drip *n* Tropfen
drip *vb* tropfen;
herabtröpfeln/herabtropfen (lassen);
träufeln
**drip catcher/drip catch
(for distillation apparatus)**
Tropfenfänger
drive *n* Antrieb, Trieb
drive *vb* antreiben
drive shaft Antriebswelle
drive system/drive unit Antriebssystem
drop *n* Tropfen
drop *vb* tropfen; (drip) herabtröpfeln/
herabtropfen (lassen), träufeln
drop bottle/dropping bottle
Tropfflasche
droplet infection Tröpfcheninfektion

dropper Tropfglas, Tropfenglas, Tropfer, Tropfflasche, Tropffläschchen; Tropfenzähler
dropping bottle/dropper vial Pipettenflasche; Tropfflasche
dropping funnel Tropftrichter
dropping mercury electrode (DME) Quecksilbertropfelektrode
dropping pipet/dropper Tropfpipette, Tropfglas
dropping point/drop point Tropfpunkt
dropwise/drop by drop tropfenweise
drought Trockenheit, Dürre
drug Droge; Arznei, Arzneimittel, Medizin; Suchtmittel, Droge
➢ **herbal drug** Pflanzendroge
➢ **non-prescription drug** nicht verschreibungspflichtiges Arzneimittel
➢ **over-the-counter drug** frei erhältliches Medikament/Medizin/Droge (nicht verschreibungspflichtig)
➢ **prescription drug** verschreibungspflichtiges Arzneimittel/Medikament
➢ **psychoactive/psychotropic drug** Rauschmittel, Rauschgift, Rauschdroge
drug design zielgerichtete 'Konstruktion' neuer Medikamente am Computer
drug resistance Arzneimittelresistenz
drug treatment medikamentöse Behandlung
drum (barrel) Trommel, Zylinder
drum mill/tube mill/barrel mill Trommelmühle
drum rotor/drum-type rotor *centrif* Trommelrotor
drum vent Fassventil (Entlüftung)
drum wrench Fassschlüssel (zum Öffnen von Fässern)
dry *adj/adv* trocken
➢ **run dry** leerlaufen, trockenlaufen
dry *vb* trocknen
dry blotting Trockenblotten

dry cell/dry cell battery Trockenbatterie
dry ice (CO_2) Trockeneis
dry mass/dry matter Trockenmasse, Trockensubstanz
dry product/dry substance Trockengut
dry storage Trockenlagerung (Lagerung mit Kaltluftkühlung)
dry weight (*sensu stricto*: dry mass) Trockengewicht (*sensu stricto*: Trockenmasse)
drying agent Trockenmittel
drying bed Trockenbeet (Kläranlage)
drying cabinet (drying oven) Trockenschrank (Trockenofen)
drying pistol Trockenpistole, Röhrentrockner
drying rack Trockengestell
drying tower/drying column Trockenturm, Trockensäule
drying tube Trockenrohr, Trockenröhrchen
dryness/drought Trockenheit, Dürre
duct Röhre, Leitung, Kanal, Gang
duct tape (polycoated cloth tape) Panzerband, Gewebeband, Gewebeklebeband, Duct Gewebeklebeband, Universalband, Vielzweckband
ductwork/airduct system Rohrleitungssystem (Lüftung)
dung/manure Dung, Mist, tierische Exkremente, Tierkot
dunk tank (for chemicals) Auffangbecken, Auffangbehälter
duplicable/duplicatable nachvollziehbar
duplicate/repeat *vb* nachvollziehen, wiederholen
durability Verwitterungsbeständigkeit
dust Staub
➢ **airborne dust** Flugstaub
➢ **fine dust** Feinstaub (alveolengängig)
➢ **flue dust** Flugstaub (von Abgasen)
➢ **inert dust** Inertstaub
➢ **sawdust** Sägemehl

dust cover Staubschutz
dust explosion Staubexplosion
dust mask (respirator)
Grobstaubmaske
➢ **particulate respirator**
 (U.S. safety levels N/R/P according to regulation 42 CFR 84)
Staubschutzmaske
(Partikelfilternde Masken)
(DIN FFP)
dust-mist mask Feinstaubmaske
dusting (dust off) Staubwischen
dustpan
Kehrschaufel, Kehrblech
dustproof staubdicht
dusty staubig
duty Pflicht; Dienst

duty cycle (machine/equipment)
Arbeitszyklus (Gerät)
duty to inform/
 obligation to provide information
Mitteilungspflicht
dye *vb* **(add color/add pigment)**
färben, einfärben; (stain) anfärben
dye *n* **(colorant/pigment/dyestuff)**
Farbstoff, Pigment
➢ **supravital dye/supravital stain**
Supravitalfarbstoff
➢ **vital dye/vital stain**
Vitalfarbstoff, Lebendfarbstoff
dyeability/stainability Anfärbbarkeit
dyeable/stainable anfärbbar
dyeing/staining Anfärbung
dying Sterben

earmuffs ('muffs')/
hearing protectors
Gehörschützer
(speziell auch: Kapselgehörschützer)
earplugs Ohrenstöpsel
Earth/World Erde, Welt
earthenware Tonwaren
eat into (corrode) *vb chem*
ätzen, korrodieren
ebullition Sieden, Aufwallen
ebullition tube Siederöhrchen
ecological ökologisch
ecological balance/
ecological equilibrium
ökologisches Gleichgewicht
ecological genetics
ökologische Genetik, Ökogenetik
economic plant/useful plant/crop plant
Nutzpflanze
ectopic verlagert, ektopisch (an
unüblicher Stelle liegend)
eczema Ekzem
eddy (swirl) Strudel
eddy current
Wirbelstrom (Vortex-Bewegung)
eddy diffusion
Turbulenzdiffusion, Wirbeldiffusion
edge/margin Rand
edibility/edibleness Essbarkeit
edible/eatable essbar
➢ **inedible/uneatable** nicht essbar
educational facility/institution
Lehranstalt
effect (action) Wirkung; (impact)
Einwirkung
effective effektiv, wirksam, erfolgreich;
tatsächlich, wirklich
efferent ausführend, wegführend,
ableitend (Flüssigkeit)
efficiency Wirkungsgrad
efficient effizient, wirksam; tüchtig;
gründlich, leistungsstrak
effluent Ablauf, Ausfluss
(herausfließende Flüssigkeit)
efflux Ausstrom
effusion Ausströmen, Effusion (Gas)
egest/excrete ausscheiden
(Exkrete/Exkremente)

egestion/excretion
Ausscheidung (Exkretion)
egg medium/egg culture medium
Eiermedium, Eiernährmedium,
Eiernährboden
egg white/egg albumen
Eiweiß (Ei)
➢ **native egg white**
natives Eiweiß, Eiklar
egress *n* Fluchtweg
elasticity Elastizität
elbow/elbow fitting/ell
(lab glass/tube fittings)
Winkelrohr, Winkelstück, Krümmer
(Glas/Metal etc. zur Verbindung)
electric appliance Elektrogerät
electric circuit/electrical circuit
Stromkreis
electric meter Stromzähler
electric power supply/power supply
Stromversorgung
electric tape/
insulating tape/friction tape
Elektro-Isolierband
electrical appliance/electrical device
Elektrogerät
electrician Elektriker
electricity (power/juice)
Strom, Elektrizität
➢ **static electricity**
statische Elektrizität
electricity failure/power failure
Stromausfall
electrocardiogram
Elektrokardiogramm (EKG)
electrochromatography (EC)
Elektrochromatographie (EC)
electrode Elektrode
➢ **dropping electrode**
Tropfelektrode
➢ **dropping mercury electrode (DME)**
Quecksilbertropfelektrode
➢ **hydrogen electrode**
Wasserstoffelektrode
➢ **ion-selective electrode (ISE)**
ionenselektive Elektrode
➢ **reference electrode**
Bezugselektrode

electrodeposition
elektrolytische Abscheidung, Galvanotechnik
electroencephalogram
Elektroencephalogramm (EEG)
electro-endosmosis/
electro-osmotic flow (EOF)
Elektroosmose (Elektroendosmose)
electrogenic elektrogen
electroimmunodiffusion/
counter immunoelectrophoresis
Elektroimmunodiffusion
electrolysis Elektrolyse
- **molten-salt electrolysis**
Schmelzelektrolyse, Schmelzflusselektrolyse
electrolyte Elektrolyt
electrolytic bath
Elektroysebad, elektrolytischer Trog
electrolytic separation
elektrolytische Trennung, elektrolytische Dissoziation
electrolytic trough
Elektroysebad, elektrolytischer Trog
electromagnetic spectrum
elektromagnetisches Spektrum
electromotive force (emf/E.M.F.)
elektromotorische Kraft (EMK)
electron (s) Elektron(en)
- **binding electron** Bindungselektron
- **free electron** freies Elektron
- **inner electron** Rumpfelektron
- **lone pair**
(free/unshared/nonbonding)
einsames (freies) Elektronenpaar
- **odd electron (unpaired/lone)**
ungepaartes Elektron
- **outer electron** Außenelektron
- **paired electron** gepaartes Elektron
- **single electron** Einzelelektron
- **twin electrons/electron pair**
Elektronenpaar
- **valence electron/valency electron**
Valenzelektron
electron acceptor
Elektronenakzeptor, Elektronenraffer, Elektronenempfänger

electron capture detector (ECD)
Elektroneneinfangdetektor
electron carrier
Elektronenüberträger
electron configuration
Elektronenkonfiguration
electron density Elektronendichte
electron donor
Elektronendonor, Elektronenspender
electron energy loss spectroscopy (EELS)
Elektronen-Energieverlust-Spektroskopie
electron micrograph
elektronenmikroskopisches Bild, elektronenmikroskopische Aufnahme
electron microscopy (EM)
Elektronenmikroskopie
- **high voltage electron microscopy (HVEM)** Hochspannungselektronenmikroskopie
- **scanning electron microscopy (SEM)** Rasterelektronenmikroskopie (REM)
- **transmission electron microscopy (TEM)** Transmissionselektronenmikroskopie, Durchstrahlungselektronenmikroskopie
electron pair Elektronenpaar
- **lone pair**
(free/unshared/nonbonding)
einsames (freies) Elektronenpaar
electron shell
Elektronenschale, Elektronenhülle
electron spin resonance spectroscopy (ESR)/
electron paramagnetic resonance (EPR) Elektronen-Spinresonanzspektroskopie (ESR), elektronenparamagnetische Resonanz (EPR)
electron transfer
Elektronenübertragung
electron transport
Elektronentransport
electroneutral (electrically silent)
elektoneutral
electronic elektronisch

el

electron-impact ionization (EI)
Elektronenstoß-Ionisation
electron-impact spectrometry (EIS)
Elektronenstoß-Spektrometrie
electron-transport chain
Elektronentransportkette
electrophilic attack
electrophiler Angriff
electrophoresis
Elektrophorese
- **alternating field gel electrophoresis**
Wechselfeld-Gelelektrophorese
- **capillary electrophoresis (CE)**
Kapillarelektrophorese
- **capillary zone electrophoresis (CZE)**
Kapillar-Zonenelektrophorese
- **carrier electrophoresis**
Trägerelektrophorese,
Elektropherografie
- **countercurrent electrophoresis**
Gegenstromelektrophorese,
Überwanderungselektrophorese
- **direct blotting electrophoresis/ direct transfer electrophoresis**
Blotting-Elektrophorese,
Direkttransfer-Elektrophorese
- **disk electrophoresis**
Diskelektrophorese,
diskontinuierliche Elektrophorese
- **field inversion gel electrophoresis (FIGE)**
Feldinversions-Gelelektrophorese
- **free electrophoresis (carrier-free electrophoresis)**
freie Elektrophorese
- **gel electrophoresis**
Gelelektrophorese
- **gradient gel electrophoresis**
Gradienten-Gelelektrophorese
- **isotachophoresis (ITP)**
Isotachophorese,
Gleichgeschwindigkeits-
Elektrophorese
- **paper electrophoresis**
Papierelektrophorese
- **pulsed field gel electrophoresis (PFGE)** Puls-Feld-Gelelektrophorese,
Wechselfeld-Gelelektrophorese
- **temperature gradient gel electrophoresis**
Temperaturgradienten-
Gelelektrophorese
- **zone electrophoresis**
Zonenelektrophorese

electrophoretic
elektrophoretisch
electrophoretic mobility
elektrophoretische Mobilität
electroplaque
Elektroplaque (*pl* Elektroplaques, *slang:* Elektroplaxe)
electroplating
elektroplatieren, galvanisieren
electropolish elektropolieren
electropolishing
elektrolytisches Polieren
electroporation
Elektroporation
electroprecipitation
Elektroabscheidung
electroretinogram (ERG)
Elektroretinogramm
electrospray Elektrospray
electrowinning
elektrolytische Metallgewinnung (Elektrometallurgie)
element Element
- **half element/half cell (single-electrode system)**
Halbelement (galvanisches),
Halbzelle
- **trace element/ microelement/micronutrient**
Spurenelement,
Mikroelement
- **transition elements (transition metals)**
Übergangselemente
(Übergangsmetalle)

elevator Aufzug (Personen~)
eliminate *chem* eliminieren; (eradicate/extirpate) ausmerzen, ausrotten
elimination *chem* Elimination; (eradication/extirpation) Ausmerzung, Ausrottung

ELISA (enzyme-linked immunosorbent assay) ELISA (enzymgekoppelter Immunadsorptionstest, enzymgekoppelter Immunnachweis)
ell/elbow/elbow fitting (bend/bent tube/angle connector) Krümmer (gebogenes Rohrstück), Winkelrohr, Winkelstück (Glas/Metal etc. zur Verbindung)
ellagic acid/gallogen Ellagsäure
elongate/extend strecken (in die Länge ziehen), verlängern
elongation (extension) Streckung, Verlängerung
eluate *n* Eluat
eluate *vb* eluieren
elucidate verdeutlichen, klar machen; (Strukturen/Zusammenhänge) aufklären
elucidation Verdeutlichung, Aufklärung; (Strukturen/Zusammenhänge) Aufklärung
eluent/eluant Elutionsmittel, Eluens (Laufmittel)
eluotropic series eluotrope Reihe (Lösungsmittelreihe)
eluting strength (eluent strength) Elutionskraft
elution *chromat* Elution (herauslösen adsorbierter Stoffe aus stationärer Phase)
elution rate Elutionsgeschwindigkeit, Durchlaufgeschwindigkeit
elutriation Elutriation, Aufstromklassierung
emanate hervorquellen
embed *micros* einbetten
embedded specimen Einbettungspräparat
embedding *micros* Einbettung
embedding machine/ embedding center *micros* Einbettautomat, Einbettungsautomat
embolism Embolie (Obstruktion der Blutbahn)
embolus Embolus
emboly/invagination Embolie, Invagination, Einfaltung, Einstülpung
embryo transfer Embryotransfer
embryotoxic embryotoxisch
emerge hervorkommen, herauskommen, auftauchen
emergence Auftauchen; Emergenz, Auswuchs
emergency Notfall, Notstand, Notlage
emergency call Notruf
emergency escape mask Fluchtgerät, Selbstretter (Atemschutzgerät)
emergency evacuation plan Notfall-Evakuierungsplan, Notfall-Fluchtplan
emergency evacuation route/emergency escape route Notfall-Fluchtweg
emergency exit Notausgang
emergency generator/ standby generator Notstromaggregat
emergency level/alert level Alarmstufe
emergency lighting Notbeleuchtung
emergency number Notruf, Notfallnummer
emergency provisions Notfallvorkehrungen
emergency response Notfalleinsatz
emergency response plan (ERP) Notfalleinsatzplan
emergency response team Notfalleinsatztruppe
emergency room Ambulanz, Notaufnahme
emergency service Notdienst, Hilfsdienst
emergency shower/safety shower Notdusche
➤ **quick drench shower/deluge shower** 'Schnellflutdusche'
emergency shutdown Notabschaltung
emergency ward (clinic) Notaufnahme, Unfallstation (Krankenhaus)

emery Schmirgel
emery cloth Schmirgelleinen
emery paper (*Br*)/sandpaper (*US*)
Schmirgelpapier
emission
Emission, Ausstoss, Ausstrahlung
emissivity
Strahlungsvermögen,
Emissionsvermögen
(Wärmeabstrahlvermögen)
**emissivity coefficient
(absorptivity coefficient)**
Emissionskoeffizient
emit emittieren, aussenden;
ausstrahlen, verströmen, ausstoßen
emphysema Emphysem, Aufblähung
empiric(al) empirisch
empirical formula
empirische Formel, Summenformel,
Elementarformel, Verhältnisformel
employee Angestellter, Bediensteter,
Mitarbeiter, Betriebszugehöriger
➤ **administrative employee**
Verwaltungsangestellte(r)
➤ **civil servant/public service officer**
staatlicher Bediensteter, 'Beamter'
➤ **full-time employee**
Vollzeitbeschäftigte(r)
➤ **staff member**
Mitarbeiter, Betriebszugehöriger
employer Arbeitgeber
empty leer; ausleeren
empty out (pour out)
entleeren, ausleeren, auskippen
emptying out (pouring out)
Entleeren, Entleerung
(eines Gefäßes; allgemein)
emulsifier/emulsifying agent
Emulgator
emulsify emulgieren
emulsion Emulsion
enamel Email, Emaille,
Schmelzglas; Glasur
enantiomere Enantiomer
enantiomeric separation
Enantiomerentrennung,
Racemattrennung, Racematspaltung
enbalm einbalsamieren

encapsulation Einkapselung
encode/code kodieren, codieren
encrusting krustenbildend
encyst zystieren, enzystieren
end nippers
Monierzange, Rabitzzange
**end-group analysis/
terminal residue analysis**
Endgruppenanalyse,
Endgruppenbestimmung
end point (point of neutrality: titration)
Äquivalenzpunkt
end-point determination
Endpunktsbestimmung
end-point dilution technique
Endpunktverdünnungsmethode
(Virustitration)
**end-product inhibition/
feedback inhibition**
EndproduktHemmung,
Rückkopplungshemmung
endanger/imperil gefährden
endangered (in danger/at risk)
gefährdet
endangerment/imperilment
Gefährdung
endergonic endergon,
energieverbrauchend
endocrine gland endokrine Drüse
endothermic endotherm
**endurance/
persistence/hardiness/perseverance**
Ausdauer, Dauerhaftigkeit
endure/persist
ausdauern, ausharren;
ertragen; überstehen
energetics Energetik
energy Energie
➤ **appearance energy (MS)**
Auftrittsenergie (MS)
➤ **binding energy/bond energy**
Bindungsenergie
➤ **ignition energy** Zündenergie
➤ **lattice energy** Gitterenergie
➤ **law of conservation of energy**
Energieerhaltungssatz
➤ **maintenance energy**
Erhaltungsenergie

- **nuclear energy/atomic energy**
  Atomenergie
- **potential energy** Lageenergie
- **radiant energy** Strahlungsenergie
- **solar energy**
  Solarenergie, Sonnenenergie
- **thermic energy**
  thermische Energie,
  Wärmeenergie
- **useful energy**
  Nutzenergie, nutzbare Energie

energy balance/energy budget
Energiebilanz
energy barrier Energiebarriere
energy charge Energieladung
energy efficiency
Energiewirkungsgrad;
Energieausbeute
energy-efficient energieeffizient
(mit hohem Energiewirkungsgrad)
energy flux/energy flow Energiefluss
energy metabolism
Energiestoffwechsel
energy profile Energieprofil
energy requirement Energiebedarf
energy-rich energiereich
energy-saving lightbulb
Energiesparlampe
energy source Energiequelle
energy supply Energiezuführung
energy transfer
Energieübergang, Energietransfer
engulf vertilgen; einverleiben
enhance verstärken
enhancer Verstärker; *gen* (sequence)
Enhancer, Verstärker(sequenz)
enlarging adapter/expansion adapter
Expansionsstück (Laborglas)
enology Weinbaukunde, Önologie
enrich (concentrate/accumulate/fortify)
anreichern
enrichment (concentration/
accumulation/fortification)
Anreicherung
- **filter enrichment**
  Anreicherung durch Filter

enrichment culture
Anreicherungskultur
enrichment medium
Anreicherungsmedium
enter hineingehen, eintreten;
eintragen (z.B. Daten ins
Laborbuch);
eingeben (Daten/in den Computer)
enthalpy Enthalpie
entomology
Insektenkunde, Entomologie
entrainer/separating agent *dist*
Schleppmittel
entrance Eingang, Zugang, Zutritt
- **No Entrance!/**
  Do Not Enter!/
  No Trespassing!
  Zutritt verboten!, Betreten verboten!

entropy Entropie
entry Eintritt, Eingang
- **route of entry** Eintrittspforte

enucleate (cell) entkernt (Zelle)
envelop *vb* einhüllen, einpacken,
einschlagen, einwickeln
envelope *n* (jacket)
Hülle, Umschlag, Umhüllung
envenom vergiften durch Tiergift
envenomation/envenomization
Vergiftung (durch Tiergift)
environment
Umwelt, Umgebung, Gegend
- **dangerous for the environment**
  (N = nuisant)
  umweltgefährlich

environmental analysis
Umweltanalyse
environmental analytics
Umweltanalytik
environmental audit
Umweltaudit, Öko-Audit
environmental burden/
environmental load
Umweltbelastung
environmental chemistry
Umweltchemie
environmental compatibility
Umweltverträglichkeit
environmental conditions
Umweltbedingungen,
Umweltverhältnisse

environmental crime
Umweltkriminalität
environmental degradation
Umweltzerstörung
environmental factor Umweltfaktor
environmental impact assessment (EIA)
Umweltverträglichkeitsprüfung (UVP)
environmental law Umweltrecht
environmental medicine
Umweltmedizin
environmental monitoring technology
Umweltmesstechnik
environmental physics Umweltphysik
environmental politics Umweltpolitik
environmental pollution
Umweltverschmutzung
environmental process engineering
Umweltverfahrenstechnik
environmental protection
Umweltschutz
environmental requirements
Umweltansprüche
environmental resistance
Umweltwiderstand
environmental science
Umweltwissenschaft
environmental variance
Umweltvarianz
environmentalist Umweltschützer
environmentally compatible/ environmentally friendly
umweltverträglich, umweltgerecht
enzymatic coupling Enzymkopplung
enzymatic degradation
enzymatischer Abbau
enzymatic digestion
enzymatischer Abbau
enzymatic inhibition/ repression of enzyme/ inhibition of enzyme
Enzymhemmung
enzymatic pathway
enzymatische Reaktionskette
enzymatic reaction Enzymreaktion
enzymatic specificity/ enzyme specificity Enzymspezifität

enzyme Enzym, Ferment
➢ **core enzyme**
Kernenzym (RNA-Polymerase)
➢ **digestive enzyme** Verdauungsenzym
➢ **holoenzyme** Holoenzym
➢ **isozyme/isoenzyme**
Isozym, Isoenzym
➢ **key enzyme**
Schlüsselenzym, Leitenzym
➢ **multienzyme complex/ multienzyme system**
Multienzymkomplex, Multienzymsystem, Enzymkette
➢ **processive enzyme**
progressiv arbeitendes Enzym
➢ **proenzyme/zymogen**
Proenzym, Zymogen
➢ **repair enzyme** Reparaturenzym
➢ **restriction enzyme**
Restriktionsenzym
➢ **tracer enzyme** Leitenzym
enzyme activity (*katal*)
Enzymaktivität (*katal*)
enzyme-immunoassay/ enzyme immunassay (EIA)
Enzymimmunoassay, Enzymimmuntest (EMIT-Test)
enzyme kinetics Enzymkinetik
enzyme-linked immunosorbent assay (ELISA) enzymgekoppelter Immunadsorptionstest, enzymgekoppelter Immunnachweis (ELISA)
enzyme-linked immunotransfer blot (EITB) enzymgekoppelter Immunoelektrotransfer
EPA
(Environmental Protection Agency)
U.S. Umwelt und Naturschutzbehörde
(entspricht in etwa UBA + BFN)
epidemic Epidemie, Seuche
epidemiologic(al) epidemiologisch
epidemiology Epidemiologie
epidermal/cutaneous
epidermal, Haut..,
die Haut betreffend
epidermis Epidermis, Oberhaut

**epiillumination/
incident illumination**
Auflicht, Auflichtbeleuchtung
epimerization Epimerisierung
epithelium Epithel (*pl* Epithelien)
epitope/antigenic determinant
Epitop, Antigendeterminante
epizooic disease
Tierseuche, Viehseuche
**Epsom salts/epsomite/
magnesium sulfate**
Bittersalz, Magnesiumsulfat
equal/same/identical
gleich, identisch
(völlig gleich, ein und dasselbe)
**equalization/
adjustment/balancing/balance**
Abgleich
equate *math* gleichen
equation Gleichung
➢ **balanced equation** *chem*
'eingerichtete' Gleichung
➢ **chemical equation**
chemische Gleichung,
Reaktionsgleichung
equation of the *x*th order
Gleichung *x*ten Grades
equilibrium Gleichgewicht
➢ **imbalance/disequilibrium**
Ungleichgewicht
➢ **ion equilibrium/ionic steady state**
Ionengleichgewicht
➢ **steady-state equilibrium**
Fließgleichgewicht,
dynamisches Gleichgewicht
**equilibrium centrifugation/
equilibrium centrifuging**
Gleichgewichtszentrifugation
equilibrium constant
Gleichgewichtskonstante
equilibrium dialysis
Gleichgewichtsdialyse
equilibrium distillation
Gleichgewichtsdestillation
equilibrium potential
Gleichgewichtspotential
equilibrium state
Gleichgewichtszustand

equipment (appliances/device)
Gerät, Ausrüstung, Ausstattung;
Gegenstand; Einrichtung, Anlage,
Maschine, Apparat
equipment probe Gerätesonde
equipment room Geräteraum
eradicate/eliminate/extirpate
ausrotten, ausmerzen
eradication/elimination/extirpation *med*
Ausrottung, Ausmerzung
(z.B. Schädlinge)
Erlenmeyer flask Erlenmeyer Kolben
erroneous/mistaken/flawed
fehlerhaft, falsch
error (mistake/defect)
Fehler (Defekt)
➢ **random error**
zufälliger Fehler, Zufallsfehler
➢ **statistical error** statistischer Fehler
**error in measurement/
measuring mistake** Messfehler
error of estimation *stat* Schätzfehler
erucic acid/(Z)-13-docosenoic acid
Erucasäure, Δ^{13}-Docosensäure
erythrocyte ghost
Erythrozytenschatten, Schatten
(leeres/ausgelaugtes rotes
Blutkörperchen)
escape fliehen, entkommen;
chem entweichen (Gas etc.)
escape hatch Fluchtluke,
Ausstiegsluke, Rettungsluke
escape route/egress Fluchtweg
escape shaft Fluchtschacht,
Notschacht, Rettungsschacht
ESR (electron spin resonance)
ESR (Elektronenspinresonanz)
essence *chem/pharm* Essenz
essential essentiell
➢ **essential for life/vital**
lebenswichtig,
lebensnotwendig, vital
essential amino acids
essentielle Aminosäure
essential oil/ethereal oil ätherisches Öl
establish gründen, einrichten;
micb (start a culture) anzüchten
(einer Kultur)

established cell line etablierte Zellinie
establishing growth/
  starting growth *micb* **(of a culture)**
  Anzüchtung (einer Kultur)
establishment phase
  Eingewöhnungsphase
esterification Veresterung
esterify verestern
estimate *n* **(estimation/assumption)**
  Schätzung, Annahme; Schätzwert
estimate *vb* **(assume)**
  schätzen, annehmen
etch *vb* m*etall/tech/micros*
  ätzen (*siehe:* Gefrierätzen)
etchant *metall/tech/micros* Ätzmittel
etching *metall/tech/micros*
  Ätzen, Ätzung, Ätzverfahren
  (*siehe:* Gefrierätzen)
➢ **deep etching** Tiefenätzung
ethanol/ethyl alcohol/alcohol
  Äthanol, Ethanol, Äthylalkohol,
  Ethylalkohol, 'Alkohol'
➢ **graded ethanol series**
  Alkoholreihe,
  aufsteigende Äthanolreihe
ether Ether, Äther
ether trap Etherfalle
ethereal oil/essential oil ätherisches Öl
ethylene Ethylen, Äthylen
etiological agent
  Krankheitsverursacher
  (Wirkstoff, Agens, Mittel)
etiology Krankheitsursache, Ätiologie
eupnea Eupnoe
eutectic point eutektischer Punkt
eutrophicate eutrophieren
evacuate entleeren,
  luftleer pumpen, herauspumpen;
  evakuieren, räumen
evacuation plan Evakuierungsplan
evaluate (e.g., results)
  auswerten (z.B. von Ergebnissen)
evaluation (e.g., of results)
  Auswertung (z.B. von Ergebnissen);
  Beurteilung
evaporate/vaporize
  verdunsten, abdampfen, abdunsten;
  eindampfen

evaporating dish
  Abdampfschale,
  Eindampfschale
evaporating flask
  Verdampferkolben
evaporation (vaporization)
  Verdunstung, Verdampfung,
  Abdampfen, Eindampfen
➢ **reduce by evaporation**
  (evaporate completely)
  eindampfen (vollständig)
evaporation burner
  Verdunstungsbrenner
evaporative cooling
  Verdunstungskälte,
  Verdunstungsabkühlung
evaporative light scattering detector
  (ELSD)
  Verdampfungs-Lichtstreudetektor
evaporator/concentrator
  Evaporator, Verdampfer,
  Abdampfvorrichtung
evaporimeter/
  evaporation gauge/
  evaporation meter
  Evaporimeter, Verdunstungsmesser
evert/evaginate/
  protrude/turn inside out
  ausstülpen
exalbuminous eiweißlos
examination Untersuchung
examination couch *med*
  Untersuchungsliege
examination under a microscope/
  usage of a microscope
  Mikroskopieren
examine untersuchen
➢ **examine under a microscope/**
  use a microscope
  mikroskopieren
exceed überschreiten
exception (special case)
  Ausnahme, Sonderfall
exceptional permission/
  special permission
  Ausnahmegenehmigung,
  Sondergenehmigung
excess Überfluss, Überschuss (Menge)

exchange Austausch
exchange reaction Austauschreaktion
excise herausschneiden, exzidieren
excision Excision, Exzision, Herausschneiden
excitability/irritability/sensitivity Erregbarkeit
excitable/irritable/sensitive erregbar
excitation/irritation Erregung, Irritation
excitatory exzitatorisch, erregend
excite (irritate) erregen; (stimulate) reizen, anregen, stimulieren
excited state *chem/med/physiol* erregter Zustand, angeregter Zustand
exciter (fluorescence microscopy) Erreger
exciter filter (fluorescence microscopy) Erregerfilter
exclusion Exclusion, Exklusion, Ausschluss
exclusion of air (air-tight) Luftausschluss, Luftabschluss
excreta/excretions Ausscheidungen, Exkrete, Exkremente
excretion Exkret, Exkretion; pl Ausscheidungen, Exkrete, Exkremente
excursion/field trip Exkursion
exergonic exergon, energiefreisetzend
exhalation Ausatmung, Ausatmen, Expiration, Exhalation
exhalation valve (respirator/mask) Ausatemventil (an Atemschutzgerät)
exhale (breathe out) ausatmen
exhaust (exhaust air/waste air/extract air) Abluft
exhaust duct Abluftschacht
exhaust fumes Abgase
exhaust stack Abzugschornstein
exhaust system/off-gas systemAblufteinrichtung, Abluftsystem
exhaust vapor/fuel-laden vapor Brüden
existing (extant) bestehend, existierend
existing chemicals/ existing substances Altstoffe

exit *n* Ausgang; Austritt
exit pupil *micros* Austrittspupille
exit slit Austrittsspalt
exit velocity (hood) Ausströmgeschwindigkeit, Austrittsgeschwindigkeit (Sicherheitswerkbank)
exocytosis Exocytose
exogenic/exogenous exogen
exothermic exotherm
expand expandieren, ausbreiten, entfalten; ausdehnen, erweitern
expansion Expansion, Ausdehnung, Erweiterung; (dilation/dilatation) Dilatation, Ausweitung
expansion adapter/enlarging adapter Expansionsstück (Laborglas)
expansivity Dehnbarkeit
experiment *vb* experimentieren
experiment *n* (test/trial) Versuch
➢ long-term experiment Langzeitversuch
➢ performing an experiment/ performance of an experiment Versuchsdurchführung
➢ pretrial/preliminary experiment Vorversuch
experiment setup Versuchsanordnung, Versuchsaufbau
experimental conditions Versuchsbedingungen
experimental physics Experimentalphysik
experimental procedure/ experimental method Versuchsverfahren
experimental series/trial series Versuchsreihe
expert (specialist/authority) Sachkundiger, Sachverständiger
expertise (expert opinion) Gutachten, Expertise, Sachverständigengutachten; Fachkenntnis, Sachkenntnis, Expertenwissen

expiration Verfall;
(exhalation) *med* Ausatmung, Ausatmen, Expiration, Exhalation
expiration date Verfallsdatum
expire ablaufen, verfallen;
(exhale/breathe out) *med* ausatmen
expired/outdated
abgelaufen, verfallen (Haltbarkeitsdatum)
explant *n* Explantat
explode explodieren
exploit ausbeuten (Rohstoffe)
explorative data analysis
explorative Datenanalyse
explosion Explosion
➤ **dust explosion** Staubexplosion
explosion hazard
Explosionsgefahr
explosion limit
Exposionsgrenze
(untere=UEG, obere=OEG)
explosionproof
explosionsgeschützt, explosionssicher
explosive *adj/adv* explosiv;
(E) explosionsgefährlich (Gefahrenbezeichnungen)
explosive *n*
Explosivstoff (*siehe:* Sprengstoff)
➤ **high energy explosive (HEX)**
hochbrisanter Sprengstoff
➤ **high explosive**
brisanter Sprengstoff
➤ **low explosive**
verpuffender Sprengstoff, Schießstoff, Schießmittel
explosive flame/sudden flame
Stichflamme
explosive force/explosive power
Sprengkraft
exponential growth phase
exponentielle Wachstumsphase, exponentielle Entwicklungsphase
export regulations
Ausfuhrbestimmungen
expose (to chemicals/radiation)
aussetzen, exponieren;
(film/plants) belichten

exposure *med/chem* Exposition, Aussetzen, Ausgesetztsein, Gefährdung (Strahlung/Chemikalie etc.), Bestrahlung;
(to light) Belichtung
➤ **permissible workplace exposure**
zulässige/maximale Arbeitsplatzkonzentration
exposure level of air pollutants
Immission
(Belastung durch Luftschadstoffe)
exposure time/duration of exposure
Einwirkungsdauer, Einwirkungszeit
express exprimieren, ausdrücken
expression Expression, Ausdruck
extension Ausdehnung, Verlängerung
extension cable *electr*
Verlängerungskabel
extension clamp
Verlängerungsklemme
extension cord *electr*
Verlängerungsschnur
external (extrinsic)
äußerlich, von außen, extern
external thread/
 male thread (pipe/fittings)
Außengewinde
extinction
Extinktion; (dying out) Aussterben
extinction coefficient/absorptivity
Extinktionskoeffizient
extinguish/put out (fire)
löschen (Feuer)
extracellular
extrazellulär, außerzellulär
extract *n* Extrakt, Auszug
➤ **cell extract** Zellextrakt
➤ **cell-free extract** zellfreier Extrakt
➤ **crude extract** Rohextrakt
➤ **meat extract** Fleischextrakt
extract *vb*
extrahieren, herauslösen
➤ **extract with ether/**
 shake out with ether
ausethern
extraction Extraktion, Herausziehen; Auszug; Ausscheidung, Gewinnung
extraction flask Extraktionskolben

extraction forceps
Extraktionszange (Zähne)
extraction thimble Extraktionshülse
extractive distillation
Extraktivdestillation,
extrahierende Destillation
extrapolate
extrapolieren (hochrechnen)
extremely flammable (F+)
hochentzündlich
extremely toxic (T+)
sehr giftig
extrude
extrudieren; spritzen;
strangpressen
extrusion Extrusion, Extrudieren;
Ausstoßen, Spritzen;
Extrusionsverfahren, Strangpressen
extrusion die
Extrudierdüse, Extruderdüse,
Pressdüse

extrusion molding
Extrudieren, Strangpressen
exudate/exudation/secretion
Exsudat, Absonderung, Abscheidung
exude/secrete/discharge
absondern, abscheiden
(Flüssigkeiten)
eyepiece/ocular Okular
  ➢ **pointer eyepiece** *micros*
  Zeigerokular
  ➢ **spectacle eyepiece/**
  high-eyepoint ocular *micros*
  Brillenträgerokular
eyepiece diaphragm/
eyepiece field stop/
ocular diaphragm *micros*
Okularblende,
Gesichtsfeldblende des Okulars
eyesight Sehkraft, Sehvermögen
eye-wash (station/fountain)
Augendusche

fabric/cloth/tissue Gewebe
fabricate fabrizieren, fertigen; (faking/falsification) fälschen, 'erfinden'
fabricated data gefälschte Daten
fabrication Fabrikation, Fertigung; (faking/falsification) Fälschung, 'Erfindung'
face mask Gesichtsmaske
face seal (stirrer/impeller shaft) Gleitringdichtung (Rührer)
face value Nennwert, Nominalwert
face velocity (not same as 'air speed' at face of hood) Lufteintrittsgeschwindigkeit, Einströmgeschwindigkeit (Sicherheitswerkbank)
faceshield Gesichtsschutz, Gesichtsschirm
facies Fazies
FACS (fluorescence-activated cell sorting) FACS (fluoreszenzaktivierte Zelltrennung, Zellsortierung)
factory (plant/manufacturing plant) Werk, Fabrik
facultative/optional fakultativ
fade verblassen, ausbleichen *vb* (*passiv*/z.B. Fluoreszenzfarbstoffe) (*siehe* bleichen)
fading Verblassen, Ausbleichen *n* (*passiv*/z.B. Fluoreszenzfarbstoffe) (*siehe* bleichen)
faint (become unconscious/ pass out/black out) ohnmächtig werden, in Ohnmacht fallen
fake *vb* **(falsify/forge/fabricate)** fälschen
fall ill (get sick/sicken/ contract a disease) erkranken
falling liquid film Rieselfilm
false (spurious) falsch
false report Fehlermeldung, Falschmeldung
false-positive (false-negative) falschpositiv (falschnegativ)

fan (blower/ventilator) Lüfter, Ventilator; Fächer
Faraday cage Faradaykäfig
fasciation Fasziation, Verbänderung
fast *vb* fasten
fast-atom bombardment (FAB) *spectr* Beschuss mit schnellen Atomen (MS)
fast-growing/rapid-growing schnellwachsend
fast-scanning detector (FSD)/ fast-scan analyzer Schnellscan-Detektor
fasten (to) befestigen, fest machen, anschließen, verbinden
fastener Verschluss, Klemme; Halter, Schließer
fasting Fasten
fat Fett
fat droplet Fetttröpfchen, Fett-Tröpfchen
fat-soluble fettlöslich
fat storage/fat reserve Fettspeicher, Fettreserve
fatigue *vb* **(tire/become tired)** ermüden
fatigue *n* **(tiring)** Ermüdung
  ➤ **material fatigue** Materialermüdung
fatten/cram/stuff mästen (z.B. Geflügel)
fatty fettig; Fett..., fettartig, fetthaltig
fatty acid Fettsäure
  ➤ **monounsaturated fatty acid** einfach ungesättigte Fettsäure
  ➤ **polyunsaturated fatty acid** mehrfach ungesättigte Fettsäure
  ➤ **saturated fatty acid** gesättigte Fettsäure
  ➤ **unsaturated fatty acid** ungesättigte Fettsäure
faucet Wasserhahn, Zapfen (z.B. Fass~); Muffe (Röhrenleitung)
fecal matter (incl. urin) (*see:* Fäzes, Kot) Fäkalien (Kot & Harn)
feces Kot, Fäkalien
  ➤ **human feces** Stuhl
fecund/prolific fruchtbar, produktiv
fecundity Fekundität, Fruchtbarkeit

fed-batch culture/fedbatch culture
Zulaufkultur, Fedbatch-Kultur
(semi-diskontinuierlich)
fed-batch process/fed-batch procedure
Zulaufverfahren, Fedbatch-Verfahren
(semi-diskontinuierlich)
fed-batch reactor/fedbatch reactor
Fedbatch-Reaktor,
Fed-Batch-Reaktor, Zulaufreaktor
feed *n* Futter; (inlet) Zuleitung
feed *vb* füttern;
(feed on something/ingest)
etwas zu sich nehmen, fressen,
sich von etwas ernähren, leben von
feed pump *tech* Förderpumpe
feed tube *tech* Zulaufschlauch
feedback Rückkopplung
feedback inhibition/
 end-product inhibition
negative Rückkopplung,
Rückkopplungshemmung,
Endprodukthemmung
feedback loop Rückkopplungsschleife
feedback system/
 feedback control system Regelkreis
feeding/nourishing Fütterung, Füttern
(z.B. eines Tieres); Ernährung
feel (sense/perceive) empfinden,
fühlen, spüren; (touch/palpate) tasten
feeler gage/feeler gauge (*Br***)**
Fühlerlehre
feeling/sensation Gefühl
Fehling's solution Fehlingsche Lösung
feign death/play dead totstellen
felt-tip pen/felt-tipped pen
Filzstift, Filzschreiber
felty/felt-like/tomentose filzig
female weiblich;
tech/mech Hülse, Minus...
female joint Minus-Verbindung
(Rohrverbindungen etc.)
female Luer hub (lock) Luerhülse
ferment *vb*
fermentieren, gären, vergären
➢ **bottom fermenting (beer brewing)**
untergärig
➢ **top fermenting (beer brewing)**
obergärig

fermentation Fermentation,
Gärung, Vergärung
➢ **anaerobic fermentation**
anaerobe Dissimilation,
anaerobe Gärung
➢ **bottom fermenting** untergärig
➢ **heterolactic fermentation**
heterofermentative Milchsäuregärung
➢ **homolactic fermentation**
homofermentative Milchsäuregärung
➢ **lactic acid fermentation/**
 lactic fermentation
Laktatgärung, Milchsäuregärung
fermentation chamber reactor
 (compartment reactor/
 cascade reactor/stirred tray reactor)
Rührkammerreaktor
fermentation tank Gärbottich
fermentation tube/bubbler
Gärröhrchen, Einhorn-Kölbchen
fermenter/fermentor Fermenter,
Gärtank (*siehe auch:* Reaktor)
Fernbach flask Fernbachkolben
ferrule *chromat*
Dichtkonus, Schneidring
fertile fertil, fruchtbar,
fortpflanzungsfähig
➢ **infertile/sterile**
unfruchtbar, steril
fertility Fertilität, Fruchtbarkeit,
Fortpflanzungsfähigkeit
➢ **infertility/sterility**
Unfruchtbarkeit, Sterilität
fertilization Düngung
fertilize (fecundate) fruchtbar machen,
befruchten; (manure) düngen
fertilizer/plant food/manure
Dünger, Düngemittel
ferulic acid Ferulasäure
fetal calf serum (FCS)
fetales Kälberserum
fetid/smelly/smelling bad/
 malodorous/stinking
übelriechend, stinkend
fetotoxic fetotoxisch
fiber Faser
➢ **dietary fiber** Ballaststoffe (diätätisch)
➢ **hollow fiber** Hohlfaser

fi

fiber optic illumination
Kaltlichtbeleuchtung
fiber optics Fiberoptik,
Faseroptik, Glasfaseroptik
fiberboard Faserstoffplatte
fiberglass
Fiberglas, Faserglas, Glasfaser
fiberscope Fibroskop,
Faserendoskop, Fiberendoskop
fibrous (stringy) faserig, fasrig
Fick diffusion equation
Ficksche Diffusionsgleichung
field capacity/
  field moisture capacity/
  capillary capacity
  Feldkapazität (Boden)
field desorption (FD)
Felddesorption (FD)
field diaphragm *opt/micros*
Feldblende, Leuchtfeldblende,
Kollektorblende
field inversion gel electrophoresis (FIGE)
Feldinversions-Gelelektrophorese
field ionization Feldionisation
field lens *micros* Feldlinse
field of vision/field of view/
  scope of view/range of vision/
  visual field
  Gesichtsfeld, Sehfeld, Blickfeld
field representative/field rep
Außendienstmitarbeiter
field stop (a field diaphragm) *micros*
Sehfeldblende, Gesichtsfeldblende
field study/field investigation/field trial
Freilanduntersuchung,
Freilandversuch, Feldversuch,
vor-Ort-Untersuchung
figure/design Maserung, Fladerung
filament (thread) Filament, Faden;
(of light bulb etc.) Glühwendel
filament tape Filamentband
file *n* Ordner, Akte; Liste, Verzeichnis;
Datei; (tool) Feile
  ➢ **needle file** Nadelfeile
filings (metal) Feilspäne (Metall~)
fill level Füllstand
  (z.B. Flüssigkeit eines Gefäßes)

fill up auffüllen
filler Füllstoff
  (auch: Füllmaterial/Verpackung)
filling Schüttung
filling funnel Fülltrichter
film badge *rad* Strahlenschutzplakette
film reactor Filmreaktor
film water/retained water Haftwasser
film wrap (transparent film/foil)
Klarsichtfolie (Einwickelfolie)
filter *vb* (pass through) filtrieren,
passieren; (percolate/strain) kolieren
filter *n* Filter
  ➢ **ashless quantitative filter**
    aschefreier quantitativer Filter
  ➢ **ceramic filter** Tonfilter
  ➢ **cut-off filter** Sperrfilter
  ➢ **exciter filter** Erregerfilter
    (Fluoreszenzmikroskopie)
  ➢ **folded filter/plaited filter/fluted filter**
    Faltenfilter
  ➢ **fritted glass filter** Glasfritte
  ➢ **HEPA-filter (high-efficiency particulate and aerosol air filter)**
    HOSCH-Filter
    (Hochleistungsschwebstoffilter)
  ➢ **membrane filter** Membranfilter
  ➢ **noise filter** Rauschfilter
  ➢ **particle filter** Partikelfilter
  ➢ **polarizing filter/polarizer**
    Polarisationsfilter, 'Pol-Filter',
    Polarisator
  ➢ **prefilter** Vorfilter
  ➢ **pressure filter** Druckfilter
  ➢ **ribbed filter/fluted filter** Rippenfilter
  ➢ **rotary vacuum filter**
    Vakuumdrehfilter,
    Vakuumtrommeldrehfilter
  ➢ **round filter/filter paper disk/'circles'**
    Rundfilter
  ➢ **selective filter/barrier filter/**
    stopping filter/selection filter *micros*
    Sperrfilter
  ➢ **sterile filter** Sterilfilter
  ➢ **suction filter/suction funnel/**
    vacuum filter (Buechner funnel)
    Filternutsche,
    Nutsche (Büchner-Trichter)

fi

- **syringe filter**
  Spritzenvorsatzfilter, Spritzenfilter
- **trickling filter (sewage treatment)**
  Tropfkörper (Tropfkörperreaktor, Rieselfilmreaktor)
- **filter adapter** Filterstopfen; (Guko) Filtermanschette, Guko
- **filter aid**
  Filterhilfsmittel, Filtrierhilfsmittel
- **filter cake/filtration residue/sludge**
  Filterkuchen, Filterrückstand
- **filter cartridge** Filterkartusche
- **filter crucible** Filtertiegel
- **filter disk** Filterblättchen
- **filter disk method**
  Filterblättchenmethode
- **filter enrichment**
  Anreicherung durch Filter, Filteranreicherung
- **filter feeder** Filtrierer, Filterer
- **filter flask/filtering flask/vacuum flask**
  Filtrierkolben, Filtrierflasche, Saugflasche
- **filter funnel/suction funnel/ suction filter/vacuum filter**
  Filternutsche, Nutsche
- **filter holder** *micros* Filterträger
- **filter mask** Filtermaske
- **filter paper** Filterpapier
- **filter paper disk/round filter/'circles'**
  Rundfilter
- **filter press** Filterpresse
- **filter pump** Filterpumpe
- **filter screen** Filterblende (Schirm)
- **filtering** Filtrierung, Filtrieren
- **filtering rate** Filtrierrate, Filtrationsrate
- **filtrate** *n* Filtrat
- **filtrate** *vb* filtrieren, klären
- **filtration**
  Filtration, Filtrierung, Klärung
- **clarifying filtration** Klärfiltration
- **cross-flow filtration**
  Kreuzstrom-Filtration, Querstromfiltration
- **dead-end filtration** Kuchenfiltration
- **gravity filtration**
  Schwerkraftsfiltration (gewöhnliche F.)

- **pressure filtration** Druckfiltration
- **sterile filtration** Sterilfiltration
- **suction filtration** Saugfiltration
- **ultrafiltration** Ultrafiltration
- **vacuum filtration/suction filtration**
  Vakuumfiltration
- **filtration residue/filter cake/sludge**
  Filterrückstand, Filterkuchen
- **final image** *micros* Endbild
- **findings/result** Befund
- **fine** *n* Bußgeld
- **fine adjustment**
  (fine focus adjustment) *micros*
  Feinjustierung, Feineinstellung
- **fine adjustment knob** *micros*
  Feinjustierschraube, Feintrieb; (micrometer screw) Mikrometerschraube
- **fine chemicals** Feinchemikalien
- **fine dust/mist** Feinstaub (alveolengängig)
- **fine structure** Feinstruktur, Feinbau
- **finger cot** Fingerling (Schutzkappe)
- **fingerprint** Fingerabdruck
- **fingerprinting/**
  genetic fingerprinting/
  DNA fingerprinting
  Fingerprinting, genetischer Fingerabdruck
- **finish** *n* (in painting) Deckanstrich; (coating of a surface) Verputz (innen/außen)
- **fire** *vb* (firing) feuern
- **fire** *n* Feuer (*siehe auch:* Flamm...)
- **put out a fire/quench a fire**
  Feuer löschen
- **fire alarm** Feueralarm; Feuermelder
- **fire axe** Brandaxt
- **fire blanket**
  Löschdecke, Feuerlöschdecke
- **fire brigade/fire department** Feuerwehr
- **fire classification** Brandarten
- **fire code** Feuerschutzvorschriften
- **fire control** Brandschutz
- **fire drill** Feueralarmübung, Feuerwehrübung
- **fire engine/fire truck**
  Feuerlöschfahrzeug

fi

fire extinguisher
Feuerlöscher, Feuerlöschgerät, Löschgerät
fire-extinguishing agent
Löschmittel, Feuerlöschmittel
fire fighting
Brandbekämpfung, Feuerbekämpfung
fire foam Feuerlöschschaum
fire hazard Brandrisiko, Feuergefahr
fire hose Feuerwehrschlauch
fire-polished feuerpoliert
fire protection (fire prevention)
Feuerschutz
fire protection association (U.S.: National Fire Protection Association NFPA)
Feuerwehrvereinigung
fire resistance class
Feuerwiderstandsklasse
fire-resistant feuerbeständig
fire-retardant/flame-retardant
feuerhemmend, flammenhemmend
fire sprinkler system
Sprinkleranlage (Beregnungsanlage, Berieselungsanlage: Feuerschutz)
fire wall/fire barrier Brandmauer, Feuerschutzwand
fireclay Schamotte
firefighter/fireman Feuerwehrmann
fireproof feuerfest, feuersicher
fireproofing agent/fire retardant
Feuerschutzmittel
first aid
Erste Hilfe, Erstbehandlung, Nothilfe
first-aid attendant/nurse Sanitäter
first-aid box/first-aid kit
Verbandskasten
first-aid cabinet/medicine cabinet
Erste-Hilfe-Kasten, Verbandsschrank, Medizinschrank, Medizinschränkchen
first-aid kit
Erste-Hilfe-Kasten, Erste-Hilfe-Koffer, Sanitätskasten
first-aid supplies
Erste-Hilfe Ausrüstung

first-aider Ersthelfer
first run/forerun *dist* Vorlauf
Fischer projection/ Fischer formula/ Fischer projection formula
Fischer-Projektion, Fischer-Formel, Fischer-Projektionsformel
FISH (fluorescence activated *in situ* hybridization)
FISH
(*in situ* Hybridisierung mit Fluoreszenzfarbstoffen)
fish hook (in chemical equation)
Halbpfeil
(in chem. Reaktionsgleichungen)
fission Fission, Spaltung; Teilung
fissure
Fissur, Furche, Einschnitt; Riss
fitness (suitability) Fitness, Eignung
fitted
(well fitted) passend (gut passend)
(z.B. Verschluss/Stopfen etc.)
fitter's hammer/ locksmith's hammer
Schlosserhammer
fitting/fittings Passstück, Zubehörteil(e) (Kleinteile/Passteile: an Geräten etc.); Armatur(en) (Hähne im Labor/an der Spüle etc.); (coupler/coupling) Kupplung, Verbinder (z.B. Schlauch); Verbindungsmuffe
(Kupplung: Rohr/ Schlauch etc.)
➢ **compression fitting**
Hochdruck-Steckverbindung
➢ **quick-disconnect fitting**
Schnellkupplung
(z.B. Schlauchverbinder)
fix fixieren (mit Fixativ härten)
fixation Fixierung, Fixieren
fixative Fixiermittel, Fixativ
fixed-angle rotor *centrif*
Festwinkelrotor
fixed bed reactor/solid bed reactor
Festbettreaktor (Bioreaktor)
fixed coupling
starre Kupplung
fixing bolt Haltebolzen

fixture Anschluss, Versorgungsanschluss (Zubehörteil/Armatur); (mounting/support/holding) Halterung
➢ **electrical fixture(s)/electricity outlet** elektrische(r) Anschluss
➢ **service fixtures/service outlets** Versorgungsanschlüsse (Wasser/Strom/Gas)
flakeboard/chipboard Pressspan, Spanplatte
flame *vb* abflammen, 'flambieren' (sterilisieren)
flame *n* Flamme (*siehe auch:* Feuer...)
➢ **pilot flame** Sparflamme
flame arrestor Flammensperre, Flammenrückschlagsicherung
flame atomic emission spectroscopy (FES)/ flame photometry Flammenemissionsspektroskopie (FES)
flame coloration Flammenfärbung
flame-ionization detector (FID) Flammenionisationsdetektor (FID)
flame point/ kindling temperature/ ignition point/flame temperature/ spontaneous-ignition temperature (SIT) Zündpunkt, Zündtemperatur, Entzündungstemperatur
flame-resistant flammbeständig, flammwidrig
flame retardant/flame retarder Flammschutzmittel
flame-retardant *adj/adv* flammenhemmend, feuerhemmend
flame spectroscopy Flammenspektroskopie
flame spectrum Flammenspektrum
flame test Leuchttest, Leuchtprobe
flameproof feuerfest, feuersicher, flammsicher, flammfest (schwer entflammbar)

flammability Entflammbarkeit, Brennbarkeit, Entzündbarkeit
flammable (R10) entzündlich
➢ **inflammable** entflammbar, brennbar
➢ **nonflammable/incombustible** nicht entflammbar, nicht brennbar
flange *vb* flanschen
flange *n* Flansch
➢ **lap-joint flange** Bördelflansch
flange connection/ flange coupling/flanged joint Flanschverbindung
flare *n* (flaring off/burning off) Abfackelung
flare *vb* (flare off/burn off) abfackeln
flare-up Aufflackern, Auflodern, Aufflammen
flash *n* (light/lightning/spark) Blitz, Lichtblitz
flash *vb* blitzen
flash arrestor Flammschutzfilter
flash chromatography Flash-Chromatographie
flash distillation Entspannungs-Destillation, Flash-Destillation
flash photolysis Blitzlichtphotolyse
flash point Flammpunkt
flash vaporizer (GC) Schnellverdampfer
flashlight/torch (*Br*) Taschenlampe
flask Kolben
➢ **culture flask** Kulturkolben
➢ **culture media flask** Nährbodenflasche
➢ **distilling flask/distillation flask/'pot'** Destillierkolben, Destillationskolben
➢ **Erlenmeyer flask** Erlenmeyer Kolben
➢ **evaporating flask** Verdampferkolben
➢ **extraction flask** Extraktionskolben
➢ **filter flask/filtering flask/ vacuum flask** Filtrierkolben, Filtrierflasche, Saugflasche
➢ **Fernbach flask** Fernbachkolben
➢ **Florence boiling flask/Florence flask (boiling flask with flat bottom)** Stehkolben, Siedegefäß
➢ **ground-jointed flask** Schliffkolben

fl

- **Kjeldahl flask**
  Birnenkolben, Kjeldahl-Kolben
- **narrow-mouthed flask/ narrow-necked flask**
  Enghalskolben
- **pear-shaped flask (small/pointed)**
  Spitzkolben
- **recovery flask/ receiving flask/receiver flask (collection vessel)**
  Vorlagekolben
- **rotary evaporator flask**
  Rotationsverdampferkolben
- **round-bottomed flask/ round-bottom flask/ boiling flask with round bottom**
  Rundkolben
- **saber flask/ sickle flask/ sausage flask**
  Säbelkolben, Sichelkolben
- **shake flask** Schüttelkolben
- **sidearm flask** Seitenhalskolben
- **spinner flask**
  Spinnerflasche, Mikroträger
- **suction flask/filter flask/ filtering flask/vacuum flask/ aspirator bottle**
  Saugflasche, Filtrierflasche
- **sulfonation flask** Sulfierkolben
- **swan-necked flask/ S-necked flask/ gooseneck flask**
  Schwanenhalskolben
- **three-neck flask** Dreihalskolben
- **tissue culture flask**
  Gewebekulturflasche, Zellkulturflasche
- **two-neck flask** Zweihalskolben
- **volumetric flask**
  Messkolben, Mischzylinder
- **wide-mouthed flask/ wide-necked flask**
  Weithalskolben

flask brush Kolbenbürste
flask clamp Kolbenklemme;
(four-prong) Federklammer
(für Kolben: Schüttler/Mischer)
flask tongs Kolbenzange
flat-bed gel/horizontal gel *electrophor*
horizontal angeordnetes Plattengel
flat-blade impeller
Scheibenrührer, Impellerrührer
flat-flange ground joint/ flat-ground joint/ plane-ground joint
Planschliff (glatte Enden), Planschliffverbindung
flat-nosed pliers Flachzange
flat plug Flachstecker

- **flat-plug connector**
  Flachsteckverbinder
- **flat-plug socket**
  Flachsteckhülse

flavor *vb* würzen, schmackhaft machen, Geschmack geben
flavor *n* (flavoring)
Geschmackstoff(e);
(pleasant taste) Aroma;
Wohlgeschmack

- **artificial flavor/artificial flavoring**
  künstlicher Geschmackstoff

flavor enhancer
Aromazusatzstoff
flavoring/aromatic substance
Aromastoff
flea/bar magnet/stir bar/ stirrer bar/stirring bar
Magnetstab, Magnetstäbchen, Magnetrührstab, 'Fisch', Rühr'fisch'
fleaker (Corning: Pyrex)
Becherglaskolben
(spezielles Produkt von Corning/Pyrex: mit Ausgussöffnung)
fleece Vlies(stoff)
fleshing knife Fleischmesser
fleshy fleischig
flexibility/pliability Biegsamkeit
flexible/pliable biegsam
flicker flackern, flimmern
flint (flint stone)
Zündstein, Feuerstein, Flintstein, Flint
flint glass Flintglas
flipping (NMR) Spinumkehr

float *n* Schwimmer
  (z.B. am Flüssigkeitsstandregler)
float *vb* **(floating)/**
  suspend (suspended)
  schweben (schwebend)
float switch Schwimmerschalter
floating rack (for ice bath)
  Schwimmständer, Schwimmgestell,
  Schwimmer (für Eiswanne)
flocculate ausflocken
flocculation Flockulation, Ausflockung
flocking Flockung
flood *vb* fluten, gründlich spülen
floor Boden, Fußboden
➢ **monolithic floor**
  monolithischer Fußboden
  (Labor: Stein/Beton aus einem Guß)
floor drain Bodenabfluss, Bodenablauf
floor tile Bodenfliese
Florence boiling flask/Florence flask
  (boiling flask with flat bottom)
  Stehkolben, Siedegefäß
floridean starch Florideenstärke
flour Mehl
flourish/thrive florieren, gedeihen
flow *vb* fließen
flow *n* Fluss, Strom, Strömung
➢ **direction of flow** Fließrichtung
➢ **turbulent flow** turbulente Strömung
flow cytometry Durchflusscytometrie
flow injection Fließinjektion
flow injection analysis (FIA)
  Fließinjektionanalyse (FIA)
flow-injection titration
  Fließinjektions-Titration
flow pattern Strömungsmuster
flow rate Fließgeschwindigkeit;
  Durchflussrate,
  Durchflussgeschwindigkeit;
  Förderleistung, Saugvermögen
  (Pumpe); (volume per time) Strom;
  chromat (mobile-phase velocity)
  Durchlaufgeschwindigkeit (Säule)
flow reactor
  Durchflussreaktor (Bioreaktor)
flow regulator Flussregler
flow resistance/resistance to flow
  Strömungswiderstand

flower *vb* **(bloom)** blühen
flower pot Blumentopf
flowerbed Blumenbeet
flowers of sulfur Schwefelblüte
fluctuate fluktuieren, schwanken
fluctuation Fluktuation, Schwankung
fluctuation analysis/noise analysis
  Fluktuationsanalyse,
  Rauschanalyse
fluctuation of temperature
  Temperaturschwankung
fluctuation test Fluktuationstest
flue dust Flugstaub (von Abgasen)
flue gases/fumes Rauchgase
fluence Flussrate
fluffy flockig, locker
fluid/liquid *adj/adv* flüssig
fluid/liquid *n* Flüssigkeit
fluid bed reactor
  Fließbettreaktor
fluid extract Flüssigextrakt,
  flüssiger Extrakt, Fluidextrakt
fluidity Fluidität, Fließfähigkeit
fluidize verflüssigen; fluidisieren,
  in den Fließbettzustand überführen
fluidized bed reactor/
  moving bed reactor
  Wirbelschichtreaktor,
  Wirbelbettreaktor,
  Fließbettreaktor
fluoresce fluoreszieren
fluorescence Fluoreszenz
fluorescence-activated cell sorter
  fluoreszenzaktivierter Zellsorter,
  Zellsortierer
fluorescence-activated cell sorting
  (FACS)
  fluoreszenzaktivierte Zellsortierung,
  Zelltrennung
fluorescence analysis/fluorimetry
  Fluoreszenzanalyse, Fluorimetrie
fluorescence-*in-situ*-hybridization
  (FISH)
  Fluoreszenz-*in-situ*-Hybridisation
  (FISH)
fluorescence marker
  Fluoreszenzsonde,
  Fluoreszenzmarker

fl

fluorescence photobleaching recovery / fluorescence recovery after photobleaching (FRAP) Fluoreszenzerholung nach Lichtbleichung
fluorescence quenching Fluoreszenzlöschung
fluorescence spectroscopy Fluoreszenzspektroskopie, Spektrofluorimetrie
fluorescent fluoreszierend
fluorescent dye Farbstoff
fluorescent tube Leuchtstoffröhre, Leuchtstofflampe ('Neonröhre')
fluoridation Fluoridierung
fluorinate fluorieren
fluorinated hydrocarbon Fluorkohlenwasserstoff
fluorine (F) Fluor
fluorosulfonic acid Fluoroschwefelsäure, Fluorsulfonsäure
flush *vb* spülen, fluten
flutings (e.g., tube connections) Riffelung, Riefen (z.B. Schlauchverbinder)
flux (light/energy) Fluss (Licht/Energie; Volumen pro Zeit pro Querschnitt); *electr* Strömung
➢ **fluxing agent/fusion reagent** Flussmittel, Schmelzmittel, Zuschlag
fly ash Flugstaub
foam *n* Schaum; (froth) feiner Schaum (z.B. auf Flüssigkeiten); (lather) Seifenschaum
foam *vb* schäumen
foam fire extinguisher Schaumlöscher
foam inhibitor Schaumhemmer, Schaumdämpfer, Schaumverhütungsmittel
foam rubber/plastic foam/foam Schaumgummi
foamed plastic/plastic foam Schaumstoff
foamer/foaming agent Schäumer, Schaumbildner
focal depth Abbildungstiefe
focal length Brennweite
focal plane Brennebene
focal point (focus) Brennpunkt
focus *vb* **(focussing)** fokussieren, scharf stellen
focus *n* Fokus, Brennpunkt, Sammelpunkt
➢ **in focus** *micro/photo* scharf
➢ **not in focus/out of focus/ blurred** *micro/photo* unscharf
focus control (focussing) Scharfeinstellung, Brennpunkteinstellung, Fokussierung
focussing Fokussierung, Bündelung, Sammlung; Scharfeinstellung, Brennpunkteinstellung
fodder/forage (plant) Futterpflanze
fog Nebel
foggy nebelig
fold *n* **(plication/wrinkle)** Falte
fold *vb* **(ply/wrinkle)** falten, zusammenfalten, zusammenlegen
folded (pleated/plicate) gefaltet, faltig
folded filter/plaited filter/fluted filter Faltenfilter
folding rule (tool) Gliedermaßstab
foliate *adj/adv tech* blättrig, blattförmig
foliate *vb tech* mit Folie/Blattmetall belegen
foliated gypsum/selenite/spectacle stone Marienglas (Gips)
food Nahrung, Essen; (feed) Fressen; (diet/nourishment/nutrition) Ernährung, Nahrung, Nahrungsmittel
food additive Lebensmittelzusatzstoff
food chain *ecol* Nahrungskette
food chemistry Lebensmittelchemie
food crop (forage plant/food plant) Nahrungspflanze
food inspection Lebensmittelüberwachung, Lebensmittelkontrolle
food poisoning Lebensmittelvergiftung, Nahrungsmittelvergiftung

food preservation
Lebensmittelkonservierung, Nahrungsmittelkonservierung
food preservative
Lebensmittelkonservierungsstoff
food quality control
Lebensmittelkontrolle, Lebensmittelprüfung
food quantity Nahrungsmenge
food shortage Nahrungsmangel
food source/nutrient source
Nahrungsquelle
food value/nutritive value Nährwert
food web *ecol*
Nahrungsgefüge, Nahrungsnetz
foodstuff/nutrients Lebensmittel, Nahrungsmittel, Nahrung, Nährstoffe
footprinting Fußabdruckmethode
forbidden (prohibited)
verboten, untersagt
➤ **strictly forbidden/strictly prohibited**
strengstens verboten
force microscopy (FM)
Kraftmikroskopie
force open *vb* gewaltsam öffnen
forced air/
recirculating air (air circulation)
Umluft
forced-air oven Umluftofen
forced-draft hood Saugluftabzug
forceps Pinzette, Zange, Klemme; (*siehe:* tweezers)
➤ **artery forceps/**
artery clamp (hemostat)
Arterienklemme
➤ **bone-cutting forceps/**
bone-cutting shears
Knochenzange
➤ **cartilage forceps**
Knorpelpinzette
➤ **cover glass forceps**
Deckglaspinzette
➤ **dissection tweezers/**
dissecting forceps
Sezierpinzette, anatomische Pinzette
➤ **extraction forceps** *dent*
Extraktionszange
➤ **hemostatic forceps/artery clamp**
Gefäßklemme, Arterienklemme, Venenklemme
➤ **membrane forceps**
Membranpinzette
➤ **microdissection forceps/**
microdissecting forceps
anatomische Mikropinzette, Splitterpinzette
➤ **sponge forceps** Tupferklemme
➤ **tissue forceps** Gewebepinzette
➤ **watchmaker forceps/**
jeweler's forceps
Uhrmacherpinzette
forcing bed/hotbed *bot/agr*
Frühbeet, Mistbeet, Treibbeet (beheizt)
forensic
forensisch, gerichtsmedizinisch
forensics/forensic medicine
Forensik, forensische Medizin, Gerichtsmedizin, Rechtsmedizin
forerun/foreshot (alcohol) *dist*
Vorlauf (Destillation)
forklift Gabelstapler, Hubstapler
formaldehyde/methanal
Formaldehyd, Methanal
formic acid (formate)
Ameisensäure (Format)
formula Formel; *pharm* Rezeptur
➤ **chain formula/**
open-chain formula
Kettenformel
➤ **empirical formula**
empirische Formel, Summenformel, Elementarformel, Verhältnisformel
➤ **ionic formula** Ionenformel
➤ **molecular formula**
Molekularformel, Molekülformel
➤ **ring formula** Ringformel
➤ **structural formula**
Strukturformel
formulary Formelsammlung; (pharmacopoeia) Pharmakopöe, Arzneimittel-Rezeptbuch, amtliches Arzneibuch
fossil fuel(s) fossile(r) Brennstoff(e)

fossilization
Fossilisierung, Versteinerung
fossilize fossilisieren, versteinern
foul *adj/adv*
(rotten/decaying/decomposing)
faul, modernd;
(decomposed/decayed) verfault, zersetzt
foul *vb* (rot/decompose/decay)
verfaulen, zersetzen
fountain (for drinking water)
Trinkbrunnen, Trinkfontäne
fraction Fraktion
fraction collector Fraktionssammler
fraction cutter Wechselvorlage
fractional precipitation
fraktionierte Fällung
fractionate fraktionieren
fractionating column Fraktioniersäule
fractionation Fraktionierung
fragile zerbrechlich;
(handle with care!)
Vorsicht, zerbrechlich!
fragment Bruchstück, Fragment
fragment ion (MS) Bruchstückion
fragmentation pattern
Fragmentierungsmuster
fragrance (scent/pleasant smell)
angenehmer Duft/Geruch;
(perfume: stronger scent)
angenehmer Geruchsstoff
fragrant duftend (angenehm)
frame raster mapping/
grid mapping *ecol/biogeo*
Rasterkartierung
frameshift *gen* Rasterverschiebung
fraud Betrug, Schwindel,
arglistige Täuschung
free-floating/pendulous frei schwebend
free from streaks/free from reams
schlierenfrei
free from water
(moisture-free/anhydrous)
wasserfrei
free-living freilebend
free radical freies Radikal
freeze einfrieren, gefrieren, erstarren
➤ **quickfreeze** schnellgefrieren

freeze-dry/lyophilize
gefriertrocknen, lyophilisieren
freeze-drying/lyophilization
Gefriertrocknung, Lyophilisierung
freeze-etch gefrierätzen
freeze-etching Gefrierätzung
freeze-fracture/freeze-
fracturing/cryofracture *micros*
Gefrierbruch
freeze preservation/cryopreservation
Gefrierkonservierung,
Kryokonservierung
freeze storage Gefrierlagerung
freezer Gefriertruhe, Gefrierschrank
➤ **upright freezer** Gefrierschrank
➤ **chest freezer** Gefriertruhe
freezer compartment/
freezing compartment/'freezer'
Kühlfach, Gefrierfach
(im Kühlschrank)
freezing Gefrieren, Frieren, Einfrieren
➤ **rapid freezing** Schnellgefrieren
freezing microtome/cryomicrotome
Gefriermikrotom
freezing point Gefrierpunkt
freezing point depression
Gefrierpunktserniedrigung
freight (load/cargo/goods) Fracht
freight company/shipping company/
shipper Spedition
freight container Frachtcontainer
frequency Frequenz; (of occurrence;
abundance) Häufigkeit
frequency distribution (FD)
Häufigkeitsverteilung
frequency histogram
Häufigkeitshistogramm
frequency ratio *stat*
relative Häufigkeit
frequent/abundant häufig
fresh mass (fresh weight)
Frischmasse (Frischgewicht)
freshwater Süßwasser
frit *n* Fritte
frit *vb* (sinter) fritten (sintern)
fritted glass Sinterglas
fritted glass filter Glasfritte
frock Kittel, Arbeitskittel

front side (front/face)
  Vorderseite (Gerät etc.)
front-side *adj/adv* **(ventral)**
  vorderseitig (bauchseitig)
fronting *chromat* Bartbildung,
  Signalvorlauf, Bandenvorlauf
frost (rime frost/white frost) Frost
frost damage/
  frost injury/freezing injury
  Frostschaden, Frostschädigung
frost-resistant/frost hardy
  frostbeständig, frostresistent
frost-tender/susceptible to frost
  frostempfindlich
frosted-end slide *micros*
  Mattrand-Objektträger
frostproof frostsicher
frozen gefroren
frozen section Gefrierschnitt
fruit essence Fruchtessenz
fruit press/juice press
  (e.g., for making juice) Kelter
fruit pulp
  Fruchtmark, Obstpulpe, Fruchtmus
fruit sugar/fructose
  Fruchtzucker, Fruktose
fruity taste Fruchtgeschmack
fucose/6-deoxygalactose
  Fukose, Fucose,
  6-Desoxygalaktose
fuel *vb*
  auftanken, mit Brennstoff versehen
fuel *n*
  Treibstoff, Brennstoff, Kraftstoff;
  Benzin
  ➢ **fossil fuel** fossiler Brennstoff
fuel cell Brennstoffzelle
fuel consumption Kraftstoffverbrauch
fuel equivalence
  Brennäquivalent;
  Brennstoffäquivalenz
fuel injection Kraftstoffeinspritzung
fueled *adj* **(powered)**
  betrieben, getrieben
fulcrum *phys*
  Drehpunkt, Gelenkpunkt;
  Angelpunkt
fulcrum pin Drehbolzen, Drehzapfen

full blast
  voll aufdrehen (Wasserhahn etc.)
full-face respirator
  Atemschutzvollmaske,
  Gesichtsmaske
full-facepiece respirator
  Vollsicht-Atemschutzmaske
full-mask (respirator)
  Vollmaske, Atemschutz-Vollmaske
full width at half-maximum (fwhm)/
  half intensity width *math/stat*
  Halbwertsbreite
fumaric acid (fumarate)
  Fumarsäure (Fumarat)
fume *vb* rauchen, dampfen,
  Dämpfe von sich geben
fume cupboard (*Br***)**
  Abzug, Dunstabzugshaube
fume extraction
  Rauchabzug (Raumentlüftung)
fume hood/hood Rauchabzug,
  Dunstabzugshaube, Abzug
fumes *n* Dämpfe, Dunst, Rauch;
  (flue gases) Rauchgase
  ➢ **exhaust fumes** Abgase
fumigant Ausräucherungsmittel
fumigate begasen
fumigation Begasung
fuming *adj/adv* **(acid)** rauchend (Säure)
fuming *n* Abrauchen
function Funktion
  ➢ **distribution function**
  Verteilungsfunktion
functional group *chem*
  funktionelle Gruppe
functional unit/module
  Funktionseinheit, Modul
functionality Funktionalität
fund(s) Gelder, Geldmittel
  (z.B. für Forschung)
funding/financing
  Finanzierung, zur Verfügungstellung
  von Geldmitteln
fungal infestation Pilzbefall
fungicide Fungizid,
  Pilzbekämpfungsmittel;
  (treatment of seeds) Beizmittel
  (zur Saatgutbehandlung)

fu

funnel Trichter
- addition funnel Zulauftrichter
- analytical funnel Analysentrichter
- dropping funnel Tropftrichter
- filling funnel Fülltrichter
- filter funnel/suction funnel/
  suction filter/vacuum filter
  Filternutsche, Nutsche
- Hirsch funnel Hirsch-Trichter
- hot-water funnel (double-wall funnel)
  Heißwassertrichter
- powder funnel Pulvertrichter
- separatory funnel Scheidetrichter
- short-stem funnel/
  short-stemmed funnel
  Kurzhalstrichter, Kurzstieltrichter
- suction funnel/suction filter/
  vacuum filter (Buechner funnel)
  Filternutsche,
  Nutsche (Büchner-Trichter)

funnel brush Trichterbürste
funnel tube Trichterrohr
furnace
  Ofen; Hochofen, Schmelzofen;
  Heizkessel
- annealing furnace Glühofen
- arc furnace Lichtbogenofen
- blast furnace Hochofen
- combustion furnace
  Verbrennungsofen
- crucible furnace Tiegelofen
- heating furnace Wärmeofen
- induction furnace/
  inductance furnace
  Induktionsofen
- muffle furnace Muffelofen
- reverberatory furnace
  Flammofen
- roasting furnace/
  roasting oven/roaster
  Röstofen
- smelting furnace Schmelzofen
- tube furnace Rohrofen

furnishings Ausstattung, Mobiliar, Einrichtung (Möbel etc.), Einrichtungsgegenstände
fuse *vb* fusionieren, verschmelzen
fuse *n* Zündschnur;
  (circuit breaker *electr*) Sicherung
- blow/kick a fuse
  Sicherung 'rausfliegen' lassen

fuse box/fuse cabinet/cutout box
  Sicherungskasten
fused (fusible/molten) schmelzflüssig;
  (coalescent) verwachsen,
  angewachsen
fused quartz Quarzgut
fused-salt electrolysis
  Schmelzelektrolyse
fusel oil Fuselöl
fusible schmelzbar
fusible plug *electr* Schmelzplombe
fusible wire *electr* Schmelzdraht
fusion Fusion, Verschmelzung
fusion point (melting point)
  Fließpunkt (Schmelzpunkt)
fusion tube/melting tube
  Abschmelzrohr
futile cycle *biochem*
  Leerlauf-Zyklus, Leerlaufcyclus

gage/gauge (*Br*) Messgerät
gain *n* (increase/increment)
Zunahme, Steigerung, Vergrößerung
gain *vb* (increase)
zunehmen, steigern, vergrößern
galactose Galaktose
galactosemia Galaktosämie
galacturonic acid Galakturonsäure
galena PbS Bleiglanz
gallic acid Gallussäure
gallipot Salbentopf, Medikamententopf (Apotheke)
gamete/sex cell
Gamet, Keimzelle, Geschlechtszelle
gap Lücke
gape klaffen, offen stehen
garbage Müll, Abfall
garbage can/dustbin (*Br*)
Mülleimer
garbage chute/waste chute
Müllschacht
garden hose Gartenschlauch
gas Gas; (gasoline) Benzin
➢ **carrier gas**
(an inert gas) *chromat* **(GC)**
Trägergas, Schleppgas
➢ **compressed gas/pressurized gas**
Druckgas
➢ **evolution of gas** Gasentwicklung
➢ **flue gases/fumes** Rauchgase
➢ **inert gas/rare gas** Edelgas
➢ **irritant gas** Reizgas
➢ **laughing gas/nitrous oxide**
Lachgas, Distickstoffoxid, Dinitrogenoxid
➢ **liquid gas/liquefied gas**
Flüssiggas
➢ **natural gas** Erdgas
➢ **producer gas** Generatorgas
➢ **protective gas/shielding gas (in welding)** Schutzgas
➢ **purge gas** Spülgas
➢ **reference gas (GC)** Vergleichsgas
➢ **sewer gas** Faulschlammgas
➢ **sludge gas/sewage gas**
Faulgas, Klärgas (Methan)
➢ **tracer gas/probe gas** Prüfgas
➢ **water gas** Wassergas

gas balance Gaswaage
gas bottle (gas cylinder/ compressed-gas cylinder)
Gasflasche
gas bottle cart/gas cylinder trolley (*Br*)
Gasflaschen-Transportkarren
gas bubble Gasblase
gas burner Gasbrenner, Gaskocher
gas chromatography
Gaschromatographie
gas cock/gas tap Gashahn
gas collecting tube/ gas sampling bulb/ gas sampling tube
Gassammelrohr, Gasprobenrohr
gas constant Gaskonstante
gas counter Gaszählrohr
gas cylinder Druckgasflasche
gas cylinder pressure regulator
Gasdruckreduzierventil, Druckminderventil, Druckminderungsventil, Reduzierventil (für Gasflaschen)
gas detector Gasdetektor, Gasspürgerät
gas-discharge tube
Gasentladungsröhre
gas exchange/ gaseous interchange/ exchange of gases
Gasaustausch
gas flowmeter
Gasdurchflusszähler, Gasströmungsmesser
gas leak/gas leakage Gasleck
gas leak detector
Gasdetektor, Gasspürgerät
gas lighter Gasanzünder
gas line (natural gas line)
Gasleitung (Erdgasleitung)
gas-liquid chromatography
Gas-Flüssig-Chromatographie
gas mask Gasmaske
gas measuring bottle Gasmessflasche
gas outlet Gasaustritt, Gasausgang, Gasabgang (aus Geräten)
gas poisoning Gasvergiftung
gas purifier Gasreiniger

ga

gas regulator/
  gas cylinder pressure regulator
  Gasdruckreduzierventil,
  Druckminderventil,
  Druckminderungsventil,
  Reduzierventil (für Gasflaschen)
gas sampling bulb/
  gas sampling tube
  Gassammelrohr,
  Gas-Probenrohr, Gasmaus
gas scrubbing Gaswäsche
gas separator Gasabscheider
gas supply Gaszufuhr
gas tap Gashahn
gas thermometer Gasthermometer
gas washing bottle Gaswaschflasche
gaseous gasförmig, gasartig, Gas...
gaseous state
  Gaszustand, gasförmiger Zustand
gasket Dichtung, Dichtungsmanschette
➤ **rubber gasket**
  Gummidichtung(sring)
gasohol Treibstoffalkohol, Gasohol
gasoline Benzin, Kraftstoff
gasoline canister
  Benzinkanister, Kraftstoffkanister
gasproof gasdicht
gastight (impervious to gas)
  gasundurchlässig
gastric lavage/gastric irrigation
  Magenspülung
gate Gitter, Tor, Pforte;
  Sperre, Schranke;
  electrophor (gel-casting)
  Verschluss-Scheibe
gate impeller Gitterrührer
gated ion channel Ionenschleuse
gating current *neuro*
  Torstrom (*pl* Torströme)
gauntlets Schutzhandschuh
➤ **sleeve gauntlets**
  Ärmelschoner, Stulpen
Gaussian curve
  Gauß-Kurve, Gauß'sche Kurve
Gaussian distribution (Gaussian curve/normal probability curve)
  Gauß-Verteilung, Normalverteilung,
  Gauß'sche Normalverteilung

gauze Gaze
gauze bandage Mullbinde, Gazebinde
GC (gas chromatography)
  GC (Gaschromatographie)
gear pump Zahnradpumpe
Geiger counter Geiger-Zähler
Geiger-Müller counter
  Geiger-Müller-Zähler
gel *vb* gelieren
gel *n* Gel
➤ **denaturing gel** denaturierendes Gel
➤ **flat bed gel/horizontal gel**
  horizontal angeordnetes Plattengel
➤ **native gel** natives Gel
➤ **running gel/separating gel** Trenngel
➤ **slab gel**
  Plattengel (hochkant angeordnetes)
➤ **stacking gel** *electrophor* Sammelgel
gel caster *electrophor*
  Gelgießstand, Gelgießvorrichtung
gel chamber *electrophor* Gelkammer
gel comb *electrophor* Gelkamm
gel electrophoresis Gelelektrophorese
gel filtration/
  molecular sieving chromatography/
  gel permeation chromatography
  Gelfiltration,
  Molekularsiebchromatographie,
  Gelpermeations-Chromatographie
gel permeation chromatography/
  molecular sieving chromatography
  Gelpermeationschromatographie,
  Molekularsiebchromatographie
gel retention analysis/band shift assay
  Gelretentionsanalyse
gel retention assay/
  electrophoretic mobility shift assay
  (EMSA) Gelretentionstest
gel-sol-transition Gel-Sol-Übergang
gel tray *electrophor*
  Gelträger, Geltablett
gelatin/gelatine Gelatine
gelatinizing agent Gelbildner
gelatinous/gel-like
  gallertartig, gelartig, gelatinös
gelation Gelieren
gelling agent Geliermittel
gelling point Gelierpunkt

geminal coupling (NMR)
   geminale Kopplung
gene Gen, Erbfaktor
gene dosage effect Gendosiseffekt
gene manipulation Genmanipulation
gene map/genetic map Genkarte
gene mapping/genetic mapping
   Genkartierung
gene technology Gentechnologie,
   Gentechnik, Genmanipulation
gene therapy/gene surgery
   Gentherapie
generate (develop)
   bilden (entwickeln)
   (z.B. Gase, Dämpfe)
generation Generation; Erzeugung
generation period Generationsdauer
generation time (doubling time)
   Generationszeit (Verdopplungszeit)
generator Erzeuger, Produzent
generic drug
   Generica, Generika,
   Fertigarzneimittel
generic name
   Sammelbegriff, Sammelname;
   ungeschützter Name (einer Substanz)
genetic analysis Erbanalyse
genetic counsel(l)ing
   genetische Beratung
genetic diagnostics/genotyping
   Gendiagnostik,
   Bestimmung des Genotyps
genetic engineering/gene technology
   Gentechnik, Gentechnologie,
   Genmanipulation
genetic hazard Erbschaden,
   genetischer Schaden
genetic screening genetischer Suchtest
genetically engineered
   gentechnisch verändert
genetically engineered organism/
   genetically modified organism
   (GMO)
   gentechnisch veränderter
   Organismus (GVO)
genetically modified microorganism
   (GMM) gentechnisch veränderter
   Mikroorganismus

genetics/transmission genetics
   (study of inheritance)
   Genetik, Vererbungslehre
genome Genom
genomic blotting genomisches Blotting
gentisic acid Gentisinsäure
gentle/mild (careful)
   schonend (vorsichtig)
genus (*pl* genera) Gattung
geranic acid Geraniumsäure
germ Keim (Mikroorganismus);
   (embryo) Keim, Keimling, Embryo
germ count Keimzahl
➢ **total germ count/**
   total cell count
   Gesamtkeimzahl
germ-free (sterile) keimfrei (steril)
germinability Keimfähigkeit
germinate (sprout) keimen
➢ **pregerminate** vorkeimen
germination Keimung
➢ **dark germination** Dunkelkeimung
germination percentage
   Keimzahl (Samenkeimung)
gibberellic acid Gibberellinsäure
gilding vergolden
gimlet bit
   Schneckenbohrer,
   Windenschneckenbohrer
girdle/cingulum
   Gürtel, Gurt, Cingulum
girth Umfang
glacial acetic acid Eisessig
glass Glas
➢ **acrylic glass** Acrylglas
➢ **alkali-lime glass**
   (e.g., soda-lime glass)
   Kalk-Alkali-Glas
➢ **amber glass** Braunglas
➢ **bits of broken glass** Glassplitter
➢ **borosilicate glass** Borosilikatglas
➢ **bulletproof glass** Panzerglas
➢ **clear glass** Klarglas
➢ **crown glass** Kronglas
➢ **fiberglass**
   Fiberglas, Glasfaser, Faserglas
➢ **flint glass** Flintglas
➢ **fritted glass** Sinterglas

- laminated safety glass
  Verbundsicherheitsglas
- magnifying glass/magnifier/lens
  Vergrößerungsglas
- milk glass Milchglas
- packaging glasses
  Verpackungsgläser
- quartz glass Quarzglas
- ribbed glass
  Rippenglas,
  geripptes Glas, geriffeltes Glas
- safety glass/laminated glass
  Schutzglas, Sicherheitsglas
- shatterproof (safety glass)
  splitterfrei (Glas)
- sheet of glass/pane
  Glasplatte, Glasscheibe
- soda-lime glass
  (alkali-lime glass)
  (see also: crown glass)
  Kalk-Soda-Glas
- tempered glass/
  resistance glass
  Hartglas
- tempered safety glass (toughened)
  Einscheibensicherheitsglas (ESG)
- toughened gehärtet
- watch glass/clock glass
  Uhrglas, Uhrenglas
- water glass/
  soluble glass $M_2O\times(SiO_2)_x$
  Wasserglas
- window glass Fensterglas

glass bead Glasperle, Glaskügelchen
glass ceramics Glaskeramik
glass cutter Glasschneider
glass cylinder Glaszylinder
glass homogenizer
  (Potter-Elvehjem homogenizer;
  Dounce homogenizer)
  Glashomogenisator ('Potter'; Dounce)
glass marker
  Glasschreiber, Glasmarker
glass pestle
  Glasstößel, Glaspistill
  (Homogenisator)
glass pressure vessel
  Druckbehälter aus Glas
glass rod Glasstab
glass scrap/
  shattered glass/
  broken glass
  Glasbruch
glass stirring rod Glasrührstab
glass stopcock Glashahn
glass transition *polym*
  Glasübergang
glass-transition temperature (T_g)
  polym Glasübergangstemperatur
glass tube/glass tubing
  Glasrohr, Glasröhre, Glasröhrchen
glass tubing cutter
  Glasrohrschneider;
  (glass-tube cutting pliers)
  Glasrohrschneider (Zange)
glass vessel Glasbehälter
glass wool Glaswolle
glassblower Glasbläser
glassblower's workshop ('glass shop')
  Glasbläserei
glassine paper/glassine Pergamin
  (durchsichtiges festes Papier)
glasslike/glassy/vitreous
  glasartig, glasig
glassmaker Glashersteller
glassware (glasswork)
  Glasgeschirr, Glaswaren, Glassachen
- standard-taper glassware
  Normschliffglas (Kegelschliff)
glasswork/glazing Glaserei
  (Handwerk)
glassy/made out of glass/vitreous
  gläsern, aus Glas
glazed paper
  Glanzpapier (glanzbeschichtetes
  Papier), satiniertes Papier
glazier (one who sets glass)
  Glaser
glazier's workshop/glass shop
  Glaserei (Werkstatt)
GLC (gas-liquid chromatography)
  GFC (Gas-Flüssig-Chromatographie)
glide angle/gliding angle *aer*
  Gleitwinkel
global radiation Globalstrahlung
global warming globale Erwärmung

glove(s) Handschuh(e)
- **acid gloves/ acid-resistant gloves** Säureschutzhandschuhe
- **cold-resistant gloves** Kälteschutzhandschuhe
- **cotton gloves** Baumwollhandschuhe
- **cut-resistant gloves** Schnittschutz-Handschuhe
- **deep-freeze gloves** Tiefkühlhandschuhe, Kryo-Handschuhe
- **disposable gloves/ single-use gloves** Einweg-, Einmalhandschuhe
- **gauntlets** Schutzhandschuhe
- **heat defier gloves/ heat-resistant gloves** Hitzehandschuhe
- **insulated gloves** Isolierhandschuhe
- **medical gloves** medizinische Handschuhe, OP-Handschuhe
- **oven gloves** Hoch-Hitzehandschuhe, Ofenhandschuhe
- **protective gloves** Schutzhandschuhe
- **single-use gloves/disposable gloves** Einmalhandschuhe
- **sleeve gauntlets** Ärmelschoner, Stulpen
- **work gloves** Arbeitshandschuhe

glove box/dry-box Handschuhkasten, Handschuhschutzkammer
glove liners Handschuhinnenfutter
glucaric acid/saccharic acid Glucarsäure
gluconic acid (gluconate)/ dextronic acid Gluconsäure (Gluconat)
glucose (grape sugar) Glukose, Glucose (Traubenzucker)
glucosuria/glycosuria Glukosurie, Glycosurie
glucuronic acid (glucuronate) Glucuronsäure (Glukuronat)

glue *n* **(gum)** Leim, Klebstoff, Kleber
- **wood glue** Holzleim

glue *vb* leimen
- **glue together** verleimen

glutamic acid (glutamate)/ 2-aminoglutaric acid Glutaminsäure (Glutamat), 2-Aminoglutarsäure
glutaraldehyde/1,5-pentanedione Glutaraldehyd, Glutardialdehyd, Pentandial
glutaric acid Glutarsäure
glyceraldehyde/ dihydroxypropanal Glyzerinaldehyd, Glycerinaldehyd
glycerol/glycerin/1,2,3-propanetriol Glyzerin, Glycerin, Propantriol
glycine/glycocoll Glycin, Glyzin, Glykokoll
glycol/ ethylene glycol/1,2-ethanediol Glykol, Glycol, Ethylenglykol
glycol aldehyde/ glycolal/hydroxyaldehyde Glykolaldehyd, Hydroxyacetaldehyd
glycolic acid (glycolate) Glykolsäure (Glykolat)
glycosidic bond/glycosidic linkage glykosidische Bindung
glycosuria/glucosuria Glykosurie, Glukosurie
glycyrrhetinic acid Glycyrrhetinsäure
glyoxalic acid (glyoxalate) Glyoxalsäure (Glyoxalat)
glyoxylic acid (glyoxylate) Glyoxylsäure (Glyoxylat)
goggles (safety goggles) Schutzbrille, Augenschutzbrille (ringsum geschlossen)
gold (Au) Gold
- **gilding** vergolden

gold foil/gold leaf Blattgold
gold-labelling Goldmarkierung
Golgi staining method Golgi-Anfärbemethode
Gooch crucible Gooch-Tiegel

Good Laboratory Practice (GLP)
Gute Laborpraxis
Good Manufacturing Practice (GMP)
Gute Industriepraxis
(Produktqualität)
Good Work Practices (GWP)
Gute Arbeitspraxis
gooseflesh/
goose pimples/goose bumps
Gänsehaut
gooseneck Schwanenhals
gowning room
(cleanroom/containment level)
Umkleidekabine
graded ethanol series
Alkoholreihe,
aufsteigende Äthanolreihe
gradient
Gradient; Neigung;
chem Gefälle, Gradient
gradient gel electrophoresis
Gradienten-Gelelektrophorese
graduate *n* Mensur
(Messbehälter/Messgefäß)
graduate *vb* graduieren, mit einer
Maßeinteilung versehen,
in Grade einteilen; *chem* gradieren;
(from school) einen (Schul~)
Abschluss machen (eine
Abschlussprüfung bestehen)
graduated graduiert,
mit einer Gradeinteilung versehen
graduated cylinder Messzylinder
graduated pipet/
measuring pipet
Messpipette
graduation (of glassware)
Ringmarke (Laborglas etc.)
graft *n*
Pfropfen, Pfropfung; Transplantat
graft *vb* pfropfen
graft rejection *med*
Transplantatabstoßung
graft-versus-host reaction (GVH)
Transplantat-anti-Wirt-Reaktion
grain Korn, Körnung;
Faserorientierung
grain-size class Kornklasse

gram/gramme (*Br*)
Gramm (Maßeinheit für Masse)
gram equivalent Grammäquivalent
gram-negative gramnegativ
gram-positive grampositiv
Gram stain/Gram's method
Gram-Färbung
grant *n* Zuschuss, Beihilfe;
(for educational purposes)
Ausbildungs-/Studienbeihilfe;
(scholarship) Stipendium
granular granulär, gekörnt, körnig
granulate granulieren, körnen
granulation
Granulieren, Körnen, Körnigkeit
granulator Granulierapparat (Mühle)
granule Körnchen
grape sugar/glucose/dextrose
Traubenzucker,
Glukose, Glucose, Dextrose
graph *n* **(plot/chart/diagram)**
graphische Darstellung
graph paper/
metric graph paper Millimeterpapier
graphite Grafit
graphite furnace Grafitofen
grasping claws/clasper(s)/clasps
Haltezange, Klasper
grate *n*
Gitter, Rost, Rätter; Fangrechen
grate *vb* reiben, raspeln; vergittern,
mit einem Rost versehen
grater Reibe, Reibeisen, Raspel
gravel Kies
gravid/pregnant trächtig, schwanger
gravimetry/gravimetric analysis
Gravimetrie, Gewichtsanalyse
gravitation Gravitation, Schwerkraft
gravitational field Schwerefeld
gravitational pull Anziehungskraft
gravitational sense Schweresinn
gravity (gravitational force)
Schwerkraft, Gravitationskraft
➤ **specific gravity** spezifisches Gewicht
gravity column chromatography
Normaldruck-Säulenchromatographie
gravity filtration Schwerkraftsfiltration
(gewöhnliche F.)

grease *vb* schmieren, einfetten
grease *n* Schmierfett, Schmiere
➢ **apiezon grease** Apiezonfett
➢ **silicone grease** Silikon-Schmierfett
greasy schmierig, verschmiert, fettig
greenhouse (hothouse/forcing house)
  Gewächshaus, Treibhaus
greenhouse effect Treibhauseffekt
grid (screen/raster) Raster;
  micros Gitter, Netz, Gitternetz,
  Probenträger(netz)
  (für Elektronenmikroskop);
  (grate/bar screen: sewage treatment
  plant) Rechen (der Kläranlage)
➢ **power grid** *electr* Verteilungsnetz
grid method Rastermethode
grid sampling *ecol*
  Rasterstichprobenerhebung
grind/crush/pulverize
  mahlen, zerkleinern
grinder/grinding machine
  Mühle (*mittel*); Schleifer,
  Schleifmaschine
grinding Zermahlen
grinding balls Mahlkugeln (Mühle)
grinding jar Mahlbecher (Mühle)
grip (grasp/handle) Griff
  (zupackend/festhaltend)
grippers Greifzange
grommet Öse, Metallöse
gross potential Summenpotential
gross weight Bruttogewicht
ground *electr* Erde, Erdung
ground fault Erdfehler, Erdschluss
ground fault current (leakage current)
  Erdschlussstrom, Fehlerstrom
ground glass
  geschliffenes Glas, Glasschliff
ground-glass equipment
  Schliffgerät
ground-glass joint
  Schliff, Schliffverbindung,
  Glasschliffverbindung
➢ **flat-flange ground joint/
  flat-ground joint/
  plane-ground joint**
  Planschliffverbindung,
  Planschliff (glatte Enden)

➢ **spherical ground joint**
  Kugelschliffverbindung,
  Kugelschliff
➢ **tapered ground joint
  (S.T. = standard taper)**
  Kegelschliff
  (N.S. = Normalschliff)
**ground-glass stopper/
  ground-in stopper/
  ground stopper** Schliffstopfen
ground joint (ground-glass joint)
  Schliffverbindung,
  Glasschliffverbindung, Schliff
➢ **flat-flange ground joint/
  flat-ground joint/
  plane-ground joint**
  Planschliffverbindung,
  Planschliff (glatte Enden)
➢ **ground-joint clip/
  ground-joint clamp**
  Schliffklammer,
  Schliffklemme (Schliffsicherung)
➢ **ground-jointed flask** Schliffkolben
➢ **spherical ground joint**
  Kugelschliffverbindung, Kugelschliff
➢ **tapered ground joint**
  Kegelschliffverbindung, Kegelschliff
ground state Grundzustand
ground water Grundwasser
grounded/earthed (*Br*) *electr* geerdet
groundwater Grundwasser
**group leader/
  principal investigator (lab/research)**
  Gruppenleiter
grow wachsen
➢ **slow-growing** langsam wachsend
grow up aufwachsen
growth Wachstum; (habit) Wuchs;
  (cover/stand) Bewuchs
growth curve Wachstumskurve
growth factor Wachstumsfaktor
**growth form/
  appearance/habit** Wuchsform,
  Wachstumsform, Habitus
growth inhibitor
  Wachstumshemmer,
  Wuchshemmer, Wuchshemmstoff
growth period Wachstumsperiode

growth phase Wachstumsphase
> **exponential growth phase**
 exponentielle Wachstumsphase,
 exponentielle Entwicklungsphase
> **logarithmic phase (log-phase)**
 logarithmische Phase
growth rate (vigor)
 Wachstumsgeschwindigkeit,
 Wachstumsrate, Zuwachsrate;
 (vigor) Wachstumsleistung
**growth regulator/
phytohormone/growth substance**
 Wuchsstoff (Pflanzenwuchsstoff),
 Phytohormon
growth-retarding/growth-inhibiting
 wachstumshemmend
growth-stimulating
 wachstumsfördernd
growth vigor Wuchskraft
guano Guano
guanylic acid (guanylate)
 Guanylsäure (Guanylat)
guar gum/guar flour
 Guarmehl, Guar-Gummi
guar meal/guar seed meal
 Guar-Samen-Mehl
guarantee *n* Garantie, Gewährleistung,
 Bürgschaft, Sicherheit

guarantee *vb* **(warrant)**
 garantieren, gewährleisten,
 verbürgen, sicherstellen
guard *n* (security guard) Wächter,
 Wachmann; (protective device)
 Schutzvorrichtung
guard *vb* bewachen
guard column/precolumn (HPLC)
 Schutzsäule, Vorsäule
guard tube Sicherheitsrohr (Laborglas)
guide bushing (of shaft/impeller)
 Führungsbuchse
guideline Leitlinie, Richtlinie
gulonic acid (gulonate)
 Gulonsäure (Gulonat)
gum arabic/acacia gum
 Gummi arabicum,
 Arabisches Gummi, Acacia Gummi
gunpowder Schießpulver
gush *n* Guss, Erguss, Strom, Schwall
gush *vb* entströmen,
 (heftig) herausströmen,
 (schwallartig) hervorsprudeln,
 sich heftig ergießen
gypsum (selenite) $CaSO_4 \times 2H_2O$
 Gips
gypsum board (ceiling)
 Gipsplatte (Deckenbeschalung)

hacksaw Bügelsäge
hair bundle (hygrometer) Haarharfe(-Messelement)
half cell/half element (single-electrode system) Halbelement (galvanisches), Halbzelle
half-cell potential Halbzellenpotential
half-life Halbwertszeit; (Enzyme) Halblebenszeit
half-mask (respirator) Halbmaske
half-reaction (electrode potentials) Teilreaktion
hallucinogen Halluzinogen
hallucinogenic halluzinogen
hallway/hall/corridor Flur, Korridor
halve *vb* halbieren
hammer Hammer
➢ **ball pane hammer/ ball peen hammer/ ball pein hammer** Schlosserhammer mit Kugelfinne
➢ **bush hammer** Scharrierhammer
➢ **carpenter's roofing hammer** Latthammer
➢ **claw hammer** Klauenhammer, Splitthammer
➢ **club hammer/mallet** Fäustel, Handfäustel
➢ **fitter's hammer/ locksmith's hammer** Schlosserhammer
➢ **sledge hammer** Vorschlaghammer
hand cart/barrow (*Br* trolley) Sackkarre
hand mill Handmühle
hand motion (handshaking motion) Handbewegung
hand-operated vacuum pump manuelle Vakuumpumpe
hand pump Handpumpe
handgrips Griffe, Tragegriffe
handicapped behindert
➢ **physically handicapped** körperbehindert
handing over/delivery Übergabe
handle *n* Griff, Schaft
handling Handhabung, Hantieren, Gebrauch, Umgang (Verhalten)
handsaw Handsäge
handtooled handgearbeitet (Glas etc.)
hard coal/anthracite Glanzkohle, Anthrazit
hard hat/hardhat/safety helmet Schutzhelm
harden härten
hardness (toughness) Härte
➢ **permanent hardness** bleibende Härte, permanente Härte
➢ **temporary hardness** temporäre Härte, Carbonathärte
➢ **total hardness** Gesamthärte (Wasser)
➢ **water hardness** Wasserhärte
hardy/persistent/enduring ausdauernd (wiederstandsfähig), winterfest, winterhart
hardware dealer Metallwarenhändler
harmful (causing damage/damaging) schädlich
harmful/nocent (Xn) gesundheitsschädlich
harmful organism/harmful lifeform Schadorganismus
harmless (not dangerous: safe) ungefährlich (sicher); (not harmful/inactive) unschädlich
hartshorn salt/ammonium carbonate Hirschhornsalz, Ammoniumcarbonat
harvest *n* Ernte
harvest *vb* **(a crop)** ernten
hatchet Beil
Haworth projection/Haworth formula Haworth-Projektion, Haworth-Formel
hay infusion Heuaufguss
hazard (source of danger) Gefahrenquelle
➢ **biohazard** biologische Gefahrenquelle, biologische Gefahr; biologisches Risiko
➢ **occupational hazard** Berufsrisiko; Gefahr am Arbeitsplatz

hazard bonus Gefahrenzulage
hazard code
 Gefahrencode,
 Gefahrenkennziffer
hazard diamond Gefahrendiamant
hazard icon/hazard symbol/
 hazard warning symbol
 Gefahrensymbol,
 Gefahrenwarnsymbol
hazard label Gefahrzettel
hazard rating/
 hazard class/hazard level
 Gefahrenstufe,
 Gefahrenklasse, Risikostufe
hazard warning(s)
 Gefährlichkeitsmerkmale
hazard warning sign/
 warning sign/danger signal
 Gefahrenwarnzeichen
hazard warning symbol
 Gefahrensymbol
hazardous gefährlich,
 gesundheitsgefährdend;
 unfallträchtig
➤ **nonhazardous** ungefährlich,
 nicht gesundheitsgefährdend
hazardous material
 gefährlicher Stoff
hazardous material class
 Gefahrenstoffklasse
hazardous materials regulations
 Gefahrenstoffverordnung,
 Gefahrgutbestimmungen
hazardous materials safety cabinet
 Gefahrstoffschrank
hazardous waste
 Sondermüll, Sonderabfall,
 Problemabfall
hazardous waste disposal
 Sondermüllentsorgung
hazardous waste dump
 Sondermülldeponie
hazardous waste incineration plant
 Sondermüllverbrennungsanlage
hazardous waste treatment plant
 Sondermüllentsorgungsanlage
head (e.g., fat molecule) Kopf
head cover Kopfbedeckung

headspace
 Gasraum, Dampfraum, Headspace
headspace gas chromatography
 Dampfraum-Gaschromatographie
health Gesundheit; Gesundheitszustand
➤ **detrimental to one's health**
 ungesund, der Gesundheit abträglich
➤ **injurious to one's health**
 gesundheitsschädlich
health care medizinische Versorgung,
 Gesundheitsfürsorge
health center (health-care center)
 Ärtzezentrum, Gesundheitszentrum
health certificate
 Gesundheitszeugnis, ärtzliches Attest
health education Gesundheitserziehung
health hazard Gesundheitsrisiko
health insurance Krankenversicherung
health resort Kurort
health-threatening
 gesundheitsbedrohend
healthy gesund
➤ **unhealthy**
 (detrimental to one's health)
 ungesund
hearing Gehör;
 (sense of hearing) Hörfähigkeit
hearing limit/auditory limit/
 limit of audibility
 Hörgrenze
hearing threshold/auditory threshold
 Hörschwelle
heartburn/acid indigestion Sodbrennen
heat *vb* heizen; (heat up) erhitzen,
 aufheizen; (warm up) erwärmen
heat *n* Hitze
➤ **specific heat** spezifische Wärme
heat capacity/thermal capacity
 Wärmekapazität
heat conductivity/thermal conductivity
 Wärmeleitfähigkeit
heat defier gloves/
 heat-resistant gloves
 Hitzehandschuhe
heat dissipation Wärmeabstrahlung
heat evolution Hitzeentwicklung
heat exchanger Wärmetauscher
heat gun Heißluftpistole

heat loss Wärmeverlust
heat of formation Bildungswärme
heat of reaction/heat effect
 Reaktionswärme, Wärmetönung
heat of solution/heat of dissolution
 Lösungswärme
heat of vaporization
 Verdampfungswärme,
 Verdunstungswärme
heat pump Wärmepumpe
heat-resistant/heat-stable
 hitzebeständig, hitzestabil
heat seal *vb* heißsiegeln
heat shock Hitzeschock
heat shock reaction/
 heat shock response
 Hitzeschockreaktion
heat-shrinkable film Schrumpffolie
heat shrinking Wärmeschrumpfen
heat source Wärmequelle
heat supply Wärmezufuhr
heat-tolerant hitzeverträglich
heat tone/heat tonality Wärmetönung
heat transfer Wärmeübergang
heat transmission Wärmeübertragung
heat transport Wärmetransport
heat treatment/baking
 Hitzebehandlung, Backen
heater/heating system Heizung
heating (warming)
 Erhitzung (Erwärmung)
heating bath Heizbad
heating coil Heizschlange, Heizwendel
heating mantle
 Heizhaube, Heizmantel, Heizpilz
heating oven/
 heating furnace (more intense)
 Wärmeofen
heating system
 Heizung, Heizungsanlage
heating tape/heating cord
 Heizband, Heizbandage
heatstroke Hitzeschlag
heavy schwer; (less volatile) schwer
 flüchtig (höhersiedend)
heavy-duty/superior performance
 superstark, verstärkt,
 Hochleistungs...

heavy metal
 Schwermetall
heavy metal contamination
 Schwermetallbelastung
heavy metal poisoning
 Schwermetallvergiftung
heavyweight schwergewicht(ig)
Heimlich maneuver *med*
 Heimlich-Handgriff
helical ribbon impeller Wendelrührer
helice (column packing) *dist* Wendel
helix (*pl* helices or helixes)/spiral
 Helix, Spirale (*pl* Helices)
help (aid/assistance/support)
 Hilfe; Aushilfe
hemadsorption inhibition test
 (HAI test)
 Hämadsorptionshemmtest (HADH)
hemagglutination inhibition test
 (HI test)
 Hämagglutinationshemmtest (HHT)
hematocyte Hämatocyt, Blutzelle
hematopoiesis
 Hämatopoese, Haematopoese,
 Blutbildung, Blutzellbildung
hemiacetal Halbacetal
hemicyclic hemicyclisch
hemorrhagic
 hämorrhagisch, blutzersetzend
hemostatic forceps/artery clamp
 Gefäßklemme, Arterienklemme,
 Venenklemme
HEPA-filter (high-efficiency particulate
 and aerosol air filter)
 HOSCH-Filter
 (Hochleistungsschwebstofffilter)
hepar reaction/hepar test
 Heparreaktion, Heparprobe
hepatotoxic
 leberschädigend, hepatotoxisch
herb Kraut, Krautpflanze;
 (herbs: medicinal) Heilkräuter
➢ **herbal drug** Pflanzendroge
herbarium Herbar
herbicide/weed killer
 Herbizid,
 Unkrautvernichtungsmittel,
 Unkrautbekämpfungsmittel

hereditary (heritable)
　erblich, hereditär
hereditary disease/
　genetic disease/inherited disease/
　heritable disorder/genetic defect/
　genetic disorder
　Erbkrankheit, erbliche Erkrankung
hereditary information/
　genetic information
　Erbinformation
hereditary material
　Erbträger, Erbsubstanz;
　(genome) Genom, Erbgut
hereditary trait Erbmerkmal
heredity/inheritance/transmission
　(of hereditary traits)
　Vererbung
heritability
　Erblichkeitsgrad, Heritabilität
hermaphroditic contact *electr*
　Zwitterkontakt
heterocycle Heterocyclus
heterocyclic heterocyclisch
heterogeneity
　Heterogenität, Ungleichartigkeit,
　Verschiedenartigkeit,
　Andersartigkeit
heterogeneous
　(consisting of dissimilar parts)
　heterogen, ungleichartig,
　verschiedenartig, andersartig
heterogenetic
　heterogenetisch, genetisch
　unterschiedlichen Ursprungs
heterogenous (of different origin)
　heterogen,
　unterschiedlicher Herkunft
heterogeny Heterogenie,
　unterschiedlicher Herkunft
heterologous heterolog
heterologous probing
　mit Hilfe einer heterologen Sonde
heterologous vaccine
　heterologer Impfstoff,
　heterologe Vakzine
heteropolar bond
　heteropolare Bindung
heteropolymer Heteropolymer

heteroscedasticity *stat*
　Heteroskedastizität,
　Varianzheterogenität
heterotroph/heterotrophic heterotroph
heterotypic heterotypisch
hex-head stopper/hexagonal stopper
　Sechskantstopfen
hex nutdriver Sechskant-Steckschlüssel
hex screwdriver
　(hexagonal screwdriver)
　Sechskantschraubenzieher
hex socket wrench
　Sechskant-Stiftschlüssel
high density lipoprotein (HDL)
　Lipoprotein hoher Dichte
high energy bond
　energiereiche Bindung
high energy compound
　energiereiche Verbindung
high energy explosive (HEX)
　hochbrisanter Sprengstoff
high explosive
　brisanter Sprengstoff
high-field shift (NMR)
　Hochfeldverschiebung
high fructose corn syrup
　Isomeratzucker, Isomerose
high-grade steel/high-quality steel
　Edelstahl
high-molecular hochmolekular
high-pressure liquid chromatography/
　high performance liquid
　chromatography (HPLC)
　Hochdruckflüssigkeits-
　chromatographie,
　Hochleistungschromatographie
high-pressure tubing
　(mit größerem Durchmesser: hose)
　Hochdruckschlauch
high-resolution ...
　hochauflösend, hoch aufgelöst
high-speed centrifuge/
　high-performance centrifuge
　Hochgeschwindigkeitszentrifuge
high voltage Hochspannung
high voltage electron microscopy
　(HVEM) Höchstspannungs-
　elektronenmikroskopie

highly flammable (F) leicht entzündlich
highly ignitable hochentzündlich
highly pure (superpure/ultrapure) reinst
hinge Gelenk, Scharnier, Schloss, Schlossleiste
Hirsch funnel Hirsch-Trichter
histocompatibility Histokompatibilität, Gewebeverträglichkeit
histogram/strip diagram Histogramm, Streifendiagramm
histoincompatibility Histoinkompatibilität, Gewebeunverträglichkeit
histology Gewebelehre, Histologie
Hofmeister series/lyotropic series Hofmeistersche Reihe, lyotrope Reihe
hoist (lifting platform) Hebebühne
holdup/retention Retention
holdup time (GC) Durchflusszeit, Totzeit
holdup volume Durchflussvolumen, Totvolumen
hollow fiber Hohlfaser
hollow impeller shaft Hohlwelle (Rührer)
hollow stirrer Hohlrührer
hollow stopper Hohlstopfen, Hohlglasstopfen
homogeneity
  (with same kind of constituents) Homogenität, Einheitlichkeit, Gleichartigkeit
homogeneous
  (having same kind of constituents) homogen, einheitlich, gleichartig
homogenization Homogenisation, Homogenisierung
homogenize homogenisieren
homogenizer Homogenisator
homogenous (of same origin) homogen, gleicher Herkunft
homogentisic acid Homogentisinsäure
**homograft/syngraft
  (syngeneic graft)/allograft** Homotransplantat, Allotransplantat

homoiosmotic/homeosmotic homoiosmotisch
homologize homologisieren
homologous homolog, ursprungsgleich
homology Homologie
homopolar bond/nonpolar bond homopolare Bindung
homopolymer Homopolymer
homoscedasticity Varianzhomogenität, Varianzgleichheit, Homoskedastizität
hood Abzug
  ➤ **forced-draft hood** Saugluftabzug
  ➤ **fume hood** Rauchabzug, Dunstabzugshaube, Abzug
  ➤ **laminar flow hood/
    laminar flow unit/
    laminar flow workstation** Querstrombank
  ➤ **sash** Schiebefenster, Frontschieber, verschiebbare Sichtscheibe (Abzug, Werkbank)
  ➤ **vertical flow hood/workstation/unit** Fallstrombank
  ➤ **walk-in hood** begehbarer Abzug
hook clamp Hakenklemme (Stativ)
hook up/wire to/(make) contact *electr* anschließen
horizontal gel/flat bed gel horizontal angeordnetes Plattengel
hormonal hormonal, hormonell
hormone Hormon
hose Schlauch
  ➤ **garden hose** Gartenschlauch
hose attachment socket Schlauchtülle (z.B. am Gasreduzierventil)
hose barb Olive; Schlauch-Ansatzstutzen
hose bundle/tubing bundle Schlauchschelle (Abrutschsicherung, Befestigung an Verbindungsstück)
hose clamp/hose clip Schlauchklemme, Schlauchschelle (Installationen: zur Schlauchbefestigung)

hose connection Schlauchverbindung;
(barbed/male) Olive;
(flask) Schlauch-Ansatzstutzen
hose connection gland Schlauchtülle
hose pump Schlauchpumpe
host range *ecol* Wirtsspektrum
host specificity Wirtsspezifität
hot air Heißluft
hot-air gun Föhn, Heißluftgebläse, Labortrockner
hot plate
  Heizplatte, Kochplatte
➢ **double-burner hot plate**
  Doppelkochplatte
➢ **single-burner hot plate**
  Einfachkochplatte
➢ **stirring hot plate**
  Magnetrührer mit Heizplatte
hot-water funnel (double-wall funnel)
  Heißwassertrichter
household Haushalt
household brush
  Handbesen, Handfeger
household waste/trash
  Haushaltsmüll, Haushaltsabfälle
housing (shell/case/casing) Gehäuse
hue Farbton, Tönung, Schattierung
human *adj/adv*
  menschlich,
  den Menschen betreffend
human biology Humanbiologie
human ecology Humanökologie
human genetics
  Humangenetik, Anthropogenetik
Human Genome Project (HUGO)
  menschliches Genomprojekt
human resources
  Arbeitskräfte(potential)
human resources department
  Personalabteilung
humane menschlich
  (wie ein guter Mensch handelnd: hilfsbereit, selbstlos)
humane society Gesellschaft zur Verhinderung von Grausamkeiten an Tieren
humic acid Huminsäure
humid/damp/moist feucht

humidifier/mist blower
  Wasserzerstäuber, Sprühgerät
humidify/prewet anfeuchten
humidity/dampness/moisture
  Feuchtigkeit
humidity-gradient apparatus *ecol*
  Feuchte-Orgel, Feuchtigkeitsorgel
hunch/guess/assumption
  Vermutung, Annahme
hunger Hunger
hungry hungrig
hurt/be painful schmerzen
hutch/coop kleiner Tierkäfig,
  Verschlag (z.B. Geflügelstall)
hyaluronic acid Hyaluronsäure
hybrid *adj/adv* (crossbred)
  hybrid, durch Kreuzung erzeugt
hybrid *n* (crossbreed) Hybride
hybridization Hybridisierung
➢ **in** *situ* **hybridization**
  in situ Hybridisierung
hybridize hybridisieren
hydrate Hydrat
hydration/solvation
  Hydratation, Hydratisierung,
  Solvation (Wassereinlagerung/ Wasseranlagerung)
hydration shell
  Hydrathülle, Wasserhülle,
  Hydratationsschale
hydric hydrisch
hydrocarbon Kohlenwasserstoff
➢ **chlorinated hydrocarbon**
  chlorierter Kohlenwasserstoff
➢ **chlorofluorocarbons/ chlorofluorinated hydrocarbons (CFCs)**
  Fluorchlorkohlenwasserstoffe (FCKW)
➢ **fluorinated hydrocarbon**
  Fluorkohlenwasserstoff
hydrochloric acid
  Salzsäure, Chlorwasserstoffsäure
hydrofluoric acid/phthoric acid
  Fluorwasserstoffsäure, Flusssäure
hydrogen (H) Wasserstoff
hydrogen bond Wasserstoffbrücke,
  Wasserstoffbrückenbindung

hydrogen cyanide/
hydrocyanic acid/prussic acid
  Blausäure, Cyanwasserstoff
hydrogen electrode
  Wasserstoffelektrode
hydrogen fluoride
  Fluorwasserstoff, Fluoran,
  Hydrogenfluorid
hydrogen ion (proton)
  Wasserstoffion (Proton)
hydrogen peroxide
  Wasserstoffperoxid
hydrogen sulfide H_2S
  Schwefelwasserstoff
hydrogenate
  hydrieren, hydrogenieren
hydrogenation Hydrierung
  (Wasserstoffanlagerung)
hydroiodic acid/hydrogen iodide
  Iodwasserstoffsäure
hydrology Hydrologie
hydrolysis Hydrolyse, Wasserspaltung
hydrolytic hydrolytisch, wasserspaltend
hydrophilic
  (water-attracting/water-soluble)
  hydrophil (wasseranziehend/
  wasserlöslich)
hydrophilic bond
  hydrophile Bindung
hydrophilicity
  (water-attraction/water-solubility)
  Hydrophilie (Wasserlöslichkeit)
hydrophobic
  (water-repelling/water-insoluble)
  hydrophob (wasserabweisend/
  wasserabstoßend, wasser-unlöslich)
hydrophobic bond
  hydrophobe Bindung
hydrophobicity (water-insolubility)
  Hydrophobie (Wasserabweisung/
  Wasserunlöslichkeit)

hydroponics
  (soil-less culture/solution culture)
  Hydrokultur
hydrostatic pressure (turgor)
  hydrostatischer Druck (Turgor)
hydroxyapatite Hydroxyapatit
hydroxylation Hydroxylierung
hygiene Hygiene
➢ **industrial hygiene**
  Arbeitshygiene
➢ **occupational hygiene**
  Arbeitsplatzhygiene
hygienic hygienisch
hygienic conditions
  Hygienebedingungen
hygroscopic
  hygroskopisch, wasseranziehend,
  Feuchtigkeit aufnehmend
hyperchromicity/
hyperchromic effect/
hyperchromic shift
  Hyperchromizität
hyperfunction/hyperactivity
  Überfunktion
hypersensitivity
  Überempfindlichkeit;
  (allergy) Allergie,
  Hypersensibilität
hypertension
  Hochdruck, Bluthochdruck
hypodermic needle
  Nadel, Kanüle, Hohlnadel (Spritze)
hypodermic syringe
  Injektionsspritze
hypofunction/insufficiency
  Unterfunktion, Insuffizienz
hypothesis Hypothese
hypothetic/hypothetical
  hypothetisch
hypoxia
  Hypoxie, Sauerstoffmangel

ice Eis
- **crushed ice** zerstoßenes Eis, Eisschnee (fürs Eisbad)

ice bath/ice-bath Eisbad
ice bucket Eisbehälter
ice nucleating activity *micb* Eiskernaktivität
identical identisch, genau gleich
identification Identifikation, Bestimmung, Erkennung, Feststellung; Legitimation; Kennung
identity Identität, völlige Gleichheit
identity by state (IBS) identisch aufgrund von Zufällen
ignitability Zündbarkeit
ignitable entzündbar
- **spontaneously ignitable/ self-ignitable/autoignitable** selbstentzündlich

ignite (inflame) anbrennen, entzünden, entflammen; (start a fire) anzünden
igniter/primer Zünder
ignition Zündung, Anzünden, Entzünden
ignition energy Zündenergie
- **minimum ignition energy** Mindestzündenergie

ignition loss/loss of ignition Glühverlust
ignition point/kindling temperature/ flame temperature/flame point/ spontaneous-ignition temperature (SIT) Zündpunkt, Zündtemperatur, Entzündungstemperatur
ignition tube Zündröhrchen, Glühröhrchen
illness/sickness/disease/disorder (Störung) Erkrankung
illuminance Beleuchtungsstärke
illuminate beleuchten, erleuchten, erhellen
illumination Beleuchtung, Licht
- **epiillumination/incident illumination** Auflicht, Auflichtbeleuchtung
- **fiber optic illumination** Kaltlichtbeleuchtung
- **Koehler illumination** *micros* Köhlersche Beleuchtung
- **transillumination/ transmitted light illumination** Durchlicht, Durchlichtbeleuchtung

illuminator Strahler, Beleuchtung; (lamp/light) Leuchte, Licht
illustration Illustration, Erläuterung, Veranschaulichung; Bebilderung; (figure) Abbildung (z.B. in einer Fachzeitschrift/Buch)
image Bild
- **final image** *micros* Endbild
- **intermediate image** *micros* Zwischenbild
- **real image** *opt* reelles Bild
- **virtual image** *opt* scheinbares Bild

image point *opt* Bildpunkt
imbalance/disequilibrium Ungleichgewicht
imbibe/hydrate imbibieren, hydratieren
imbibition/hydration Imbibition, Hydratation
IMDG (Intl. Maritime Dangerous Goods Code) Internat. Code für die Beförderung von gefährlichen Gütern mit Seeschiffen
imino acid Iminosäure
immature unreif
immaturity/immatureness Unreife
immerse eintauchen, untertauchen
immersed slot reactor Tauchkanalreaktor
immersing surface reactor Tauchflächenreaktor
immersion Immersion, Eintauchen, Untertauchen
immersion bath Tauchbad
immersion circulator Tauchpumpen-Wasserbad, Einhängethermostat
immersion heater/'red rod' (*Br*) Tauchsieder
imminent danger drohende Gefahr
immiscibility Unvermischbarkeit
immiscible unvermischbar

immission
(injection/admission/introduction)
Immission, Einwirkung
immobile/fixed/motionless
immobil, fixiert, bewegungslos
immobility/motionlessness
Immobilität,
Bewegungslosigkeit
immobilization
Immobilisation, Immobilisierung,
Ruhigstellung, Unbeweglichmachen
immobilize (to make immobile)
immobilisieren, ruhigstellen,
unbeweglich machen
immortal unsterblich
immortality
Immortalität, Unsterblichkeit
immune immun; unempfänglich
immune deficiency
(immunodeficiency)
Immunschwäche, Immundefekt
> **severe combined immune deficiency**
(SCID)
schwerer kombinierter Immundefekt
immune deficiency syndrome
Immunschwächesyndrom,
Immunmangel-Syndrom
immune reaction Immunreaktion
immune response Immunantwort
immunity Immunität
> **active immunity**
aktive Immunität
> **concomitant immunity/premunition**
Prämunität, Präimmunität,
Prämunition, begleitende Immunität
> **cross protection**
Kreuzimmunität
(übergreifender Schutz)
> **passive immunity**
passive Immunität
immunization/vaccination
Immunisierung, Impfung
immunize/vaccinate
immunisieren, impfen
immunoaffinity chromatography
Immunaffinitätschromatographie
immunoblot/Western blot
Immunoblot, Western-Blot

immunodiffusion
Immundiffusion,
Gelpräzipitationstest,
Immunodiffusionstest
> **double diffusion/**
double immunodiffusion
(Ouchterlony technique)
Doppeldiffusion,
Doppelimmundiffusion
> **double radial immunodiffusion (DRI)**
(Ouchterlony technique)
doppelte radiale Immundiffusion
(Ouchterlony-Methode)
immunoelectron microscopy (IEM)
Immun-Elektronenmikroskopie
(IEM)
immunoelectrophoresis
Immunelektrophorese
> **charge-shift immunoelectrophoresis**
Tandem-Kreuzimmunelektrophorese
> **countercurrent immunoelectro-**
phoresis/counterelectrophoresis
Überwanderungsimmun-
elektrophorese,
Überwanderungselektrophorese
> **crossed immunoelectrophoresis/**
two-dimensional
immunoelectrophoresis
Kreuzimmunelektrophorese
> **rocket immunoelectrophoresis**
Raketenimmunelektrophorese
immunofluorescence chromatography
Immunfluoreszenzchromatographie
immunofluorescence microscopy
Immunfluoreszenzmikroskopie
immunogenetics Immungenetik
immunogenicity
Immunisierungsstärke,
Immunogenität
immunogold-silver staining (IGSS)
Immunogold-Silberfärbung (IGSS)
immunolabeling
Immunmarkierung
immunology Immunologie
immunopathy Immunkrankheit,
Immunopathie
immunoprecipitation
Immunpräzipitation

immunoradiometric assay (IRMA)
immunoradiometrischer Assay (IRMA)
immunosuppression/immune suppression Immunsuppression
immunosurveillance/ immunologic(al) surveillance
Immunüberwachung, immunologische Überwachung
immurement technique *biotech*
Einschlussverfahren
impact *n* Aufprall, Auftreffen; Stoß, Schlag, Wucht; Druck, Belastung
impact-resistant stoßfest
impact sound Trittschall
impact sound-reduced
trittschallgedämpft
impaction Einkeilung
impeller Rührer, Rührwerk
➢ **anchor impeller** Ankerrührer
➢ **crossbeam impeller**
Kreuzbalkenrührer
➢ **disk turbine impeller**
Scheibenturbinenrührer
➢ **flat-blade impeller**
Scheibenrührer, Impellerrührer
➢ **four flat-blade paddle impeller**
Kreuzblattrührer
➢ **gate impeller** Gitterrührer
➢ **helical ribbon impeller**
Wendelrührer
➢ **hollow stirrer** Hohlrührer
➢ **screw impeller** Schraubenrührer
➢ **multistage impulse countercurrent impeller**
Mehrstufen-Impuls-Gegenstrom (MIG) Rührer
➢ **off-center impeller**
exzentrisch angeordneter Rührer
➢ **marine blade impeller**
Schraubenblattrührer
➢ **pitch screw impeller**
Schraubenspindelrührer
➢ **pitched blade impeller/ pitched-blade fan impeller/ pitched-blade paddle impeller/ inclined paddle impeller**
Schrägblattrührer

➢ **profiled axial flow impeller**
Axialrührer mit profilierten Blättern
➢ **propeller impeller** Propellerrührer
➢ **rotor-stator impeller/ Rushton-turbine impeller**
Rotor-Stator-Rührsystem
➢ **screw impeller** Schneckenrührer
➢ **self-inducting impeller with hollow impeller shaft**
selbstansaugender Rührer mit Hohlwelle
➢ **stator-rotor impeller/ Rushton-turbine impeller**
Stator-Rotor-Rührsystem
➢ **turbine impeller** Turbinenrührer
➢ **two flat-blade paddle impeller**
Blattrührer
➢ **two-stage impeller**
zweistufiger Rührer
➢ **variable pitch screw impeller**
Schraubenspindelrührer mit unterschiedlicher Steigung
impeller pump
Kreiselpumpe, Kreiselradpumpe
impeller shaft Rührerwelle
impermeability/imperviousness
Impermeabilität, Undurchlässigkeit
impermeable/impervious
impermeabel, undurchlässig
impingement Stoß, Aufprall, Aufschlag
implant einpflanzen
implantation Einpflanzung
implosion Implosion
impregnate imprägnieren, tränken, durchtränken
improper disposal
unsachgemäße Entsorgung
impulse Impuls, Erregung
impure/contaminated
verunreinigt, schmutzig, unsauber, kontaminiert
impurity/contamination
Verunreinigung, Kontamination
in-process verification
Inprozesskontrolle
in-vitro fertilization (IVF)
In-vitro-Fertilisation, Reagensglasbefruchtung

inactivated inaktiviert;
  außer Betrieb gesetzt, ausgeschaltet
inactive inaktiv; unwirksam;
  untätig, träge, faul
inanimate (lifeless/nonliving) unbelebt
incendiary Zündstoff, Brandstoff
incidence Auftreten, Vorkommen,
  Auftreffen; Häufigkeit
incident *adj/adv* auftreffend, einfallend
incident *n* Vorfall, Zwischenfall
  (Unfall: accident)
incident light
  einfallendes Licht, Auflicht
incidental release
  versehentliche Freisetzung
  (Austritt bei üblichem Betrieb)
incinerate (combust/burn) verbrennen;
  (reduce to ashes) veraschen;
  einäschern
incinerating tube
  Verbrennungsrohr (Glas)
incineration
  Verbrennung, Veraschung,
  Einäscherung, Verglühen
incineration dish Glühschälchen
incinerator
  Verbrennungsanlage,
  Verbrennungsofen
➢ **waste incineration plant**
  Müllverbrennungsanlage
incision (cut) Einschnitt
inclination Neigung, Neigungswinkel;
  Schräge, geneigte Fläche
inclusion Einschluss;
  (intercalation) Einlagerung
inclusion body Einschlusskörperchen
inclusion compound
  Einschlussverbindung
incompatibility (intolerance)
  Inkompatibilität,
  Unverträglichkeit (Intoleranz)
incompatibility reaction
  Inkompatibilitätreaktion,
  Unverträglichkeitreaktion
incompatible (intolerant)
  inkompatibel,
  unverträglich (intolerant)
incorrect/wrong/false falsch, fehlerhaft

increase *n* Zunahme, Steigerung,
  Vergrößerung, Vermehrung;
  (increment) Zuwachs
increased safety erhöhte Sicherheit
incubate/brüten/bebrüten
  inkubieren, brood, breed
incubation
  Inkubation, Bebrütung, Bebrüten
incubation period Inkubationszeit
incubation room Brutraum
incubator Brutschrank, Wärmeschrank
➢ **shaking incubator/**
  incubating shaker/
  incubator shaker
  Inkubationsschüttler
indentation Einkerbung, Einbuchtung,
  Zahnung; (notch) Kerbe
index number/index figure/
  guiding figure Richtwert, Richtzahl;
  (indicator) *stat* Kennzahl, Kennziffer
indicator Indikator, Anzeiger;
  (recording instrument) Anzeigegerät
indicator value Zeigerwerte
indirect end-labeling *gen*
  indirekte Endmarkierung
indolyl acetic acid/indoleacetic acid
  (IAA) Indolessigsäure
induce induzieren, veranlassen,
  bewirken, auslösen, fördern
induced vomiting
  provoziertes Erbrechen
inducible induzierbar
induction
  Induktion, Auslösung, Herbeiführung
induction furnace/inductance furnace
  Induktionsofen
induction stroke Saughub, Ansaughub
inductively coupled plasma (ICP)
  induktiv gekoppeltes Plasma
industrial accident directive/
  statutory order on hazardous
  incidents Störfallverordnung
industrial hygiene Arbeitshygiene
industrial waste
  Industriemüll, Industrieabfall
industry
  Industrie, Gewerbe, Gewerbezweig
inedible/uneatable nicht essbar

inert *chem* träg, träge, reaktionsträge
inert dust Inertstaub
inert gas/rare gas Edelgas
inertia Trägheit
inertial force Trägheitskraft
infect infizieren, anstecken
infection Infektion, Ansteckung
> **double infection** Doppelinfektion
> **incomplete infection**
  unvollständige Infektion
> **latent infection** latente Infektion
> **lytic infection** lytische Infektion
> **productive infection**
  produktive Infektion
> **silent infection**
  stumme Infektion, stille Feiung
> **smear infection**
  Schmierinfektion
> **source of infection**
  Ansteckungsherd,
  Ansteckungsquelle
> **superinfection**
  Superinfektion, Überinfektion
infectious infektiös, ansteckend
infectious disease
  Infektionskrankheit
infectious dose
  (ID_{50} = 50% infectious dose)
  Infektionsdosis
infectious waste infektiöser Abfall
infectivity
  Infektionsvermögen,
  Ansteckungsfähigkeit
inferiority (zahlen-/mengenmäßige)
  Unterlegenheit
infertile (sterile) unfruchtbar (steril)
infertility (sterility)
  Unfruchtbarkeit (Sterilität)
infest (pests/parasites)
  befallen (Schädlingsbefall)
infestation Verseuchung (mit
  Mikroorganismen/Ungeziefer etc.);
  (with pests/parasites) Befall
  (Schädlingsbefall)
> **degree/level/rate of infestation**
  Befallsrate
> **reinfestation**
  Wiederbefall

infested befallen;
  (with pests) verseucht
inflame/ignite
  entzünden, entflammen, anbrennen
inflammation *med* Entzündung
inflammed/inflammatory *med*
  entzündet, entzündlich
inflate aufblasen, mit Luft/Gas füllen
influx Einstrom, Einströmen,
  Zustrom, Zufluss
informed consent
  Einverständniserklärung nach
  ausführlicher Aufklärung
infrared absorbance detector (IAD)
  Infrarot-Absorptionsdetektor
infrared detector (ID) Infrarotdetektor
infrared spectroscopy
  Infrarot-Spektroskopie,
  IR-Spektroskopie
infusion Infusion;
  Aufguss, Aufgießen, Ziehenlassen
ingest aufnehmen;
  einnehmen, etwas zu sich nehmen
ingestion/food intake
  Ingestion, Nahrungsaufnahme
ingot (zone melting) Schmelzling
ingredient Bestandteil;
  pl Inhaltsstoffe, Zutaten
> **active ingredient/**
  active principle/active component
  Wirkstoff, Wirksubstanz
ingress einströmen; *embr* einwandern
ingression
  Einströmen; *embr* Einwanderung
inhalable atembar
inhalation Inhalation, Inspiration,
  Einatmung, Einatmen
inhale (breathe in) inhalieren, einatmen
inherit erben, ererben
inheritable disease Erbkrankheit
inhibit inhibieren, hemmen
inhibition Inhibition, Hemmung
> **competitive inhibition**
  kompetitive Hemmung,
  Konkurrenzhemmung
inhibition zone Hemmzone
inhibitor (inhibitory substance)
  Hemmstoff

inhibitory hemmend,
inhibierend, inhibitorisch
inhibitory concentration
Hemmkonzentration
initial distribution *stat*
Ausgangsverteilung
initial magnification Maßstabzahl
initial pressure/
initial compression/high pressure
Vordruck, Eingangsdruck
(Hochdruck: Gasflasche)
initial velocity (vector)/initial rate
Anfangsgeschwindigkeit
(v_0: Enzymkinetik)
initial weight/amount weighed/
weighed amount/weighed quantity
Einwaage
inject (shoot)
spritzen, einspritzen, injizieren
injection Einspritzung, Injektion;
(shot) Spritze
(eine S. geben/bekommen)
injection molding *polym*
Spritzgießen, Spritzguss
injection valve/
injection port/syringe port
Einspritzventil (Einspritzblock)
injector Einspritzer
injure verletzen
injurious schädlich;
(i. to health) gesundheitsschädlich
injury Verletzung
➢ **chilling injury**
Kälteschaden, Kälteschädigung
➢ **frost injury/freezing injury**
Frostschaden, Frostschädigung
➢ **needle stick injury**
Nadel-Stichverletzung
➢ **occupational injury**
Berufsverletzung
➢ **radiation injury**
Strahlenverletzung,
Strahlenschaden, Verstrahlung
ink Tinte
inlet Zuleitung, Zulauf;
Eingang, Einlass
inlet opening Zufuhröffnung
inlet pipe Zuleitungsrohr

inlet system Einlasssystem
inlet valve Einlassventil
inlet velocity (hood)
Einströmgeschwindigkeit,
Eintrittsgeschwindigkeit
(Sicherheitswerkbank)
inoculate *med* inokulieren,
einimpfen, impfen;
micb beimpfen
inoculating loop Impföse, Impfschlinge
inoculating needle Impfnadel
inoculating wire Impfdraht
inoculation *med* Impfung,
Inokulation, Einimpfung;
(vaccination) Vakzination;
(immunization) Immunisierung;
micb Beimpfung
inoculation method/inoculation
technique Beimpfungsverfahren
inoculum/vaccine
Impfstoff, Inokulum,
Inokulat, Vakzine
inorganic anorganisch
inorganic chemistry
Anorganische Chemie
input Eingabe;
electr Eingang; *ecol* Eintrag
input air Zuluft
insatiability Unersättlichkeit
insatiable unersättlich
insect pest Insektenplage
insecticide Insektizid,
Insektenbekämpfungsmittel,
Insektenvernichtungsmittel
insensitive unempfindlich
insert *n* Einsatz, Einlage; (leaflet/slip:
package) Beipackzettel
insert *vb* (inserted)
inserieren (inseriert),
hinein stecken, einlegen, einschieben
insertion reaction *chem*
Insertionsreaktion,
Einschiebungsreaktion
insolation Sonneneinstrahlung
insolubility Unlöslichkeit
insoluble unlöslich
➢ **insoluble in water**
wasserunlöslich

in

inspection Inspektion,
Begehung, Besichtigung
(z.B. Geländebegehung);
(checking of goods) Warenkontrolle
inspection log Inspektions-Logbuch
inspection of corpse/
postmortem examination
Leichenschau
inspiration/inhalation Inspiration,
Einatmen, Einatmung, Inhalation
inspire inspirieren, einatmen
install installieren; anschließen;
(set up) einrichten (Experiment etc.)
installation(s) Installation(en),
Installierung, Anschluss, Einbau;
Anlage, Einrichtung,
Betriebseinrichtung
instruct anweisen;
(train/teach) unterweisen;
(brief/advise) belehren
instructions
(training/teaching/guidance)
Anleitung, Anweisung,
Einarbeitung, Betreuung;
(briefing/advice) Unterweisung,
Belehrung; Ratschlag
➢ **regulations/directions/rules/policy**
Richtlinie(n)standards
➢ **specifications/**
directions/prescription
Anweisungen, Vorschriften
instrument Instrument, Gerät,
Werkzeug, Apparat,
techn. Vorrichtung
instrument display/
(abgelesener Wert)
instrument reading
Instrumentenanzeige
instrumental error Gerätefehler
instrumentation and control
Mess- und Regeltechnik
insulate isolieren,
abschirmen, dämmen
insulated gloves Isolierhandschuhe
insulating panel Dämmplatte
insulating tape
(see also: duct tape)
Isolierband

integrated pest management (IPM)
integrierte Schädlingsbekämpfung/
Schädlingskontrolle,
integrierter Pflanzenschutz
interaction Interaktion,
Wechselwirkung
intercalation agent/
intercalating agent
interkalierendes Agens
interchangeable (replaceable)
austauschbar (ersetzbar)
intercom (intercom system)
Sprechanlage
➢ **two-way intercom/**
two-way radio
Wechselsprechanlage,
Gegensprechanlage
interdisciplinary research
interdisziplinäre Forschung
interface
Grenzfläche, Trennungsfläche;
electr Schnittstelle; Nahtstelle
interfacial surface tension
Grenzflächenspannung
interference assay Interferenzassay
interference microscopy
Interferenzmikroskopie
intergrade (intermediary form/
transitory form/transient)
Zwischenstufe, Übergangsform
interim storage/
temporary storage
Zwischenlager
interlock/fail safe circuit
Riegel, Verriegelung
(elektr. Sicherung)
interlocking relay *electr* Sperrrelais
intermediate (intermediate product/
intermediate form)
Zwischenstadium,
Zwischenprodukt,
Zwischenform
intermediate bulk container (IBC)
Großpackmittel
intermediate density lipoprotein (IDL)
Lipoprotein mittlerer Dichte
intermediate state/intermediate stage
Zwischenstadium, Zwischenstufe

internal/intrinsic
intern, innerlich, von innen
internal fittings (built-in elements/ structural additions) Einbauten
internal thread/female thread (tubes/pipes/fittings) Innengewinde
international standards
internationale Richtlinie
International Unit (IU)/ SI unit (*fr:* Système Internationale)
Internationale Maßeinheit, SI Einheit
interpolate interpolieren; einfügen
interpretation
Interpretation, Auslegung; Auswertung (von Ergebnissen)
interrelation/interrelationship
Wechselbeziehung
interval Intervall; Abstand
interval scale *stat* Intervallskala
interwoven/intertwined/entangled
verflochten
introduce einführen; vorstellen; (import) importieren
introduction (to)
Einführung; Vorstellung; Anleitung; (import) Importieren
invariant residue *math*
unveränderter Rest, invarianter Rest
invent erfinden
invention Erfindung
inventory Inventar, Bestand; Inventur
➢ **to make an inventory/ to take inventory** eine Bestandsaufnahme/Inventur machen
inverse PCR
inverse Polymerasekettenreaktion
invert sugar Invertzucker
inverted invers, invertiert, umgekehrt
inverted image Kehrbild
inverted microscope
Umkehrmikroskop, Inversmikroskop
inverter *electr* Wechselrichter
investigate
(examine/test/try/assay/analyze)
prüfen, untersuchen, testen, probieren, analysieren

investigation
(examination/exam/study/ search/test/trial/assay/analysis)
Untersuchung, Prüfung, Test, Probe, Analyse
invisible unsichtbar
iodination Iodierung
(mit Jod reagieren/substituieren)
iodine (I) Iod (*früher:* Jod)
iodine number/iodine value Iodzahl
iodization Iodierung
(mit Iod/Iodsalzen versehen)
iodize iodieren
(mit Iod/Iodsalzen versehen)
iodized salt Iodsalz
iodoacetic acid Iodessigsäure
ion Ion
➢ **counterion** Gegenion
➢ **daughter ion** Tochterion
➢ **fragment ion (MS)** Bruchstückion
➢ **molecular ion (MS)** Molekülion
➢ **parent ion (MS)**
Mutterion, Ausgangsion
➢ **radical ion** Radikalion
➢ **zwitterion** Zwitterion
ion channel (membrane channel)
Ionenkanal (Membrankanal)
ion equilibrium/ionic steady state
Ionengleichgewicht
ion-exchange chromatography (IEX)
Ionenaustauschchromatographie
ion-exchange resin
Ionenaustauscherharz
ion exchanger Ionenaustauscher
➢ **anion exchanger**
(*strong:* SAX/*weak:* WAX)
Anionenaustauscher (starke/schwacher)
➢ **cation exchanger**
(*strong:* SCX/*weak:* WCX)
Kationenaustauscher (starker/schwacher)
ion pair Ionenpaar
ion pore Ionenpore
ion product Ionenprodukt
ion pump Ionenpumpe
ion spray Ionenspray
ion transport Ionentransport

ion trap detector (ITD)
  Ioneneinfangdetektor (MS)
ion trap spectrometry
  Ionen-Fallen-Spektrometrie
ion-selective electrode (ISE)
  ionenselektive Elektrode
ionic ionisch
ionic bond Ionenbindung
ionic conductivity Ionenleitfähigkeit
ionic coupling Ionenkopplung
ionic current/ion current
  Ionenstrom
ionic formula Ionenformel
ionic radius Ionenradius
ionic strength Ionenstärke
ionization Ionisation
ionization chamber
  Ionisationskammer
ionize ionisieren
ionizing radiation
  ionisierende Strahlen,
  ionisierende Strahlung
ionophore Ionophor
ionophoresis
  Ionophorese, Iontophorese
ion-pair chromatography (IPC)
  Ionenpaarchromatographie (IPC)
ion-selective electrode (ISE)
  ionenselektive Elektrode
iris diaphragm *micros*
  Irisblende
iron (Fe) Eisen
> **cast iron** Gusseisen
> **malleable iron/**
  malleable cast iron/
  wrought iron
  Tempereisen; Temperguss
> **sheet iron** Eisenblech
> **wrought iron**
  Schmiedeeisen
IRP (island rescue PCR)
  IRP (inselspezifische PCR)
irradiance/
  fluence rate/
  radiation intensity/
  radiant-flux density
  Bestrahlungsintensität,
  Bestrahlungsdichte

irradiate bestrahlen
irradiation Bestrahlung
irregular irregulär, anomal;
  (anomalous) unregelmäßig;
  (non-uniform) ungleichmäßig
irregularity Irregularität, Anomalie;
  (anomaly) Unregelmäßigkeit;
  (non-uniformity) Ungleichmäßigkeit
irreversible inhibition
  irreversible Hemmung
irrigate
  bewässern, beregnen (künstlich)
irrigation
  Bewässerung, Beregnung
irritability Reizbarkeit
irritable reizbar;
  (excitable/sensitive)
  reizempfänglich;
  (sensible) empfindlich
irritant *n* Reizstoff
irritant (Xi) reizend
irritant gas Reizgas
irritate *med/physio/chem*
  reizen, irritieren, reizen
irritation Irritation; (stimulus)
  Stimulus, Reiz; (stimulation)
  Stimulation, Reizung
irritation of the mucosa
  Schleimhautreizung
isoelectric focusing
  isoelektrische Fokussierung,
  Isoelektrofokussierung
isoelectric point
  isoelektrischer Punkt
isolate isolieren, darstellen,
  rein darstellen;
  (separate) abtrennen
isolating mechanism *ecol*
  Isolationsmechanismus
isolation medium
  Isolationsmedium
isomer *n* Isomer
isomeric isomer
isomerism/isomery
  Isomerie
isomerization Isomerisation
isomerize isomerisieren
isomerous isomer, gleichzählig

**isopycnic centrifugation/
isodensity centrifugation**
isopyknische Zentrifugation
isosmotic isosmotisch
isotachophoresis (ITP)
Isotachophorese,
Gleichgeschwindigkeits-
Elektrophorese
isothiocyanic acid Isothiocyansäure

isotonic isotonisch
isotonicity Isotonie
isotope assay Isotopenversuch
isotopic dilution Isotopenverdünnung
isotropic isotrop, einfachbrechend
isovaleric acid Isovaleriansäure
issue point
  (issueing/supplies issueing)
Ausgabe (Materialausgabe)

ja

jack Heber, Hebevorrichtung, Hebebock; *electr* Klinke
› **lab jack** Hebestativ, Hebebühne, höhenverstellbare Plattform (fürs Labor)
jack knife Klappmesser
jacket (insulation) Mantel, Ummantelung, Umhüllung, Hülle, Wicklung, Umwicklung, Verkleidung; Manschette, Tropfschutz
jacketed (insulated) ummantelt, verkleidet
jacketed coil condenser Intensivkühler
jacketing Ummantelung, Verkleidung; Mantelmaterial
jacklift/lifting truck Hubwagen
jammed/ seized-up/stuck/'frozen'/caked verbacken, festgebacken, festgesteckt (Schliff, Hahn)
jasmonic acid Jasmonsäure
jaw crusher/jaw breaker Backenbrecher
jelly Gelee; (gelatin/gel) Gallerte, Gelatine
jet Strahl; (nozzle) Düse
jet flame (jetting flame) Stichflamme
jet loop reactor Düsenumlaufreaktor, Strahl-Schlaufenreaktor
jet of water Wasserstrahl
jet reactor Strahlreaktor
jigsaw machine Schweifsäge (Maschine)
jimmy/crowbar Brecheisen
job Arbeitsstelle
joint Verbindung; Fuge; Naht, Nahtstelle; Gelenk; Verbindungsstück
› **ball-and-socket joint/ spheroid joint** Kugelgelenk
› **butt joint** Stoßverbindung
› **flanged joint** Flanschverbindung
› **ground-glass joint** Glasschliffverbindung, Schliffverbindung (N.S. = Normalschliff)
› **jammed joint/ stuck joint/caked joint/ 'frozen' joint** festgebackener Schliff
› **male/female joint** Plus-/Minus-Verbindung (Rohrverbindungen etc.)
› **plane-ground joint (flat-flange ground joint/ flat-ground joint)** Planschliffverbindung,
› **spherical ground joint** Kugelschliffverbindung, Kugelschliff
› **tapered ground joint/ tapered joint (S.T. = standard taper)** Kegelschliffverbindung, Kegelschliff (N.S. = Normalschliff)
joint clip/ ground-joint clip/ ground-joint clamp Schliffklammer, Schliffklemme (Schliffsicherung)
joint sleeve (for ground-glass joints) Manschette für Schliffverbindungen
journal *tech/mech* Achsenlager, Achslager, Zapfenlager (z.B. beim Kugellager)
journal box Lagerbüchse (Kugellager)
judgement Bewertung, Beurteilung
jug/pitcher Krug, Kanne, Kännchen
juicy saftig
jump (spring/bound/leap) springen
jumper cable/coupling cable Überbrückungskabel, Starterkabel

keep clear! Abstand halten!
keeper/warden
  (caretaker of animals)
  Tierpfleger, Tierwärter
keratinize (cornify)
  keratinisieren (verhornen)
kerosene Kerosin
keto acid Ketosäure
ketoaldehyde/
  aldehyde ketone Ketoaldehyd
ketone Keton
➢ **acetone/**
  dimethyl ketone/2-propanone
  Aceton (Azeton), Propan-2-on,
  2-Propanon, Dimethylketon
ketone body (acetone body)
  Ketonkörper
ketonuria/acetonuria Ketonurie
key Schlüssel;
  (button/knob/push-button) Taste;
  (stopcock key/plug) Hahnküken,
  Küken; *biol* Bestimmungsschlüssel
key stimulus/sign stimulus
  (release stimulus)
  Schlüsselreiz, Auslösereiz
keyboard (große) Tastatur
keypad (kleine) Tastatur
kick over (knock over)
  (mit dem Fuß/Bein/Körper)
  umstoßen, umkippen, umwerfen
kieselguhr
  (loose/porous diatomite;
  diatomaceous/infusorial earth)
  Kieselgur
kiln
  (kiln oven:
  for drying grain/lumber/tobacco)
  Darre, Darrofen
kiln-dry darren
Kimwipes
  (Kimberley-Clark cleanroom wipes)
  Kimwipes (Kimberley-Clark
  Reinraum Wischtücher)

kindling temperature
  (ignition point/
  flame temperature/flame point/
  spontaneous-ignition temperature
  SIT)
  Zündpunkt, Zündtemperatur,
  Entzündungstemperatur
kinetics (zero-/first-/second-order...)
  Kinetik (nullter/erster/zweiter
  Ordnung)
➢ **reaction kinetics** Reaktionskinetik
➢ **reassociation kinetics**
  Reassoziationskinetik
Kipp generator
  Kippscher Apparat, 'Kipp',
  Gasentwickler
kitchen tissue (kitchen paper towels)
  Küchenrolle, Haushaltsrolle,
  Tücherrolle, Küchentücher,
  Haushaltstücher
kitchen towel Küchenhandtuch
Kjeldahl flask
  Birnenkolben,
  Kjeldahl-Kolben
knife Messer
➢ **amputating knife** Amputiermesser
➢ **cable stripping knife** Kabelmesser
➢ **cartilage knife** Knorpelmesser
➢ **fleshing knife** Fleischmesser
➢ **jack knife** Klappmesser
➢ **pocket knife** Taschenmesser
➢ **putty knife**
  Spachtelmesser, Kittmesser
➢ **safety cutter** Sicherheitsmesser
knife holder *micros* Messerhalter
knife switch *electr* Messerschalter
knurled nut Rändelmutter
knurled screw/knurled thumbscrew
  Rändelschraube
Koch's postulate
  Koch's Postulat, Koch'sches Postulat
Koehler illumination *micros*
  Köhlersche Beleuchtung

LD

LD$_{50}$ (median lethal dose)
  LD$_{50}$ (mittlere letale Dosis)
LDL (low density lipoprotein)
  LDL
  (Lipoproteinfraktion niedriger Dichte)
lab (laboratory) Labor (*pl* Labors), Laboratorium (*pl* Laboratorien)
lab aide Laborgehilfe
lab apron Laborschürze
lab balance/lab scales Laborwaage
lab bench Laborarbeitstisch
lab bottle
  Laborstandflasche, Standflasche
lab chemical Laborchemikalie
**lab chewing gum
  (sticks to glass/metal/wood!)**
  sog. 'Laborkaugummi'
lab cleanup Laborreinigung
lab coat Laborkittel
lab conditions Laborbedingungen
lab counter Laborbank
lab courtesy Labor-Anstandsregeln
lab diary/lab manual/log book
  Labortagebuch
lab etiquette Laboretikette,
  Laborgepflogenheiten,
  Laborbenimmregeln,
  Labor'knigge'
lab experiment/lab test
  Laborversuch, Labortest
lab furniture Labormöbel
lab gossip Labortratsch
lab grade (chem grade)
  technisch rein (Laborchemikalie)
lab head Laborleiter
lab jack Hebestativ, Hebebühne, höhenverstellbare Plattform (fürs Labor)
lab manual/lab diary/log book
  Labortagebuch
lab notes/lab documentation
  Laboraufzeichnungen
lab procedure Laborverfahren
lab protocol Laborprotokoll
lab pushcart (*Br* trolley)
  Laborwagen,
  Laborschiebewagen
lab reagent/bench reagent
  Laborreagens
lab report Laborbericht
lab roach Laborkakerlake
lab safety Laborsicherheit
lab scale *n* Labormaßstab
lab-scale *adj/adv*
  labortechnisch,
  im Labormaßstab
lab space/lab working space
  Laborplatz, Laborarbeitsplatz
lab standard Laborstandard
lab stool Laborhocker
lab supplies Laborbedarf
lab technician/technical lab assistant
  technischer Assistent
  (technische Assistentin),
  Laborassistent (Laborassistentin),
  Laborant (Laborantin)
lab technique Labortechnik
lab tray Laborschale
lab unit Laboratoriumseinheit
lab worker
  Laborant(in), Laborarbeiter
labcoat Laborkittel, Labormantel
label *n* Markierung, Marke;
  Kennzeichen;
  (tag) Etikett, Beschriftungsetikett
  ➢ **warning label** Warnetikett
label *vb* markieren; kennzeichnen;
  beschriften; (tag) etikettieren
labeling (labelling)
  Markierung, Kennzeichnung;
  Beschriftung;
  (tagging) Etikettierung
labeling requirement
  Kennzeichnungspflicht
labile
  labil, instabil, unbeständig
laboratory (lab) Labor (*pl* Labors),
  Laboratorium (*pl* Laboratorien)
  ➢ **animal laboratory/animal lab**
    Tierlabor
  ➢ **research laboratory**
    Forschungslabor
  ➢ **teaching laboratory/
    educational laboratory**
    Lernlabor

laboratory aide Laborgehilfe
laboratory animal Versuchstier
laboratory apron/lab apron
  Laborschürze
laboratory balance Laborwaage
laboratory bench Labor-Werkbank
laboratory bottle
  Laborstandflasche, Standflasche
laboratory brush Laborbürste
laboratory cart/
  lab pushcart (*Br* trolley)
  Laborwagen, Laborschiebewagen
laboratory chemical Laborchemikalie
laboratory cleanup Laborreinigung
laboratory coat/labcoat
  Laborkittel, Labormantel
laboratory conditions
  Laborbedingungen
laboratory diary/lab manual/log book
  Labortagebuch
laboratory equipment Laborgerät
laboratory facilities
  Laboreinrichtung, Laborausstattung
laboratory findings/laboratory results
  Laborbefund
laboratory furniture Labormöbel
laboratory gossip Labortratsch
laboratory jack/lab-jack
  Hebestativ, Hebebühne,
  höhenverstellbare Plattform
  (fürs Labor)
laboratory notebook
  Laborjournal, Protokollheft
laboratory personnel Laborpersonal
laboratory procedure Laborverfahren
laboratory protection plate
  Laborschutzplatte (Keramikplatte)
laboratory protocol Laborprotokoll
laboratory reagent/bench reagent
  Laborreagens
laboratory results
  Laborergebnisse; Laborbefund
laboratory safety Laborsicherheit
laboratory safety officer
  Laborsicherheitsbeauftragter
laboratory scale *n* Labormaßstab
laboratory-scale *adj/adv*
  im Labormaßstab, labortechnisch

laboratory suite Labortrakt,
  Laboratoriumstrakt
laboratory supplies Laborbedarf
laboratory table/laboratory workbench
  Labortisch, Labor-Werkbank
laboratory technician/
  lab technician/
  technical lab assistant
  technischer Assistent
  (technische Assistentin),
  Laborassistent (Laborassistentin),
  Laborant (Laborantin)
laboratory technique Labortechnik
laboratory tray Laborschale
laboratory unit Laboratoriumseinheit
laboratory worker
  Laborant(in), Laborarbeiter
laborious mühselig, schwer, arbeitsam
labware/
  laboratory supplies/lab supplies
  Laborbedarf
lachrymatory
  tränend (Tränen hervorrufend)
lacking/missing/wanting fehlend
lacquer Lack, Firnis, Farblack
lactic acid (lactate) Milchsäure (Laktat)
lactic acid fermentation/
  lactic fermentation
  Laktatgärung, Milchsäuregärung
lactose (milk sugar)
  Laktose, Lactose (Milchzucker)
ladder Leiter
lading Beladen, Verladen, Befrachten;
  Ladung, Fracht
lading form Frachtbrief-Formular
ladle Schöpfkelle, Gießlöffel
lag phase
  (latent phase/incubation phase/
  establishment phase)
  Anlaufphase, Latenzphase,
  Inkubationsphase,
  Verzögerungsphase,
  Adaptationsphase, lag-Phase
laminar flow
  laminare Strömung, Schichtströmung
laminar flow workstation/
  laminar flow hood/laminar flow unit
  Querstrombank

laminate (laminated plastic) Laminat
laminated paper Hartpapier
lancet Lanzette
landfill Deponie
➢ **sanitary landfill**
   Müllgrube, Mülldeponie (geordnet)
landing net/aquatic net
   (collecting net for fish) *ecol*
   Kescher, Käscher
   (Fangnetz für Fische)
lap-joint flange Bördelflansch
lard Schmalz,
   Schweineschmalz, Schweinefett
large scale Großmaßstab
large-scale *adj*
   im Großmaßstab, großtechnisch
lashing strap Zurrgurt
latch Raste, Riegel, Schnappriegel,
   Schnappschloss, Schnäpper,
   Rastklinke; Verriegelung, Sperre
latency Latenz
latency period/latent period
   Latenzzeit, Suchzeit,
   Verzögerungszeit;
   (incubation period) Inkubationszeit
latent latent,
   verborgen, unsichtbar, versteckt
latent phase
   (incubation phase/
   establishment phase/lag phase)
   Latenzphase, Adaptationsphase,
   Anlaufphase, Inkubationsphase,
   lag-Phase
lateral lateral, seitlich
lateral magnification *micros*
   Lateralvergrößerung,
   Seitenverhältnis, Seitenmaßstab,
   Abbildungsmaßstab
latex Latex, Milchsaft
lath/plank Latte (aus Holz)
lathe Drehbank, Drehmaschine
lattice Gitter
lattice energy Gitterenergie
lattice sampling/grid sampling *stat*
   Gitterstichprobenverfahren
laughing gas/nitrous oxide
   Lachgas,
   Distickstoffoxid, Dinitrogenoxid

laundrette Schnellwäscherei
laundry Wäsche; Wäscherei
laundry hutch/laundry hamper
   Wäschekorb
lauric acid/decylacetic acid/
   dodecanoic acid
   (laurate/dodecanate)
   Laurinsäure, Dodecansäure
   (Laurat/Dodecanat)
lavatory Waschraum, Toilette
law (act/statute) Gesetz; Regel
   (*siehe auch bei:* Verordnung)
➢ **against the law/illegal/unlawful**
   gesetzeswidrig
law of conservation of energy
   Energieerhaltungssatz
law of conservation of matter
   Massenerhaltungssatz
law of mass action
   Massenwirkungsgesetz
lawn *micb* Rasen
lawn culture Rasenkultur
laxative Abführmittel
leach *vb* auslaugen (Boden)
leachate Lauge (Bodenauslaugung)
leaching Auswaschung, Auslaugung
   (gelöste Bodenmineralien)
lead *electr* Kontakt
➢ **pigtail lead** *electr* Anschlussleitung
lead *vb* führen, anführen, leiten
lead (Pb) Blei
lead citrate Bleicitrat
lead dioxide/brown lead oxide/
   lead superoxide PbO_2
   Bleidioxid, Blei(IV)oxid
lead oxide Bleioxid
➢ **lead dioxide/lead superoxide**
   (brown) PbO_2
   Blei(IV)oxid, Bleidioxid
➢ **lead protooxide**
   (yellow monoxide) PbO
   Bleioxid, Bleiglätte, Massicot
➢ **lead suboxide (yellow)** Pb_2O
   Bleioxid
➢ **lead tetraoxide (red)/**
   minium Pb_3O_4
   Blei(II,IV)oxid, Blei(II)plumbat(IV),
   Mennige, Minium

lead ring (for Erlenmeyer)
  Gewichtsring, Stabilisierungsring,
  Beschwerungsring,
  Bleiring (für Erlenmeyerkolben)
lead sulfate Bleisulfat
lead sulfide/galena PbS
  Bleisulfid, Bleiglanz, Galenit
leader/head ('boss')
  Leiter, Führungskraft
  (Vorgesetzter/'Chef')
leaflet (notice/instructions) Merkblatt
leak *n* (leakage) Leck;
  (leakiness) Undichtigkeit
leak *vb* **(doesn't close tightly)**
  undicht sein
➢ **leak out/bleed**
  auslaufen (Flüssigkeit)
leak rate Leckagerate
leakage
  Leck, Auslauf, Austritt, Leckage;
  electr Sreuverlust
leakage current (creepage) *electr*
  Kriechstrom
**leakage current circuit breaker/
  surge protector (fuse)**
  FI-Schalter
  (Fehlerstromschutzschalter)
leaking/leaky
  undicht, leck; läuft aus
leakproof/leaktight (sealed tight)
  dicht, leckfrei, lecksicher
leaky undicht, leck; läuft aus
**least significant difference/
  critical difference** *stat*
  Grenzdifferenz
leavening/raising agent
  Treibmittel (Gärmittel/Gärstoff)
**leaving group/
  coupling-off group** *chem*
  Austrittsgruppe, Abgangsgruppe,
  austretende Gruppe
left-handed linksgängig;
  (sinistral) linkshändig
legal requirements
  rechtliche/gesetzliche Auflage(n)
lens (*also:* **lense**) Linse;
  (magnifying glass) Lupe,
  Vergrößerungsglas

lens tissue/lens paper *micros*
  Linsenpapier,
  Linsenreinigungspapier
lesion Läsion,
  Schädigung, Verletzung, Störung
lethal/deadly letal, tödlich
➢ **conditional lethal**
  bedingt letal, konditional letal
lethality Letalität
level *adj/adv* eben;
  waagrecht, horizontal;
  gleich, gleichmäßig
level *n* Ebene, ebene Fläche; Niveau;
  Horizontale, Waagrechte; Höhe;
  (tool) Wasserwaage; Libelle
level switch Niveauschalter
leveling nivellieren, einebnen,
  planieren, gleichmachen
lever *tech/mech* Hebel
lever ratchet Hebelknarre
leverage
  Hebelübersetzung,
  Hebelkraft, Hebelwirkung
leverage mechanism
  Hebelmechanismus
levulinic acid Lävulinsäure
liability
  Haftung, Haftpflicht, Haftbarkeit;
  Verpflichtung, Verbindlichkeit
liability insurance
  Haftpflichtversicherung
liable
  haftpflichtig, haftbar, verantwortlich;
  ausgesetzt, unterworfen
liberate (release/set free)
  freisetzen
  (Wärme/Energie/Gase etc.)
librarian Bibliothekar(in)
library Bibliothek
➢ **bank (clone bank)**
  Bank (Klonbank)
➢ **departmental library**
  Institutsbibliothek
➢ **unit library** Bereichsbibliothek
licence *n* Lizenz, Zulassung, Erlaubnis,
  Genehmigung, Konzession
lichen acid Flechtensäure
lick lecken

li

lid (cover/top) Deckel
Liebig condenser Liebigkühler
life Leben
  ➤ **danger of life/life threat**
    Lebensgefahr
  ➤ **essential for life/vital**
    lebenswichtig, lebensnotwendig,
    vital
  ➤ **half-life** Halbwertszeit;
    (Enzyme) Halblebenszeit
  ➤ **pot life** *chem*
    Topfzeit, Verarbeitungsdauer
  ➤ **service life (of a machine/equipment)**
    Laufzeit, Lebenszeit, Lebensdauer
  ➤ **shelf life**
    Haltbarkeit, Lagerfähigkeit;
    Verfallsdatum
**life cycle assessment/
  life cycle analysis (LCA)** *ecol*
    Ökobilanz
life expectancy Lebenserwartung
life science/biology
  Naturkunde, Biologie
life size Lebensgröße
life span Lebensdauer
life zone/biotope
  Lebensraum, Lebenszone, Biotop
lifeform/organism
  Lebewesen, Organismus
lifeless/inanimate/dead
  leblos, tot
life-threatening
  lebensgefährlich
life-time/lifetime Lebenszeit;
  (of membrane channels)
  Öffnungsdauer
lift *n* Heben, Hochhalten; Hub,
  Hubhöhe, Förderhöhe, Steighöhe;
  Aufzug, Fahrstuhl;
  (buoyancy) Auftrieb
lift gate
  Hubladebühne, Hubschranke,
  Hubtor, Hubklappe
lifting jack Hebevorrichtung,
  (Hebe)Winde, Bock
lifting platform Hebebühne
lifting truck/jacklift Hubwagen
ligament Ligament, Band

ligand Ligand
ligand blotting Liganden-Blotting
ligase chain reaction
  Ligasekettenreaktion
ligation Ligation, Verknüpfung
ligation-mediated PCR *gen*
  ligationsvermittelte
  Polymerasekettenreaktion
light Licht
  ➤ **beam of light**
    Lichtstrahl, Lichtbündel
  ➤ **emergent light** ausgestrahltes Licht
  ➤ **incident light** einfallendes Licht
  ➤ **plane-polarized light**
    linear polarisiertes Licht
  ➤ **point of light** Lichtpunkt
  ➤ **polarized light** polarisiertes Licht
  ➤ **scattered light/stray light**
    Streulicht
**light bulb/lightbulb/
  incandescent lamp**
  Glühbirne, Glühlampe
light fastness
  Lichtechtheit, Lichtbeständigkeit
**light microscope (compound
  microscope)** Lichtmikroskop
light microscopy Lichtmikroskopie
light permeability Lichtdurchlässigkeit
light pipe (fiberoptics)
  Lichtleiter (Kaltlicht)
  ➤ **gooseneck** Schwanenhals
light scattering Lichtstreuung
**light-sensitive/
  photosensitive/sensitive to light**
  lichtempfindlich (leicht reagierend)
light sensitivty Lichtempfindbarkeit
light source Lichtquelle
light stimulus Lichtreiz
lightweight leichtgewicht(ig)
lignification/sclerification
  Verholzung, Lignifizierung
lignified verholzt, lignifiziert
lignite Lignit
  (Weichbraunkohle &
  Mattbraunkohle)
lignoceric acid/tetracosanoic acid
  Lignocerinsäure, Tetracosansäure
ligroin/petroleum spirit Ligroin

likelihood function
  Wahrscheinlichkeitsfunktion
lime *vb* **(calcify)** kalken
lime *n* Kalk
➢ **caustic lime CaO** Branntkalk
➢ **slaked lime Ca(OH)$_2$**
  Ätzkalk, Löschkalk,
  gelöschter Kalk
➢ **soda lime** Natronkalk
lime deposit Kalkablagerung
limestone Kalkstein
limestone deposit Kalkablagerung
liming Kalkung
limit *n* Grenze, Begrenzung, Grenzwert;
  (limiting value) Grenzwert,
  Schwellenwert
limit of detection (LOD)/
  detection limit
  Bestimmungsgrenze,
  Nachweisgrenze
limit of resolution Auflösungsgrenze
limit valve/limiting valve
  Begrenzungsventil
limited capacity control system (LCCS)
  limitiertes Kapazitätskontrollsystem
limiting concentration
  Grenzkonzentration
limiting factor *ecol*
  Grenzfaktor, begrenzender Faktor,
  limitierender Faktor
limiting value
  Grenzwert, Schwellenwert
limp schlaff (welk)
limy/limey/calcareous
  kalkig, kalkartig, kalkhaltig
line/lines *tech/mech/electr*
  Leitung(en); Anschluss/Anschlüsse
  (Gas~/Strom~/Wasser~)
line *vb* **(coat/cover/laminate)**
  füttern, überziehen, beschichten
line diagram
  Strichdiagramm
line spectrum
  Linienspektrum, Atomspektrum
line transect method
  Linienstichprobenverfahren
linesman pliers
  Telefonzange, Kabelzange

lining (coat/coating/
  covering/lamination)
  Futter, Futterstoff, Fütterung,
  Auskleidung; Beschichtung,
  Isolationsschicht
link (up to) verbinden, anschließen;
  verketten, verknüpfen; sich
  zusammenfügen
linolenic acid Linolensäure
linolic acid/linoleic acid Linolsäure
linseed oil Leinöl
lint *n* Lint; Fussel(n)
lip seal/lip-type seal
  Lippendichtung
  (Wellendurchführung)
lipid Lipid
lipoic acid (lipoate)/thioctic acid
  Liponsäure, Dithiooctansäure,
  Thioctsäure, Thioctansäure (Liponat)
lipophilic lipophil
lipoteichoic acid Lipoteichonsäure
liquefaction Verflüssigung
liquefaction of air Luftverflüssigung
liquefied natural gas (LNG)
  Flüssiggas (verflüssigtes Erdgas)
liquefier Verflüssiger
liquefy/liquify verflüssigen
liquid *adj/adv* flüssig, liquid
liquid *n* Flüssigkeit
liquid air flüssige Luft
liquid chromatography (LC)
  Flüssigkeitschromatographie
liquid crystal display
  Flüssigkristallanzeige
liquid gas/liquefied gas
  Flüssiggas
liquid nitrogen
  Flüssigstickstoff,
  flüssiger Stickstoff
liquid oxygen
  Flüssigsauerstoff,
  flüssiger Sauerstoff
liquid soap/liquid detergent
  Flüssigseife
liquid state flüssiger Zustand
liquidus temperature
  Liquidustemperatur
liquify/liquefy verflüssigen

li

**litharge/massicot/
lead protooxide/lead oxide
(yellow monoxide) PbO**
Bleioxid
litmus paper Lackmuspapier
litocholic acid Litocholsäure
litter *n*
herumliegende Sachen/Abfall;
Streu
litter *vb* Abfall herumliegen lassen
litter bag Abfalltüte
live *vb* leben
live culture/living culture
Lebendkultur
live germ count Lebendkeimzahl
live vaccine
Lebendimpfstoff, Lebendvakzine
live weight Lebendgewicht
load *tech/mech* Last (Beladung),
Traglast; (freight) Fracht
(Flüssigkeit/Abwasser)
loading dock Laderampe
local anesthetic Lokalanästhetikum
locate orten
location Lage (Ort)
lock *n* **(closure)** Schloss (Verschluss)
lock *vb* verschließen
lock-and-key principle
Schlüssel-Schloss-Prinzip,
Schloss-Schlüssel-Prinzip
locker Spind, Schließfach
locomotion
Lokomotion, Bewegung
(Ortsveränderung)
locust bean gum/carob gum
Johannisbrotkernmehl,
Karobgummi
lod score
('logarithm of the odds ratio')
Lod-Wert
log off ausloggen
log on einloggen
log paper
Logarithmuspapier,
Logarithmenpapier
logarithmic phase (log-phase)
logarithmische Phase
logbook Arbeitstagebuch

**lognormal distribution/
logarithmic normal distribution**
Lognormalverteilung,
logarithmische Normalverteilung
lone pair (of electrons)
freies Elektronenpaar,
einsames Elektronenpaar
long-chain langkettig
long-distance heat(ing) Fernwärme
long-distance transport Ferntransport
long-lived/long-living langlebig
long-term Langzeit..., Dauer...
long-term run/operation
Dauerbetrieb, Dauerleistung,
Non-Stop-Betrieb
longevity Langlebigkeit
**longisection/
longitudinal section/long section**
Längsschnitt
**longnose pliers/
long-nose pliers**
Spitzzange
loop Schlaufe, Schlinge
loop conformation/coil conformation
Schleifenkonformation,
Knäuelkonformation
**loop reactor/
circulating reactor/recycle reactor**
Umlaufreaktor, Umwälzreaktor,
Schlaufenreaktor
loss on drying
Trocknungsverlust
loss on ignition Glühverlust
lot Posten, Partie (Waren);
(unit) Charge (Produktions-
menge/-einheit)
lot number/unit number
Chargen-Bezeichnung (Chargen-B.)
lounge
Gemeinschaftsraum, Pausenraum
low density lipoprotein (LDL)
Lipoprotein niedriger Dichte
low-field shift (NMR)
Tieffeldverschiebung
low-molecular niedermolekular
**low-voltage lamp/
low-voltage illuminator (spotlight)**
Niedervoltleuchte

LSE (least squares estimation)
MSQ-Schätzung (Methode der kleinsten Quadrate)
lubricant Gleitmittel, Schmiermittel, Schmierstoff, Schmiere, 'Fett'; (for ground joints) Schliff-Fett
lubricate (grease/oil)
schmieren (einfetten/einölen)
lubricating oil/lube oil Schmieröl
lubrication
Schmierung, Einfetten, Einschmieren
Luer Luer
➢ **female Luer hub (lock)** Luerhülse
➢ **lock** Luerlock, Luerverschluss
➢ **male Luer hub (lock)** Luerkern
➢ **tee** Luer T-Stück
➢ **tip** Luerspitze
lug *electr*
Kabelschuh, Ansatz, Öhr
lukewarm lauwarm
luminescence Lumineszenz
➢ **bioluminescence**
Biolumineszenz
luminescent
lumineszent, lumineszierend
luminescent paint Leuchtfarbe
luminescent screen Leuchtschirm
luminiferous
leuchtend, Licht erzeugend
luminophore (phosphor)
Leuchtstoff, Luminophor ('Phosphor')

luminosity Leuchtkraft; (light intensity) Lichtstärke, Lichtintensität
luminous
leuchtend, strahlend, Leucht...
luminous paint
Leuchtfarbe
luster Glanz; Lüster
➢ **metallic luster** Metallglanz
luster terminal (insulating screw joint)
Lüsterklemme
lute Kitt, Dichtungskitt, Dichtungsmasse; Gummiring (für Flaschen etc.)
lye Lauge
lyophilization/freeze-drying
Lyophilisierung, Gefriertrocknung
lyotropic series/Hofmeister series
Hofmeistersche Reihe, lyotrope Reihe
lysate Lysat
lyse lysieren
lysergic acid Lysergsäure
lysigenic/lysigenous lysigen
lysis Lyse
lysogenic (temperate)
lysogen (temperent)
lytic plaque/plaque
Lysehof, lytischer Hof, Aufklärungshof, Hof, Plaque

macerate mazerieren
maceration Mazeration
machine-washable
  waschmaschinenfest
➤ **dishwasherproof**
  spülmaschinenfest
macromolecule Makromolekül
macronutrients Kernnährelemente
macroscopic makroskopisch
magic acid (HSO_3F/SbF_5)
  Magische Säure
magic angle spinning (MAS)
  Rotation um den magischen Winkel
  (NMR)
magnesia/mangesium oxide
  Magnesia, Magnesiumoxid
magnesium (Mg) Magnesium
magnetic field Magnetfeld
magnetic resonance imaging (MRI)/
  nuclear magnetic resonance
  imaging
  Magnetresonanztomographie (MRT),
  Kernspintomographie (KST)
magnetic stirrer Magnetrührer
magnetism Magnetismus
magnification (enlargement)
  Vergrößerung
magnification at x diameters
  x-fache Vergrößerung
magnify (enlarge)
  vergrößern
magnifying glass/
  magnifier/lens
  Vergrößerungsglas
main band *chromat/electrophor*
  Hauptbande
mains (*Br*) Hauptstromleitung
mains cable (*Br*)/power cable
  Netzkabel
mains connection (*Br*)/
  power supply (electric hookup)
  Netzanschluss
maintenance (servicing)
  Instandhaltung, Wartung
maintenance coefficient (m)
  Erhaltungskoeffizient
maintenance costs
  Instandhaltungskosten

maintenance culture Erhaltungskultur
maintenance energy Erhaltungsenergie
maintenance medium
  Erhaltungsmedium
maintenance personnel
  Wartungspersonal
maintenance worker/maintenance man
  Wartungsmonteur
maintenance-free wartungsfrei
male (plug/couplings etc.) männlich
  (Stecker/Kupplung/Verbinder),
  Plus...; *tech/mech* Kern
male joint (plug/couplings etc.)
  Plus-Verbindung
  (Rohrverbindungen etc.)
male Luer hub (lock) Luerkern
maleic acid (maleate)
  Maleinsäure (Maleat)
malformation Fehlbildung
malfunction
  (functional disorder) *tech/med*
  Funktionsstörung; Dysfunktion
malfunction *vb tech/mech*
  schlecht funktionieren, versagen
malfunction report Fehleranzeige
malic acid (malate) Äpfelsäure (Malat)
malignancy/
  malignant nature
  Malignität, Bösartigkeit
malignant maligne, bösartig
malleable (kalt)hämmerbar, streckbar,
  dehnbar; verformbar; formbar
mallet Handfäustel
➤ **rubber mallet**
  Gummihammer (Fäustel)
malnourished fehlernährt;
  (undernourished) unterernährt
malnutrition Fehlernährung
malonic acid (malonate)
  Malonsäure (Malonat)
malt Malz
malt sugar/maltose Malzzucker,
  Maltose
mandatory report/
  mandatory registration
  (compulsory registration/
  obligation to register)
  Meldepflicht, Anmeldpflicht

mandelic acid/
  phenylglycolic acid/amygdalic acid
  Mandelsäure, Phenylglykolsäure
manganese (Mn) Mangan
manganese dioxide
  Braunstein, Manganoxid
manifest Manifest;
  Frachtliste, Frachtdokument
manifest document
  Ladeverzeichnis, Ladungsdokument
  (Warenverzeichnis)
manifold *n* Verteiler
man-made (artificial/synthetic)
  von Menschen gemacht
  (künstlich, naturfern, synthetisch)
mannitol Mannit
mannuronic acid Mannuronsäure
manual *adj/adv* manuell, von Hand,
  mit der Hand, handbetrieben
manual *n* (handbook/guide) Leitfaden,
  Handbuch; *biol* Bestimmungsbuch;
  (instructions) Handbuch, Anleitung,
  Gebrauchsanweisung
manual operation
  Handbedienung (Gerät)
manually controlled handgesteuert
manually operated
  mit Hand bedient
manufacture (manufacturing/
  preparation/production)
  Herstellung, Fertigung,
  Erzeugung, Produktion
manufacturer
  (producer) Hersteller, Erzeuger,
  Produzent; (manufacturing
  company/firm) Herstellerfirma
manufacturer catalog
  Herstellerkatalog
manufacturer's specifications
  Herstellerangaben
manufacturing process/procedure
  Herstellungsverfahren
manure/dung
  Mist, Dung; (droppings) Tierkot
map Karte, Landkarte
  (auch Stadtplan etc.)
map unit Karteneinheit
mapping/plotting Kartierung

marginal distribution *stat*
  Randverteilung
mark *vb* **(brand/earmark)**
  markieren, kennzeichnen
marker (genetic/radioactive) Marker,
  Markersubstanz
  (genetische/radioaktive); Markierstift
➤ **permanent marker (water-resistant)**
  wischfester/wasserfester Markierstift
marking/labeling Kennzeichnung
Marsh test Marshsche Probe
mask *vb* maskieren; verhüllen,
  verkleiden, verschleiern, verbergen;
  überdecken (z.B. Geschmack)
mask *n* Maske; Mundschutz
➤ **cartridge mask**
  Patronen-Filtermaske
➤ **dust mask (respirator)**
  Grobstaubmaske
➤ **dust-mist mask**
  Feinstaubmaske
➤ **emergency escape mask**
  Fluchtgerät, Selbstretter
  (Atemschutzgerät)
➤ **face mask/protection mask**
  Atemschutzmaske
➤ **filter cartridge** Filterpatrone
➤ **filter mask** Filtermaske
➤ **full-face respirator**
  Atemschutzvollmaske,
  Gesichtsmaske
➤ **full-facepiece respirator**
  Vollsicht-Atemschutzmaske
➤ **gas mask** Gasmaske
➤ **half-mask (respirator)** Halbmaske
➤ **mist mask/mist respirator mask**
  Feinstaubmaske
➤ **particulate respirator (U.S. safety
  levels N/R/P according to regulation
  42 CFR 84)** Staubschutzmaske
  (Partikelfilternde Masken)
  (DIN FFP)
➤ **protection mask/**
  face mask/
  respirator mask/respirator
  Atemschutzmaske
➤ **surgical mask** Operationsmaske,
  chirurgische Schutzmaske

ma

masking tape Kreppband, Maler-Krepp
mass Masse
> **biomass** Biomasse
> **dry mass/dry matter**
  Trockenmasse, Trockensubstanz
> **'fresh mass' (fresh weight)**
  'Frischmasse' (Frischgewicht)
> **molar mass ('molar weight')**
  Molmasse, molare Masse
  ('Molgewicht')
> **molecular mass ('molecular weight')**
  Molekülmasse ('Molekulargewicht')
> **molecular weight/**
  relative molecular mass (M_r)
  Molekulargewicht, relative
  Molekülmasse (M_r)
> **nominal mass**
  Nennmasse, Nominalmasse
mass action constant
  Massenwirkungskonstante
mass exchange/substance exchange
  Stoffaustausch
mass filter Massenfilter
mass flow/bulk flow
  Massenströmung (Wasser)
mass fraction
  Massenanteil (Massenbruch)
mass reproduction
  (mass spread/outbreak)
  Massenvermehrung
mass spectrometer/mass spec
  Massenspektrometer
mass spectrometry (MS)
  Massenspektrometrie (MS)
mass transfer Stoffübergang,
  Massenübergang, Stofftransport,
  Massentransport, Massentransfer
mass transfer coefficient
  Stoffübergangszahl,
  Stofftransportkoeffizient,
  Massentransferkoeffizient
mass-selective detector
  massenselektiver Detektor
mass-to-charge ratio *m/z* **(MS)**
  Masse-Ladungsverhältnis
massicot/litharge/
  lead protooxide/lead oxide
  (yellow monoxide) PbO Bleioxid

mat Matte
> **step mat/foot mat** Fußmatte;
  (doormat) Türmatte, Abstreifer
  (Fußmatte vor der Tür)
material Material, Werkstoff
material flow/chemical flow
  Stofffluss
Material Safety Data Sheet (MSDS)
  Sicherheitsdatenblatt (Merkblatt)
maturation Reifung
mature *adj/adv* **(ripe)** reif
mature *vb* **(ripen)** *vb* reifen
maturing/ripening Reifen
maturity (ripeness) Reife
> **immaturity/immatureness**
  Unreife
maximum permissible workplace
  concentration/
  maximum permissible exposure
  MAK-Wert (maximale Arbeitsplatz-
  Konzentration)
maximum rate
  Maximalgeschwindigkeit
  (V_{max} Enzymkinetik, Wachstum)
maximum tolerated dose (MTD)
  maximal verträgliche Dosis
maximum yield Höchsterträge
mealy/farinaceous mehlig
mean (average)
  Mittel, Durchschnittswert
  (*siehe auch:* Mittelwert)
> **adjusted mean**
  bereinigter Mittelwert,
  korrigierter Mittelwert
mean value/mean/
  arithmetic mean/average *stat*
  Mittelwert, Mittel,
  arithmetisches Mittel,
  Durchschnittswert
measurable messbar
measure *vb* messen, abmessen
measure *n* Maß
> **immediate measure (instant action)**
  Sofortmaßnahme
measurement
  (test/testing/reading/recording)
  Messung, Messen, Maß
> **range of measurement** Messbereich

measuring apparatus/
  measuring instrument
  Messgerät
measuring cup Messbecher
measuring procedure Messverfahren
measuring scoop Messschaufel
measuring unit/
  measuring device *math*
  Messglied (Größe)
meat extract Fleischextrakt
meat infusion
  (**meat digest/tryptic digest**) *micb*
  Fleischwasser, Fleischbrühe,
  Fleischsuppe
mechanic *n* Mechaniker
mechanical stage *micros*
  Kreuztisch
median longitudinal plane
  Sagittalebene
  (parallel zur Mittellinie)
median value *stat*
  Medianwert, Zentralwert
mediate vermitteln
mediator Mediator, Vermittler
medical examination/
  medical exam/
  medical checkup/
  physical examination/physical
  medizinische, ärztliche Untersuchung
medical gloves
  medizinische Handschuhe,
  OP-Handschuhe
medical lab technician/
  medical lab assistant
  MTLA
  (medizinisch-technische(r)
  LaborassistentIn)
medical personnel
  medizinisches Personal,
  Sanitätspersonal
medical service Sanitätsdienst
medical student Medizinstudent
medical supplies
  Medizinalbedarf, Sanitätsbedarf
medical surveillance/
  health surveillance
  medizinische Überwachung,
  ärztliche Überwachung

medical technician/
  medical assistant
  (*auch:* **Sprechstundenhilfe:**
  doctor's assistant)
  MTA
  (medizinisch-technische(r)
  AssistentIn)
medication
  Medizin, Medikament,
  Arznei, Arzneimittel
medicinal plant
  Drogenpflanze,
  Arzneipflanze, Heilpflanze
medicine Medizin; (drug/medicament)
  Droge, Arznei, Arzneimittel,
  Medizin, Medikament
medium (culture/nutrient medium)
  Medium
  (Kulturmedium, Nährmedium)
  ➢ **basal medium**
  Basisnährboden, Basisnährmedium
  ➢ **complete medium/rich medium**
  Vollmedium, Komplettmedium
  ➢ **complex medium**
  komplexes Medium
  ➢ **conditioned medium**
  konditioniertes Medium
  ➢ **deficiency medium** Mangelmedium
  ➢ **defined medium**
  synthetisches Medium
  (chem. definiertes Medium)
  ➢ **differential medium**
  Differenzierungsmedium
  ➢ **egg medium/egg culture medium**
  Eiermedium, Eiernährmedium,
  Eiernährboden
  ➢ **enrichment medium**
  Anreicherungsmedium
  ➢ **isolation medium** Isolationsmedium
  ➢ **maintenance medium**
  Erhaltungsmedium
  ➢ **minimal medium** Minimalmedium
  ➢ **rich medium/complete medium**
  Vollmedium, Komplettmedium
  ➢ **selective medium**
  Elektivmedium, Selektivmedium
  ➢ **test medium** Testmedium,
  Prüfmedium (zur Diagnose)

medulla/pith/core Mark
medullation Verkernung
melt *n* Schmelze
melt *vb* schmelzen, aufschmelzen
melted geschmolzen
melting curve Schmelzkurve
melting furnace/smelting furnace
  Schmelzofen
melting point Schmelzpunkt
melting temperature
  Schmelztemperatur
meltwater Schmelzwasser
membrane Membran
membrane conductance
  Membranleitfähigkeit
membrane filter Membranfilter
membrane forceps Membranpinzette
membrane ghost
  Membran-Ghost (künstlich
  hergestellte leere Membran)
membrane reactor
  Membranreaktor (Bioreaktor)
membraneous membranös
meniscus Meniskus
mercerize *text*
  mercerisieren, laugen
mercuric/mercury(II) ...
  Quecksilber-(II), zweiwertiges Q.
mercuric chloride/
  mercury dichloride/sublimate/
  corrosive mercury chloride
  Quecksilber-(II)-chlorid,
  Sublimat
mercurous/mercury(I) ...
  Quecksilber-(I), einwertiges Q.
mercurous chloride/calomel/
  mercury subchloride
  Quecksilber-(I)-chlorid, Kalomel
mercury (Hg) Quecksilber
mercury poisoning
  Quecksilbervergiftung,
  Merkurialismus
mercury trap/mercury well
  Quecksilberfalle
mercury vapor lamp
  Quecksilberdampflampe
mercury-in-glass thermometer
  Quecksilberthermometer

mesh Masche (Netz/Sieb),
  Drahtgeflecht; Gitterstoff;
  Maschenweite
mesh screen Maschensieb
mesh size/mesh Siebnummer
meshy maschig
mesomerism Mesomerie
mesophile Mesophile
mesophilic mesophil (20–45°C)
mesotrophic mesotroph
  (mittlerer Nährstoffgehalt)
metabolic derangement/
  metabolic disturbance
  Stoffwechselstörung
metabolic pathway/metabolic shunt
  Stoffwechselweg
metabolic rate
  Metabolismusrate,
  Stoffwechselrate,
  Energieumsatzrate
metabolic turnover
  Stoffwechselumsatz
metabolism
  Metabolismus, Stoffwechsel
➢ **basal metabolism**
  Grundstoffwechsel,
  Ruhestoffwechsel
➢ **cellular metabolism**
  Zellstoffwechsel
➢ **energy metabolism**
  Energiestoffwechsel
➢ **intermediary metabolism**
  intermediärer Stoffwechsel,
  Zwischenstoffwechsel
➢ **maintenance metabolism**
  Betriebsstoffwechsel
➢ **secondary metabolism**
  Sekundärstoffwechsel
metabolite
  Metabolit,
  Stoffwechselprodukt
metal Metall
➢ **heavy metal** Schwermetall
➢ **nonferrous metal** Buntmetall
➢ **precious metal** Edelmetall
➢ **semimetals** Halbmetalle
➢ **semiprecious metal** Halbedelmetall
➢ **transition metal** Übergangsmetall

metal alloy Metalllegierung
metal deposition
 chem Metallabscheidung;
 micros Metallaufdampfung
metal drill (bit)
 Metalbohrer (Bohreraufsatz)
metal science (metallurgy)
 Metallkunde (Metallurgie)
metal-cutting saw Metallsäge
metal-ore leaching Erzlaugung
➤ **microbial metal-ore leaching/**
 microbial leaching of metal ores
 mikrobielle Erzlaugung
metallic metallisch
metallic bond metallische Bindung
metallic luster/lustre
 Lüster, Metallglanz
metallization Metallbelag
metallizing Metallbeschattung
 (Schrägbedampfung bei TEM)
metallurgy
 (science & technology of metals)
 Metallurgie, Hüttenkunde
metastasis Metastase,
 Tochtergeschwulst
meter *n* Zähler, Messinstrument,
 Messgerät, Messer
meter *vb* messen
 (mit Messinstrument/Messgerät)
metering valve Dosierventil
methane Methan
methanogenic
 methanbildend, methanogen
method Methode
method of estimation *stat*
 Schätzverfahren
methylate methylieren
methylation Methylierung, Methylieren
metric scale metrische Skala
metrological messtechnisch
metrology/
 measurement techniques/
 measuring techniques
 Messtechnik
mevalonic acid (mevalonate)
 Mevalonsäure (Mevalonat)
micellation Micellierung
micelle Micelle

Michaelis constant/
Michaelis-Menten constant
 Michaeliskonstante,
 Halbsättigungskonstante (K_M)
Michaelis-Menten equation
 Michaelis-Menten-Gleichung
micro-environment Mikroumwelt
micro-forceps Mikropinzette
microbe/microorganism
 Mikrobe, Mikroorganismus
microbial mikrobiell
microbiological safety cabinet (MSC)
 mikrobiologische
 Sicherheitswerkbank (MSW)
microcarrier Mikroträger
microdissection forceps/
 microdissecting forceps
 anatomische Mikropinzette,
 Splitterpinzette
microelement/micronutrient
 (trace element)
 Mikroelement, Spurenelement
microfuge Mikrozentrifuge
micrograph/
 microscopic picture/
 microscopic image
 mikroskopische Aufnahme,
 mikroskopisches Bild
microinjection Mikroinjektion
micromanipulation
 Mikromanipulation
micromanipulator Mikromanipulator
micrometer Messschraube
➤ **outside micrometer**
 Bügelmessschraube
micrometer screw/
 fine-adjustment/
 fine-adjustment knob
 Mikrometerschraube
microorganism/microbe
 Mikroorganismus (*pl*
 Mikrorganismen), Mikrobe
micropipet
 Mikropipette, Mikroliterpipette
micropipet tip
 Mikropipettenspitze
microprobe Mikrosonde
microprocedure Mikroverfahren

mi

microscope Mikroskop
- compound microscope
  zusammengesetztes Mikroskop
- course microscope
  Kursmikroskop
- dissecting microscope
  Präpariermikroskop
- inverted microscope
  Umkehrmikroskop,
  Inversmikroskop
- light microscope
  (compound microscope)
  Lichtmikroskop
- phase contrast microscope
  Phasenkontrastmikroskop
- polarizing microscope
  Polarisationsmikroskop
- stereo microscope
  Stereomikroskop

microscope accessories
  Mikroskopzubehör
microscope illuminator
  Mikroskopierleuchte
microscope slide Objektträger
- prepared microscope slide
  Mikropräparat

microscope slide label
  Objektträgerbeschriftungsetikett
microscope stage Objekttisch
microscopic (microscopical)
  mikroskopisch
microscopic image/
  microscopic picture/micrograph
  mikroskopisches Bild,
  mikroskopische Aufnahme
microscopic procedure
  Mikroskopierverfahren
microscopical preparation/
  microscopic mount
  mikroskopisches Präparat
microscopy Mikroskopie
- atomic force microscopy (AFM)
  Rasterkraftmikroskopie
- brightfield microscopy
  Hellfeld-Mikroskopie
- confocal laser scanning microscopy
  konfokale Laser-Scanning
  Mikroskopie

- darkfield microscopy
  Dunkelfeld-Mikroskopie
- force microscopy (FM)
  Kraftmikroskopie
- high voltage electron microscopy
  (HVEM)
  Höchstspannungselektronen-
  mikroskopie
- immunoelectron microscopy (IEM)
  Immun-Elektronenmikroskopie
- immunofluorescence microscopy
  Immunfluoreszenzmikroskopie
- interference microscopy
  Interferenzmikroskopie
- light microscopy
  (compound microscopy)
  Lichtmikroskopie
- phase contrast microscopy
  Phasenkontrastmikroskopie
- polarizing microscopy
  Polarisationsmikroskopie
- scanning electron microscopy
  (SEM)
  Rasterelektronenmikroskopie (REM)
- scanning force microscopy (SFM)
  Rasterkraftmikroskopie (RKM)
- scanning tunneling microscopy
  (STM)
  Rastertunnelmikroskopie (RTM)
- tunneling microscopy
  Tunnelmikroskopie

microscopy accessories
  Mikroskopierzubehör
microscopy transmission electron
  microscopy (TEM)
  Transmissionselektronenmikroskopie, Durchstrahlungselektronenmikroskopie
microtome Mikrotom
- cryoultramicrotome
  Kryo-Ultramikrotom
- freezing microtome/cryomicrotome
  Gefriermikrotom
- rotary microtome
  Rotationsmikrotom
- sliding microtome
  Schlittenmikrotom
- ultramicrotome Ultramikrotom

mi

microtome blade Mikrotommesser
microtome chuck
  Mikrotom-Präparatehalter,
  Objekthalter (Spannkopf)
microtomy Mikrotomie
microwave oven ('microwave')
  Mikrowellenofen,
  Mikrowellengerät ('Mikrowelle')
migration *chromat/electrophor*
  Migration, Wanderung
milk glass Milchglas
milk sugar/lactose
  Milchzucker, Laktose
milky (opaque) milchig (opak)
mill Mühle
➢ **analytical mill** Analysenmühle
➢ **ball mill/bead mill** Kugelmühle
➢ **bead mill (shaking motion)**
  Schwing-Kugelmühle
➢ **centrifugal grinding mill**
  Rotormühle, Zentrifugalmühle,
  Fliehkraftmühle
➢ **coffee mill/coffee grinder**
  Kaffeemühle
➢ **cutting-grinding mill/
  shearing machine**
  Schneidmühle
➢ **disk mill** Tellermühle
➢ **drum mill/tube mill/barrel mill**
  Trommelmühle
➢ **grinding jar** Mahlbecher
➢ **hand mill** Handmühle
➢ **mixer mill** Mischmühle
➢ **mortar grinder mill** Mörsermühle
➢ **plate mill/disk mill** Scheibenmühle
➢ **pulverizer** Pulverisiermühle
mineral (minerals) Mineral
  (*pl* Mineralien); Mineralstoffe
mineral cycle Mineralstoffkreislauf
mineral fertilizer/inorganic fertilizer
  Mineraldünger
mineral oil Mineralöl
mineral soil Mineralboden
mineral spring Mineralquelle
mineral water Mineralwasser
mineral wool (mineral cotton)
  Mineralfasern
  (speziell: Schlackenfasern)

mineralization
  Mineralisation, Mineralisierung
minimal medium
  Minimalmedium
miniprep/minipreparation
  Miniprep, Minipräparation
minute respiratory volume
  Atemminutenvolumen (AMV)
miscibility
  Mischbarkeit, Vermischbarkeit
➢ **immiscibility**
  Unvermischbarkeit
miscible mischbar, vermischbar
➢ **immiscible** unvermischbar
misfire/backfire
  Fehlzünden, Fehlzündung
mismatch (mispairing) *gen*
  Fehlpaarung (Basenfehlpaarung)
mist feiner Nebel, leichter Nebel
➢ **fine dust/fines** Feinstaub
misty leicht nebelig
miter box Gehrungsschneidlade
miter-box saw Gehrungssäge
mix *n* Mischung;
  (mixing) Vermischung
mix *vb* michen, vermischen
mixed culture Mischkultur
**mixed-bed filter/
  mixed-bed ion exchanger**
  Mischbettfilter,
  Mischbettionenaustauscher
mixer Mischer, Mixer
➢ **barrel mixer/drum mixer**
  Trommelmischer
➢ **blade mixer** Schaufelmischer
➢ **blender (vortex)**
  Küchenmaschine (Vortex)
➢ **nutator/nutating mixer/
  'belly dancer' (shaker with
  gyroscopic, i.e., threedimensional
  circular/orbital & rocking motion)**
  Taumelschüttler
➢ **roller wheel mixer** Drehmischer
➢ **shaker with spinning-rotating
  motion (vertically rotating 360°)**
  Überkopfmischer
➢ **tumbling mixer/tumbler**
  Fallmischer

mixer mill Mischmühle
mixing
  Mischen, Durchmischung
mixing drum Mischtrommel
mixing ratio
  Mischungsverhältnis
mixotropic series
  mixotrope Reihe
mixture Mischung, Gemenge
  ➢ **binary mixture**
    Zweistoffgemisch
mobile mobil, beweglich;
  (vagile/wandering) vagil
  (Ortsveränderung des
  Gesamtorganismus)
mobility Mobilität, Beweglichkeit;
  (vagility) Vagilität (Ortsveränderung
  des Gesamtorganismus)
mobility shift experiment
  Gelretardationsexperiment
mock-up *n*
  Attrape, Nachbildung, Modell
modal value *stat* Modalwert
mode (style/status)
  Modus, Art und Weise, Regel;
  Einstellung
  (einstellbare Betriebsart);
  stat Modalwert;
  (method) Methode
mode of action/mechanism
  Wirkungsweise, Mechanismus
model Modell, Bauart, Ausführung;
  Muster, Vorlage, Typ;
  Vorbild
model building Modellbau
modeling clay Modellierknete
module Modul,
  Funktionseinheit
Mohr's salt/
  ammonium iron(II) sulfate
  hexahydrate
  (ferrous ammonium sulfate)
  Mohrsches Salz
moiety/part/section
  Teil (des Ganzen), Anteil; Hälfte
moist feucht
moisten (humidify/dampen)
  befeuchten; benetzen

moistening
  (humidification/dampening)
  Befeuchtung; Benetzung
moistness Feuchte
moisture Feuchtigkeit, Feuchte
moisture capacity
  (water-holding capacity,
  e.g., of soil)
  Wasserkapazität;
  Wasserhaltevermögen
moisture-proof
  feuchtigkeitsundurchlässig
molar mass ('molar weight')
  Molmasse, molare Masse
  ('Molgewicht')
molar volume Molvolumen
mold/mould (*Br***)**
  (frame/cavity/matrix)
  tech/mech Gießform;
  biol (mildew) Moder (Schimmel)
moldy/putrid/musty
  moderig, faulend,
  verfaulend (Geruch)
mole Mol
mole fraction
  Molenbruch,
  Stoffmengenanteil
molecular biology
  Molekularbiologie
molecular formula
  Molekularformel, Molekülformel
molecular genetics
  Molekulargenetik
molecular ion (MS) Molekülion
molecular leak Molekularleck
molecular mass ('molecular weight')
  Molekülmasse, Molmasse
  ('Molekulargewicht')
molecular peak Molekülpeak
molecular sieve
  Molekularsieb,
  Molekülsieb, Molsieb
molecular sieving chromatography/
  gel permeation chromatography/
  gel filtration
  Molekularsiebchromatographie,
  Gelpermeationschromatographie,
  Gelfiltration

molecular weight (molecular mass)
Molekulargewicht, Molgewicht
(Molmasse)
➢ **number average molecular mass (M_n)**
Zahlenmittel des Molekulargewichts,
zahlenmittlere Molmasse
➢ **relative molecular mass (M_r)**
relatives Molekulargewicht,
relative Molmasse/Molekülmasse
➢ **weight average molecular mass (M_w)**
Gewichtsmittel des
Molekulargewichts,
gewichtsmittlere Molmasse,
Durchschnitts-Molmasse
molecular-weight distribution
Molmassenverteilung
molecule Molekül
➢ **carrier molecule** Trägermolekül
➢ **macromolecule** Makromolekül
➢ **parent molecule (backbone)**
Grundkörper (Strukturformel)
➢ **tagged molecule**
markiertes Molekül
molten geschmolzen, schmelzflüssig
molten metal Metalschmelze
molten salt/salt melt
Salzschmelze,
geschmolzenes Salz
molten-salt electrolysis
Schmelzelektrolyse,
Schmelzflusselektrolyse
molybdenum (Mo) Molybdän
monitor *n* Monitor,
Mess-/Anzeige-/Kontrollgerät;
Anzeige
monitor *vb* **(survey/supervise/control)**
überwachen; abhören, mithören;
kontrollieren
monitoring (surveillance/surveying/ supervision/surveyance)
Überwachung, Supervision
monitoring camera
Überwachungskamera
monitoring protocol
Arbeitsvorschrift/Arbeitsanweisung
für die Überwachung
monobasic einbasig

monoclonal antibody
monoklonaler Antikörper
monolayer (monomolecular layer)
einlagige Schicht,
monomolekulare Schicht
monolayer cell culture
Einschichtzellkultur
monolithic floor
monolithischer Fußboden
(Labor: Stein/Beton aus einem Guß)
monoprotic acid
einwertige/einprotonige Säure
monounsaturated
einfach ungesättigt
monounsaturated fatty acid
einfach ungesättigte Fettsäure
mop Mop, Aufwischer
mop up (the floor)
aufputzen, aufwischen (den Boden)
mop wringer Auswringer, Wringer
morbidity Morbidität
(Häufigkeit der Erkrankungen)
mordant Beize, Beizenfärbungsmittel
morphologic/morphological
morphologisch
morphology Morphologie
mortal sterblich
➢ **immortal** unsterblich
mortality/death rate
Sterblichkeit, Sterberate, Mortalität
➢ **immortality**
Unsterblichkeit, Immortalität
mortar Mörser, Reibschale
➢ **agate mortar** Achatmörser
➢ **alumina mortar**
Aluminiumoxid-Mörser
➢ **apothecary mortar**
Apotheker-Mörser
➢ **glass mortar** Glasmörser
➢ **pestle** Pistill
➢ **porcelain mortar**
Porzellanmörser
mortar grinder mill Mörsermühle
mother board Hauptplatine
mother liquor Mutterlauge
motile beweglich, motil,
bewegungsfähig
(Bewegung eines Körperteils)

motility Beweglichkeit, Motilität, Bewegungsvermögen (Bewegung eines Körperteils)
motion Bewegung
> **hand motion (handshaking motion)** Handbewegung
> **rocking motion** (up-down/see-saw motion) Wippbewegung (rauf und runter); (side-to-side: fast) Rüttelbewegung (schnell hin und her)
> **see-saw motion/rocking motion** Wippbewegung, Schaukelbewegung
> **spinning/rotating motion** Drehbewegung (rotierend)
> **vibrating motion** Vibrationsbewegung
> **vibrational motion** Schwingungsbewegung
> **vortex motion/ whirlpool motion (shaker)** Vortex-Bewegung, kreisförmig-vibrierende Bewegung (Schüttler)
motion sensor Bewegungsmelder, Bewegungssensor
motionless ruhend, unbewegt
motor vehicle Kraftfahrzeug
mount *vb* fixieren, präparieren; einspannen; arrangieren, anbringen, befestigen
mount *n* m*icros* Präparat (Objektträger); Einbettung
> **microscopical preparation/ microscopic mount** mikroskopisches Präparat
> **scraping (mount)** Schabepräparat
> **squash mount** Quetschpräparat
> **wet mount** Nasspräparat (Frischpräparat, Lebendpräparat, Nativpräparat)
> **whole mount** Totalpräparat
mountant/mounting medium Einbettungsmittel, Einschlussmittel
mouth (opening/orifice) Mund, Öffnung; Mündung; Eingang, Zugang
mouth mirror Mundspiegel
mouth wash Mundspülung
mouth-to-mouth resuscitation/ respiration Mund-zu-Mund Beatmung (Wiederbelebung)
mouthpiece Mundstück, Ansatz, Tülle
movement/motion/locomotion Bewegung, Fortbewegung, Lokomotion
MS (mass spectroscopy) MS (Massenspektroskopie)
mucic acid Schleimsäure, Mucinsäure
mucilage Schleim (speziell pflanzlich)
mucosa (mucous membrane) Schleimhaut, Schleimhautepithel
> **irritation of the mucosa** Schleimhautreizung
mucous membrane/mucosa Schleimhaut, Schleimhautepithel
mucus/slime/ooze Schleim
muff Muffe, Flanschstück
muffle furnace Muffelofen
muffler Dämpfer, Schalldämpfer
muffs/earmuffs/hearing protectors Gehörschützer (speziell auch: Kapselgehörschützer)
mull (IR/Raman) Aufschlämmung
mull technique (IR spectroscopy) Suspensionstechnik
multicellular mehrzellig, vielzellig
multichannel instrument Vielkanalgerät
multichannel pump Mehrkanal-Pumpe
multicomponent adhesive (or cement) Mehrkomponentenkleber
multifunctional vector/ multipurpose vector multifunktioneller Vektor, Vielzweckvektor
multilayer film Mehrschichtfolie
multilayered vielschichtig, mehrschichtig

multi-limb vacuum receiver adapter/ cow receiver adapter/'pig' (receiving adapter for three/four receiving flasks) *dist* Wechselvorlage, Spinne, Eutervorlage, Verteilervorlage
multimeter *electr* Multimeter, Universalmessgerät, Vielfachmessgerät
multiple bond *chem* Mehrfachbindung
multiple sugar/polysaccharide Vielfachzucker, Polysaccharid
multiple-component adhesive (or cement) Mehrkomponentenkleber
multiplet signal (NMR) Multiplett-Signal
multiplication Vermehrung, Vervielfältigung, Multiplikation
multistage impulse countercurrent impeller Mehrstufen-Impuls-Gegenstrom (MIG) Rührer
multi-tray *micb* Wannen-Stapel
multiwell plate *micb* Vielfachschale, Multischale
municipal solid waste (MSW) kommunaler Müll
muramic acid Muraminsäure
mushroom poisoning/mycetism Pilzvergiftung
mustard oil Senföl
mutability Mutabilität, Mutierbarkeit, Mutationsfähigkeit
mutagen *n* Mutagen, mutagene Substanz
mutagenesis Mutagenese
mutagenic (T) erbgutverändernd, mutagen; mutationsauslösend, erbgutverändernd
mutagenicity Mutagenität
mutant Mutante
mutarotation Mutarotation
mutate mutieren
mutation Mutation
mutation rate Mutationsrate
mutualist Symbiont (in gegenseitiger Lebensgemeinschaft)
mutualistic symbiotisch (gemeinnützig)
mycoplasma (*pl* **myoplasmas)** Mykoplasma (*pl* Mykoplasmen)
mycosis Mykose
mycotoxin Mykotoxin
myeloma Myelom
myristic acid/tetradecanoic acid (myristate/tetradecanate) Myristinsäure (Myristat), Tetradecansäure

nacre (mother of pearl)
  Perlmutter, Perlmutt
nail Nagel
nail bit Nagelbohrer
nail extractor Nagelzieher
nail scissors Nagelschere
naked/nude nackt
naked flame(s) offenes Feuer
name/term Name;
  (designation/nomenclature)
  Bezeichnung, Benennung, Name,
  Namensgebung (Nomenklatur)
name tag Namensetikett,
  Namensschildchen
naming/designation/nomenclature
  Benennung, Bezeichnung,
  Namensgebung
naphthalene Naphthalin
narrow-mouthed
  (narrow-mouth/narrowmouthed/
  narrow-neck/narrownecked)
  Enghals ...
narrow-mouthed bottle
  Enghalsflasche
narrow-mouthed flask
  (narrow-necked flask)
  Enghalskolben
nasal mucosa/olfactory epithelium
  Nasenschleimhaut
National Institute of Occupational
  Safety and Health (NIOSH)
  [part of CDC] Amerikanisches
  Bundesamt für Arbeitsplatzsicherheit
  und Gesundheitsschutz
National Pipe Taper (NPT)
  U.S. Rohrgewindestandard
native (not denatured)
  nativ (nicht-denaturiert)
native gel natives Gel
natural natürlich
> **near-natural** naturnah
> **unnatural** unnatürlich
natural balance natürliches
  Gleichgewicht (Naturhaushalt)
natural colors/natural coloring
  natürliche Farbstoffe
natural flavor/natural flavoring
  natürlicher Geschmackstoff

natural gas Erdgas
natural immunity
  natürliche Immunität
natural product Naturstoff
natural product chemistry
  Naturstoffchemie
natural rubber (NR)
  Naturkautschuk
natural sciences/science
  Naturwissenschaften
natural scientist/scientist
  Naturwissenschaftler(in),
  Naturforscher
nature protection/
  nature conservation/
  nature preservation
  Naturschutz
nausea/sickness/illness
  Übelkeit, Übelsein
neat/pure *chem* rein, pur
neatness (in cleaning-up)
  Ordentlichkeit, Aufräumen
nebulizer
  Verneblér, Nebelgerät
neck *micros* Hals, Tubusträger
necrosis Nekrose
necrotic nekrotisch
needle Nadel;
  (syringe needle) Kanüle, Hohlnadel
> **cemented needle (syringe needle)**
  geklebte Injektionsnadel
> **removable needle (syringe needle)**
  abnehmbare Injektionsnadel/Nadel
> **suture needle** chirurgische Nadel
needle file Nadelfeile
needle valve
  Nadelventil, Nadelreduzierventil
  (Gasflasche)
needle-nose pliers/
  snipe-nose pliers/
  snipe-nosed pliers
  Storchschnabelzange
negative pressure
  Unterdruck
negative staining/
  negative contrasting *micros*
  Negativkontrastierung
negligible vernachläßigbar

**neon screwdriver (*Br*)/
 neon tester (*Br*)/
 voltage tester screwdriver**
 Spannungsprüfer (Schraubenzieher)
nephelometry
 Nephelometrie, Streulichtsmessung
nerve Nerv
nerve poison Nervengift
net Netz
net primary production (NPP)
 Nettoprimärproduktion
net production Nettoproduktion
net weight Nettogewicht
netted/meshy/reticulate
 vernetzen, vernetzt
network Netzwerk, Netz;
 Geflecht; Maschenwerk
 ➢ **power network**
 Versorgungsnetz
neuraminic acid Neuraminsäure
neurosecretory neurosekretorisch
neurotoxic neurotoxisch
neutron activation analysis (NAA)
 Neutronenaktivierungsanalyse
 (NAA)
neutron diffraction
 Neutronenbeugung,
 Neutronendiffraktometrie
neutron scattering
 Neutronenstreuung
**new chemicals/
 new substances**
 Neustoffe
Newtonian fluid
 Newton'sche Flüssigkeit
nick Kerbe, Schlitz;
 gen Bruchstelle, Einzelstrangbruch
nickel (Ni) Nickel
nicotinic acid (nicotinate)/niacin
 Nikotinsäure, Nicotinsäure
 (Nikotinat)
**NIOSH (National Institute for
 Occupational Safety and Health)**
 U.S. Institut für Sicherheit und
 Gesundheit am Arbeitsplatz
nitrate Nitrat
nitration/nitrification Nitrierung
nitric acid Salpetersäure

nitrification
 Nitrifikation, Nitrifizierung
nitrifier/nitrifying bacteria
 Nitrifikanten
nitrify nitrieren
nitrile rubber (NBR) Nitrilkautschuk
 (Acrylnitril-Butadien-Kautschuk)
nitrite Nitrit
nitrobenzene Nitrobenzol
nitrocotton/guncotton (12.4–13% N)
 Schießbaumwolle
nitrogen (N) Stickstoff
 ➢ **liquid nitrogen**
 Flüssigstickstoff,
 flüssiger Stickstoff
nitrogen deficiency Stickstoffmangel
nitrogenous (nitrogen-containing)
 stickstoffhaltig,
 stickstoffenthaltend, Stickstoff...
nitrogenous base
 stickstoffhaltige Base,
 'Base' (Purine/Pyrimidine)
**nitrogenous compound/
 nitrogen-containing compound**
 Stickstoffverbindung
nitroglycerin/glycerol trinitrate
 Nitroglycerin, Glycerintrinitrat
nitrous acid salpetrige Säure
No Smoking! Rauchverbot!
NOAEL (no adverse effect level)
 Wirkschwelle
NOEL (no observed effect level)
 höchste Dosis
 ohne beobachtete Wirkung
noise *tech/electro/neuro* Rauschen
noise analysis/fluctuation analysis
 Rauschanalyse, Fluktuationsanalyse
noise filter Rauschfilter
noise level Geräuschpegel
noise pollution
 Lärmverschmutzung
noise protection Lärmschutz
noise reduction
 Rauschminderung
noise thermometer
 Rauschthermometer
nomenclature
 Bezeichnungssystem, Nomenklatur

nominal frequency
 Sollfrequenz
nominal mass
 Nennmasse, Nominalmasse
nominal output/rated output
 Soll-Leistung
nominal scale
 Nominalskala
nominal value/rated value/ desired value/set point
 Sollwert
nominal volume
 Nennvolumen
nonbreakable/unbreakable/crashproof
 bruchsicher
noncombustible/nonflammable
 nicht brennbar
noncompetitive inhibition
 nichtkompetitive Hemmung
nonessential nichtessentiell
nonflammable/incombustible
 nicht entflammbar, nicht brennbar
nonhazardous ungefährlich,
 nicht gesundheitsgefährdend
nonidentity (immundiffusion)
 Verschiedenheit (Nicht-Identität)
nonmotile/immotile/immobile/ motionless/fixed
 unbeweglich, bewegungslos, fixiert
nonpersistent transmission
 nicht-persistente Übertragung
 (z.B. Krankheit)
nonrandom disjunction
 nicht-zufallsgemäße Verteilung
nonsaturation kinetics
 Nichtsättigungskinetik
nonskid/skip-proof
 nicht-rutschend, Antirutsch...
 (Gerät auf Unterlage)
nonsmoking rauchfrei
nonspecific unspezifisch
nonvolatile
 nicht flüchtig;
 schwerflüchtig
norm of reaction
 Reaktionsnorm
normal distribution
 Normalverteilung

nosepiece (nosepiece turret) *micros*
 Objektivrevolver, Revolver
 ➤ **double nosepiece** Zweifachrevolver
 ➤ **triple nosepiece** Dreifachrevolver
 ➤ **quadruple nosepiece** Vierfachrevolver
 ➤ **quintuple nosepiece** Fünffachrevolver
nosocomial infection/ hospital-acquired infection
 Nosokomialinfektion,
 nosokomiale Infektion,
 Krankenhausinfektion
notation/scoring *stat* Bonitur
notched/nicked kerbig, gekerbt
notice *n* Notiz, Anzeige,
 Benachrichtigung, Bericht,
 Mitteilung, Hinweis
notice of approval
 Genehmigungsbescheid
notification
 Benachrichtigung,
 Inkenntnissetzung
 ➤ **obligation to notify/ notifiable/reportable**
 anzeigepflichtig
nozzle (spout)
 Tülle (ausgießen),
 Ausgussstutzen (Kanister);
 (socket) Stutzen
nozzle loop reactor/ circulating nozzle reactor
 Düsenumlaufreaktor,
 Umlaufdüsen-Reaktor
NTP (s.t.p./STP) (standard temperature 0°C/ pressure 1 bar)
 Normzustand
 (Normtemperatur & Normdruck)
nuclear nukleär, nucleär
nuclear energy/atomic energy
 Atomenergie
nuclear magnetic resonance (NMR)
 kernmagnetische Resonanz,
 Kernspinresonanz
nuclear magnetic resonance spectroscopy (NMR spectroscopy)
 kernmagnetische
 Resonanzspektroskopie,
 Kernspinresonanz-Spektroskopie

nuclear Overhauser effect (NOE)
Kern-Overhauser-Effekt
nuclear physics Kernphysik
nuclear power/atomic power
Atomkraft
nuclear radiation Kernstrahlung
nuclear transfer/
  nuclear transplantation *biol*
Kerntransplantation
nuclear waste Atommüll
nucleic acid
Nucleinsäure, Nukleinsäure
nucleophilic attack *chem*
nukleophiler Angriff
nucleoside Nucleosid, Nukleosid
nucleotide Nukleotid, Nucleotid
nude mouse Nacktmaus
numb taub, gefühllos
numbness
Taubheit, Gefühllosigkeit
nurse Krankenschwester,
Sanitäter, Krankenpfleger, Pfleger
➢ **company nurse**
Betriebssanitäter
nurture/feed
ernähren, nähren, füttern
nut (and bolt) *tech/mech*
Mutter (und Schraube)
➢ **acorn nut** Hutmutter
➢ **blind rivet nut**
Blindnietmutter
➢ **swivel nut/coupling nut/**
  mounting nut/cap nut
Überwurfmutter,
Überwurfschraubkappe
(z.B. am Rotationsverdampfer)
➢ **wing nut** Flügelmutter
nutate (gyroscopic motion) taumeln
nutation/gyroscopic motion
  (threedimensional circular/
  orbital & rocking motion)
dreidimensionale Taumelbewegung
nutator/nutating mixer/'belly dancer'
  (shaker with gyroscopic, i.e.,
  threedimensional circular/orbital &
  rocking motion) Taumelschüttler

nutdriver (wrench or screwdriver)
Steckschlüssel
➢ **hex nutdriver**
Sechskant-Steckschlüssel
nutrient Nahrung, Nährstoff
nutrient agar Nähragar
nutrient broth
Nährbouillon, Nährbrühe
nutrient budget
Nährstoffhaushalt
nutrient deficiency
Nährstoffarmut
➢ **food shortage**
Nahrungsmangel
nutrient demand/
  nutrient requirement
Nährstoffbedarf
nutrient medium (solid and liquid)/
  culture medium/substrate
Nährboden, Nährmedium,
Kulturmedium, Medium, Substrat
nutrient salt Nährsalz
nutrient solution/culture solution
Nährlösung
nutrient table/
  food composition table
Nährwert-Tabelle
nutrient uptake
Nährstoffaufnahme
nutrient-deficient/oligotroph(ic)
nährstoffarm, oligotroph
nutrient-rich/eutroph(ic)
nährstoffreich, eutroph
nutrition Nahrung, Ernährung
nutrition science/
  nutrition studies (dietetics)
Ernährungswissenschaft
(Diätetik)
nutritional deficit Nährstoffmangel
nutritional requirements
Nahrungsbedarf
(*pl* Nahrungsbedürfnisse)
nutritious/nutritive
nahrhaft, nährend, nutritiv
nutritive ratio/nutrient ratio
Nährstoffverhältnis

object Objekt, Ding, Gegenstand
object stage *micros*
　Objekttisch
objective *micros* Objektiv
objective lens Objektivlinse
obligation to provide welfare services
　Fürsorgepflicht
**obligatory/binding/
　mandatory/compulsory**
　verbindlich
observance (compliance)
　Einhaltung (Vorschrift)
obtuse/blunt stumpf
occupation Beruf, Beschäftigung
occupational
　beruflich, Berufs...,
　betrieblich, Betriebs...
occupational accident
　Arbeitsunfall
occupational disease
　Berufskrankheit
occupational exposure limit (OEL)
　maximale Arbeitsplatzkonzentration
　(MAK)
occupational hazard
　Berufsrisiko;
　Gefahr am Arbeitsplatz
occupational hygiene
　Arbeitsplatzhygiene
occupational injury
　Berufsverletzung
occupational medicine
　Arbeitsmedizin
**occupational protection/
　workplace protection/
　safety provisions
　(for workers)** Arbeitsschutz
occupational safety (workplace safety)
　Arbeitsplatzsicherheit
**Occupational Safety and Health
　Administration (OSHA)
　[Dept. of Labor]**
　Amerikanische
　Bundesministrialbehörde für
　Arbeitsplatzsicherheit und
　Gesundheitsschutz
occupational safety code
　Arbeitsplatzsicherheitsvorschriften

**occupational trainee (professional
　school & on-the-job training)**
　Auszubildende(r), Azubi
occurrence Ereignis, Vorfall;
　(presence) Vorkommen, Auftreten
octa-head stopper/octagonal stopper
　Achtkantstopfen
ocular (eyepiece) *micros* Okular
　➢ **binocular head** Binokularaufsatz
　➢ **spectacle eyepiece/
　high-eyepoint ocular**
　Brillenträgerokular
　➢ **trinocular head**
　Trinokularaufsatz, Tritubus
**ocular diaphragm/
　eyepiece diaphragm/
　eyepiece field stop** *micros*
　Gesichtsfeldblende des Okulars,
　Okularblende
ocular lens *micros*
　Okularlinse, Augenlinse
ocular micrometer
　Okularmikrometer
odd electron ungepaartes Elektron
odor (*Br* odour) Geruch, Duft
**odor threshold/
　olfactory threshold**
　Riechschwelle,
　Geruchsschwellenwert
odorfree geruchsfrei
odoriferous riechend,
　einen Geruch ausströmend;
　wohlriechend, duftend
odorless geruchlos
off-limits verboten
off-limits to unauthorized personnel
　Zutritt/Zugang für Unbefugte
　verboten!
office (bureau)
　Büro, Sekretariat; Amt, Behörde
office supplies Bürobedarf
official dress Dienstkleidung
official orders Dienstanweisung(en)
official uniform Dienstuniform
offset adapter (joint-glass adapter)
　Übergangsstück mit seitlichem
　Versatz
offset screwdriver Winkelschrauber

oil Öl
- **ben oil/benne oil** Behenöl
- **bitter almond oil** Bittermandelöl
- **canola oil (rapeseed oil)** Speise-Rapsöl, Rüböl
- **castor oil/ricinus oil** Rizinusöl
- **coconut oil** Kokosöl
- **cod-liver oil** Lebertran
- **corn oil** Maisöl
- **cotton oil** Baumwollsaatöl
- **crude oil/petroleum** Erdöl
- **drying oil** trocknendes Öl
- **essential oil/ethereal oil** ätherisches Öl
- **fusel oil** Fuselöl
- **linseed oil** Leinöl
- **lubricating oil/lube oil** Schmieröl
- **mineral oil** Mineralöl
- **mustard oil** Senföl
- **nondrying oil** nicht trocknendes Öl
- **olive kernel oil** Olivenkernöl
- **olive oil** Olivenöl
- **palm oil** Palmöl
- **peanut oil** Erdnussöl
- **pumpkinseed oil** Kürbiskernöl
- **safflower oil** Safloröl
- **sesame oil** Sesamöl
- **soybean oil** Sojaöl
- **sperm oil (whale)** Walratöl
- **sunflower seed oil** Sonnenblumenöl
- **turpentine oil** Terpentinöl
- **vegetable oil** Pflanzenöl
- **virgin oil (olive)** Jungfernöl
- **waste oil/used oil** Altöl

oil bath Ölbad
oil crops/oil seed crops Ölsaaten (ölliefernde Pflanzen)
oil paint Ölfarbe
oil pollution Ölverschmutzung, Ölpest
oil slick Ölteppich (auf Wasseroberfläche)
oil spill Ölkatastrophe
oiliness Fettigkeit, fettig-ölige Beschaffenheit
oilseed Ölsaat
oilskin(s) Ölzeug
oily ölig
oleic acid/ (Z)-9-octadecenoic acid (oleate) Ölsäure, Δ^9-Octadecensäure (Oleat)
olfactory epithelium/ nasal mucosa Nasenschleimhaut
olfactory sense Geruchssinn, olfaktorischer Sinn
oligomer Oligomer
oligomerous oligomer
oligonucleotide Oligonucleotid, Oligonukleotid
oligosaccharide Oligosaccharid
oligotrophic/nutrient-deficient oligotroph, nährstoffarm
olive kernel oil Olivenkernöl
olive oil Olivenöl
oncogene (onc gene) Onkogen
oncogenic (oncogenous) onkogen, oncogen, krebserzeugend
oncogenicity Onkogenität
oncology Onkologie
oncotic pressure onkotischer Druck, kolloidosmotischer Druck
one-pot reaction *chem* Eintopfreaktion
onset/start (of a reaction) Einsetzen, Beginn (einer Reaktion)
opalesce schillern
open öffnen, aufmachen
- **force open** gewaltsam öffnen

open-end wrench/ open-end spanner (*Br*) Gabelschlüssel, Maulschlüssel
opening (aperture/orifice/mouth/ perforation/entrance) Öffnung, Mund, Mündung; (mouth) Öffnung (Flasche/Gläschen)

op

operate *tech*
(work) arbeiten, in Betrieb sein, funktionieren, laufen; handhaben; (effect/bring about) betätigen, bedienen, betreiben, in Gang setzen; *med* (perform surgery) operieren
operating conditions
(of a piece of equipment)
funktionsfähiger Zustand (Geräte)
operating electrode
Arbeitselektrode
operating instructions
Bedienungsanleitung, Gebrauchsanleitung; Betriebsanleitung; Betriebsvorschrift; (manual) Handbuch
operating mode
Arbeitsweise, Funktionsweise; Funktionszustand
operating pressure Betriebsdruck
operating procedure
Arbeitsverfahren; Arbeitsanweisung
➢ **standard operating procedure (SOP)**
Standard-Arbeitsanweisung
operating range (Geräte)
Funktionsbereich, Arbeitsbereich
operating temperature
Arbeitstemperatur
operation *tech* Betrieb, Tätigkeit, Lauf; Wirkungsweise, Vefahren, Prozess, Inbetriebsetzung; Bedienung; *med* Operation
operational betriebsbereit; Betriebs..., Funktions..., Arbeits...
operational condition
funktionsfähiger Zustand
operational mode
Betriebsmodus
operational permission
Betriebserlaubnis
operations manager/plant manager
Betriebsleiter
operations personnel
Bedienungspersonal (Arbeiter/Handwerker/Mechaniker)
operations worker
Handwerker, Arbeiter

operator Maschinist, Bediener, Bedienungsperson, Durchführender
optical density/absorbance
optische Dichte, Absorption
optical diffusion/dispersion/ dissipation/scattering (light)
Streuung (Lichtstreuung)
optical pyrometer
Pyropter, optisches Pyrometer
optical refraction Lichtbrechung, optische Brechung, Refraktion
optical resolution optische Auflösung
optical specificity optische Spezifität
optics Optik
orbital shaker/ rotary shaker/circular shaker
Rundschüttler, Kreisschüttler
order *n*
Ordnung; Bestellung, Auftrag
order *vb*
bestellen, in Auftrag geben
order confirmation
Auftragsbestätigung
order statistics Ordnungsstatistik
ordinal scale Ordinalskala
ordinance/decree
Verordnung, Verfügung, Erlass
ore Erz
organic organisch
organic chemistry
organische Chemie, 'Organik'
organic matter
organische Substanz, organisches Material
organism/lifeform Organismus
origin Ursprung; (descent/provenance) Herkunft, Abstammung (Provenienz)
original (basic/simple/primitive)
originär, ursprünglich; einfach, primitiv
orotic acid Orotsäure
orsellic acid/orsellinic acid
Orsellinsäure
oscillate/vibrate
oszillieren, schwingen, vibrieren
oscillation/vibration
Oszillation, Schwingung, Vibration

oscillator Oszillator
oscillometry/
high-frequency titration
Oszillometrie,
oszillometrische Titration,
Hochfrequenztitration
osmic acid Osmiumsäure
osmiophilic osmiophil
(färbbar mit Osmiumfarbstoffen)
osmium tetraoxide Osmiumtetroxid
osmolality Osmolalität
osmolarity (osmotic concentration)
Osmolarität,
osmotische Konzentration
osmosis Osmose
➢ **reverse osmosis**
Reversosmose, Umkehrosmose
osmotic osmotisch
osmotic concentration/osmolarity
Osmolarität,
osmotische Konzentration
osmotic pressure osmotischer Druck
osmotic shock osmotischer Schock
ossification Ossifikation,
Verknöcherung, Knochenbildung
OTA
(Office of Technology Assessment)
US-Büro für
Technikfolgenabschätzung
otoscope/ear speculum
Otoskope, Ohrenspiegel,
Ohrenspekulum
outfit *n* Ausrüstung, Ausstattung;
Geräte, Werkzeuge, Utensilien
outflow *n* **(efflux/draining off)**
Abfluss, Ausfluss
outlet Ablauf, Ausfluss, Auslauf,
Austritt (Austrittsstelle einer
Flüssigkeit), Ableitung (von
Flüssigkeiten; Zulauf von
Flüssigkeit/Gas); (socket/wall
socket/receptacle/jack/mains
electricity supply *Br*) Steckdose
➢ **wall outlet** Wandsteckdose
outlet pressure Hinterdruck
outlet strip *electr*
Mehrfachsteckdose, Steckdosenleiste
outlier *stat* Ausreißer

output Output, Ertrag, Leistung,
Produktion; Ausgabe (z.B. Daten:
Computer); *electr* Ausgang
➢ **nominal output/rated output**
Soll-Leistung
outside facility Außenanlage
outside micrometer
Bügelmessschraube
oven Ofen, Backofen; (furnace)
Hochofen, Schmelzofen
➢ **convection oven** Konvektionsofen
➢ **drying oven** Trockenofen
➢ **gravity convection oven**
Konvektionsofen mit natürlicher
Luftumwälzung
➢ **heating oven/**
heating furnace (more intense)
Wärmeofen
➢ **hybridization oven**
Hybridisierungsofen
➢ **microwave oven**
Mikrowellenofen
oven drying/kiln drying/kilning
Ofentrocknung
oven gloves
Hoch-Hitzehandschuhe,
Ofenhandschuhe
overactivity/hyperactivity
Überfunktion, Hyperaktivität
overall(s)
Arbeitskittel, Overall (Einteiler)
overdose Überdosis
overfertilization/excessive fertilization
Überdüngung
overflow/overrun
Überfließen, Überschwemmung;
Überlauf; Überschuss
overgrow (overgrown)
zuwachsen (zugewachsen),
überwachsen (überwuchert)
overhaul (reconditioning) überholen
overhead projector
Tageslichtprojektor,
Overhead-Projektor
overheating/superheating
Überhitzen, Überhitzung
overload *n tech/electr* Überlastung
overload *vb tech/electr* überlasten

overpacking Umverpackung
overpotential Überspannung
overshoot
über etwas herausschießen; übersteuern
oversize (sieving residue)
Überkorn (Siebrückstand)
overswing
Überschwingen (aufheizen)
over-the-counter drug
frei erhältliches Medikament/Medizin/Droge (nicht verschreibungspflichtig)
overtone (IR) Oberschwingung
overwinding Überdrehung
oxalic acid (oxalate)
Oxalsäure (Oxalat)
oxaloacetic acid (oxaloacetate)
Oxalessigsäure (Oxalacetat)
oxalosuccinic acid (oxalosuccinate)
Oxalbernsteinsäure (Oxalsuccinat)
oxidation Oxidation
oxidation-reduction reaction
Redoxreaktion
oxidative oxidativ
oxidize oxidieren
oxidizing oxidierend;
(pyrophoric) brandfördernd (O)

oxidizing agent (oxidant/oxidizer)
Oxidationsmittel
oxoacid Oxosäure
oxoglutaric acid (oxoglutarate)
Oxoglutarsäure (Oxoglutarat)
oxygen Oxygen, Sauerstoff
➤ **liquid oxygen**
Flüssigsauerstoff, flüssiger Sauerstoff
oxygen debt
Sauerstoffverlust, Sauerstoffschuld, Sauerstoffdefizit
oxygen demand
Sauerstoffbedarf
oxygen partial pressure
Sauerstoffpartialdruck
oxygen transfer rate (OTR)
Sauerstofftransferrate
oxygenate
mit Sauerstoff anreichern/sättigen; oxidieren, mit Sauerstoff verbinden/behandeln
oxygenation Oxygenierung
oxygeneous
sauerstoffhaltig, Sauerstoff...
ozone Ozon
ozonization Ozonisierung
ozonolysis Ozonolyse

pacemaker Schrittmacher
pack (package) Abpackung
package Paket, Packung
➤ **bulk package** Großpackung
package insert Packungsbeilage
packaging Verpackung
➤ **in** *vitro* **packaging**
  in vitro-Verpackung
packaging bottle
  Verpackungsflasche
packaging glasses
  Verpackungsgläser
packaging material
  Packmaterial, Verpackungsmittel
packaging tape
  Verpackungsklebeband
packed bed reactor
  Packbettreaktor, Füllkörperreaktor
packed distillation column
  Füllkörperkolonne
packing Verpacken, Verpackung;
  Dichtung, Abdichtung;
  Dichtungsmaterial;
  Füllung, Füllmaterial
packing box (seal)
  Stopfbüchse (Dichtung)
packing nut Dichtungsmutter
packing sleeve Dichtungsmuffe
pad (gauze pad) Tupfer;
  (swab/pledget [cotton]/tampon)
  Bausch, Wattebausch,
  Tupfer, Tampon
paddle stirrer/paddle impeller
  Schaufelrührer, Paddelrührer
paddle wheel
  Schaufelrad, Laufrad
paddle wheel reactor
  Schlaufenradreaktor
padlock Vorhängeschloss
  (für Laborspind etc.)
pail Eimer (meist aus Metal);
  (bucket) *allg* Eimer
pail opener Eimeröffner
pain Schmerz
pain sensation Schmerzgefühl
painful schmerzhaft
painkiller Schmerzmittel,
  schmerzstillendes Mittel

paint *n* Farbe, Lack, Tünche
paint brush Malpinsel
palatable genießbar, schmackhaft
pale bleich, blass, fahl
paleness
  Bleiche, Blässe, bleiche Farbe
pallet Palette
palm oil Palmöl
**palmitic acid/hexadecanoic acid
  (palmate/hexadecanate)**
  Palmitinsäure, Hexadecansäure
  (Palmat/Hexadecanat)
pamphlet/brochure
  Pamphlet, Broschüre,
  Informationsschrift
pan Pfanne
pan balance Tafelwaage
pandemic *adj/adv* pandemisch
pandemic *n* Pandemie
panel *tech* Paneel; Frontplatte
  (eines Geräts/Instruments)
**panel board/
  switchboard** *electr*
  Instrumentenbrett,
  Schalttafel, Armaturenbrett
pantoic acid Pantoinsäure
pantothenic acid (pantothenate)
  Pantothensäure (Pantothenat)
PAP stain/Papanicolaou's stain
  PAP-Färbung,
  Papanicolaou-Färbung
paper Papier
➤ **absorbent paper (bibulous paper)**
  Saugpapier ('Löschpapier')
➤ **bibulous paper (for blotting dry)**
  Löschpapier
➤ **bond paper/stationery**
  Schreibpapier
➤ **brown paper/kraft** Packpapier
➤ **chart paper**
  Registrierpapier,
  Aufzeichnungspapier;
  Tabellenpapier
➤ **construction paper**
  Bastelpapier
➤ **filter paper** Filterpapier
➤ **glassine paper/glassine** Pergamin
  (durchsichtiges festes Papier)

pa

- glazed paper Glanzpapier (glanzbeschichtetes Papier), satiniertes Papier
- graph paper/metric graph paper Millimeterpapier
- laminated paper Hartpapier
- lens paper *micros* Linsenpapier, Linsenreinigungspapier
- litmus paper Lackmuspapier
- log paper Logarithmuspapier, Logarithmenpapier
- metric graph paper Millimeterpapier
- parchment paper Pergamentpapier
- photographic paper Fotopapier
- recycled paper Umweltschutzpapier
- tracing paper Pauspapier
- waste paper Altpapier
- wax paper Wachspapier
- weighing paper Wägepapier
- wrapping paper Einpackpapier

paper chromatography Papierchromatographie
paper clip Büroklammer, Heftklammer
paper electrophoresis Papierelektrophorese
paper napkin Papierserviette
paper towel Papierhandtuch
paperwork Schreibarbeit(en), 'Papierkram'
paramedic Sanitäter, Rettungssanitäter; ärztliche(r) Assistent(in)
paramedical nichtärztlich
parameter Parameter
paraphernalia Utensilien, Ausrüstung, Zubehör
parasite Parasit, Schmarotzer
parasitic parasitär, parasitisch, schmarotzend
parasitism Parasitismus, Schmarotzertum
parasitize parasitieren, schmarotzen

parathion Parathion (E 605)
parcel Paket (Postpaket)
parcel service Paketdienst
parchment Pergament; Pergamentpapier
parent compound/ parent molecule (backbone) Grundkörper (Strukturformel)
parent ion (MS) Mutterion, Ausgangsion
parent material/raw material Ausgangsstoff
parent substance Muttersubstanz
partial correlation coefficient *stat* Teilkorrelationskoeffizient
partial digest Partialverdau
partial identity Teilidentität, partielle Übereinstimmung
partial pressure Partialdruck
partial reaction Teilreaktion
partial survey *stat* Teilerhebung
particle Partikel, Teilchen
particle filter Partikelfilter, Teilchenfilter
particle physics Teilchenphysik
particle size (grain size) Korngröße; (soil texture) Teilchengröße (Bodenpartikel)
particulate respirator Partikelfilter-Atemschutzmaske
partition Abtrennung, Teilung, Abteilung, Verteilung; Trennwand, Scheidewand, Querwand (räumlich)
partition chromatography/ liquid-liquid chromatography (LLC) Verteilungschromatographie, Flüssig-flüssig-Chromatographie
partition coefficient/ distribution constant *chromat* Verteilungskoeffizient
partition wall (of a building) Trennwand (Gebäude)
PAS stain (periodic acid-Schiff stain) PAS-Anfärbung (Periodsäure, Schiff-Reagens)

passage (opening/outlet/port/ conduit/duct) Durchlass; (walkthrough) Durchgang; (subculture) Passage, Subkultivierung
passageway (duct) Ausführgang, Ausführkanal; (passage/walkthrough) Durchgang
Pasteur effect Pasteur-Effekt
Pasteur pipet Pasteurpipette
pasteurization Pasteurisierung, Pasteurisieren
pasteurize pasteurisieren
pasteurizing/pasteurization Pasteurisierung, Pasteurisieren
patch *n* Flicken
patch clamp technique Patch-Clamp Verfahren
paternity test Vaterschaftsbestimmung, Vaterschaftstest
path Weg, Pfad, Gang; (course/trend) Verlauf (einer Kurve)
path difference *opt* Gangunterschied
pathogen/ disease-causing agent Krankheitserreger
pathogenic (causing/capable of causing disease) pathogen, krankheitserregend
pathogenicity Pathogenität
pathological (altered or caused by disease) pathologisch, krankhaft
pathology Pathologie, Lehre von den Krankheiten
pattern (sample/model) Muster, Vorlage, Modell; (design) Musterung, Zeichnung
payment Zahlung (einer Rechnung)
➢ **conditions of payment** Zahlungsbedingungen

PCR (polymerase chain reaction) PCR (Polymerasekettenreaktion)
➢ **bubble linker PCR** Blasen-Linker-PCR
➢ **DOP-PCR (degenerate oligonucleotide primer PCR)** DOP-PCR (PCR mit degeneriertem Oligonucleotidprimer)
➢ **inverse PCR** inverse Polymerasekettenreaktion
➢ **IRP (island rescue PCR)** IRP (inselspezifische PCR)
➢ **ligation-mediated PCR** ligationsvermittelte Polymerasekettenreaktion
➢ **RACE-PCR (rapid amplification of cDNA ends)** RACE-PCR (schnelle Vervielfältigung von cDNA-Enden)
➢ **RT-PCR (reverse transcriptase-PCR)** RT-PCR (PCR mit reverser Transcriptase)
peak Peak, Spitze, Scheitel, Höhe, Maximum
peak broadening *chromat/spectr* Peakverbreiterung
peak value/maximum (value) Scheitelwert, Höchstwert, Maximum
peanut oil Erdnussöl
pear-shaped flask (small/pointed) Spitzkolben
peat Torf
➢ **granulated peat/garden peat** Torfmull
peat humus Torfhumus
pebble Kieselstein
pectic acid (pectate) Pektinsäure (Pektat)
peer review Begutachtungsverfahren (durch Kollegen)
pellet *n* Pellet; Kügelchen, Körnchen; Pille, Mikrodragée, Granulatkorn; *spectr* Pressling, Tablette
pellet *vb* pelletieren
pelletize pelletisieren, garnulieren; zu Pellets formen; körnen

penalty Strafe
pencil marking Bleistiftmarkierung
pendant überhängend, überstehend
pendant group *chem* Seitengruppe
penicillanic acid Penicillansäure
pentavalent fünfwertig
peptide bond/peptide linkage
  Peptidbindung
peptone water Peptonwasser
peptonize peptonisieren
perceive
  wahrnehmen, empfinden (Reiz);
  perzipieren, sinnlich wahrnehmen
percentage
  Prozentsatz, prozentualer Anteil
perceptible
  wahrnehmbar, empfindbar
perception Wahrnehmung,
  Empfindung, Perzeption (Reiz)
perchloric acid Perchlorsäure
percolate (flow through) durchfließen;
  (seep through) durchsickern
percolation
  (flowing through/flux) Durchfluss;
  (seepage) Durchsickern
percussion drill Schlagbohrer
percussion hammer/
  plexor/plessor/percussor
  Reflexhammer
perennial mehrjährig, ausdauernd
perforate(d)
  perforieren (perforiert, löcherig)
performance
  Auftritt; Leistung; Verhalten;
  (realization/completion/
  implementation) Durchführung
  (z.B. eines Experiments)
performance audit
  Leistungsaudit, Leistungsprüfung,
  Tauglichkeitsprüfung
performance criteria
  Leistungskriterien (Geräte etc.)
performance factor Gütefaktor
performance range Leistungsbereich
performance value/
  performance coefficient
  Leistungszahl
performic acid Perameisensäure

perfusion culture
  Perfusionskultur
periodic (periodical) periodisch
periodic acid-Schiff stain (PAS stain)
  Periodsäure, Schiff-Reagens
  (PAS-Anfärbung)
periodic table (of the elements)
  Periodensystem (der Elemente)
periodicity Periodizität
perish verderben;
  zugrunde gehen, vergehen
perishable verderblich
➤ **highly perishable**
  leicht verderblich
peristalsis Peristaltik
peristaltic peristaltisch
peristaltic pump
  peristaltische Pumpe
perlite Perlit, Perlstein
permanent hardness
  bleibende Härte,
  permanente Härte
permanent marker (water-resistant)
  wischfester/wasserfester Markierstift
permanent mount/slide *micros*
  Dauerpräparat
permanent run/operation
  Dauerbetrieb, Dauerleistung,
  Non-Stop-Betrieb
permanent wilting percentage
  permanenter Welkungsgrad
permeability
  Permeabilität, Durchlässigkeit
➤ **impermeability/imperviousness**
  Undurchlässigkeit, Impermeabilität
➤ **semipermeability**
  Halbdurchlässigkeit,
  Semipermeabilität
permeable (pervious)
  permeabel, durchlässig
➤ **impermeable/impervious**
  impermeabel, undurchlässig
➤ **semipermeable**
  semipermeabel, halbdurchlässig
permissible exposure limit (PEL)
  zulässige/erlaubte Belastungsgrenze
permissible radiation
  zulässige Strahlung

permissible workplace exposure
zulässige/maximale Arbeitsplatzkonzentration
permission Erlaubnis
permissivity/permissive conditions
Permissivität
permit *n* Zulassung, Lizenz, Erlaubnis
➤ **requiring official permit**
genehmigungsbedürftig
persist
persistieren, verharren, ausdauern
persistence
Persistenz, Beharrlichkeit, Ausdauer; (survival) Überdauerung, Überleben
persisting infection
persistierende Infektion, anhaltende Infektion
pervious/permeable
durchlässig, permeabel
perviousness/permeability
Durchlässigkeit, Permeabilität
pest Schädling, Ungeziefer
pest control
Schädlingsbekämpfung, Schädlingskontrolle
➤ **biological pest control**
biologische Schädlingsbekämpfung
pest infestation Schädlingsbefall
pest insect Schadinsekt
pesticide (biocide)
Pestizid, Biozid, Schädlingsbekämpfungsmittel
pesticide accumulation
Pestizidanreicherung
pesticide residue
Pestizidrückstand
pesticide resistance
Schädlingsbekämpfungsmittelresistenz, Pestizidresistenz
pestle (and mortar)
Stößel, Pistill (und Mörser)
PET (positron emission tomography)
PET (Positronenemissionstomographie)
Petri dish Petrischale
petroleum (crude oil)
Erdöl (Rohöl)
petroleum ether Petroläther

petroleum jelly/vaseline
Petrolatum, Vaseline
petroleum spirit/ligroin Ligroin
pharmaceutic/
pharmaceutical chemist
Arzneimittelchemiker
pharmaceutical pharmazeutisch
pharmacognosy
Pharmakognosie, Drogenkunde, pharmazeutische Biologie
pharmacology Pharmakologie
pharmacopeia/pharmacopoeia/
formulary
Pharmakopöe, amtliches Arzneibuch, Arzneimittel-Rezeptbuch
pharmacy
Pharmazie, Arzneilehre, Arzneikunde
phase (layer) Phase (nicht mischbare Flüssigkeiten)
➤ **lower phase**
Unterphase (flüssig-flüssig)
➤ **transition phase**
Übergangsphase
➤ **upper phase**
Oberphase (flüssig-flüssig)
phase boundary Phasengrenze
phase contrast Phasenkontrast
phase contrast microscope
Phasenkontrastmikroskop
phase contrast microscopy
Phasenkontrastmikroskopie
phase diagram Phasendiagramm
phase down
stufenweise verringern/runterstellen
phase in
stufenweise einführen/hochstellen/ heraufstellen
phase out stufenweise abbauen, stufenweise außer Kraft setzen/ auslaufen lassen
phase ring/phase annulus
Phasenring
phase shifting Phasenverschiebung
phase transition Phasenübergang
phase transition temperature
Phasenübergangstemperatur
phase variation Phasenveränderung

Phillips®-head screwdriver/ Phillips® screwdriver
Kreuzschraubenzieher, Kreuzschlitzschraubenzieher
phosgene Phosgen
phosphate Phosphat
phosphatidic acid
Phosphatidsäure
phosphodiester bond
Phosphodiesterbindung
phosphoric acid (phosphate)
Phosphorsäure (Phosphat)
phosphorous *adj/adv*
phosphorhaltig, phosphorig, Phosphor...
phosphorous acid
phosphorige Säure
phosphorus (P) *n* Phosphor
phosphorylation Phosphorylierung
photo-ionization detector (PID)
Fotoionisations-Detektor (PID)
photoacoustic spectroscopy (PAS)
fotoakustische Spektroskopie (PAS), optoakustische Spektroskopie
photoallergenic fotoallergen
photobleaching Lichtbleichung
photocopier/copy machine
Fotokopierer, Kopiergerät, Kopierer
photoelectron spectrometry (PES)
Fotoelektronenspektrometrie
photographic paper Fotopapier
photographic plate Fotoplatte
photometric titration
fotometrische Titration
photometry
Fotometrie, Lichtmessung
photomultipier
Fotovervielfacher, Fotomultiplier
photon Photon, Strahlungsquant
photoperception Lichtwahrnehmung
photoreactivation Fotoreaktivierung
photoresist/photoresistor
Fotowiderstand, lichtelektrischer Widerstand
photorespiration Fotorespiration, Fotoatmung, Lichtatmung
photosensibilization
Fotosensibilisierung

photosensitive/ photoresponsive
fotoempfindlich, lichtempfindlich
photosensitivity
Fotoempfindlichkeit, Lichtempfindlichkeit
photostability (lightfastness)
Lichtbeständigkeit, Lichtechtheit
photostable (lightfast/nonfading)
lichtbeständig, lichtecht
photosynthesis
Photosynthese, Fotosynthese
photosynthesize
photosynthetisieren, fotosynthetisieren
photosynthetic photon flux (PPF)
Photonenstromdichte
photosynthetically active radiation (PAR)
photosynthetisch aktive Strahlung
phthalic acid Phthalsäure
physical *adj/adv*
physisch, körperlich; physikalisch
physical *n*
(physical/medical examination/ medical exam)
medizinische Untersuchung, ärztliche Untersuchung
physical containment
physikalische/technische Sicherheit(smaßnahmen)
physical containment level
Sicherheitsstufe (Laborstandard), Laborsicherheitsstufe
physical examination (medical examination/ medical exam/physical)
medizinische Untersuchung, ärztliche Untersuchung
physical exercises
Leibesübungen, körperliche Ertüchtigung
physical map physikalische Karte
physical state (solid, liquid, gas)
Aggregatzustand
physical work körperliche Arbeit
physically handicapped
körperbehindert

physician Arzt, Mediziner
physician's white coat/ white coat
Medizinerkittel
physicist Physiker
physics Physik
physiological physiologisch
physiologist Physiologe
physiology Physiologie
phytanic acid Phytansäure
phytic acid Phytinsäure
phytochemical/plant chemical
Pflanzeninhaltsstoff
phytotoxic
pflanzenschädlich, phytotoxisch
pickle *vb* pökeln
(sauer einlegen: Gurken/Hering etc.)
pickling Pökeln (in Salzlake oder Essig einlegen: Gurken/Hering etc.)
pickup Abholung (Lieferung etc.);
electr Geber; Greifer;
kleiner Pritschenwagen
pickup point Abholstelle; Haltestelle
picric acid (picrate)
Pikrinsäure (Pikrat)
pictograph (for hazard labels)
Bilddiagramm, Begriffszeichen
picture/image Aufnahme, Bild
PID control
Proportional-Integral-Differential-Regelung
pie chart Kreisdiagramm
pierce durchstechen, durchbohren, durchstoßen
piercer Bohrer, Locher
pig (cow receiver adapter: receiving adapter for three/four receiving flasks)
'Spinne', Eutervorlage, Verteilervorlage;
rad (outermost container of lead for radioactive materials) Bleiblock
pigment *n* Pigment; Farbe, Farbstoff
pigmentation
Pigmentation, Färbung;
Pigmentierung
pigtail lead *electr*
Anschlussleitung

pilot flame/pilot light (from a pilot burner)
Sparflamme; *auch:* Zündflamme
pilot plant
Pilotanlage, Versuchsanlage
pilot scale Pilotmaßstab
pilot wire *electr*
Messader, Prüfader, Prüfdraht;
Steuerleitung; Hilfsleiter
pilot-operated (valve)
hydraulisch vorgesteuert (Ventil)
pimelic acid Pimelinsäure
pin *n* (lead) *electr*
Stift (Stecker/Anschluss);
(dowel/wall plug) Dübel
pinch kneifen, klemmen, quetschen
➤ **pinch off/tip** pinzieren, entspitzen
pinch clamp Schraubklemme
pinch valve Quetschventil
pinchcock Quetschhahn
pinchcock clamp
Schlauchklemme
pinewood chip/chip of pinewood
Kienspan
pipe (tube) Rohr, Röhre;
(pipes/plumbing) Rohre, Rohrleitungen
➤ **downpipe** Fallrohr
pipe clamp/pipe clip
Rohrschelle
pipe cleaner
Pfeifenreiniger, Pfeifenputzer
pipe fitting(s)/fittings
Rohrverbinder, Rohrverbindung(en)
pipe wrench (rib-lock pliers/ adjustable-joint pliers)
Rohrzange
pipe-to-tubing adapter
Schlauch-Rohr-Verbindungsstück
pipet *vb* pipettieren
pipet *n* (**pipette** *Br*) Pipette
➤ **blow-out pipet** Ausblaspipette
➤ **capillary pipet/capillary pipette**
Kapillarpipette
➤ **dropping pipet/dropper**
Tropfpipette, Tropfglas
➤ **filter pipet** Filterpipette

- ➤ graduated pipet/measuring pipet
  Messpipette
- ➤ micropipet (pipettor)
  Mikropipette, Mikroliterpipette (Kolbenhubpipette)
- ➤ **Pasteur pipet** Pasteurpipette
- ➤ **piston-type pipet** Saugkolbenpipette
- ➤ serological pipet
  serologische Pipette
- ➤ suction pipet (patch pipet)
  Saugpipette
- ➤ transfer pipet/volumetric pipet
  Vollpipette, volumetrische Pipette

pipet aid/pipet helper
Pipettierhilfe
pipet ball Pipettierball
- ➤ safety pipet filler/
  safety pipet ball
  Peleusball (Pipettierball)

pipet brush Pipettenbürste
pipet bulb/rubber bulb
Saugball, Pipettierball, Pipettierbällchen
pipet filler/pipet aspirator
Pipettensauger
pipet pump Pipettierpumpe
pipet rack Pipettenständer
pipet tip Pipettenspitze
pipeting nipple/
rubber nipple/teat (*Br*)
Pipettierhütchen, Pipettenhütchen, Gummihütchen
pipettor/micropipet
Pipette, Mikropipette
piston/plunger
(e.g., of a syringe/pump)
Kolben (Stempel/Schieber: Spritze/Pumpe etc.)
piston pump/reciprocating pump
Kolbenpumpe
piston stroke Kolbenhub
piston valve Kolbenventil
pitch *n* Neigung, Gefälle; Höhe; Grad, Stufe; (DNA: helix periodicity) Ganghöhe (*DNA-Helix:* Anzahl Basenpaare pro Windung); (resin from conifers) Terpentinharz

pitch angle
Steigungswinkel, Steigwinkel
pitch screw impeller
Schraubenspindelrührer
pitched-blade fan impeller/
pitched-blade paddle impeller/
inclined paddle impeller
Schrägblattrührer
pitcher
Krug; Becher(glas) mit Griff
pivot Spindel, Zapfen, Stift, Achse; Drehpunkt, Drehzapfen, Drehbolzen
pivoted drehbar
pixel Bildpunkt, Rasterpunkt
placard
Kennzeichen für Fahrzeuge/ Container
placebo
Placebo, Plazebo, Scheinarznei
plain stage *micros*
Standardtisch
plane *n* (flat/level surface)
Ebene, ebene Fläche
plane mirror/plano-mirror
Planspiegel
plane-polarized light
linear polarisiertes Licht
plano-concave mirror
Plan-Hohlspiegel, Plankonkav
plant *n* *bot* Pflanze;
tech Betriebseinrichtung, Werk, Anlage
plant *vb* pflanzen, einpflanzen, bepflanzen; anlegen
plant pest
Pflanzenschädling
plant pigment
Pflanzenfarbstoff
plant protection
Pflanzenschutz
plant-protective agent (pesticide)
Pflanzenschutzmittel (Pestizid)
plaque Plaque (auch: Zahnbelag), Aufklärungshof, Lysehof, Hof
plaque assay Plaque-Test
plasmenic acid Plasmensäure

plaster Mörtel, Verputz, Tünche;
 med Pflaster
plaster cast *med* Gipsverband;
 Gipsabdruck, Gipsabguss
plaster of Paris (POP) *med*
 Gips (für Gipsverband)
plaster splint *med*
 Gipsschiene
plastic wrap (household wrap)
 Plastikfolie (Frischhaltefolie)
plasticine Plastilin
plasticity Plastizität
plastination Plastination
➤ **whole mount plastination**
 Ganzkörperplastination
plate *vb micb* plattieren
plate *n* Teller;
 chromat (HPLC) Trennstufe;
 dist/chromat Boden
➤ **agar plate** Agarplatte
➤ **baffle plate**
 Prallblech, Prallplatte, Leitblech,
 Ablenkplatte (Strombrecher z.B. an
 Rührer von Bioreaktoren)
➤ **counting plate** Zählplatte
➤ **hot plate** Heizplatte, Kochplatte
➤ **laboratory protection plate**
 Laborschutzplatte (Keramikplatte)
➤ **multiwell plate** *micb*
 Vielfachschale, Multischale
➤ **photographic plate** Fotoplatte
➤ **pour-plate method/technique** *micb*
 Plattengussverfahren,
 Gussplattenmethode
➤ **precoated plate** *chromat*
 Fertigplatte
➤ **sieve plate (perforated plate)**
 Siebplatte
➤ **spot plate** *micb* Tüpfelplatte
➤ **spread-plate method/technique** *micb*
 Spatelplattenverfahren
➤ **stirring hot plate**
 Magnetrührer mit Heizplatte
➤ **streak-plate method/technique** *micb*
 Plattenausstrichmethode
➤ **theoretical plates** *dist/chromat*
 theoretische Böden
➤ **well plate** *gen/micb* Lochplatte

plate assay/plating
 Platten-Test
plate column *dist*
 Bodenkolonne
plate count *micb*
 Plattenzählverfahren
plate efficiency *dist*
 Bodenwirkungsgrad
plate height *dist/chromat*
 Bodenhöhe
plate mill/disk mill
 Scheibenmühle
plate number/number of plates
 dist/chromat Bodenzahl
platform Plattform
platform truck
 Plattformwagen/-karren
➤ **dolly**
 kleines/rundes Schiebegestell auf
 Rollen (Kistenroller/Fassroller etc.)
plating (plating out) *micb*
 Plattierung, Plattieren
➤ **efficiency of plating**
 Plattierungseffizienz
platinum (Pt) Platin
pleated sheet (α-sheet)
 Faltblatt (α-Faltblatt)
plenum chamber (*pl* **plena)**
 Luftkammer (Schacht: z.B. Abzug)
plier/pliers (*Br* **nippers)**
 Zange; Beißzange, Kneifzange
➤ **bent longnose pliers/**
 bent long-nose pliers
 gebogene Spitzzange
➤ **combination pliers/**
 linesman pliers
 Kombizange
➤ **cutting pliers/pincers**
 Kneifzange
➤ **diagonal pliers**
 Seitenschneider
➤ **dip needle-nose pliers**
 gebogene Storchschnabelzange
➤ **end nippers**
 Monierzange, Rabitzzange
➤ **flat-nosed pliers** Flachzange
➤ **glass-tube cutting pliers**
 Glasrohrschneider (Zange)

pl

- **griplock pliers (US)/ channellock pliers (US)**
  Rohrzange
- **grippers** Greifzange
- **linesman pliers**
  Telefonzange, Kabelzange
- **needle-nose pliers/ snipe-nose pliers/ snipe-nosed pliers**
  Storchschnabelzange
- **punch pliers** Lochzange
- **revolving punch pliers**
  Revolverlochzange
- **rib joint pliers/rib-lock pliers**
  Eckrohrzange
- **snap-ring pliers/circlip pliers**
  Sicherungsringzange
- **utility pliers**
  Mehrzweckzange
- **Vise-Grip® pliers**
  Gripzange, Haltezange
- **water pump pliers/ slip-joint adjustable water pump pliers (adjustable-joint pliers)**
  Pumpenzange, Wasserpumpenzange

plot *n tech* Plan, Entwurf; Diagramm, grafische Darstellung; Plotten, Auftragung; Aufzeichnung, Registrierung

plot *vb tech* planen, entwerfen; plotten, auftragen; aufzeichnen, registrieren

plotter Plotter, Kurvenzeichner, Kurvenschreiber

plug *vb* verschließen, zustopfen, stöpseln; dübeln

plug *n electr/tech* **(jack/connector/coupler)** Stecker

- **banana plug** Bananenstecker
- **female**
  weiblicher Stecker (Minus~); Hülse
- **male**
  männlicher Stecker (Plus~); Kern

plug connection/fitting
Steckverbindung

plug in (connect) *electr/tech*
einstecken, anschließen
(Stecker reinstecken/Stecker in Steckdose stecken)

- **unplug/disconnect**
  ausstöpseln, Stecker herausziehen

plug valve Auslaufventil

plug wrench (bung removal)
Spundschlüssel (für Fässer); Stopfenschlüssel

plug-flow reactor
Pfropfenströmungsreaktor, Kolbenströmungsreaktor

plumber Klempner, Installateur

plumbing Rohre, Rohrleitungen; Klempner-, Installationsarbeiten

plumbing system
Rohrleitungssystem (Wasser)

plunger Stempel, Kolben, Schieber; Stößel

- **syringe piston**
  Spritzenkolben

plunging jet reactor/ deep jet reactor/ immersing jet reactor
Tauchstrahlreaktor

plunging siphon Stechheber

plus (minus) connection *electr*
Plus~ (Minus~)Verbindung

pluviometer/rain gauge
Regenmesser

plywood Sperrholz

plywood board Sperrholzplatte

pneumatic pneumatisch

pneumatic valve Druckluftventil

pneumoconiosis Pneumokoniose, Staublunge, Staublungenerkrankung

pocket knife Taschenmesser

pointer eyepiece *micros*
Zeigerokular

poison *vb* **(intoxicate)** vergiften

poison *n* **(toxin)** Gift, Toxin

- **cumulative poison**
  Summationsgift, kumulatives Gift

poison cabinet Giftschrank

poison control center/
  poison control clinic
  Entgiftungszentrale,
  Entgiftungsklinik,
  Vergiftungszentrale
poison information center
  Giftinformationszentrale
poisoning/intoxication
  Vergiftung, Intoxikation
poisonous (toxic)
  giftig, toxisch
poisonous materials/
  poisonous substances
  Giftstoffe
poisonous plant Giftpflanze
poisonousness/toxicity
  Giftigkeit, Toxizität
polar growth polares Wachstum
polarity reverseal (Umpolung)
  Polaritätsumkehrung,
  Polwechsel, Umpolung
polarized light
  polarisiertes Licht
polarizing filter/polarizer
  Polarisationsfilter, 'Pol-Filter',
  Polarisator
polarizing microscope
  Polarisationsmikroskop
polarizing microscopy
  Polarisationsmikroskopie
polarography Polarografie
> **current-sampled polarography**
  Tastpolarografie
> **differential pulse polarography (DPP)**
  differentielle Pulspolarografie
pole *electr* Pol; (rod) Stange
policeman/
  rubber policeman/scraper
  (glass/plastic or metal rod with
  rubber or Teflon tip)
  Kolbenwischer, Gummiwischer
  (zum mechanischen Loslösen von
  Rückständen im Glaskolben)
policy/rule
  Vorschrift(en), Regel(n)
> **general policy**
  allgemeine Richtlinie

pollutant
  (harmful substance/contaminant)
  Schadstoff, Schmutzstoff
pollute (contaminate) verschmutzen,
  verunreinigen, belasten; beflecken
polluted (contaminated)
  verschmutzt
> **unpolluted/uncontaminated**
  unverschmutzt
polluter Umweltverschmutzer
pollution (contamination)
  Verschmutzung,
  Verunreinigung, Belastung
> **air pollution**
  Luftverschmutzung,
  Luftverunreinigung
> **amount of pollution/**
  degree of contamination
  Verschmutzungsgrad
> **environmental pollution**
  Umweltverschmutzung
> **water pollution**
  Wasserverschmutzung
pollution control Umweltschutz
polyacrylamide Polyacrylamid
polydispersity index (PDI)
  Polydispersitätsindex (PDI)
polymer Polymer
polymerase chain reaction
  (*see also:* PCR)
  Polymerasekettenreaktion
polymerization Polymerisation
> **degree of polymerization**
  Polymerisationsgrad
polysulfide rubber
  Polysulfid-Kautschuk
polyunsaturated
  mehrfach ungesättigt
polyunsaturated fatty acid
  mehrfach ungesättigte Fettsäure
pool *n* **(whole quantity of a**
  particular substance: body
  substance, metabolite etc)
  'Pool' (Gesamtheit einer
  Stoffwechselsubstanz)
pool *vb* **(combine/accumulate)**
  poolen, vereinigen,
  zusammenbringen, zusammenfassen

population Population, Bevölkerung
population crash
  Populationszusammenbruch,
  Bevölkerungszusammenbruch
population curve Populationskurve,
  Bevölkerungskurve
population genetics
  Populationsgenetik
porcelain Porzellan
porcelain dish Porzellanschale
porcelain enamel Email, Emaille
pore size/mesh size
  Porenweite (Filter/Gitter etc.)
porosity Porosität, Durchlässigkeit
porous porös, porig, durchlässig
port *tech/mech/electr*
  Eingang, Anschluss (Gerät)
portion/fraction Portion, Anteil,
  Teilmenge, Fraktion
position *n* Position,
  Lage (in Bezug), Stellung, Standort
position *vb* positionieren,
  in die gewünschte Lage bringen
  (in Bezug), aufstellen;
  einstellen, anbringen
positive displacement pump
  Direktverdrängerpumpe
positive pressure Überdruck
positive-displacement valve
  Verdrängerventil
positron emission tomography (PET)
  Positronenemissionstomographie
  (PET)
post-emergence treatment *agr*
  Nachauflaufbehandlung
posttreatment examination/
  follow-up (exam)/
  reexamination after treatment *med*
  Nachuntersuchung
pot *n* Topf, Kanne, Gefäß
pot *vb* eintopfen (Pflanze)
pot cleaner/scouring pad
  Topfkratzer, Topfreiniger
pot life *chem*
  Topfzeit, Verarbeitungsdauer
potable water trinkbares Wasser
potash (potassium carbonate)
  Pottasche (Kaliumcarbonat)
potassium (K) Kalium
potassium cyanide
  Kaliumcyanid, Cyankali, Zyankali
potassium hydroxide solution
  Kalilauge, Kaliumhydroxidlösung
potassium permanganate
  Kaliumpermanganat
potential *adj/adv* potentiell
potential *n* Potential
➤ **gross potential**
  Summenpotential
potential barrier
  Potentialwall, Potentialbarriere
potential difference (voltage) *electr*
  Potentialdifferenz, Spannung
potential energy Lageenergie
potholder Topflappen
Potter-Elvehjem homogenizer
  (glass homogenizer)
  'Potter' (Glashomogenisator)
potting soil (potting mixture:
  soil & peat a.o.) Topferde
pour gießen
➤ **pour off/pour out/decant**
  abgießen, ausgießen, dekantieren;
  (drain) ablassen
pouring ring Ausgießring
pouring spout
  Gießschnauze (an Gefäß)
pour-plate method/technique *micb*
  Plattengussverfahren,
  Gussplattenmethode
powder Puder, Pulver
powder funnel Pulvertrichter
powder spatula Pulverspatel
power *n*
  Leistung; Elektrizität, Strom
power *vb* betreiben, antreiben,
  versorgen, mit Strom versorgen
➤ **power down** ausschalten, abschalten;
  (computer/reactor) herunterfahren
➤ **power up** einschalten, anschalten;
  (computer/reactor) hochfahren
power control Leistungsregelung
power cord/electric cord/
  power cable/electric cable
  Stromkabel
power grid *electr* Verteilungsnetz

power input Aufnahmeleistung
power lead Stromkontakt
power network Versorgungsnetz
power output/rated power output
  Nennleistung, Nominalleistung
power plug Netzstecker
power screwdriver Elektroschrauber
power supply *electr*
  Stromquelle, Stromzufuhr
power supply unit *electr*
  Netzgerät, Netzteil, Stromgerät
power switch
  Netzschalter, Stromschalter
preamplifier Vorverstärker
precaution (safety warning)
  Vorkehrung, Vorsichtsmaßnahme,
  Vorsichtsmaßregel
➢ **take precautions**
  (take precautionary measures)
  Vorkehrungen treffen
precautionary measure
  (protective measure)
  Schutzmaßnahme,
  Vorsichtsmaßnahme (Vorkehrung)
precious metal Edelmetall
precipitant/precipitating agent
  Fällungsmittel
precipitate *n chem*
  Präzipitat, Fällung, Ausfällung
precipitate *vb chem*
  präzipitieren, fällen, ausfällen;
  (crystals) ausschieden
precipitation *chem*
  Präzipitation, Ausfällung, Ausfällen,
  Fällung, Fällen; *meteo* Niederschlag
➢ **fractional precipitation**
  fraktionierte Fällung
precipitation titration
  Fällungstitration
precise (exact/accurate)
  präzis, genau
precision (exactness/accuracy)
  Präzision, Genauigkeit
precision balance
  Präzisionswaage, Feinwaage
precision of measurement/
  measurement precision
  Messgenauigkeit

precleaned vorgereinigt
precleaning Vorreinigung
precoated plate *chromat*
  Fertigplatte
preculture Vorkultur
precursor
  Präkursor, Vorläufer
prediction
  Vorraussage, Vorhersage
predictive
  vorraussagend, vorhersagend
predictive medicine
  vorhersagende Medizin
predictive model
  Vorraussagemodell
predisposition
  Prädisposition,
  Veranlagung
predominate vorherrschen
pre-emergence treatment *agr*
  Vorauflaufbehandlung
prefilter Vorfilter
pregerminate vorkeimen
pregermination Vorkeimung
preheat vorwärmen, anheizen
preheater Vorwärmer
preheating time/rise time (autoclave)
  Anheizzeit, Steigzeit (Autoklav)
preimplantation testing
  Präimplantationstest (Untersuchung
  vor Einnistung des Eis)
preliminary vorläufig
preliminary test/crude test
  Vorversuch, Vorprobe
prenatal diagnostics
  pränatale Diagnostik
preparation Vorbereitung,
  Zubereitung, Herstellung;
  (preserved specimen) Präparat
  (*Lebewesen*)
preparation process/procedure
  Herstellungsverfahren,
  Vorbereitungsverfahren
preparative präparativ
preparative centrifugation
  präparative Zentrifugation
preparative chromatography
  präparative Chromatographie

preparatory school (prep school)
(auf ein College)
vorbereitende Schule
preparatory work Vorarbeiten
prepare
präparieren, vorbereiten, richten;
anfertigen, herstellen, zubereiten
prepared microscope slide
Mikropräparat
prephenic acid (prephenate)
Prephensäure (Prephenat)
prepurify vorreinigen
prescribe vorschreiben, vorgeben;
med verschreiben
**prescribed work procedure/
prescribed operating procedure**
Arbeitsanweisung, Arbeitsvorschrift
prescription
Vorschrift, Verordnung;
med Rezept
prescription drug
verschreibungspflichtiges
Arzneimittel/Medikament
preservation Bewahrung,
Erhaltung, Konservierung;
Aufbewahrung, Lagerung
preservative *n*
Konservierungsstoff
**preserve/
keep/maintain**
bewahren, erhalten, schützen,
vor dem Verderben schützen,
konservieren, präservieren
preset *adj/adv* voreingestellt,
vorgewählt
press *n* Presse, Druckmaschine
press *vb* pressen,
ausdrücken, zusammendrücken;
(fruit/grapes) keltern
pressure Druck
➢ **air pressure** Luftdruck
➢ **ambient pressure**
Umgebungsdruck
➢ **atmospheric pressure**
atmosphärischer Luftdruck
➢ **blood pressure** Blutdruck
➢ **breaking pressure**
Öffnungsdruck (Ventil)
➢ **counterpressure** Gegendruck
➢ **high pressure** Hochdruck
➢ **hydrostatic pressure**
hydrostatischer Druck
➢ **hydrostatic pressure (turgor)**
hydrostatischer Druck (Turgor)
➢ **initial pressure/
initial compression/
high pressure**
Vordruck, Eingangsdruck
(Hochdruck: Gasflasche)
➢ **loss of pressure/pressure drop**
Druckverlust
➢ **low pressure** Niederdruck
➢ **negative pressure** Unterdruck
➢ **normal pressure** Normaldruck
➢ **oncotic pressure**
onkotischer Druck,
kolloidosmotischer Druck
➢ **osmotic pressure**
osmotischer Druck
➢ **outlet pressure** Hinterdruck
➢ **oxygen partial pressure**
Sauerstoffpartialdruck
➢ **partial pressure** Partialdruck
➢ **positive pressure** Überdruck
➢ **reduced pressure**
erniedrigter Druck
➢ **selective pressure/
selection pressure**
Selektionsdruck
➢ **standard pressure**
Normaldruck, Normdruck
➢ **supply pressure (HPLC)**
Eingangsdruck
➢ **total pressure** Gesamtdruck
➢ **turgor pressure** Turgordruck
➢ **vapor pressure** Dampfdruck
➢ **working pressure/
delivery pressure**
Hinterdruck, Arbeitsdruck
(Druckausgleich)
pressure bandage *med*
Druckverband
pressure control valve
Druckregelventil
pressure cooker
Dampfkochtopf, Schnellkochtopf

pressure cycle reactor
  Druckumlaufreaktor
pressure drop
  Druckabfall
pressure equalization
  Druckausgleich
pressure filtration Druckfiltration
pressure fluctuation
  Druckschwankung
pressure gauge/
  pressure gage/gauge/gage
  Druckmesser, Manometer
pressure head Staudruck, Druckhöhe,
  Fließdruck, Druckgefälle;
  Förderhöhe
pressure protection device
  Druckentlastungseinrichtung
pressure regulator
  Druckminderer (Gasflasche),
  Druckregler
pressure resistant druckfest
pressure rise/pressure increase
  Druckanstieg
pressure tubing Druckschlauch
pressure valve/
  pressure relief valve
  Druckventil, Überdruckventil
pressure vessel Druckbehälter
pressure-flow theory/hypothesis
  Druckstromtheorie,
  Druckstromhypothese
pressure-relief valve
  (gas regulator/
  gas cylinder pressure regulator)
  Gasdruckreduzierventil,
  Druckminderventil,
  Druckminderungsventil,
  Reduzierventil (für Gasflaschen)
pressure-tight druckdicht
pressurize unter Druck setzen
  (unter Überdruck halten)
pressurizer
  Druckerzeuger; Druckanlage
presymptomatic diagnostics
  präsymptomatische Diagnostik
pretreatment Vorbehandlung
pretrial (preliminary experiment)
  Vorversuch

prevalence/prevalency
  Prävalenz
prevention Prävention;
  (provision) Verhinderung
  (Unfälle/Vorsorge)
prevention of accidents
  Unfallverhütung
preventive medical checkup
  Vorsorgeuntersuchung
preventive medicine
  Präventivmedizin
primary product (initial product)
  Ausgangsprodukt
primary structure (proteins)
  Primärstruktur
prime *n* Anfang, Beginn
prime *vb* vorbereiten; (Farbe)
  grundieren; (pump) vorpumpen,
  anlassen (auch: selbstansaugend)
prime conductor *electr* Hauptleiter
primer Zündvorrichtung;
  Grundiermasse, Spachtelmasse;
  gen Primer
primitive form
  (basic form/parent form)
  Stammform, Urform
printout (from a printer)
  Ausdruck (Drucker)
priority rule Prioritätsregel
prism Prisma
pristine ursprünglich, urtümlich
probability Wahrscheinlichkeit
probe *n* Sonde, Fühler;
  (microprobe) Mikrosonde
▷ **proton microprobe**
  Protonensonde
probe *vb* prüfen, testen,
  untersuchen, analysieren
probing head *micros* Tastkopf
procedure/technique Verfahren
process *n* Prozess, Vorgang,
  Verlauf, Arbeitsgang
process *vb* **(treat)** prozessieren,
  verarbeiten, weiterverarbeiten;
  (metabolize) umsetzen
process control
  Prozess-Kontrolle,
  Prozesssteuerung

process engineering
Verfahrenstechnik
➢ **environmental process engineering**
Umweltverfahrenstechnik
process water/service water/ industrial water (nondrinkable water)
Brauchwasser, Betriebswasser (nicht trinkbares Wasser)
processing/treatment/finishing
Prozessierung, Verarbeitung, Aufbereitung, Weiterverarbeitung
procreate/reproduce/propagate
zeugen, fortpflanzen
procreation/reproduction/propagation
Zeugung, Fortpflanzung
produce *vb* **(manufacture/make)**
produzieren, erzeugen, herstellen
producer
Produzent, Erzeuger, Hersteller
producer gas Generatorgas
product Produkt; Erzeugnis, Ware; Ergebnis, Resultat
product inhibition Produkthemmung
product purity Produktreinheit
production costs/ manufacturing costs
Herstellungskosten
productivity Produktivität; Ertragsfähigkeit, Ergiebigkeit, Rentabilität
product-moment correlation coefficient
Maßkorrelationskoeffizient, Produkt-Moment-Korrelationskoeffizient
profession
Profession, Beruf, Erwerbstätigkeit
professional association (organization)
Berufsverband
prognosis Prognose, Vorhersage
programmed cell death (apoptosis)
programmierter Zelltod (Apoptose)
progressing cavity pump
Schneckenantriebspumpe
prohibition/ban Verbot
project *vb opt* projizieren, abbilden
projection Projektion, Abbildung

proliferate proliferieren
proliferation Proliferation
proof/check *n chem* Probe, Versuch, Untersuchung, Test, Prüfung
proove/check *vb chem*
beweisen, versuchen, untersuchen, die Probe machen
prop up stützen
propagate (reproduce) fortpflanzen, vermehren, reproduzieren; propagieren;
neuro weiterleiten, fortleiten
propagation (reproduction)
Fortpflanzung, Vermehrung, Reproduktion;
neuro Weiterleitung, Fortleitung
propellant *n* **(e.g., in pressure cans)**
Treibmittel, Treibgas (z.B. für Sprühflaschen)
propeller impeller Propellerrührer
property management (custodian)
Hausverwaltung
prophylactic prophylaktisch
prophylaxis Prophylaxe
propionic acid (propionate)
Propionsäure (Propionat)
propionic aldehyde/ propionaldehyde
Propionaldehyd
proportional truncation
proportionaler Schwellenwert
proportional valve (P valve)
Proportionalventil
propositus Proband, Propositus
propulsion Antrieb, Voranbringen (Fortbewegung)
propulsive force
Antriebskraft, Triebkraft
prostanoic acid Prostansäure
protect (protected)
schützen (geschützt)
protection Schutz
protection assay/ protection experiment
Schutzversuch, Schutzexperiment
protection mask/ face mask/respirator mask/respirator
Atemmaske, Atemschutzmaske

protective clothing Schutzkleidung
➢ **workers' protective clothing**
Arbeitsschutzkleidung
protective coat/protective gown
Schutzkittel, Schutzmantel
protective coating
Schutzüberzug (z.B. Anstrich)
protective curtain Schutzvorhang
protective gas/shielding gas
(**in welding**) Schutzgas
protective gloves
Schutzhandschuhe
protective group/protecting group
Schutzgruppe (*chem* Synthese)
protective hood Schutzhaube
protective measure
(**precautionary measure**)
Schutzmaßnahme
protective screen/shield
Schutzscheibe, Schutzschirm
protein Protein, Eiweiß
protein engineering
gezielte Konstruktion von Proteinen
protein synthesis Proteinsynthese
protein tagging
Proteinmarkierung, Protein-Tagging
proteinaceous proteinartig,
proteinhaltig, Protein...,
aus Eiweiß bestehend, Eiweiß...
proteolytic
proteolytisch, eiweißspaltend
protocol (**record/minutes**)
Protokoll, Aufzeichnungen;
Sitzungsbericht;
genormte Verfahrensvorschrift
proton gradient Protonengradient
proton microprobe Protonensonde
proton motive force
protonenmotorische Kraft
proton pump Protonenpumpe
proton shift (**NMR**)
Protonenverschiebung
protraction (**delay/procrastination:**
through neglect)
Verschleppung, Übertragung
protrude (**project/stand out/**
stick out/rise over)
herausragen, hervorstehen; überragen

provision Vorsorge
provisional measure
(**precautionary measure**)
Vorsorgemaßnahme
provisions (**furnishings/equipment/**
outfit/supplies) Ausstattung;
jur Bestimmungen
proximal proximal, ursprungsnah
pruners/pruning shears
(*Br* secateurs)
Gartenschere
prussiat Blutlaugensalz,
Kaliumhexacyanoferrat
psychoactive/psychotropic drug
Rauschmittel, Rauschgift,
Rauschdroge
psychrometer/
wet-and-dry-bulb hygrometer
Psychrometer
(ein Luftfeuchtigkeitsmessgerät)
psychrophilic
(**thriving at low temperatures**)
psychrophil
public danger
öffentliche Gefahr
public servant/civil servant
Staatsbedienstete(r)
public service officer
(*Br* civil servant)
Beamter, Beamtin
pulley Flaschenzug
pulmonary edema
Lungenödem
pulpwood Papierholz
pulsate/throb/beat pulsieren
pulse *n* Puls, Pulsieren;
electr Stromstoß, Impuls
pulse *vb* pulsieren;
impulsweise ausstrahlen/senden
pulse current *electr*
Impulsstrom, Stoßstrom
pulse labeling/pulse chase
Pulsmarkierung
pulse polarography
Pulspolarografie
➢ **differential pulse polarography**
(**DPP**)
differentielle Pulspolarografie

pu

pulsed field gel electrophoresis (PFGE)
Puls-Feld-Gelelektrophorese, Wechselfeld-Gelelektrophorese
pulsed laser
Impulslaser, gepulster Laser
pulverization
Pulverisierung; Zerstäubung
pulverize pulverisieren, fein zermahlen; zerstäuben
pulverizer
Zerkleinerer, Pulverisiermühle; Zerstäuber
pumice Bims
pumice rock Bimsstein
pump Pumpe
- **aspirator pump/vacuum pump**
  Absaugpumpe, Saugpumpe
- **barrel pump/drum pump**
  Fasspumpe
- **bellows pump** Balgpumpe
- **centrifugal pump** Kreiselpumpe
- **circulation pump** Umwälzpumpe
- **diaphragm pump** Membranpumpe
- **dispenser pump/dispensing pump**
  Dispenserpumpe
- **displacement pump**
  Verdrängungspumpe, Kolbenpumpe (HPLC)
- **dosing pump/ proportioning pump/ metering pump**
  Dosierpumpe
- **double-acting pump**
  Druckpumpe, Saugpumpe, doppeltwirkende Pumpe
- **feed pump** Förderpumpe
- **filter pump** Filterpumpe
- **gear pump** Zahnradpumpe
- **hand pump** Handpumpe
- **hand-operated vacuum pump**
  manuelle Vakuumpumpe
- **heat pump** Wärmepumpe
- **hose pump** Schlauchpumpe
- **impeller pump**
  Kreiselpumpe, Kreiselradpumpe
- **ion pump** Ionenpumpe
- **multichannel pump**
  Mehrkanal-Pumpe
- **peristaltic pump**
  peristaltische Pumpe
- **pipet pump**
  Pipettierpumpe
- **piston pump/ reciprocating pump**
  Kolbenpumpe
- **positive displacement pump**
  Direktverdrängerpumpe
- **prime** selbstansaugend
- **progressing cavity pump**
  Schneckenantriebspumpe
- **proton pump**
  Protonenpumpe
- **rotary piston pump**
  Drehkolbenpumpe
- **rotary vane pump**
  Drehschieberpumpe
- **squeeze-bulb pump (hand pump for barrels)**
  Quetschpumpe (Handpumpe für Fässer)
- **suction pump/ aspirator pump/ vacuum pump**
  Saugpumpe, Vakuumpumpe
- **syringe pump**
  Spritzenpumpe
- **total static head**
  Gesamtförderhöhe
- **tubing pump**
  Schlauchpumpe
- **vacuum pump**
  Vakuumpumpe
- **vane-type pump**
  Propellerpumpe
- **water pump/filter pump/ vacuum filter pump**
  Wasserstrahlpumpe

pump drive Pumpenantrieb
pump head Pumpenkopf
pumpkinseed oil Kürbiskernöl
punch pliers Lochzange
puncture *n* **(needle biopsy)**
Punktion
puncture *vb med* **(tap)** punktieren

pungency
Schärfe; stechender Geruch
pungent (taste/smell/pain)
scharf, stechend, beißend,
ätzend (Geruch)
pupil dilatation *med/opt*
Pupillenerweiterung
purchase *n* Einkauf, Erwerb
purchase order Kauforder, Bestellung
pure rein;
chem (purissimum/puriss.) reinst ;
(not denatured) unvergällt
pure culture/axenic culture
Reinkultur
pure substance Reinstoff
purge *n*
Reinigung, Säuberung, Befreiung;
chem Klärung; Klärflasche;
med Entschlackung, Darmentleerung
purge *vb* reinigen, säubern, befreien;
chem klären; *med* entschlacken,
entleeren (Darm)
purge assembly/purge device
Spülvorrichtung (z.B. Inertgas)
purge gas Spülgas
purge valve/
pressure-compensation valve
Entlüftungsventil
purification
Reinigung, Klärung, Reindarstellung
purification procedure/purification
technique
Reinigungsverfahren (Aufreinigung)
purified water gereinigtes Wasser,
aufgereinigtes Wasser,
aufbereitetes Wasser
purify reinigen, aufreinigen

purity Sauberkeit;
Reinheit (ohne Zusätze)
purity grades (chemical grades)
Reinheitsgrade (chemische R.)
purity of variety/variety purity
Sortenreinheit
pus Eiter
push button Drucktaste, Bedienknopf
push-pull current *electr*
Gegentaktstrom
putrefaction (rotting/decomposition)
Verwesung, Zersetzung
putrefactive bacteria Fäulnisbakterien
putrefy (rot/decompose)
verwesen, zersetzen
putty *vb* kitten, verkitten,
spachteln, verspachteln
putty *n* Kitt (Fensterkitt etc.)
putty knife
Spachtelmesser, Kittmesser
pycnometer
Pyknometer, Messflasche,
Wägeflasche
pyrethric acid Pyrethrinsäure
pyrite Eisenkies, Schwefelkies
pyrolysis/thermolysis
Pyrolyse, Thermolyse
pyrometer
Pyrometer, Hitzemessgerät
➢ **optical pyrometer**
Pyropter,
optisches Pyrometer
pyrometry Pyrometrie
pyroxylin (11.2–12.4% N)
Schießbaumwolle
pyruvic acid (pyruvate)
Brenztraubensäure (Pyruvat)

quad Vierer
quadrangle connection
  Viereckschaltung
quadrat method/quadrat sampling *ecol*
  Quadratmethode
quadratic mean *stat*
  Quadratmittel (Mittelwert)
quadruple nosepiece *micros*
  Vierfachrevolver
quadrupod (for burner)
  Vierfuß (für Brenner)
qualitative analysis
  qualitative Analyse
quality assessment
  Qualitätsbeurteilung, Qualitätsbewertung
quality assurance (QA)
  Qualitätssicherung
quality control (QC)
  Qualitätskontrolle, Qualitätsprüfung, Qualitätsüberwachung
quality factor
  Qualitätsfaktor, Bewertungsfaktor
quality indicator
  Qualitätskennzeichen
quality manual
  Qualitätssicherungshandbuch (EU-CEN)
quantification/quantitation *med/chem*
  Quantifizierung
quantify/quantitate *med/chem*
  quantifizieren
quantile/fractile *stat* Quantil, Fraktil
quantitative analysis
  quantitative Analyse
quantity (amount/number)
  Quantität, Menge (Anzahl), Größe
➢ **physical quantity**
  physikalische Größe
quantity to be measured
  Messgröße
quantization
  Quantisierung, Quantelung

quantum state Quantenzustand
quantum yield Quantenausbeute
quarantine Quarantäne
quarternary structure (proteins)
  Quartärstruktur
quartile *stat* Quartil, Viertelswert
quartz glass Quarzglas
quartz thermometer
  Quarzthermometer
quaternary quartär, quaternär
quench (put out/extinguish: fire/flame) löschen, stillen, ablöschen; abschrecken; rasch abkühlen; abdämpfen; *polym* härten
quencher/quenching agent
  Löscher, Quencher; Abschreckmittel, Ablöschmittel
quenching gas Löschgas
quick connect/quick connection
  Schnellverbindung
quick drench shower/ deluge shower
  'Schnellflutdusche'
quick section *micros/med*
  Schnellschnitt
quick-disconnect fitting
  Schnellkupplung (z.B. Schlauchverbinder)
quick-fit connection
  Schnellverbindung (Rohr/Glas/Schläuche etc.)
quick-release clamp
  Schnellspannklemme, Schnellspannverschluss
quick-stain *micros* Schnellfärbung
quickfreeze schnellgefrieren
quiescent ruhend, untätig, unterdrückt; ruhig, still
quiescent current *neuro* Ruhestrom
quill (bobbin/spool) *text*
  Hülse, Buchse; Spule; *tech/mech* (hollow shaft) Hohlwelle
quintuple nosepiece *micros*
  Fünffachrevolver

R phrases (Risk phrases)
R-Sätze (Gefahrenhinweise)
R$_F$-value *chromat*
(retention factor/ratio of fronts)
R$_F$-Wert
race (of ball bearing)
Laufring (beim Kugellager)
RACE-PCR (rapid amplification of cDNA ends) RACE-PCR (schnelle Vervielfältigung von cDNA-Enden)
racemate
Racemat, racemische Verbindung
racemization Racemisierung
rack Gestell, Ständer (Sammlung/ Aufbewahrung etc.); *tech* Zahnstange
rack-and-pinion gear
Zahnstangengetriebe
radial immunodiffusion (RID)
radiale Immundiffusion
radiant strahlend
radiant energy Strahlungsenergie
radiant heat Strahlungswärme
radiate strahlen, ausstrahlen, leuchten
radiation Strahlung
➤ **background radiation**
Hintergrundsstrahlung
➤ **corpuscular radiation**
Teilchenstrahlung
➤ **electromagnetic radiation**
elektromagnetische Strahlung
➤ **global radiation** Globalstrahlung
➤ **ionizing radiation**
ionisierende Strahlen,
ionisierende Strahlung
➤ **nuclear radiation** Kernstrahlung
➤ **permissible radiation**
zuläßige Strahlung
➤ **photosynthetically active radiation (PAR)**
photosynthetisch aktive Strahlung
➤ **polarized radiation**
polarisierte Strahlung
➤ **radioactive radiation**
radioaktive Strahlung
➤ **scattered radiation/diffuse radiation**
Streustrahlung
➤ **solar radiation** Sonnenstrahlung
➤ **thermal radiation** Wärmestrahlung

radiation biology
Strahlenbiologie
**radiation control/
radiation protection/
protection from radiation**
Strahlenschutz
radiation dosage/irradiation dosage
Bestrahlungsdosis
radiation hazards/radiation injury
Strahlenschäden
radiation incident Strahlenvorfall
radiation injury
Stahlenschädigung,
Strahlenschaden
radiation intensity
Strahlungsintensität
radiation level Strahlenbelastung
radiation protection Stahlenschutz
radiation sickness
Strahlenkrankheit
radiation therapy/radiotherapy
Bestrahlungstherapie,
Strahlentherapie
radiator (heater)
Strahler; Heizkörper;
(des Motors) Kühler
radiator coil Kühlschlange
radical Radikal
➤ **free radical** freies Radikal
radical ion Radikalion
radical scavenger Radikalfänger
radioactive (nuclear disintegration)
radioactiv (Atomzerfall)
**radioactive decay/
radioactive disintegration**
radioaktiver Zerfall
radioactive marker
radioaktiver Marker
radioactive radiation
radioaktive Strahlung
radioactive waste/nuclear waste
radioaktive Abfälle
radioactively contaminated
radioaktiv/atomar verstrahlt,
radioaktiv/atomar verseucht
radioactivity Radioaktivität
radioallergosorbent test (RAST)
Radio-Allergo-Sorbent Test

radiocarbon method
Radiokarbonmethode,
Radiokohlenstoffmethode,
Radiokohlenstoffdatierung
radioimmunoassay
Radioimmunassay,
Radioimmunoassay
radioimmunoelectrophoresis
Radioimmunelektrophorese
radiolabelling radioaktive Markierung
radionuclide
Radionuklid, Radionuclid
rag Lappen, Lumpen;
Wischtuch, Putzlumpen
rain gauge Niederschlagsmesser
rainwater Regenwasser
rake *n* Rechen, Harke; Schürhaken
rake *vb* rechen, harken;
scharren, kratzen
rancid ranzig
random zufällig, wahllos, willkürlich
random deviation
Zufallsabweichung
random distribution
Zufallsverteilung
random error
zufälliger Fehler, Zufallsfehler
random event Zufallsereignis
random number Zufallszahl
random sample/
  sample taken at random
Zufallsstichprobe, Zufallsprobe
random sampling
Zufallsstichprobenerhebung
random screening Zufallsauslese
random variable Zufallsvariable
random-access memory *comp*
Arbeitsspeicher
randomization Randomisierung
randomize randomisieren
range Spanne, Messspanne,
Reichweite (Strahlung);
stat Spannweite
range of measurement
Messbereich
range of saturation/
  zone of saturation
Sättigungsbereich, Sättigungszone

range of variation/
  range of distribution *stat*
Variationsbreite
range of vision/visual distance
Sehweite
rank *n* Rang, Stufe, Einstufung
▸ **order of rank/ranking/hierarchy**
Stufenfolge, Rangordnung,
Rangfolge, Hierarchie
rank *vb* **(classify)**
einordnen, einstufen, klassifizieren
rank correlation coefficient *stat*
Rangkorrelationskoeffizient
rank statistics/
  rank order statistics
Rangmaßzahlen
rapid freezing Schnellgefrieren
Raschig ring (column packing)
Raschig-Ring (Glasring)
rash (skin rash/skin eruptions)
Ausschlag (Hautausschlag)
rasp Raspel; (grater) Haushaltsraspel
ratchet Knarre;
(ratchet wrench) Ratsche, Rätsch
▸ **change-over ratchet** Umschaltknarre
▸ **lever ratchet** Hebelknarre
ratchet clamp Ratschen-Klemme,
Ratschen-Absperrklemme
(Schlauchklemme)
rate Rate, Ziffer, Quote;
Tarif; Preis, Gebühr; Klasse, Grad;
Geschwindigkeit, Tempo
rate constant (enzyme kinetics)
Geschwindigkeitskonstante
rated output (rated amperage output)
Nennstrom, Nominalstrom;
(rated power output) Nennleistung,
Nominalleistung
rate-determining step/reaction
geschwindigkeitsbestimmende(r)
Schritt/Reaktion
rate-limiting step/reaction
geschwindigkeitsbegrenzende(r)
Schritt/Reaktion
ratio (quotient/proportion/relation)
Verhältnis, Quotient, Proportion
ratio scale *stat*
Verhältnisskala, Ratioskala

raw material/resource Rohstoff
raw sewage Rohabwasser
raw sludge Rohschlamm
raw sugar/crude sugar (unrefined sugar) Rohzucker
ray Strahl; (of sunshine/sunbeam) Sonnenstrahl
ray diagram Strahlendiagramm
razor blade Rasierklinge
react reagieren
reactant Reaktand, Reaktionsteilnehmer, Ausgangsstoff
reaction (zero-order/first-order/second-order..) Reaktion (nullter/erster/zweiter.. Ordnung)
➢ **biosynthetic reaction (anabolic reaction)** Biosynthesereaktion
➢ **bisubstrate reaction** Zweisubstratreaktion, Bisubstratreaktion
➢ **chain reaction** Kettenreaktion
➢ **condensation reaction/ dehydration reaction** Kondensationsreaktion, Dehydrierungsreaktion
➢ **coupled reaction** gekoppelte Reaktion
➢ **coupling reaction** *chem* Kupplungsreaktion
➢ **dark reaction** Dunkelreaktion
➢ **displacement reaction** Verdrängungsreaktion
➢ **enzymatic reaction** Enzymreaktion
➢ **exchange reaction** Austauschreaktion
➢ **immune reaction** Immunreaktion
➢ **one-pot reaction** Eintopfreaktion
➢ **rate-determining reaction** geschwindigkeitsbestimmende(r) Schritt/Reaktion
➢ **rate-limiting reaction** geschwindigkeitsbegrenzende(r) Schritt/Reaktion

➢ **reduction-oxidation reaction/ redox reaction** Redoxreaktion
➢ **runaway reaction** Durchgeh-Reaktion
➢ **sequential reaction/chain reaction** sequentielle Reaktion, Kettenreaktion
➢ **side reaction** *chem* Nebenreaktion
➢ **vigorous reaction/violent reaction** heftige Reaktion
reaction distillation Reaktionsdestillation
reaction intermediate Reaktionszwischenprodukt
reaction kinetics Reaktionskinetik
reaction pathway Reaktionskette
reaction rate Reaktionsrate, Reaktionsgeschwindigkeit
reaction sequence Reaktionsfolge
reaction vessel Reaktionsgefäß
reactive force Gegenkraft, Rückwirkungskraft
reactor Reaktor
➢ **airlift loop reactor** Mammutschlaufenreaktor
➢ **airlift reactor/ pneumatic reactor** Airliftreaktor, pneumatischer Reaktor
➢ **bead-bed reactor** Kugelbettreaktor
➢ **bioreactor** Bioreaktor
➢ **bubble column reactor** Blasensäulen-Reaktor
➢ **column reactor** Säulenreaktor, Turmreaktor
➢ **fedbatch reactor/ fed-batch reactor** Fedbatch-Reaktor, Fed-Batch-Reaktor, Zulaufreaktor
➢ **film reactor** Filmreaktor
➢ **fixed bed reactor/ solid bed reactor** Festbettreaktor
➢ **flow reactor** Durchflussreaktor

- ➤ **fluidized bed reactor/ moving bed reactor** Fließbettreaktor, Wirbelschichtreaktor, Wirbelbettreaktor
- ➤ **immersed slot reactor** Tauchkanalreaktor
- ➤ **immersing surface reactor** Tauchflächenreaktor
- ➤ **jet loop reactor** Strahlschlaufenreaktor, Strahl-Schlaufenreaktor
- ➤ **jet reactor** Strahlreaktor
- ➤ **loop reactor/circulating reactor/ recycle reactor** Umlaufreaktor, Umwälzreaktor, Schlaufenreaktor
- ➤ **membrane reactor** Membranreaktor
- ➤ **nozzle loop reactor/ circulating nozzle reactor** Umlaufdüsen-Reaktor, Düsenumlaufreaktor
- ➤ **packed bed reactor** Füllkörperreaktor, Packbettreaktor
- ➤ **paddle wheel reactor** Schlaufenradreaktor
- ➤ **plug-flow reactor** Pfropfenströmungsreaktor, Kolbenströmungsreaktor
- ➤ **plunging jet reactor (deep jet reactor/immersing jet reactor)** Tauchstrahlreaktor
- ➤ **sieve plate reactor** Siebbodenkaskadenreaktor, Lochbodenkaskadenreaktor
- ➤ **solid phase reactor** Festphasenreaktor
- ➤ **stirred loop reactor** Rührschlaufenreaktor, Umwurfreaktor
- ➤ **stirred-tank reactor** Rührkesselreaktor
- ➤ **tray reactor** Gärtassenreaktor
- ➤ **trickling filter reactor** Tropfkörperreaktor, Rieselfilmreaktor
- ➤ **tubular loop reactor** Rohrschlaufenreaktor

read (off/from)/record ablesen, messen
readability Ablesbarkeit (Waage)
reading error/false reading Ablesefehler
readout Ablesung, Ablesen (Gerät/Messwerte); Ausgabe, Auslesen
ready-to-use gebrauchsfertig
ready-to-use solution/test solution Gebrauchslösung, gebrauchsfertige Lösung, Fertiglösung
reagent/ reagent-grade/ analytical reagent (AR)/ analytical grade pro Analysis (pro analysi = p.a.)
reagent *n* Reagens (*pl* Reagentien)
reagent bottle Reagentienflasche
reagent grade analysenrein, zur Analyse
reagent solution Reagenslösung
real image *micros* reelles Bild
real time Echtzeit
reaminate(d) wiederbeleben (wiederbelebt)
rearrange umlagern, umordnen
rearrangement *chem* Umlagerung, Umordnung
reassociation kinetics Reassoziationskinetik
receipt Erhalt, Entgegennahme; Quittung, Erhaltsbestätigung
receiver Empfänger, Empfangsgerät; Hörer; Behälter, Gefäß; Auffanggefäß
receiver adapter Destilliervorstoß
receiving vessel (receiver/collection vessel) Auffanggefäß
reception Rezeption, Empfang
receptive empfänglich
receptive capacity Aufnahmekapazität
receptor Rezeptor, Empfänger
rechargeable wiederaufladbar
recharge *vb* wiederaufladen, auffüllen, wiederauffüllen, nachladen

recipient (also: host) Empfänger,
Rezipient (z.B. Transplantate)
**reciprocating shaker
(side-to-side motion)**
Reziprokschüttler,
Horizontalschüttler,
Hin- und Herschüttler (rütteln)
reclamation
Reklamation, Rückforderung;
chem Wiedergewinnung,
Regenerierung
recognition Erkennung, Erkennen
**recognition site affinity
chromatography**
Erkennungssequenz-
Affinitätschromatographie
recoil (return motion)
Rückstoß, Rückprall, Abprall
recoil radiation
Rückstoßstrahlung
recombinant (cell)
Rekombinante (Zelle)
recombine rekombinieren
recommendation
Empfehlung, Fürsprache
recommended daily allowance (RDA)
empfohlener täglicher Bedarf
reconstitute rekonstituieren,
wiederherstellen; wiedereinsetzen;
in Wasser auflösen (Milch etc.)
reconstitution
Rekonstitution, Wiederherstellung;
Wiedereinsetzung,
Wiederzusammensetzen
record *n* Aufzeichnung(en),
Bericht, Dokument, Urkunde;
(registration) Registrierung
record *vb* aufzeichnen, aufschreiben,
erfassen; (register) registrieren
recorder (plotter)
Schreiber (Gerät zur Aufzeichnung)
recording/registration
Aufnahme, Aufschreiben,
Registration
recordkeeping
Protokollierung,
Verwahrung/Verwaltung von
Aufzeichnung(en)

recover erholen, wiedergewinnen,
rückgewinnen, zurückbekommen;
aufbereiten
recovery Erholung,
Rückgewinnung, Wiedererlangung
**recovery flask/
receiving flask/receiver flask
(collection vessel)**
Vorlagekolben
recrystallization
Rekristallisation;
Umkristallisation
recrystallize rekristallisieren;
umkristallisieren
rectification Rektifikation;
electr Gleichrichtung
rectifier *chem/dist* Rektifizierapparat,
Rektifiziersäule; *electr* Gleichrichter
rectify *chem* rektifizieren, destillieren;
electr gleichrichten; *tech/mech*
korrigieren, eichen, richtig einstellen
rectifying column Rektifiziersäule
recuperation Erholung
recurrence risk
Wiederholungsrisiko
recyclable wieder verwertbar
recycle recyceln, wieder verwerten
recycled paper Umweltschutzpapier
recycling
Recycling, Wiederverwertung
**recycling plant
(waste recycling plant)**
Müllverwertungsanlage
redistill (rerun)
redestillieren, erneut destillieren,
wiederholt destillieren,
umdestillieren (nochmal destillieren)
redox couple Redoxpaar
redox potential Redoxpotential
redox titration Redoxtitration
reduce reduzieren;
(to small pieces) zerkleinern;
(concentrate) einengen,
konzentrieren;
verkleinern, erniedrigen
**reduce by evaporation
(evaporate completely)**
eindampfen (vollständig)

reduced pressure
  erniedrigter Druck
**reducer/reducing adapter/
  reduction adapter**
  Reduzierstück (Laborglas/Schlauch)
reducing agent Reduktionsmittel
reduction Reduktion;
  photo (size reduction) Verkleinerung
**reduction-oxidation reaction/
  redox reaction**
  Redoxreaktion
redundancy Redundanz; Überfluss,
  Überflüssigkeit; (unnötige)
  Wiederholung
reference book Nachschlagewerk
reference electrode Bezugselektrode
reference gas (GC) Vergleichsgas
reference strain *micb* Referenzstamm
reference temperature
  Bezugstemperatur
reference value Bezugswert
refill wieder füllen, nachfüllen,
  auffüllen, wiederauffüllen
refillable nachfüllbar
**refinement (improvement/processing/
  finishing)** Veredlung
refinement process Veredlungsprozess
reflux *n* Rückfluss, Rücklauf, Reflux
reflux *vb* am Rückflusskühler kochen,
  unter Rückfluss erhitzen/kochen
reflux condenser Rückflusskühler
refract *opt* brechen; *chem* analysieren
refracting angle *opt* Brechungswinkel
refraction Refraktion, Brechung;
  Refraktionsvermögen
refractive lichtbrechend
refractive index/index of refraction
  Brechungsindex,
  Brechungskoeffizient, Brechzahl
refractivity Brechungsvermögen
refractometer Refraktometer
refractory *adj/adv* feuerfest,
  hochschmelzend, refraktär;
  widerstandsfähig, unempfindlich
refractory clay
  Schamotte, Schamotteton
refractory material feuerfester Stoff,
  feuerfestes Material

refrigerant Kältemittel,
  Kühlflüssigkeit, Kühlmittel
refrigerate kühl stellen,
  in den Kühlschrank stellen
refrigerator/fridge (icebox)
  Kühlschrank
**regenerate/
  regrow/grow back/reestablish**
  regenerieren, nachwachsen
regeneration
  Regenerierung, Regeneration
**register/announce oneself/sign in
  (schriftlich eintragen/einschreiben)**
  anmelden
registration/signing in Anmeldung
regression analysis *stat*
  Regressionsanalyse
**regression coefficient/
  coefficient of regression** *stat*
  Regressionskoeffizient
regression to the mean
  Regression zum Mittelwert
regressive regressiv,
  zurückbildend, zurückentwickelnd
regular regelmäßig
regulate/control regeln,
  kontrollieren, regulieren, steuern
regulator Regler; (control/adjustment
  knob/adjustment button) Schalter,
  Knopf
regulatory agency
  Regulierungsbehörde
regulatory mechanisms
  Regulationsmechanismen,
  Steuerungsmechanismen
regulatory procedure
  Regelungsprozess
rehydration
  Rehydratation, Rehydratisierung
reinfestation Wiederbefall
reinforce verstärken
reinforced verstärkt (fest/solide)
reinforcement/amplification
  Verstärkung
rejuvenate/regenerate
  verjüngen, regenerieren
rejuvenation/regeneration
  Verjüngung, Regeneration

relation/correlation/ interrelationship/ connection Zusammenhang, Verhältnis, Verbindung
relationship Verhältnis, Beziehung
relative frequency
 relative Häufigkeit
relative molecular mass/ molecular weight (M_r)
 relative Molekülmasse, Molekulargewicht (M_r)
relax entspannen, lockern, erschlaffen (z.b. Muskel)
relaxation Entspannung, Erschlaffung, Relaxation
relaxed (conformation) relaxiert, entspannt
relay *electr* Relais
release *n* Freisetzung, Entweichen; Abgabe; Auslösung; Ausschüttung (z.B. Hormone/Neurotransmitter); Mitteilung, Verlautbarung
release *vb* freisetzen, entweichen; abgeben; auslösen; ausschütten
release button Auslöser, Auslösetaste
release factor Freisetzungsfaktor
release time Auslösezeit
releaser Auslöser
relevé Vegetationsaufnahme
reliability Zuverläßigkeit
relic Relikt, Überbleibsel, Rest
relief Erleichterung, Entlastung
relief valve (pressure-maintaining valve) Ausgleichsventil, Überdruckventil
remote control Fernbedienung
removable needle (syringe needle) abnehmbare Injektionsnadel/Nadel
removal (withdrawal) Beseitigung, Entfernung; (taking out) Entnahme
remove beseitigen, entfernen; entnehmen
reorient/reorientate umstimmen, neu orientieren, neu ausrichten
repair *n* (restoration) Reparatur, Instandsetzung, Wiederherstellung

repair *vb* (fix/mend/restore) reparieren, instand setzen, wiederherstellen
repeat *n* (repetition) Wiederholung
repeatability Wiederholbarkeit
repeated distillation/cohobation Redestillation, mehrfache Destillation
repellent *adj/adv* (*also*: **repellant**) abstoßend
repellent *n* (*also*: **repellant**) Repellens (*pl* Repellentien)
repetition Wiederholung
replace austauschen, ersetzen; vertreten
replacement Austausch, Ersatz; Vertretung
replacement bulb (lamp) Ersatzbirnchen
replacement parts (spare parts) Ersatzteile
replacement vector Substitutionsvektor
replant verpflanzen, umpflanzen, umsetzen, versetzen
replenish nachfüllen, wiederbefüllen
replica (Oberflächenabdruck: *EM*) *micros* Abdruck
replica-plating *micb* Replikaplattierung, Stempel-Methode
report *n* Bericht, Meldung; Anzeige
report *vb* berichten, melden; anzeigen
reportable (by law)/subject to registration meldepflichtig
repot umtopfen
representative *n* ('rep') Vertreter
repress (control/suppress/subdue) reprimieren, unterdrücken, hemmen
repression (control/suppression) Reprimierung, Unterdrückung, Hemmung
reprocess wieder aufbereiten
reprocessing Wiederaufbereitung
reproduce reproduzieren, wiederholen; kopieren, nachmachen; wiedergeben
reproducibility Reproduzierbarkeit; Vergleichspräzision

reproduction Fortpflanzung, Vermehrung
➢ **asexual/vegetative reproduction** ungeschlechtliche/vegetative Fortpflanzung
➢ **sexual reproduction** geschlechtliche/sexuelle Fortpflanzung
reproductive cell Fortpflanzungszelle
repugnant unangenehmer/abweisender Geruchsstoff
repulsion Abstoßung
repulsion conformation Repulsionskonformation
rescue *n* Rettung, Bergung, Befreiung
rescue *vb* retten, bergen, befreien
rescue helicopter Rettungshubschrauber
rescue operation Rettungsaktion
rescue service/lifesaving service Rettungsdienst
research *n* **(trial/experimentation/investigation)** Forschung, Untersuchung, Erforschung
➢ **basic research** Grundlagenforschung
research *vb* forschen, untersuchen, erforschen
research assignment Forschungsauftrag
research contract Forschungsauftrag
research department Forschungsabteilung
research funding Forschungsfinanzierung
research funds Forschungsgelder
research laboratory Forschungslabor
research program Forschungsprogramm
research project Forschungsvorhaben, Forschungsprojekt

research advisory committee Forschungsbeirat
researcher/research scientist/ research worker/investigator Forscher
resemble sich gleichen, gleichartig sein
reserve material/ storage material/food reserve Reservestoff
reset *vb* zurücksetzen
residence time Verweilzeit, Verweildauer, Aufenthaltszeit, Verweildauer
residual dampness (H_2O)/ residual humidity Restfeuchte
residue Rest, Rückstand; (bottoms/heel) abgesetzte Teilchen
resin Harz
resin acids Harzsäure
resiniferous harzabsondernd
resinous harzig
resinous gum Gummiharz
resistance/resistivity/hardiness Resistenz, Beständigkeit, Widerstand, Widerstandsfähigkeit
resistance temperature detector (RTD) Widerstands-Temperatur-Detektor
resistance thermometer Widerstandsthermometer
resistant/resistive/hardy resistent, beständig, widerstandsfähig
resistive heating Widerstandsheizung
resistivity spezifischer Widerstand
resolution Auflösung; *chromat* Trennschärfe
➢ **high-resolution** *adj/adv* hoch aufgelöst
➢ **limit of resolution** Auflösungsgrenze
➢ **low-resolution** *adj/adv* niedrig aufgelöst
➢ **optical resolution** optische Auflösung
resolve *opt* auflösen; *chromat* trennen
resolving power *opt* Auflösungsvermögen

resonance/echo/reverberation
Resonanz, Schall, Widerhall
resorb resorbieren, aufsaugen
resorbent resorbierend, aufsaugend
resorption Resorption, Aufsaugung
resource Ressource, Rohstoff; Rohstoffquelle
respiration
Respiration, Atmung
➢ **aerobic respiration**
aerobe Atmung
➢ **anaerobic respiration**
anaerobe Atmung
➢ **cellular respiration**
Zellatmung
➢ **cutaneous respiration/ cutaneous breathing/ integumentary respiration**
Hautatmung
➢ **diaphragmatic respiration/ abdominal breathing**
Bauchatmung, Zwerchfellatmung
➢ **thoracic respiration/costal breathing**
Brustatmung, Thorakalatmung
respirator (breathing apparatus)
Atemschutzgerät, Atemgerät
➢ **dust mask respirator**
Grobstaubmaske
➢ **full-face respirator**
Atemschutzvollmaske, Gesichtsmaske
➢ **full-facepiece respirator**
Vollsicht-Atemschutzmaske
➢ **full-mask respirator**
Vollmaske, Atemschutz-Vollmaske
➢ **half-mask respirator**
Halbmaske
➢ **particulate respirator**
Partikelfilter-Atemschutzmaske
respiratory center Atemzentrum
respiratory poison Atmungsgift
respiratory quotient
Atmungsquotient, respiratorischer Quotient
respiratory system Atemwege
respiratory toxin/fumigants
Atemgifte, Fumigantien

respiratory tract burn/ (alkali/acid) caustic burn of the respiratory tract
Atemwegsverätzung
response Antwort (auf Reiz); (conditioned/unconditioned r.) bedingte/unbedingte Reaktion
response time
Anlaufzeit, Reaktionszeit; (metering equipment) Ansprechzeit
responsibility
Verantwortung, Haftung; Zuständigkeit
rest *n* **(residue)**
Rest (z.B. Aminosäuren-Seitenkette)
rest *vb* **(lie dormant)** ruhen
resting (quiescent/dormant)
ruhend
resting period/ quiescent period/ dormancy period
Ruhephase, Ruheperiode
restitute
restituieren, wiederherstellen
restitution
Restitution, Wiederherstellung
restock auffüllen, aufstocken, nachfüllen (Vorräte, Lager)
restricted access/ access control
Zutrittsbeschränkung
restriction
Einschränkung, Beschränkung
restriction enzyme
Restriktionsenzym
restriction fragment length polymorphism (RFLP)
Restriktionsfragmentlängen-polymorphismus
resuscitation
Wiederbelebung, Reanimation
➢ **mouth-to-mouth resuscitation/respiration**
Mund-zu-Mund Beatmung (Wiederbelebung)
resuspend wiederaufschlämmen

retail business/retail trade Einzelhandel
retail price Einzelhandelspreis
retail store Einzelhandelsgeschäft
retailer/retail dealer/retail vendor Einzelhändler
retain zurückbehalten, einbehalten, beibehalten; halten, sichern, stützen
retainment capacity/ retainability/ retention efficiency Rückhaltevermögen
retention time Retentionszeit, Verweildauer, Aufenthaltszeit
retinic acid Retinsäure
retort Retorte
retrieval Wiedergewinnung, Rückholung
retting rötten, rösten (Flachsrösten)
return *n* Rücksendung (einer Ware)
re-uptake Wiederaufnahme
reusable wiederverwendbar, Mehrweg...
reuse *n* Wiederverwendung
reuse *vb* wiederverwenden
reverberatory furnace Flammofen
reversal potential Umkehrpotential
reversal spectrum Umkehrspektrum
reverse osmosis Reversosmose, Umkehrosmose
reversed phase (reverse phase) Umkehrphase, Reversphase
reversed phase chromatography/ reverse-phase chromatography Umkehrphasenchromatographie
reversibility Reversibilität, Umkehrbarkeit
reversible reversibel, umkehrbar
reversible inhibition reversible Hemmung
reversion Reversion, Umkehrung
revolutions per minute (rpm)/ number of revolutions Umdrehungen pro Minute (UpM), Drehzahl

revolving punch pliers Revolverlochzange
Reynolds number Reynold'sche Zahl, Reynolds-Zahl, Reynoldsche Zahl
rib joint pliers/rib-lock pliers Eckrohrzange
ribbed filter/fluted filter Rippenfilter
ribbed glass Rippenglas, geripptes Glas, geriffeltes Glas
ribonucleic acid (RNA) Ribonucleinsäure, Ribonukleinsäure (RNA/RNS)
riboprobe Ribosonde, RNA-Sonde
right-handed rechtsgängig; (dextral) rechtshändig
rigor mortis Totenstarre, Leichenstarre
rim/edge Rand (eines Gefäßes)
ring (for support stand/ring stand) Stativring
ring binder Ringbuch
ring cleavage *chem* Ringspaltung
ring closure/ ring formation/cyclization *chem* Ringschluss, Ringbildung
ring compound Ringverbindung
ring form/ring conformation Ringform
ring formula Ringformel
ring spanner wrench/ box wrench/box spanner/ ring spanner (*Br*) Ringschlüssel
ring stand/ support stand/retort stand/stand Bunsenstativ, Stativ
ring structure Ringstruktur
Ringer's solution Ringerlösung, Ringer-Lösung
rinse ausspülen, ausschwenken, nachspülen
ripe reif
➢ **unripe/immature** unreif
rise time/preheating time (autoclave) Anheizzeit, Steigzeit
riser pipe/riser tube/chimney Steigrohr

risk (danger)
Risiko (*pl* Risiken), Gefahr
➢ **recurrence risk**
Wiederholungsrisiko
risk assessment
Risikoabschätzung
risk class/security level/safety level
Sicherheitsstufe, Risikostufe
risk of contamination
Verseuchungsgefahr
roast rösten; (calcine) ausglühen
roasting furnace/
  roasting oven/roaster
Röstofen
rock drill (bit) Steinbohrer
rock salt (halite)/
  common salt/table salt/
  sodium chloride NaCl
Steinsalz (Halit), Kochsalz,
Tafelsalz, Natrium chlorid
rock wool Steinwolle
rocker/rocking shaker
  (side-to-side/up-down)
Wippschüttler; Rüttler
(hin und her/rauf-runter); Schwinge
rocket immunoelectrophoresis
Raketenimmunelektrophorese
rocking motion
(up-down/see-saw motion)
Wippbewegung (rauf und runter);
(side-to-side: fast) Rüttelbewegung
(schnell hin und her)
rod Stab, Stange; (rod cell) Stäbchen,
Stäbchenzelle; (bacilli) Stäbchen,
Stäbchenbakterien, Bazillen
rod clevis Bügelschaft
roll *vb* rollen
roller Drehwalze (Roller-Apparatur)
roller bottle Rollerflasche
roller tube culture
Rollerflaschenkultur
roller wheel mixer Drehmischer
rolling step-stool
'Elefantenfuß', Rollhocker
(runder Trittschemel mit Rollen)
room temperature/
  ambient temperature
Raumtemperatur

root *vb* **(take root)** bewurzeln
rope Seil
rot *n* **(decaying matter/mold/**
  mildew/blight)
Mulm, Fäule
rot *vb* **(decay/decompose/disintegrate)**
faulen, verfaulen;
(putrefy) vermodern, modern
rotary evaporator
  ('rovap'/rotary film evaporator *Br*)
Rotationsverdampfer
rotary evaporator flask
Rotationsverdampferkolben
rotary microtome Rotationsmikrotom
rotary piston pump Drehkolbenpumpe
rotary-piston meter Drehkolbenzähler
rotary vacuum filter
Vakuumdrehfilter,
Vakuumtrommeldrehfilter
rotary vane pump
Drehschieberpumpe
rotating stage *micros*
Drehtisch
rotation Rotation,
Umdrehung, Kreislauf
rotation speed adjustment
Drehzahlregelung
rotational motion
Rotationsbewegung
rotational sense/sense of rotation
Rotationssinn, Drehsinn
rotational spectrum
Rotationsspektrum
rotor Rotor
➢ **angle rotor/angle head rotor**
Winkelrotor
➢ **swing-out rotor/**
  swinging-bucket rotor/
  swing-bucket rotor
Ausschwingrotor
➢ **vertical rotor**
Vertikalrotor
rotor-stator impeller/
  Rushton-turbine impeller
Rotor-Stator-Rührsystem
rotting/
  decaying/putrefying/decomposing
moderig, faulend, verfaulend

round filter/
 filter paper disk/'circles'
  Rundfilter
round-bottomed flask/
 round-bottom flask/
 boiling flask with round bottom
  Rundkolben
row (series) Reihe (Serie)
RT-PCR (reverse transcriptase-PCR)
  RT-PCR
  (PCR mit reverser Transcriptase)
rub/grind zerreiben
rubber Gummi
rubber band/elastic (*Br*)
  Gummiband, Gummi
rubber boots Gummistiefel
rubber gasket
  Gummidichtung(sring)
rubber mallet Gummihammer
rubber policeman (scraper rod with rubber or Teflon tip)
  Kolbenwischer,
  Gummiwischer, Gummischaber
  (zum Loslösen von festgebackenen Rückständen im Kolben)
rubber ring (e.g., flask support)
  Gummiring
rubber sleeve
 (seal for glassware joints)
  Gummimanschette
rubber stopper/rubber bung (*Br*)
  Gummistopfen, Gummistöpsel
rubber tubing Gummischlauch
rudiment (*sensu lato*: vestige)
  Rudiment
rudimentary (*sensu lato*: vestigial)
  rudimentär
rules Regeln
rules of conduct Verhaltensregeln
run dry leerlaufen, trockenlaufen
runaway reaction
  Durchgeh-Reaktion
running gel/separating gel *electrophor*
  Trenngel
rust *n* Rost
rust *vb* rosten
rust inhibitor/
 antirust agent/
 anticorrosive agent
  Rostschutzmittel
rust remover/rust-removing agent
  Rostentferner, Rostlöser,
  Rostentfernungsmittel,
  Entrostungsmittel

S phrases (Safety phrases)
S-Sätze (Sicherheitsratschläge)
s.t.p. (STP/NTP)
(standard temperature 0 °C/ pressure 1 bar) Normzustand (Normtemperatur & Normdruck)
saber flask/sickle flask/sausage flask Säbelkolben, Sichelkolben
saccharic acid/aldaric acid Zuckersäure, Aldarsäure
sacchariferous/saccharogenic zuckerbildend
saccharification Verzuckerung
saccharify verzuckern
saccharimeter Saccharimeter
saccharolytic zuckerspaltend
sachet Tütchen
saddle (column packing) *dist* Sattelkörper (Füllkörper)
safe *adj/adv* sicher; (without risk/unrisky) unbedenklich
➢ **unsafe** unsicher, gefährlich
safe handling sicherer Umgang
safelight *photo* Dunkelkammerlampe (Rotlichtlampe)
safety Sicherheit
➢ **margin of safety** Sicherheitsspielraum
safety cabinet Sicherheitsschrank
➢ **microbiological safety cabinet (MSC)** mikrobiologische Sicherheitswerkbank (MSW)
safety check/safety inspection Sicherheitsüberprüfung, Sicherheitskontrolle
safety cutter Sicherheitsmesser
safety data Sicherheitsdaten
safety data sheet Sicherheitsdatenblatt
safety device Sicherheitsvorrichtung
safety engineer Sicherheitsingenieur
safety feature Sicherheitsmerkmal
safety glass/laminated glass Schutzglas, Sicherheitsglas
safety guidelines Sicherheitsrichtlinien
safety helmet/hard hat/hardhat Schutzhelm

safety instructions/ safety protocol/ safety policy Sicherheitsvorschriften
safety labeling Sicherheitskennzeichnung
safety measures/safeguards Sicherheitsvorkehrungen, Absicherungen
safety of operation Betriebssicherheit
safety officer Sicherheitsbeauftragter
safety pipet filler/safety pipet ball Peleusball (Pipettierball)
safety policy Sicherheitsverhaltensmaßregeln
safety precaution Sicherheitsvorkehrung, Sicherheitsvorbeugemaßnahme
safety profile Sicherheitsprofil
safety regulations Sicherheitbestimmungen
safety spectacles (einfache) Schutzbrille
safety valve Sicherheitsventil
safety vessel/ safety container/ safety can Sicherheitsbehälter, Sicherheitskanne
sagittal section/median longisection Sagittalschnitt (parallel zur Mittelebene)
sale Verkauf, Vertrieb; *pl* Absatz, Umsatz; Schlussverkauf
sales account Warenausgangskonto, Verkaufskonto
sales representative Vertreter (im Verkauf)
sales talk Verkaufsgespräch
sales tax Umsatzsteuer
salesman Verkäufer; (sales representative) Handlungsreisender, Vertreter
salicic acid (salicylate) Salicylsäure (Salicylat)

saline Kochsalzlösung
- physiological saline solution
physiologische Kochsalzlösung
saline water salziges Wasser
salinity/saltiness
Salinität, Salzgehalt, Salzigkeit
salinization Versalzung (Boden)
saliva Speichel
salt *vb* salzen, einsalzen
salt Salz
- bile salts Gallensalze
- complex salt Komplexsalz
- double salt Doppelsalz
- Epsom salts/epsomite/ magnesium sulfate
Bittersalz, Magnesiumsulfat
- hartshorn salt/ammonium carbonate
Hirschhornsalz, Ammoniumcarbonat
- iodized salt Iodsalz
- Mohr's salt/ ammonium irin(II) sulfate hexahydrate (ferrous ammonium sulfate)
Mohrsches Salz
- molten salt/salt melt
Salzschmelze
- nutrient salt Nährsalz
- rock salt (halite)/ common salt/table salt/ sodium chloride (NaCl)
Steinsalz (Halit), Kochsalz, Tafelsalz, Natrium chlorid
- sea salt Meersalz
- table salt/common salt (NaCl)
Kochsalz
salt beads Salzperlen
salt bridge (ion pair)
Salzbrücke (Ionenpaar); *electrolyt* Stromschlüssel
salt out *vb* aussalzen
salt water *n* Salzwasser
- saltwater *adj* Salzwasser...
saltiness Salzigkeit
salting in Einsalzen, Einsalzung
salting out Aussalzen
salting-out chromatography
Aussalzchromatographie
saltpeter Salpeter

salty/saline salzig
salvage pathway
Wiederverwertungsreaktion, Wiederverwertungsstoffwechselwege
sample Muster, Probe (Teilmenge eines zu untersuchenden Stoffes)
- spot sample/aliquot Stichprobe
sample concentrator
Probenkonzentrator
sample custody
Probenverwaltung
sample function/sample statistic
Stichprobenfunktion
sample preparation
Probenvorbereitung
sample size *stat*
Fallzahl; Stichprobenumfang
sample vial/specimen vial
Probefläschchen, Probegläschen
sample-taking/taking a sample
Probennahme, Probeentnahme
sampler Probenehmer, Probenentnahmegerät
sampling Probe, Probieren; Auswahlverfahren; Stichprobenerhebung; Prüfung, Erhebung
sampling device
Probenahmevorrichtung
sandblasting
Sandstrahlreinigung
sandblasting apparatus
Sandstrahlgebläse
sandpaper/emery paper (*Br*)
Schmirgelpapier
sanitary facilities/sanitary installations
sanitäre Einrichtungen
sanitary measure
Hygienemaßnahme
sanitary sewer Abwasserkanal
sanitary supplies (sanitary equipment/ plumbing supplies or equipment)
Sanitärzubehör
sanitation worker Müllmann
sanitize keimfrei machen, sterilisieren
saponification Verseifung
saponify verseifen

saprobes/saprobionts
  Saprobien (Organismen), Fäulnisbewohner
saprogenic saprogen, fäulniserregend
saprophage/saprotroph/saprobiont
  Saprophage, Fäulnisernährer, Fäulnisfresser
sash (*see also:* **hood**)
  Schiebefenster, Frontschieber, Frontscheibe, verschiebbare Sichtscheibe (Abzug/Sicherheitswerkbank)
satellite band Satellitenbande
saturate (saturated)
  sättigen (gesättigt)
  ➢ **monounsaturated**
    einfach ungesättigt
  ➢ **polyunsaturated**
    mehrfach ungesättigt
  ➢ **unsaturated** ungesättigt
saturated fatty acid
  gesättigte Fettsäure
saturated solution
  gesättigte Lösung
saturation Sättigung
  ➢ **range of saturation/ zone of saturation**
    Sättigungsbereich, Sättigungszone
  ➢ **unsaturation** ungesättigter Zustand
saturation deficit
  Sättigungsverlust, Sättigungsdefizit
saturation hybridization
  Sättigungshybridisierung
saturation kinetics
  Sättigungskinetik
saw Säge
  ➢ **chain saw** Kettensäge
  ➢ **compass saw (with open handle)/pad saw**
    Stichsäge
  ➢ **coping saw** Bogensäge
  ➢ **hacksaw** Bügelsäge
  ➢ **handsaw** Handsäge
  ➢ **metal-cutting saw** Metallsäge
  ➢ **miter-box saw** Gehrungssäge
  ➢ **scroll saw/jigsaw/fretsaw**
    Laubsäge (Blatt <2 mm), Dekupiersäge (Blatt: größer 2 mm)

saw blade Sägeblatt
sawdust Sägemehl
scab Schorf (Wundschorf), Grind
scab lesion (crustlike disease lesion)
  Schorfwunde
scaffold/scaffolding (framework)
  Gerüst, Gerüstmaterial, Gestell
scald/scalding
  Verbrühung, Verbrühungsverletzung
scale Skala (*pl* Skalen), Maßstab;
  bot Schuppe;
  (boilerstone) Kesselstein
  ➢ **bench-scale/lab-scale**
    im Labormaßstab, labortechnisch
  ➢ **commercial-scale**
    in kommerziellem Maßstab (in handelsüblichen Mengen)
  ➢ **industrial-scale**
    im Industriemaßstab, industrietechnisch
  ➢ **laboratory-scale/lab-scale** im Labormaßstab, labortechnisch
  ➢ **large-scale**
    im Großmaßstab, großtechnisch
  ➢ **metric scale** metrische Skala
  ➢ **small-scale**
    im Kleinmaßstab
scale-down/scaling down
  Maßstabsverkleinerung, maßstabsgerecht verkleinern; herabsetzen, herunterschrauben
scalepan/ weigh tray/weighing tray/ weighing dish
  Waagschale
scales (balance) Waage
  ➢ **bench scales**
    Tischwaage
  ➢ **checkweighing scales**
    Kontrollwaage
  ➢ **spring scales/spring balance**
    Federzugwaage, Federwaage
scale-up/scaling up
  Maßstabsvergrößerung, maßstabsgerecht vergrößern; heraufsetzen, hoch schrauben
scalpel Skalpell
scalpel blade Skalpellklinge

scan *vb* **(screen)**
absuchen, kritisch prüfen;
rastern, scannen, abtasten
scanning calorimetry
Raster-Kalorimetrie
scanning electron microscopy (SEM)
Rasterelektronenmikroskopie (REM)
scanning force microscopy (SFM)
Rasterkraftmikroskopie (RKM)
scanning tunneling microscopy (STM)
Rastertunnelmikroskopie (RTM)
scar/cicatrix/cicatrice
Narbe, Wundnarbe, Cicatricula
scarce/rare selten, rar
scarcity/rarity Seltenheit, Rarität
scatter diagram (scattergram/ scattergraph/scatterplot)
Streudiagramm
scatter *vb* (spread/distribute) streuen, verstreuen, ausstreuen, verteilen; (disperse) zerstreuen, dispergieren
scattered light Streulicht
scattered radiation (diffuse radiation)
Streustrahlung
scattering (spreading/distribution) Streuung, Verstreuen, Verteilung; (dispersion) Zerstreuung, Dispergierung
scedasticity (heterogeneity of variances) *stat* Streuungsverhalten
scent Geruch, Wohlgeruch, Duft; (odiferous substances) Duftstoffe
scented duftend; parfürmiert
scentless geruchlos
scholarship (grant-in-aid to a student) Stipendium; Gelehrtheit, Wissen; Gelehrsamkeit
science Wissenschaft; (natural science) Naturwissenschaft
scientific naturwissenschaftlich
scientist Naturwissenschaftler; (research scientist) Forscher
scintillate szintillieren, funkeln, Funken sprühen, glänzen
scintillation Szintillation; Lichtblitz
scintillation counter/scintillometer
Szintillationszähler ('Blitz'zähler)
scintillation vial Szintillationsgläschen

scissors Schere
➢ **bandage scissors**
Verbandsschere
➢ **blunt point scissors**
stumpfe Schere
➢ **dissecting scissors**
Präparierschere, Sezierschere
➢ **iris scissors**
Irisschere, Listerschere
➢ **nail scissors**
Nagelschere
➢ **sharp point scissors**
spitze Schere
➢ **surgical scissors**
chirurgische Schere
➢ **wire shears/wire cutters**
Drahtschere
sclerification
Sklerifizierung
sclerified sklerifiziert
sclerotic sklerotisch
sclerotized (hardened)
sklerotisiert
scoop
Schöpfkelle, Schöpfer, Schaufel; Löffel
➢ **scoopula**
Löffelspatel
scorch
versengen, verbrennen, anbrennen; *electr* verschmoren
scour scheuern, schrubben; säubern, polieren
scouring agent/abrasive
Scheuermittel
scouring pad/pot cleaner
Topfkratzer, Topfreiniger
scrape kratzen, schaben
scraper Kratzer (Gerät zum abkratzen), Schaber, Ziehklinge, Schabhobel
➢ **wiper blade/spreading knife/ coating knife/doctor knife**
Rakel, Rakelmesser, Schabeisen, Abstreichmesser
scraping (mount) *micros*
Schabepräparat
scraps/shavings
Krümel

screen *vb* abschirmen,
beschirmen, verdecken, tarnen;
sichten; (size) sieben, klassieren
(nach Korngröße)
screen *n* Schirm, Schutzschirm;
Abschirmung; (projection)
Leinwand; Drahtgitter, Sieb
➢ **intensifying screen
(autoradiography)**
Verstärkerfolie
(Autoradiographie)
➢ **optical screen**
Schirm, Filter, Blende
screening Durchmustern, Durchtesten;
med Rasteruntersuchung,
Reihenuntersuchung; (siftage/size
separation by screening) Siebung
screening test Suchtest
screw *vb* schrauben
screw *n* Schraube
➢ **adjusting screw/
adjustment screw/
setting screw/
adjustment knob/fixing screw**
Stellschraube;
(tuning screw) Einstellschraube
➢ **knurled screw/knurled thumbscrew**
Rändelschraube
➢ **socket screw/socket-head screw**
Inbusschraube
➢ **thumbscrew**
Daumenschraube, Flügelschraube
screw bolt Schraubenbolzen
screw cap/screw-cap/screwtop
Schraubkappe, Schraubdeckel,
Schraubkappenverschluss
screw-cap bottle Schraubflasche
screw-cap vial/screw-cap jar
Schraubengläschen,
Schraubdeckelgläschen,
Probegläschen mit Schraubverschluss
screw clamp/pinch clamp
Schraubklemme, Schraubzwinge
screw compression pinchcock
Schraub-Quetschhahn
screw impeller
Schraubenrührer,
Schneckenrührer

screw jack Schraubenwinde
screw micrometer
Schraubenmikrometer;
Messschraube
screw thread
Schraubgewinde, Schraubengewinde
screw-base socket
Gewindefassung
screwdriver
Schraubenzieher, Schraubendreher
➢ **cordless screwdriver**
Akkuschrauber
➢ **hexagonal screwdriver/
hex screwdriver**
Sechskantschraubenzieher
➢ **neon screwdriver (*Br*)/
neon tester (*Br*)/
voltage tester screwdriver**
Spannungsprüfer (Schraubenzieher)
➢ **offset screwdriver**
Winkelschrauber
➢ **Phillips®-head screwdriver/
Phillips® screwdriver**
Kreuzschraubenzieher,
Kreuzschlitzschraubenzieher
➢ **power screwdriver**
Elektroschrauber
➢ **slotted screwdriver**
Schlitzschraubenzieher
➢ **watchmaker's screwdriver/
jeweler's screwdriver**
Uhrmacherschraubenzieher
screwtop (threaded top)
Schraubverschluss,
Schraubdeckel
scroll saw/jigsaw/fretsaw
Laubsäge (Blatt <2 mm),
Dekupiersäge (Blatt >2 mm)
scrubbing brush/scrub brush
Scheuerbürste, Schrubbbürste
scurfy/scabby/furfuraceous
schorfig, Schorf...
sea salt Meersalz
seal *vb* versiegeln, plombieren;
fest verschließen
➢ **seal off** (make tight/make
leakproof/insulate) abdichten;
abriegeln

seal *n* Siegel, Verschluss, Dichtung, Abdichtung; (cap/closure) Verschlusskappe;
➤ **face seal (impeller)** Gleitringdichtung (Rührer)
➤ **lip seal/lip-type seal** Lippendichtung (Wellendurchführung)
sealability Abdichtbarkeit
sealable verschließbar
sealant (sealing compound/sealing material) Dichtungsmasse, Dichtungsmittel
sealing Dichtung, Verschluss
sealing tape Dichtungsband
sealing wax Siegellack
sealless dichtungsfrei, ohne Dichtung (Pumpe)
seam (border/edge/fringe) Saum, Rand; (suture/raphe) Fuge, Naht, Verwachsungslinie
seam sealant/joint filler Fugendichtungsmasse
season *vb* (store/keep: e.g. wood) lagern (Holz)
seawater/salt water Meerwasser
sebaceous (tallowy) Talg..., talgig
sebaceous matter/sebum Talg
secondary infection Sekundärinfekt, Sekundärinfektion
secondary response Sekundärantwort
secondary settling tank Nachklärbecken
secondary structure (proteins) Sekundärstruktur
secrecy agreement Geheimhaltungsvereinbarung
secretarial assistant (secretarial help/typist) Sekretariatsgehilfe, Schreibkraft
secretary Sekretär(in)
secretary's office ('office') Sekretariat
secrete ausscheiden; (excrete) sezernieren, abgeben (Flüssigkeit)
secretion Sekretion, Freisetzung; Ausscheidung; Sekret
secretory sekretorisch

section *n* Abschnitt, Teil; *micros* Schnitt
➤ **cross section** Querschnitt
➤ **frozen section** Gefrierschnitt
➤ **quick section** *micros/med* Schnellschnitt
➤ **sagittal section/ median longisection** Sagittalschnitt (parallel zur Mittelebene)
➤ **semithin section** Semidünnschnitt
➤ **serial sections** *micros/anat* Serienschnitte
➤ **thin section/microsection** Dünnschnitt
➤ **ultrathin section** *micros* Ultradünnschnitt
secure *adj/adv* (*personal protection*) sicher; geschützt, in Sicherheit
secure *vb* sichern, absichern
security (*personal protection*) Sicherheit
security measures/ safety measures/containment Sicherheitsmaßnahmen, Sicherheitsmaßregeln
security personnel/security Sicherheitspersonal
security valve/ security relief valve Sicherheitsventil
sedentary sedentär, niedergelassen
sediment *n* (deposit/precipitate) Sediment, Präzipitat, Niederschlag, Fällung
sedimentation analysis Sedimentationsgeschwindigkeitsanalyse
sedimentation coefficient Sedimentationskoeffizient
seed *n* Same, Samen, Saatgut
seed *vb* säen, besäen, aussäen
seed repository Samenbank
see-saw motion/rocking motion Wippbewegung, Schaukelbewegung
segmentation Segmentierung

segregate segregieren, aufspalten;
(separate out/reseparate) entmischen
segregation Segregation, Aufspaltung;
(separation/reseparation)
Entmischung
select selektieren, auswählen, auslesen
selection Selektion, Auswahl, Auslese
selection coefficient/
coefficient of selection
Selektionswert,
Selektionskoeffizient
selective selektiv, auswählend
selective advantage
Selektionsvorteil
selective disadvantage
Selektionsnachteil
selective filter/barrier filter/
stopping filter/selection filter *micros*
Sperrfilter
selective medium
Elektivmedium, Selektivmedium
selective pressure/selection pressure
Selektionsdruck
selectivity
Selektivität, Unterscheidung;
Trennschärfe
selector switch
Umschalter; Umpolschalter
selenium (Se) Selen
self-adhesive/self-adhering/gummed
selbstklebend
self-assembly
Selbstzusammenbau,
Spontanzusammenbau,
Selbstassoziierung,
spontaner Zusammenbau
(molekulare Epigenese)
self-balancing selbstabgleichend
self-cleansing Selbstreinigung
self-contained
in sich geschlossen, selbständig,
autonom, kompakt, unabhängig
self-curing (resins/polymers)
selbsthärtend (Harze/Polymere)
self-fertilization/selfing/autogamy
Selbstbefruchtung,
Selbstung, Autogamie
self-healing Selbstheilung

self-igniting selbstzündend
self-incompatibility
Selbstinkompatibilität
self-locking selbstverschließend
self-organization Selbstorganisation
self-protection Selbstschutz
self-regulating/self-adjusting
selbstregulierend, selbsteinstellend
self-sealing selbstdichtend
self-tolerance
Selbsttoleranz, Eigentoleranz
semiconductor Halbleiter
semiconservative replication
semikonservative Replikation
semifinished halbfertig
semifinished product
Halbzeug
semimetals Halbmetalle
semimicro batch
Halbmikroansatz
semimicro procedure/method
Halbmikroverfahren,
Halbmikromethode
semipermeability
Halbdurchlässigkeit,
Semipermeabilität
semipermeable
halbdurchlässig,
semipermeabel
semiprecious metal
Halbedelmetall
semisynthesis Halbsynthese
semisynthetic halbsynthetisch
semithin section
Semidünnschnitt
sender Absender, Übersender
sensation/perception
Empfindung
sense of taste/
gustatory sense/
gustatory sensation
Geschmackssinn
sensibility/sensitiveness
Empfindbarkeit
sensitive empfindlich
(sensitiv/leicht reagierend)
sensitiveness/touchiness
Empfindlichkeit, Gekränktsein

sensitivity
 Sensitivität, Empfindlichkeit
sensitivity to pain
 Schmerzempfindlichkeit
sensitivity to temperature
 Temperaturempfindlichkeit
sensitization
 Sensibilisierung, Allergisierung
sensitize sensibilisieren
sensitizing sensibilisierend
 (Gefahrenbezeichnungen)
sensor (detector) Fühler, Sensor, Detektor (*tech:* z.B. Temperaturfühler); (probe) Messfühler, Sonde
sensory sensorisch
separate *vb* scheiden, trennen, abtrennen; (disconnect) trennen, lösen, entkuppeln, auskuppeln; (fractionate) auftrennen, trennen, fraktionieren
**separating column/
fractionating column** *dist*
 Trennsäule
separating gel (running gel) *chromat*
 Trenngel
separation
 Trennung, Scheidung, Abtrennung; (fractionation) Auftrennung, Trennung, Fraktionierung
separation accuracy *chromat*
 Trennschärfe
separation efficiency *chromat*
 Trennwirkungsgrad, Trennleistung
separation factor
 Trennfaktor, Separationsfaktor
separation method
 Trennmethode
**separation technique/
separation procedure**
 Trennverfahren
**separator (precipitator/settler/
trap/catcher/collector)**
 Abscheider
separatory funnel
 Scheidetrichter
sepsis (septicemia/blood poisoning)
 Sepsis, Septikämie, Blutvergiftung

septic tank Faulbehälter (Abwässer)
septum (*pl* **septa or septums)**
 Septum (*pl* Septen)
sequela(e) Folge, Folgeerscheinung, Folgezustand
➢ **late sequelae** Spätfolgen
sequence *n* Sequenz; Aufeinanderfolge, Folge, Reihe, Reihenfolge, Serie
sequence *vb* sequenziren
sequence of operation
 Arbeitsablauf
sequencer (apparatus)
 Sequenzierungsautomat
sequential aufeinander folgend
sequential reaction/chain reaction
 sequentielle Reaktion, Kettenreaktion
sequestration *chem* Maskierung
serial sections *micros/anat*
 Serienschnitte
serologic(al) serologisch
serological pipet
 serologische Pipette
serology Serologie
serous serös
serum (*pl* **sera or serums)**
 Serum (*pl* Seren)
service *n* Dienst, Dienstleistung, Arbeit; Betrieb, Bedienung; Wartung, Kundendienst
service *vb* bedienen, betreiben; warten
service cart/service trolley (*Br***)**
 Servierwagen
service hatch Durchreiche
service life (of a machine/equipment)
 Laufzeit (Gerät), Lebenszeit
service pipe
 Hauptanschlussrohr
**service regulations/
job regulations/
official regulations**
 Dienstvorschrift
service switch Hauptschalter
service voltage
 Betriebsspannung
servicing Wartung, Pflege
servicing schedule Wartungsplan

set *n* Satz, Garnitur;
(instrument/equipment/apparatus)
Gerät, Anlage, Apparat
set *vb* **(turn solid)** fest/steif werden,
abbinden; gerinnen, koagulieren
set point (nominal value/
rated value/desired value)
Sollwert; Bezugspunkt, Festpunkt
set-point adjuster/setting device
Sollwertgeber; Stelleinrichtung
set-point correction
Sollwertkorrektur
setting screw/setscrew Stellschraube
setting time (autoclave)
Ausgleichszeit,
thermisches Nachhinken
setting up (assemble the equipment)
aufbauen (Experiment)
settings
Einstellungen (eines Geräts)
settle (establish) besiedeln, etablieren;
(colonize) kolonisieren
settle out absetzen, ausfallen
settlement/establishment
Besiedlung, Etablierung
settling tank Klärbecken, Absetzbecken
setup *n* **(of an experiment)**
Aufbau (eines Experiments)
severe combined immune deficiency
(SCID)
schwerer kombinierter Immundefekt
sewage Abwasser
➢ **raw sewage** Rohabwasser
sewage fields/sewage farm
Rieselfelder (Abwasser-Kläranlage)
sewage sludge
(*esp.:* **excess sludge from digester)**
Faulschlamm
(*speziell:* ausgefaulter Klärschlamm)
sewage system/sewer Kanalisation
sewage treatment
Abwasseraufbereitung, Klärung
sewage treatment plant
Klärwerk, Kläranlage (Abwasser)
sewer Abwasserkanal, Kloake
sewer gas Faulschlammgas
sewer pipe Abflussrohr
sewerage system/sewer Kanalisation

sex (male/female/neuter)/gender
Geschlecht
(männlich/weiblich/neutral)
sex cell/gamete
Geschlechtszelle, Keimzelle, Gamet
sexually transmitted disease (STD)/
venereal disease (VD)
sexuell übertragbare Krankheit,
Geschlechtskrankheit,
venerische Krankheit
shade *n*
Schatten, Schattierung, Tönung
shade *vb* schattieren
shading Beschattung
shadow Schatten (eines
bestimmten Gegenstandes;
Gegensatz, *siehe:* shade)
shadowcasting
(rotary shadowing in TEM)
Beschattung
(Schrägbedampfung bei TEM)
shady schattig
shaft (spindle) Schaft, Welle
shaft seal (of stirrer/impeller etc.)
Wellendichtung
shake schütteln;
(shake out) ausschütteln;
(shake off) abschütteln
shake culture Schüttelkultur
shake flask Schüttelkolben
shaker
Schüttler; (dredger) Streuer
➢ **circular shaker/**
orbital shaker/
rotary shaker
Kreisschüttler, Rundschüttler
➢ **incubating shaker/**
incubator shaker/
shaking incubator
Inkubationsschüttler
➢ **nutator/nutating mixer/'belly dancer'**
(shaker with gyroscopic, i.e.,
threedimensional circular/orbital &
rocking motion)
Taumelschüttler
➢ **orbital shaker/**
rotary shaker/circular shaker
Rundschüttler, Kreisschüttler

- ➢ **reciprocating shaker (side-to-side motion)** Reziprokschüttler, Horizontalschüttler, Hin- und Herschüttler (rütteln)
- ➢ **rocking shaker (see-saw motion/up-down)** Wippschüttler (rauf-runter); (side-to-side) Rüttler (hin und her); Wippe, Schwinge
- ➢ **vortex shaker/vortex** Vortexmischer, Vortexschüttler, Vortexer (für Reagensgläser etc.)
- ➢ **water bath shaker/ shaking water bath** Schüttelbad, Schüttelwasserbad
- ➢ **with spinning-rotating motion** Drehschüttler (rotierend); (vertically rotating 360°) Überkopfmischer

shaker bottle/shake flask Schüttelflasche, Schüttelkolben
shaking Schütteln
shaking incubator/ incubating shaker/ incubator shaker Inkubationsschüttler
shaking out Ausschütteln, Ausschüttelung
shaking water bath/water bath shaker Schüttelbad, Schüttelwasserbad
shape *n* **(form/appearance/contour)** Gestalt
sharp scharf; (pungent/acrid) beißend (Geruch/Geschmack)
sharpen schärfen (Messer, Scheren)
sharpening stone/ grindstone/honing stone Schleifstein, Abziehstein
sharpie (permanent marker) wasserfester/wischfester Markierstift
sharpness (focus) *micro/photo* Schärfe
sharps scharfe Gegenstände (scharfkantige/spitze G.)
sharps collector Sicherheitsbehälter (Abfallbox zur Entsorgung von Nadeln, Skalpellklingen, Glas etc.)

shatter zerschmettern, zerschlagen, zerspringen, zerbrechen
shatterproof (safety glass) splitterfrei (Glas), bruchsicher
shatterproof glass Sicherheitsglas
sheaf/bundle Garbe (Licht/Funke etc.)
shear *n* Scherung, Gleitung
shear *vb* scheren, schneiden, abschneiden
shear force Scherkraft; (shear stress: shear force per unit area)
shear gradient Schergradient, Schergefälle
shear rate (rate of shear) Scherrate
shear strength/shearing strength Schubfestigkeit, Scherfestigkeit (Holz)
shear stress (shear force per unit area) Scherspannung, Schubspannung
shearing scheren
shearing action/shearing effect Scherwirkung, Schubeffekt
shears (große) Schere
- ➢ **sheet-metal shears/plate shears** Blechschere
- ➢ **trimming shears** Trimmschere
- ➢ **wire cable shears/cable shears** Drahtseilschere, Kabelschere
- ➢ **wire shears/wire cutters** Drahtschere

sheath Scheide, Umhüllung
sheathed scheidenförmig, umhüllt
sheet Bogen, Blatt, (dünne) Platte; Schicht
sheet copper Kupferblech
sheet glass Tafelglas
sheet iron Eisenblech
sheet lead Tafelblei
sheet metal Metallblech, Blech
sheet of glass (pane) Glasplatte, Glasscheibe
sheet steel Stahlblech
sheet-metal shears/plate shears Blechschere
shelf life Haltbarkeit, Lagerfähigkeit; Verfallsdatum

sh

shield *n* Schild, Abschirmung, Schutz
shield *vb* **(from radiation)**
  abschirmen (von Strahlung)
shielding (from radiation)
  Abschirmung (von Strahlung)
shift *n* Wechsel, Verschiebung,
  Veränderung; Schicht (Arbeit)
➢ **chemical shift** *spectr*
  chemische Verschiebung
➢ **frameshift** *gen*
  Rasterverschiebung
➢ **high-field shift (NMR)**
  Hochfeldverschiebung
➢ **low-field shift (NMR)**
  Tieffeldverschiebung
➢ **proton shift (NMR)**
  Protonenverschiebung
➢ **tautomeric shift**
  tautomere Umlagerung
shift work Schichtarbeit
shikimic acid (shikimate)
  Shikimisäure (Shikimat)
shim *tech*
  Ausgleichsring, Ausgleichsscheibe
shipment (dispatch)
  Versand, Warensendung, Lieferung
➢ **bulk shipment/bulk delivery**
  Großlieferung
➢ **ready for shipment/delivery**
  versandfertig
shipment costs/
  shipping charges/
  carriage charges
  Versandkosten
shipper (freight company/
  shipping company)
  Logistikdienstleister, Spedition
shipping documents
  Versandpapiere
shipping papers Frachtpapiere
shiver frösteln, vor Kälte zittern
shock absorption Stoßdämpfung
shock freezing Schockgefrieren
shock-pressure resistant druckstoßfest
shock resistance Stoßfestigkeit (Holz)
shock wave
  Druckwelle, Schockwelle, Stoßwelle
shockproof stoßfest, stoßsicher

shoe covers/
  shoe protectors (disposable)
  Überschuhe, Überziehschuhe
  (Einweg~)
short circuit *n* **(short-circuiting/short)**
  Kurzschluss
short-circuit *vb* kurzschließen
short-chain kurzkettig
short-path distillation/
  flash distillation
  Kurzwegdestillation,
  Molekulardestillation
short-stem funnel/
  short-stemmed funnel
  Kurzhalstrichter,
  Kurzstieltrichter
shot/injection
  (hypodermic injection) *med*
  Spritze, Injektion
showcase Schaukasten, Vitrine
shower
  Dusche; Duschkabine, Duschraum
➢ **emergency shower/safety shower**
  Notdusche
➢ **quick drench shower/deluge shower**
  'Schnellflutdusche'
shred
  zerfetzen, zerreißen, in Fetzen reißen
shredder Reißwolf, Aktenwolf;
  Schneidemaschine
shrink schwinden, schrumpfen;
  einlaufen (Textilien/Stoffe)
➢ **heat shrinking** Wärmeschrumpfen
shrink film/
  shrink wrap/shrink foil/shrinking foil
  Schrumpffolie (zum 'einschweißen')
shrinkage/shrinking
  Schrumpfung, Schwund; Abnahme;
  Einlaufen (Textilien)
shrinkproof schrumpffest, schrumpffrei
shunt Nebenschluss, Stromzweig,
  Nebenschaltung, Überbrückung
shunt current
  Nebenschlussstrom, Zweigstrom
shunt resistance
  Nebenschlusswiderstand,
  Nebenwiderstand
shunt switch Umgehungsschalter

shutdown Abschaltung, Abstellen
shutoff
 Abschaltung, Absperrung,
 Abschaltung
shutoff valve
 Abschaltventil, Absperrventil
shutter Klappe, Schieber;
 photo Verschluss, Blende; Jalousie,
 Rolladen, Fensterladen
shuttle vector/bifunctional vector
 Schaukelvektor,
 bifunktionaler Vektor
sialic acid (sialate) Sialinsäure (Sialat)
siccative/
 desiccant/drying agent/
 dehydrating agent
 Trockenmittel, Sikkativ
sick krank
sick-building syndrome
 Sick-Building-Syndrom
sick leave Fehlen wegen Krankheit
 (krankgeschrieben sein mit
 Lohnfortzahlung)
sick note Krankheitsattest
sickle flask/sausage flask/saber flask
 Säbelkolben, Sichelkolben
side chain (of a molecule) Seitenkette
➢ **main chain (of a molecule)**
 Hauptkette
side effect(s) Nebenwirkung(en)
side product
 Begleitprodukt, Nebenprodukt
side reaction *chem* Nebenreaktion
side tubulation/side arm
 (hose connection on flask)
 Ansatzstutzen
 (Olive für Schlauche/an Kolben)
sidearm (tubulation)
 Seitenarm, Tubus (Kolben etc.)
sidearm flask Seitenhalskolben
sideband *spectr* Nebenbande
sieve *vb* **(sift/screen)** sieben
sieve *n* **(sifter/strainer)** Sieb
➢ **molecular sieve**
 Molekularsieb,
 Molekülsieb, Molsieb
sieve analysis/screen analysis
 Siebanalyse

sieve material/
 sieving material/
 material to be sieved
 Siebgut
sieve plate (perforated plate)
 Siebplatte
sieve plate reactor
 Lochbodenkaskadenreaktor,
 Siebbodenkaskadenreaktor
sieve residue/oversize
 Siebrückstand, Siebüberlauf,
 Überkorn
sieve shaker Siebmaschine (Schüttler)
sievings/screenings/
 siftings/undersize
 Siebdurchgang, Siebunterlauf,
 Unterkorn
sift *vb* sieben
siftage (size separation by screening)
 Siebung
sifter Schüttelsieb
siftings Siebdurchgang
sign in (register)
 eintragen (bei Anmeldung)
sign out (deregister)
 austragen (bei Abmeldung)
signal substance Signalstoff
signal transducer Signalwandler
signal transduction
 Signalübertragung
signal-to-noise ratio (S/N ratio)
 Signal-Rausch-Verhältnis;
 Rauschspannungsabstand
significance level/
 level of significance (error level)
 Signifikanzniveau,
 Irrtumswahrscheinlichkeit
significance test/
 test of significance *stat*
 Signifikanztest
silent infection
 stumme Infektion,
 stille Feiung
silica (silicon dioxide)
 Siliziumdioxid
silica gel Kieselgel, Silicagel
siliceous kieselsäurehaltig
silicic acid Kieselsäure

silicon (Si) Silizium, Silicium
silicone (silicon ketone/silicoketone)
Silikon (Siliciumketon),
Poly(organylsiloxan)
silicone grease
Silikon-Schmierfett
silicone rubber
Siliconkautschuk
silicosis *med* Silicose
silk (fibroin/sericin) Seide
silk suture Seidenfaden
silken seiden, Seiden...
silky (sericeous/sericate)
seidenartig, seidenhaarig, seidig
silt *geol* Schluff
simmer (boil gently) köcheln
(auf 'kleiner' Flamme), sieden
simmering/ebullient
siedend; (boiling) kochend
simple distillation
Gleichstromdestillation
sinapic acid Sinapinsäure
sinapic alcohol Sinapinalkohol
single (solitary) einzeln, solitär
single digest einfacher Verdau
single dose Einzeldosis
single immunodiffusion (Oudin test)
einfache Immundiffusion,
lineare Immundiffusion
(Oudin-Methode)
single radial immunodiffusion (SRI) (Mancini technique)
einfache radiale Immundiffusion
(Mancini-Methode)
single sugar/monosaccharide
Einfachzucker, einfacher Zucker,
Monosaccharid
single-burner hot plate
Einfachkochplatte (Heizplatte)
single-celled/unicellular einzellig
single-use (disposable)
Einmal..., Einweg..., Wegwerf...
single-use gloves (disposable gloves)
Einmalhandschuhe
single-way cock
Einweghahn
singulet condition
Singulettzustand

sink *n* Ausguss, Spüle;
(basin) Abflussbecken, Spülbecken;
physiol (importer of assimilates)
Senke, Verbrauchsort (von
Assimilaten)
➢ **sink unit** Spültisch
sinter/sintering sintern
siphon Siphon, Saugheber
site/location Ort, Fundort, Lage
site-directed mutagenesis
ortsspezifische Mutagenese
size *n* Größe, Maß, Format, Umfang;
Abmessung(en)
size *vb* abmessen; (cut into discreet
length: glass tubing) ablängen
(mit Glasrohrschneider);
tech leimen, grundieren;
(Stoff) appretieren, schlichten
size exclusion chromatography (SEC)
Ausschlusschromatographie,
Größenausschlusschromatographie
sizing Bemessen, auf ein bestimmtes
Maß zurechtschneiden;
Größenbestimmung; Sichtung;
(textile) Schlichten, Schlichtung;
(paper) Leimen, Leimung;
(waste etc.) Sortieren, Sortierung,
Klassieren, Klassierung
sizzle brutzeln, zischen
skid rutschen
skid-proof (non-skid)
nicht-rutschend, Antirutsch ...
(Gerät auf Unterlage)
skim off/scoop off, up
abschöpfen
skin Haut; (cutis) Kutis, Cutis
(eigentliche Haut;
Epidermis & Dermis)
skin care Hautpflege
skin care product
Hautpflegemittel
skin graft/skin transplant
Hauttransplantat
skin irritation Hautreizung
skin ointment Hautsalbe
skin-irritant hautreizend
skull and crossbones
Totenkopf (Giftzeichen)

slab gel *electrophor*
Plattengel (hochkant angeordnetes)
slag Schlacke
slaked lime Ca(OH)$_2$
Ätzkalk, Löschkalk, gelöschter Kalk
slant culture/slope culture
Schrägkultur (Schrägagar)
slaughter/butcher *vb* schlachten
slaughter/slaughtering/butchering
Schlachtung, Schlachten
slaughterhouse Schlachthof
sledge hammer Vorschlaghammer
sleet (glaze/frozen rain)
Eisüberzug, überfrorene Nässe, gefrorener Regen
sleeve/collar *mech* Manschette; (joint sleeve) Manschette für Schliffverbindungen
sleeve gauntlets
Ärmelschoner, Stulpen
slide *n micros* Objektträger
➢ **frosted-end slide**
Mattrand-Objektträger
➢ **microscope depression slide/ concavity slide/cavity slide**
Objektträger mit Vertiefung
slide caliper/caliper square
Schublehre
slide rod Führungsstange
slide rule Rechenschieber
slide valve Schieberventil, Schieber
sliding microtome
Schlittenmikrotom
slimy (mucilaginous/glutinous)
schleimig
slip *vb* abrutschen, ausrutschen
slip-joint connection
Gleitverbindung
slip-resistant rutschfest
slope/slant/dip Neigung
slops Spülicht
slot blot Schlitzlochplatte
slotted screwdriver
Schlitzschraubenzieher
sludge (sewage sludge) Klärschlamm; (sapropel) Sapropel, Faulschlamm
sludge gas/sewage gas
Faulgas, Klärgas (Methan)

sluice *n* Schleuse
sluice *vb* **(channel)** schleusen
slurry *n* Schlamm, Aufschlämmung
slurry *vb* aufschlämmen
slurry-packing technique *chromat*
Einschlämmtechnik
small-angle X-ray scattering (SAXS)
Röntgenkleinwinkelstreuung
small scale *n* Kleinmaßstab
small-scale *adj/adv*
im Kleinmaßstab, in kleinem Maßstab
small-scale application
Kleinanwendung
smear *n med* Abstrich; *micb* Ausstrich
smear infection
Schmierinfektion
smell *vb* riechen
smell *n* **(odor/scent)**
Duft, Geruch
➢ **pleasant smell (fragrance)**
angenehmer Duft/Geruch
➢ **unpleasant smell**
unangenehmer Duft/Geruch
smellable/ perceptible to one's sense of smell
riechbar
smelting furnace Schmelzofen
smock/gown Arbeitskittel
smog ordinance
Smogverordnung
smoke Rauch
➢ **clouds of smoke**
Rauchschwaden
smoke barrier
Rauchschranke, Rauchschutzwand
smoke detector Rauchmelder
smoking Rauchen
➢ **ban on smoking/smoking ban**
Rauchverbot
smoldering/smouldering
Schwelen, Schwelung
smother the flames
Flammen ersticken
smudge-free
unverschmiert, schmutzfrei

snap cap (push-on cap)
Schnappdeckel,
Schnappverschluss
snap-cap bottle/snap-cap vial
Schnappdeckelglas,
Schnappdeckelgläschen
snap-ring pliers/circlip pliers
Sicherungsringzange
soak (drench/steep)
tränken, durchtränken, einweichen
(durchfeuchten), einwirken lassen
(in einer Flüssigkeit), quellen
(Wasseraufnahme)
➢ **soak up/absorb**
aufsaugen, absorbieren
soaking up/absorption
Aufsaugen, Absorption
soap Seife
➢ **a bar of soap** ein Stück Seife
➢ **curd soap (domestic soap)**
Kernseife (feste Natronseife)
➢ **liquid soap/liquid detergent**
Flüssigseife
➢ **soft soap** Schmierseife
soap dispenser (liquid soap)
Seifenspender (Flüssigseife)
soberness *med/physio*
Nüchternheit
socket (ferrule) Hülse, Ring;
(receptacle) Tülle,
electr Fassung, Steckbuchse;
(chuck/nut) Steckschlüsseleinsatz,
Stecknuss, Nuss
➢ **female (spherical joint)**
Schliffpfanne
➢ **ground socket/**
ground-glass socket/
female (ground-glass joint)
Hülse, Schliffhülse
('Futteral', Einsteckstutzen)
➢ **screw-base socket**
Gewindefassung
➢ **threaded socket (connector/nozzle)**
Gewindestutzen
socket screw/socket-head screw
Inbusschraube
socket wrench/box spanner
Stiftschlüssel, Steckschlüssel

soda Soda,
kohlensaures Natrium
soda lime
Natronkalk
soda machine
(for soft drinks/soda pop)
'Kola'maschine
soda water Selterswasser,
Sodawasser, Sprudel
soda-lime glass/
alkali-lime glass (crown glass)
Kalk-Soda-Glas (Kronglas)
sodium (Na) Natrium
sodium chloride NaCl
Natriumchlorid
sodium dodecyl sulfate (SDS)
Natriumdodecylsulfat
sodium hydroxide NaOH
Natriumhydroxid
sodium hydroxide solution
Natronlauge,
Natriumhydroxidlösung
sodium hypochlorite NaOCl
Natriumhypochlorit
soft soap Schmierseife
soft water weiches Wasser
soften weich machen, erweichen;
(Wasser) enthärten
softener Weichmacher;
Weichspülmittel, Weichspüler;
Enthärtungsmittel, Enthärter;
(plasticizer: in plastics a.o.)
Plastifikator
soggy aufgeweicht,
durchnässt, durchweicht
soil (ground/earth)
Boden, Erdreich, Erdboden, Erde
soil decontamination Bodensanierung
soil salinization Bodenversalzung
soil skeleton (inert quartz fraction)
Bodenskelett
soil texture Bodenpartikelgrößen
solar cell/photovoltaic cell
Solarzelle
solar energy
Solarenergie, Sonnenenergie
solar radiation
Sonnenstrahlung

solder *n* Lot, Lötmittel, Lötmetall
solder *vb* löten
soldering acid Lötsäure
soldering fluid/soldering liquid
Lötwasser
soldering flux/solder flux
Lötflussmittel
soldering gun Lötpistole
soldering iron Lötkolben
soldering lug Lötöse
soldering wire Lötdraht
solenoid Zylinderspule
solenoid valve
Magnetventil (Zylinderspule)
solid *adj/adv* fest
solid *n* Festkörper; Körper;
(solids) feste Bestandteile
solid body Festkörper
solid matter Feststoff
solid phase (bonded phase)
Festphase
solid phase reactor
Festphasenreaktor
solid state fester Zustand
solid waste Festmüll
solidify fest werden (lassen),
erstarren
solubility Löslichkeit
➢ **insolubility** Unlöslichkeit
➢ **of low solubility**
schwerlöslich
solubility product
Löslichkeitsprodukt
solubilization
Solubilisierung, Solubilisation,
Löslichkeitsvermittlung
solubilizer/solutizer
Lösungsvermittler,
Löslichkeitsvermittler
soluble löslich
➢ **easily soluble/readily soluble**
leichtlöslich
➢ **insoluble** unlöslich
➢ **sparingly soluble/barely soluble**
kaum löslich, wenig löslich
solute gelöster Stoff
solute potential
Löslichkeitspotential

solution Lösung
➢ **aqueous solution**
wässrige Lösung
➢ **buffer solution** Pufferlösung
➢ **dilute solution**
verdünnte Lösung
➢ **Fehling's solution**
Fehlingsche Lösung
➢ **nutrient solution/culture solution**
Nährlösung
➢ **ready-to-use solution/**
test solution
Gebrauchslösung,
gebrauchsfertige Lösung,
Fertiglösung
➢ **reagent solution** Reagenslösung
➢ **Ringer's solution**
Ringerlösung, Ringer-Lösung
➢ **saline/**
physiological saline solution
physiologische Kochsalzlösung
➢ **saturated solution** gesättigte Lösung
➢ **standard solution** Standardlösung
➢ **stock solution**
Stammlösung, Vorratslösung
➢ **test solution/**
solution to be analyzed
Untersuchungslösung, Prüflösung
➢ **volumetric solution**
(a standard analytical solution)
Maßlösung
solvable löslich, lösbar
solvate *n* solvatisierter Stoff
(Ion/Molekül)
solvate *vb* solvatisieren
solvation Solvatation, Solvatisierung
solve *math* lösen
solvent Lösemittel, Lösungsmittel;
(mobile phase) *chromat* Laufmittel,
Fließmittel
➢ **mobile solvent/eluent/eluant**
(mobile phase)
Laufmittel, Elutionsmittel,
Fließmittel, Eluent (mobile Phase)
solvent front
Lösemittelfront, Lösungsmittelfront;
chromat Laufmittelfront,
Fließmittelfront

solvent recovery
Lösemittelrückgewinnung,
Lösungsmittelrückgewinnung
solvent reistance
Lösemittelbeständigkeit
sonicate beschallen,
mit Schallwellen behandeln
sonification/sonication
Sonifikation, Sonikation,
Beschallung, Ultraschallbehandlung
sonogram Sonogramm
**sonography/
ultrasound/ultrasonography**
Sonographie, Ultraschalldiagnose
sooty (forming soot) rußend, rußig
sorbent Sorbens (*pl* Sorbentien),
Absorbens, absorbierender Stoff,
Absorptionsmittel, Sorptionsmittel
sorbic acid (sorbate)
Sorbinsäure (Sorbat)
sorbitol Sorbit
sort *n* **(type/kind/variety/cultivar)**
Sorte
sort *vb* sortieren
sound Schall, Geräusch, Laut, Ton,
Klang; (noise) lautes Geräusch
sound proofing
Schalldämmung, Schallisolation
sound spectrum
Klangspektrum
sound waves Schallwellen
soundproof
schalldicht, schallundurchlässig
source
Quelle, Produktionsort
source of danger (troublespot)
Gefahrenherd
source of error/mistake/defect
Fehlerquelle
source of fire Brandherd
sow *vb* säen, aussäen, einsäen
sowing/seed sowing
Säen, Aussäen, Aussaat
soybean oil Sojaöl
space
Raum, Platz; Abstand,
Zwischenraum
space heating Raumheizung

space restrictions i.s.v.
Platzbeschränkung, Platznot
space-filling model *chem*
Kalottenmodell
spacer Abstandhalter, Abstandshalter,
Distanzstück
spanner (*US* wrench)
Schlüssel, Schraubenschlüssel
spare parts (replacement parts)
Ersatzteile
sparger Gasverteiler, Luftverteiler
(Düse in Reaktor); Sprenger,
Wassersprenggerät; (in fermentation
reactors etc.) Anschwänzappart,
Anschwänzvorrichtung
sparingly soluble/barely soluble
kaum löslich, wenig löslich
spark *n* Funke, Zündfunke; Entladung
spark *vb* Funken sprühen, zünden
spark coil Zündspule
spark plug Zündkerze
spark spectrum Funkenspektrum
**spat (protective cloth/leather gaiter
covering instep and ankle)**
Gamasche (Schuh~)
spatial räumlich;
(three-dimensional) dreidimensional
spatial perception
räumliche Wahrnehmung
spatula Spatel
➤ **powder spatula** Pulverspatel
➤ **weighing spatula** Wägespatel
special license/special permit
Sondergenehmigung
specialization Spezialisierung
species Spezies, Art
specific
spezifisch, speziell, bestimmt
➤ **nonspecific**
unspezifisch, unbestimmt
specific gravity spezifisches Gewicht
specific gravity bottle
Pyknometerflasche
specific heat
spezifische Wärme
specifications (specs)
Spezifizierung, Spezifikation,
technische Beschreibung

specificity Spezifität
specificity of action
  Wirkungsspezifität
specify spezifizieren, einzeln angeben/benennen/aufführen; bestimmen, festsetzen
specimen (sample)
  Exemplar, Muster, Probe
specimen jar
  Probengefäß,
  großes Probegläschen,
  Sammelglas,
  Sammelgefäß
speckled/
  patched/spotted/spotty
  fleckig
spectacle(s)/pair of s.
  Brille
spectacle eyepiece/
  high-eyepoint ocular *micros*
  Brillenträgerokular
spectral colors
  Spektralfarben
spectrometry
  Spektrometrie
➢ **electron-impact spectrometry (EIS)**
  Elektronenstoß-Spektrometrie
➢ **ion trap spectrometry**
  Ionen-Fallen-Spektrometrie
➢ **mass spectrometry (MS)**
  Massenspektrometrie (MS)
➢ **photoelectron spectrometry (PES)**
  Photoelektronenspektrometrie
➢ **time-of-flight mass spectrometry (TOF-MS)**
  Flugzeit-Massenspektrometrie (FMS)
spectroscopy
  Spektroskopie
➢ **atomic absorption spectroscopy (AAS)** Atom-Absorptionsspektroskopie (AAS)
➢ **atomic emission spectroscopy (AES)** Atom-Emissionsspektroskopie (AES)
➢ **atomic fluorescence spectroscopy (AFS)** Atom-Fluoreszenzspektroskopie (AFS)
➢ **Auger electron spectroscopy (AES)**
  Auger-Elektronenspektroskopie (AES)
➢ **electron energy loss spectroscopy (EELS)**
  Elektronen-Energieverlust-Spektroskopie
➢ **electron spin resonance spectroscopy (ESR)/**
  electron paramagnetic resonance (EPR)
  Elektronen-Spinresonanzspektroskopie (ESR), elektronenparamagnetische Resonanz (EPR)
➢ **flame atomic emission spectroscopy (FES)/**
  flame photometry
  Flammenemissionsspektroskopie (FES)
➢ **infrared spectroscopy**
  Infrarot-Spektroskopie, IR-Spektroskopie
➢ **mass spectroscopy (MS)**
  Massenspektroskopie (MS)
➢ **microwave spectroscopy**
  Mikrowellenspektroskopie
➢ **nuclear magnetic resonance spectroscopy/**
  NMR spectroscopy
  Kernspinresonanz-Spektroskopie, kernmagnetische Resonanzspektroskopie
➢ **photoacoustic spectroscopy (PAS)**
  photoakustische Spektroskopie (PAS), optoakustische Spektroskopie
➢ **ultraviolet spectroscopy/**
  UV spectroscopy
  UV-Spektroskopie
➢ **X-ray absorption spectroscopy (XAS)**
  Röntgenabsorptionsspektroskopie
➢ **X-ray emission spectroscopy (XES)**
  Röntgenemissionsspektroskopie
➢ **X-ray fluorescence spectroscopy (XFS)**
  Röntgenfluoreszenzspektroskopie (RFS)

spectrum (*pl* **spectra/spectrums**)
  Spektrum (*pl* Spektren)
➢ **absorption spectrum/**
  dark-line spectrum
  Absorptionsspektrum
➢ **arc spectrum**
  Lichtbogenspektrum
➢ **band spectrum/molecular spectrum**
  Bandenspektrum, Molekülspektrum (Viellinienspektrum)
➢ **electromagnetic spectrum**
  elektromagnetisches Spektrum
➢ **line spectrum**
  Linienspektrum, Atomspektrum
➢ **reversal spectrum**
  Umkehrspektrum
➢ **rotational spectrum**
  Rotationsspektrum
➢ **spark spectrum**
  Funkenspektrum
➢ **vibrational spectrum**
  Schwingungsspektrum
speed (velocity: vector)
  Geschwindigkeit
spermaceti Walrat
spermaceti oil/sperm oil
  Walratöl
spherical ground joint
  Kugelschliff, Kugelschliffverbindung
spider wrench/spider spanner (*Br*)
  Kreuzschlüssel
spigot (plug of a cask) Zapfen; (faucet) Hahn, Zapfhahn, Fasshahn (Behälter/Kanister/Leitungen)
spill *n* Verschütten, Ausschütten, Überlaufen; Pfütze
spill *vb* verschütten
spill containment pillow
  Saugkissen (zum Aufsaugen von verschütteten Chemikalien)
spillage/spill
  Vergossene(s), Übergelaufene(s)
spillway Abflusskanal
spin decoupling (NMR)
  Spinentkopplung
spindle diagram
  Spindeldiagramm

spinner flask
  Spinnerflasche, Mikroträger
spinning band column
  Drehbandkolonne
spinning band distillation
  Drehband-Destillation
spinning motion (rotating)
  Drehbewegung (rotierend)
spin-spin splitting (NMR)
  Spin-Spin-Aufspaltung
spiral *n* (coil) Gewinde, Spirale; (helix) Helix, Schraube, Spirale; (column packing) Spirale (Füllkörper)
spiral movement/spiral coiling
  Windung (Bewegung)
spiral winding/coiling
  Spiralwindung
spiraled/helical/
  spirally twisted/contorted
  schraubig, spiralig, helical
spirally coiled spiralig aufgewickelt
spirilla (*sg* **spirillum**) **(spiraled forms)**
  Spirillen (*sg* Spirille)
spirit *chem* Spiritus
spirit of wine
  (rectified spirit: alcohol)
  Weingeist
spit *vb* speien
splash *n* Spritzen, Spritzfleck; Spritzer (verspritzte Chemikalie)
splash *vb* **(splatter/spatter/squirt)**
  spritzen, verspritzen, herumspritzen (auch versehentlich)
splash adapter/splash protector/
  antisplash adapter/
  splash-head adapter *dist*
  Spritzschutzadapter, Spritzschutzaufsatz, Schaumbrecher-Aufsatz, Reitmeyer-Aufsatz (Rückschlagschutz: Kühler/Rotationsverdampfer etc.)
splash-proof spritzfest
splice *gen* spleißen
splint *med* Schiene
splinter Splitter; (bits of broken glass) Glassplitter

split *n chromat* Abzweig
split *vb* spalten, aufspalten; zerlegen
split valve *chromat* Abzweigventil
splitting Zerlegen, Zerlegung, Aufspaltung
spokeshave Ziehklinge
sponge forceps Tupferklemme
sponge stopper Schwammstopfen
spontaneous decomposition/ autodecomposition Selbstzersetzung
spontaneous ignition (self-ignition/autoignition) Selbstenzündung
spontaneous ignition temperature (SIT) Selbstenzündungstemperatur
spontaneously ignitable/ self-ignitable/autoignitable selbstentzündlich
spool/coil Spule
spoon/scoop Löffel
sporadic sporadisch
spot/stain *n* Fleck
spot plate *micb* Tüpfelplatte
spot remover/stain remover Fleckenentferner
spot test Tüpfelprobe
spotlight/spot Strahler, Punktstrahler, Spot
spotted/mottled gefleckt
spout Ausguss (zum Ausgießen einer Flüssigkeit), Mundstück, Schnauze; (nozzle/lip/pouring lip) Ausgießschnauze; (pouring spout) Gießschnauze (an Gefäß)
spray bottle Sprühflasche
spray can/aerosol can Sprühdose, Druckgasdose
spray column *dist* Sprühkolonne
spray nozzle Zerstäuberdüse
spread (scatter/disseminate) verstreuen, ausstreuen; (e.g., disease/epidemic) *med* übergreifen

spread-plate method/technique *micb* Spatelplattenverfahren
spreading Spreitung; (expansion/propagation) Ausbreitung, Propagation
spring Feder, Sprungfeder; Elastizität; Quelle, Ursprung
spring balance/spring scales Federzugwaage, Federwaage, Zugwaage
spring constant Federkonstante, Federsteifigkeit
spring-loaded mit Federdruck
springwater Quellwasser
sprinkle (spray) besprühen; (irrigate) berieseln, besprengen
sprinkle irrigation Berieselung
sprinkler/
 sprinkler irrigation system Beregnungsanlage, Berieselungsanlage, Sprinkler
sprout (grow/bud) sprießen, knospen
sprouting/budding Sprossung, Knospung
spur Sporn (Immunodiffusion)
sputter (EM) *micros* sputtern, besputtern (Vakuumzerstäubung)
sputtering (EM) *micros* Sputtern, Besputtern, Kathodenzerstäubung (Metallbedampfung)
sputtering unit/appliance Besputterungsanlage
square bottle Vierkantflasche
squash (mount) *micros* Quetschpräparat
squeegee Abstreicher, Rakel (Gummi), Abzieher, Gummiwischer; (for floors) Wasserschieber, Wasserabzieher (Bodenwischer); (for windows) Fensterwischer, Fensterabzieher
squeeze-bulb pump (hand pump for barrels) Quetschpumpe (Handpumpe für Fässer)

St. Andrew's cross
 Andreaskreuz (Gefahrenzeichen)
stab culture
 Stichkultur,
 Einstichkultur (Stichagar)
stabilization Stabilisierung
stabilize stabilisieren
stabilizer Stabilisator
stable stabil
➢ **unstable (instable)**
 instabil, nicht stabil
stack *vb* **(stacked)**
 stapeln (gestapelt)
stack *n* Stapel
➢ **smokestack** Schornstein
stacking forces Stapelkräfte
stacking gel *electrophor*
 Sammelgel
staff/employees/personnel
 Belegschaft
stage Stadium (*pl* Stadien);
 Stufe; Bühne
➢ **mechanical stage** *micros*
 Kreuztisch
➢ **microscope stage** Objekttisch
➢ **plain stage** *micros* Standardtisch
➢ **rotating stage** *micros* Drehtisch
stage clip *micros*
 Objekttisch-Klammer
stage micrometer *micros*
 Objektmikrometer
stain *vb tech/micros* kontrastieren,
 färben, einfärben;
 (e.g., wood) beizen;
 (bleed) abfärben
stain *n tech/micros* Kontrastierung,
 Farbstoff, Pigment;
 (staining) Färben, Färbung,
 Einfärbung
➢ **vital stain/vital dye**
 Vitalfarbstoff, Lebendfarbstoff
stainability *micros* Färbbarkeit
staining *micros*
 Färbung (durch Farbstoffzugabe)
➢ **supravital staining**
 Supravitalfärbung
➢ **vital staining**
 Lebendfärbung, Vitalfärbung

staining dish/
 staining jar/staining tray *micros*
 Färbeglas, Färbetrog, Färbekasten,
 Färbewanne
staining method/technique
 Färbemethode, Färbetechnik
staining tray Färbegestell
stainless steel rostfreier Stahl
stainless-steel sponge
 Edelstahlschrubber
staircase Treppe, Treppenaufgang;
 Treppenhaus
stairs Treppe
stairway Treppenaufgang; Treppenhaus
stairway entry/exit
 Treppenhauseingang/~ausgang
stairwell Treppenschacht, Treppenhaus
stance phase Stemmphase
stand/rack Ständer
standard condition Standardbedingung
standard deviation/
 root-mean-square deviation *stat*
 Standardabweichung
standard electrode potentials
 (tabular series)/
 standard reduction potentials/
 electrochemical series (of metals)
 Spannungsreihe (der Metalle),
 Normalpotentiale
standard error
 (standard error of the means) *stat*
 Standardfehler, mittlerer Fehler
standard hydrogen electrode
 Normalwasserstoffelektrode
standard measure Normalmaß
standard operating procedure (SOP)
 Standard-Arbeitsanweisung
standard potential/
 standard electrode potential
 Standardpotential, Normalpotential
standard pressure
 Normaldruck, Normdruck
standard procedure Standardverfahren
standard solution Standardlösung
standard taper (S.T.)
 Normalschliff (NS)
standard-taper glassware
 Normschliffglas (Kegelschliff)

standardization
  Standardisierung, Normierung; Vereinheitlichung, Normung
standardize
  standardisieren, normen, vereinheitlichen
standby Bereitschaft; Not..., Hilfs..., Reserve..., Ersatz...
standby duty/standby service
  Bereitschaftsdienst
standby mode
  Bereitschaftsstellung, Wartestellung
standby unit Notaggregat
standstill Stillstand
staple *n* **(U-shaped metal loop)**
  Heftklammer
staple food/basic food
  Grundnahrungsmittel
staple remover
  Enthefter, Heftklammern-Entferner
stapler Hefter (Büro~)
starch Stärke
starch granule Stärkekorn
star-crack (in glass) Sternriss
start/
  prepare/mix/make/set up (solution/experiment etc.)
  ansetzen (z.B. eine Lösung)
starter culture (growth medium)
  Starterkultur (Anzuchtmedium)
starter medium (growth medium)
  Anzuchtmedium
starting material/
  basic material/base material/
  source material/primary material
  Ausgangsmaterial, Ausgangsstoff
starvation *n micb* Hungern
starvation phase Auszehrphase
starve *vb micb*
  hungern, aushungern
state *n* Lage, Stand; (condition) Zustand
state equation/equation of state
  Zustandsgleichung
statement Erklärung, Verlautbarung, Aussage, Angabe
static *adj/adv* **(static charge)**
  statisch, elektrostatisch

static *n* **(static charge)**
  statische Elektrizität, Ladung
static culture
  statische Kultur
static current Ruhestrom
static electricity
  statische Elektrizität
statics (in construction)
  Statik, Baustatik
stationary
  stationär, feststehend
stationary equilibrium
  stationäres Gleichgewicht
stationary phase/stabilization phase/adsorbent *chromat*
  stationäre Phase
stationary wave *phys*
  stehende Welle, Stehwelle
stationery
  Schreibwaren; Briefpapier
  ➢ **letter-head**
  Briefpapier mit Briefkopf
statistic/statistic value
  Kennzahl, statistische Maßzahl
statistical deviation
  statistische Abweichung
statistical distribution
  statistische Verteilung
statistical error
  statistischer Fehler
statistics Statistik
stator-rotor impeller/
  Rushton-turbine impeller
  Stator-Rotor-Rührsystem
steady state
  stationärer Zustand, gleichbleibender Zustand
steady-state equilibrium
  Fließgleichgewicht, dynamisches Gleichgewicht
steam bath Dampfbad
steam distillation
  Trägerdampfdestillation
stearic acid/octadecanoic acid (stearate/octadecanate)
  Stearinsäure, Octadecansäure (Stearat/Octadecanat)

steel Stahl
➤ **high-grade steel/**
high-quality steel
Edelstahl
➤ **stainless steel**
rostfreier Stahl
steel cylinder (gas cylinder)
Stahlflasche (Gasflasche)
steer/steering steuern
(in eine Richtung lenken)
stem cell research
Stammzellforschung
stem culture/stock culture
Stammkultur, Impfkultur
stencil Zeichenschablone
(für Formeln etc.)
step gradient Stufengradient
step-on pail Treteimer (Mülleimer)
step-stool Trittschemel
➤ **rolling step-stool**
Rollhocker, 'Elefantenfuß'
(runder Trittschemel mit Rollen)
stepladder/steps
Stehleiter, Treppenleiter, Trittleiter
stereo microscope Stereomikroskop
stereoisomer Stereoisomer
stereoselective stereoselektiv
stereospecificity Stereospezifität
steric/sterical/spacial
sterisch, räumlich
steric hindrance
sterische Hinderung,
sterische Behinderung
sterile steril; (disinfected) desinfiziert;
(infertile) unfruchtbar
sterile bench sterile Werkbank
sterile filter Sterilfilter
sterile filtration Sterilfiltration
sterility/infertility Sterilität,
Unfruchtbarkeit
sterilizability
Sterilisierbarkeit
sterilizable sterilisierbar
sterilization/sterilizing
Sterilisation, Sterilisierung
sterilization in place (SIP)
SIP-Sterilisation (ohne Zerlegung,
Öffnung der Bauteile)

sterilize sterilisieren
sterol Sterin, Sterol
stewpan
Schmorpfanne, Kasserole
stick *vb* (adhere: paste/cement) kleben;
med (by a needle) stechen
➤ **stick together** verkleben
stick injury Stichverletzung
stick-and-ball model/
ball-and-stick model *chem*
Stab-Kugel-Modell,
Kugel-Stab-Modell
sticker Aufkleber
sticky (glutinous/viscid)
klebrig, glutinös
stiffen versteifen, verstärken;
starr machen;
verdicken (Flüssigkeiten)
stifling/stuffy stickig
still *n dist* Destillierapparat;
Destillierkolben
still pot (boiler/distillation boiler flask/
reboiler) Blase, Destillierblase,
Destillierrundkolben
stillhead (distillation head)
Destillieraufsatz, Destillierbrücke
stimulate (excite)
stimulieren, anregen, beleben
stimulation (excitation)
Stimulierung, Anregung,
Antrieb, Anreiz, Belebung
stimulus
Stimulus, Reiz, Antrieb, Ansporn
➤ **external stimulus** Außenreiz
➤ **light stimulus** Lichtreiz
stimulus threshold
Reizschwelle
sting *vb* stechen, beißen, brennen
stipend Gehalt (auf Grund eines
'höheren' Dienstverhältnisses)
stir rühren, umrühren;
(agitate) schütteln, aufrühren,
aufwühlen;
(swirl) umwirbeln, herumwirbeln
stir bar/stirrer bar/
stirring bar/bar magnet/'flea'
Magnetstab, Magnetstäbchen,
Magnetrührstab, 'Fisch', Rühr'fisch'

stirred cascade reactor
Rührkaskadenreaktor
stirred loop reactor
Rührschlaufenreaktor,
Umwurfreaktor
stirred-tank reactor
Rührkesselreaktor
stirrer (impeller/agitator) Rührer,
Rührwerk;
(mixer) Mixer, Rührgerät
➤ **hollow stirrer** Hohlrührer
➤ **magnetic stirrer** Magnetrührer
➤ **paddle stirrer/paddle impeller**
Schaufelrührer, Paddelrührer
➤ **turbine stirrer/turbine impeller**
Turbinenrührer
stirrer bearing
Lagerhülse (Glasaussatz);
Rührerlager (Rührwelle)
stirrer blade Rührerblatt
stirrer gland Rührhülse
stirrer seal Rührverschluss
stirrer shaft Rührerschaft,
Rührerwelle, Rührwelle
stirring bar (stirrer bar/stir bar/'flea')
Rührstab, Rührstäbchen,
Magnetrührstab,
Magnetrührstäbchen, Rührfisch,
'Fisch'
stirring bar extractor
(stirring bar retriever/'flea' extractor)
Rührstabentferner,
Magnetrührstabentferner
(zum 'Angeln' von Magnetstäbchen)
stirring hot plate
Magnetrührer mit Heizplatte
stirring rod Rührstab (Glasstab)
stock/store/supply
(*meist* p*l* supplies)/
provisions/reserve
Vorrat; Lager
➤ **on stock** lieferbar
➤ **out of stock**
nicht lieferbar
➤ **temporarily out of stock**
derzeit nicht lieferbar
stock culture/stem culture
Stammkultur, Impfkultur

stock solution Stammlösung
stocking density Besatzdichte
stockroom/storage room
(repository/warehouse)
Warenlager
stoichiometric(al) stöchiometrisch
stoichiometry Stöchiometrie
stomach acid Magensäure
stomach juice/gastric juice
Magensaft, Magenflüssigkeit
stool/feces
Stuhl, Fäzes, Kot (Mensch)
➤ **swivel stool** Drehhocker
stool sample Stuhlprobe
stop (limit/detent) Anschlag
(Endpunkt, Sperre, Stop);
Arretierung (z.B. am Mikroskop)
stopcock
Absperrhahn, Sperrhahn
➤ **glass stopcock** Glashahn
stopper *vb* zustöpseln,
mit Stopfen verschließen
stopper (cork) *n* (*Br* bung)
Stopfen, Stöpsel (Korken)
➤ **hex-head stopper/**
hexagonal stopper
Sechskantstopfen
➤ **octa-head stopper/**
octagonal stopper
Achtkantstopfen
➤ **rubber stopper/rubber bung** (*Br*)
Gummistopfen, Gummistöpsel
storability/durability/shelf life
Haltbarkeit
storable/durable/lasting haltbar
storage Speicherung, Aufbewahrung;
Vorrat, Lager, Stauraum
➤ **interim storage/temporary storage**
Zwischenlager
storage cabinet Vorratsschrank
storage chamber Vorratskammer
storage container
Sammelbehälter, Sammelgefäß
storage pest Vorratsschädling
storage tank Speichertank, Lagertank
store *vb* (keep/save/preserve)
aufbewahren; (save/accumulate)
speichern, anreichern, akkumulieren

storeroom/storage room
Abstellraum, Abstellkammer
stowage/storage Stauraum
STP (s.t.p./NTP)
(standard temperature 0°C/ pressure 1 bar) Normzustand (Normtemperatur & Normdruck)
straight-end distillation
einfache/direkte Destillation
strain *vb* belasten;
tech deformieren, verformen, verziehen;
(filter) abseihen
strain *n* Belastungsursache;
micb Stamm
➢ **bacterial strain**
Bakterienstamm
➢ **inbred strain** Inzuchtstamm
➢ **reference strain** *micb*
Referenzstamm
straining cloth Siebtuch
strap *n* Gurt, Band, Riemen
strapping fabric Bindevlies
stratification (act/process of stratifying)
Schichtenbildung;
(state of being stratified: layering)
Schichtung
stray light Streulicht
streak *vb* **(smear)** *micb*
ausstreichen (z.B. Kultur)
streak culture/smear culture
Ausstrichkultur, Abstrichkultur
streak formation/streaking/striation
Schlierenbildung
streak-plate method/technique *micb*
Plattenausstrichmethode
streaky/streaked schlierig
stream *n* **(flow: liquid)** Strom
stream *vb* **(flow)** strömen
stress *n* Stress,
Belastungszustand, Spannung
stress *vb* stressen, belasten
stressful stressig, anstrengend
stretch *tech/mech*
spannen, dehnen, strecken
stretch film (stretch foil) Stretchfolie
stretcher Trage, Krankentrage
stretching vibration Streckschwingung

strictly forbidden/ strictly prohibited
strengstens verboten
striker (e.g., ignite gas)
Anzünder (Gas)
string Schnur
stringency
(of reaction conditions)
Stringenz
(von Reaktionsbedingungen)
stringent conditions
stringente Bedingungen, strenge Bedingungen
stripping column *dist*
Abtriebsäule, Abtreibkolonne
stripping section *dist*
Abtriebsteil (Unterteil der Säule)
stroboscope (strobe/strobe light)
Stroboskop
strong ion difference (SID)
Starkionendifferenz
strontium (Sr) Strontium
structural analysis
Strukturanalyse
structural formula
Strukturformel
structure Struktur
➢ **ultrastructure**
Ultrastruktur
structure elucidation
Strukturaufklärung
stuffing gland/packing box seal
Stopfbuchse (Rührer: Wellendurchführung)
stupefacient *adj/adv*
(stupefying/narcotic/anesthetic)
betäubend, narkotisch, anästhetisch
stupefacient *n*
(narcotic/narcotizing agent/ anesthetic/anesthetic agent)
Betäubungsmittel, Narkosemittel, Anästhetikum
stupefaction (narcosis/anesthesia)
Betäubung, Narkose, Anästhesie
stupefy (narcotize/anesthetize)
betäuben, narkotisieren, anästhesieren
stylet/stiletto Stilett

styptic/hemostatic (astringent) blutstillend (adstringent)
styrene Styrol
styrene-butadiene rubber (SBR) Styrol-Butadien-Kautschuk
subbituminous coal Glanzbraunkohle, subbituminöse Kohle
subculture/passage (of cell culture) Subkultur, Subkultivierung, Passage (einer Zellkultur)
subcutis Unterhaut, Unterhautbindegewebe, Subcutis, Tela subcutanea
subdivide(d) untergliedern (untergliedert), unterteilen (unterteilt)
subdivision Untergliederung, Unterteilung
suberic acid/octanedioic acid Korksäure, Suberinsäure, Octandisäure
sublethal subletal
sublimate sublimieren
sublimation Sublimation
submerged (submersed) untergetaucht, submers
submerged culture Submerskultur, Eintauchkultur
submersible (pump) tauchfähig
subordinate/submit *vb* unterordnen
subsample *stat* Teilstichprobe
subset selection *stat* Teilmengenauswahl
substage illuminator *micros* Anstechleuchte
substance mixture Substanzgemisch
substitute *n* Ersatz
substitute *vb* substituieren, einsetzen (anstelle von)
substitute name Ersatzname
substitute substance Ersatzstoff
substitution Substitution, Ersatz, Ersetzen
substitution therapy Ersatztherapie
substrate Substrat
➤ **following substrate** Folgesubstrat
➤ **leading substrate** Leitsubstrat

substrate constant Substratkonstante (K_S)
substrate inhibition Substrathemmung, Substratüberschusshemmung
substrate recognition Substraterkennung
substrate saturation Substratsättigung
substrate specificity Substratspezifität
substrate-level phosphorylation Substratkettenphosphorylierung
subtyping Subtypisierung
subunit (component) Untereinheit, Komponente
subunit vaccine Komponentenimpfstoff, Spaltimpfstoff, Spaltvakzine, Subunitimpfstoff, Subunitvakzine
succinic acid (succinate) Bernsteinsäure (Succinat)
suck saugen, ansaugen
suck in/draw in einsaugen, aufsaugen
suck-back Einsaugen (Rückschlag bei Wasserstrahlpumpe etc.)
sucrose (beet sugar/cane sugar) Saccharose, Sucrose (Rübenzucker/Rohrzucker)
suction Saugen, Sog, Absaugen, Aufsaugen;
suction-cup feet Saugfüßchen
suction disk Saugnapf, Saugscheibe
suction filter/suction funnel/ vacuum filter (Buechner funnel) Filternutsche, Nutsche (Büchner-Trichter)
suction filtration Saugfiltration
suction flask (filter flask/filtering flask/vacuum flask/aspirator bottle) Saugflasche, Filtrierflasche
suction force Saugkraft
suction funnel/suction filter/ vacuum filter (Buechner funnel) Filternutsche, Nutsche (Büchner-Trichter)

suction head Ansaughöhe
suction lift Ansaugtiefe
suction pipet (patch pipet)
  Saugpipette
suction pump/
  aspirator pump/vacuum pump
  Saugpumpe, Vakuumpumpe
suction stroke (pump)
  Ansaugpuls
suction tension
  Saugspannung
suds
  Seifenschaum, Seifenwasser
suet (from abdominal cavity of ruminants) Talg
suffocate ersticken
suffocation Ersticken
sugar Zucker
➢ **amino sugar** Aminozucker
➢ **blood sugar** Blutzucker
➢ **cane sugar/**
  beet sugar/table sugar/sucrose
  Rohrzucker, Rübenzucker,
  Saccharose, Sukrose, Sucrose
➢ **double sugar/disaccharide**
  Doppelzucker, Disaccharid
➢ **fruit sugar/fructose**
  Fruchtzucker, Fruktose
➢ **grape sugar/**
  glucose/dextrose
  Traubenzucker,
  Glukose, Glucose, Dextrose
➢ **high fructose corn syrup**
  Isomeratzucker, Isomerose
➢ **invert sugar** Invertzucker
➢ **malt sugar/maltose**
  Malzzucker, Maltose
➢ **milk sugar/lactose**
  Milchzucker, Laktose
➢ **multiple sugar/polysaccharide**
  Vielfachzucker, Polysaccharid
➢ **raw sugar/crude sugar**
  (unrefined sugar)
  Rohzucker
➢ **single sugar/**
  monosaccharide
  Einfachzucker, einfacher Zucker,
  Monosaccharid

➢ **table sugar/**
  cane sugar/beet sugar/sucrose
  Rohrzucker, Rübenzucker,
  Saccharose, Sukrose, Sucrose
➢ **wood sugar/xylose**
  Holzzucker, Xylose
sugar-containing zuckerhaltig
suicide inhibition Suizidhemmung
suicide substrate Selbstmord-Substrat
sulfa drug/sulfonamide
  Sulfonamid
sulfanilic acid/
  p-aminobenzenesulfonic acid
  Sulfanilsäure
sulfate Sulfat
sulfonation flask Sulfierkolben
sulfur (S) Schwefel
sulfur compound
  Schwefelverbindung,
  schwefelhaltige Verbindung
sulfurated/sulfuretted
  geschwefelt, Schwefel...
  (mit Schwefel verbunden)
sulfuric Schwefel...
sulfuric acid H_2SO_4
  Schwefelsäure
sulfuricants Sulfurikanten
sulfuring
  Schwefeln, Schwefelung (Fässer)
sulfurize (e.g., vats)
  schwefeln (z.B. Fässer)
sulfurous schweflig, schwefelig
  (vierwertigen Schwefel enthaltend);
  (sulfur-containing) schwefelhaltig
sulfurous acid H_2SO_3
  schweflige Säure, Schwefligsäure
sum n **(total)** Summe, Gesamtheit
sum rule Summenregel
sump Sammelbehälter, Sammelgefäß;
  (cesspit/cesspool/soakaway Br)
  Senkgrube, Sickergrube
sunscreen lotion
  Lichtschutzmittel
sunstroke Sonnenstich;
  (heatstroke) Hitzschlag
superacid Supersäure
supercharge vb
  überladen; vorverdichten

supercoiled superspiralisiert, superhelikal, überspiralisiert
supercoiling Überspiralisierung
supercool unterkühlen
supercooling Unterkühlung
supercritical (gas/fluid) überkritisch (Gas/Flüssigkeit))
supercritical fluid chromatography (SFC) überkritische Fluidchromatographie, superkritische Fluid-Chromatographie, Chromatographie mit überkritischen Phasen (SFC)
superficial (on the surface) oberflächlich
superheat/overheat überhitzen
superheating/overheating Überhitzen, Überhitzung
superinfection Superinfektion, Überinfektion
superior höher, höher stehend, besser; (dominant) überlegen, überragend, vorherrschend, dominant
superior performance überragende/hervorragende Leistung
superior quality beste Qualität
superiority/dominance Überlegenheit, Dominanz
supernatant *n* Überstand
supersaturated übersättigt
supersaturation Übersättigung
supervise beaufsichtigen, überwachen, kontrollieren
supervision Aufsicht, Überwachung, Beaufsichtigung, Kontrolle
supervisor Aufseher, Kontrolleur; leitender Beamter; Chef; Doktorvater
supervisory function Kontrollfunktion
supplier (vendor/supply house) Lieferant; (for accessories) Zubehörlieferant; (distributor) Vertrieb
supplier catalog/distributor catalog Händlerkatalog
supplies Zubehör

supplies storage/ supplies 'shop'/'supplies' Zubehörlager
supply *n* (shipment) Lieferung; (delivery) Zulieferung
supply *vb* liefern; (feed/pipe in/let in) zuleiten
➢ **supply with blood/vascularize** durchbluten
supply line/utility line/service line Versorgungsleitung
supply pressure (HPLC) Eingangsdruck
support *n* Stütze; Unterstützung; *chem* Stativ
support *vb* stützen, unterstützen, helfen
support base Stativplatte
support clamp Stativklemme
support rod Stativstab
support stand/ ring stand/retort stand/stand Stativ, Bunsenstativ
suppress supprimieren, unterdrücken, zurückdrängen
suppressible supprimierbar, unterdrückbar
suppression Suppression, Unterdrückung
supravital dye/supravital stain Supravitalfarbstoff
supravital staining Supravitalfärbung
surface Oberfläche
surface culture Oberflächenkultur
surface labeling Oberflächenmarkierung
surface runoff Oberflächenabfluss
surface tension Oberflächenspannung, Grenzflächenspannung
surface-to-volume ratio Oberflächen-Volumen-Verhältnis
surfactant oberflächenaktive Substanz, Entspannungsmittel

surge *n* Woge, Welle;
electr Spannungsstoß
surge *vb electr*
plötzlich ansteigen,
emporschnellen
surge suppressor/surge protector
Überspannungsfilter,
Überspannungsschutz
surgeon Chirurg
surgeon's gown
Operationskittel
surgeon's mask
Operationsmaske
surgery Chirurgie;
chirurgischer/operativer Eingriff
surgical instruments
Operationsbesteck, OP-Besteck,
chirurgische Instrumente
surgical mask
Operationsmaske,
chirurgische Schutzmaske
surgical scissors
chirurgische Schere
surplus Überschuss
surplus production
Überschussproduktion
surroundings/
environs/environment/vicinity
Umgebung
survey *n* Inspektion, Untersuchung,
Begutachtung, Schätzung;
math/stat Erhebung
survey *vb* untersuchen, begutachten,
betrachten; *math/stat* erheben,
eine Erhebung vornehmen
survival Überleben
survive überleben
survivorship curve
Überlebenskurve
susceptibility
Empfindlichkeit, Anfälligkeit
susceptible empfindlich, anfällig
suspected toxin Verdachtsstoff
suspend suspendieren
(schwebende Teilchen in
Flüssigkeit);
chem (slurry/slurrying)
aufschlämmen; (hang) aufhängen

suspended condenser/
cold finger
Einhängekühler, Kühlfinger
suspended particle
Schwebeteilchen
suspended substance
(suspended matter)
Schwebstoff(e)
suspension (slurry) Suspension
sustained yield
Nachhaltigkeit, nachhaltiger Ertrag
suture *n med*
Naht; Nahtmaterial, Faden
suture needle
chirurgische Nadel
swab Abstrich; Abstrichtupfer
  ➤ **buccal swab** Wangenabstrich
  ➤ **to take a swab**
    einen Abstrich machen
swallow *vb* schlucken
swallowing Schlucken
swan-necked flask/
S-necked flask/
gooseneck flask
Schwanenhalskolben
sweat *n* **(perspiration)** Schweiß
sweat *vb* **(perspire)** schwitzen
sweating/perspiration/hidrosis
Schwitzen
sweep *n* Einzeldurchlauf, Abtastung;
Abtaststrahl;
Zeitablenkung (Oszillograph)
sweep *vb* absuchen; scannen, abtasten
sweep (up) kehren, fegen
sweep coil (NMR)
Sweep-Spule,
Ablenkspule, Kippspule
sweep generator *electr*
Kippgenerator;
Frequenzwobbler
sweep voltage Kippspannung
sweet süß
sweetener Süßstoff
sweetness Süße
swell/swelling/turgescent
schwellen, anschwellen, turgeszent
swelling Schwellung;
(turgescence) Turgeszenz

swing phase/suspension phase
Schwingphase
swing-out rotor/
 swinging-bucket rotor/
 swing-bucket rotor
Ausschwingrotor
swirl schwenken
(Flüssigkeit in Kolben), wirbeln
switch *n* Schalter; Weiche;
Umstellung, Wechsel
➤ **toggle switch (rocker)**
Kippschalter
switch gear Schaltvorrichtung,
Schaltgetriebe, Schaltwerk
switchboard
Schaltanlage; Telefonzentrale;
(electrical control panel) Schalttafel
swivel sich drehen, schwenken;
drehbar, schwenkbar
swivel casters
Schwenkrollen, Lenkrollen,
Schwenkrollfüße
swivel chair Drehstuhl
swivel nut/
 coupling nut/
 mounting nut/cap nut
Überwurfmutter,
Überwurfschraubkappe
(z.B. am Rotationsverdampfer)
swivel stool Drehhocker
symmetry Symmetrie
synchronous culture
Synchronkultur
syndrome/complex of symptoms
Syndrom, Symptomenkomplex
synergist/booster Synergist
(Promoter/Aktivator)

syngas Syngas, Synthesegas
synthesis Synthese;
(preparation) Darstellung
➤ **biosynthesis** Biosynthese
➤ *de-novo*-**synthesis**
Neusynthese, *de-novo* Synthese
➤ **semisynthesis**
Halbsynthese
synthesize synthetisieren,
künstlich herstellen;
(prepare) herstellen;
chem darstellen
synthetic synthetisch
synthetic reactions
(metabolism/anabolism)
Synthesestoffwechsel, Anabolismus
syringe (hypodermic syringe)
Spritze
➤ **disposable syringe**
Einwegspritze
syringe connector
Nadeladapter
syringe filter
Spritzenvorsatzfilter, Spritzenfilter
syringe needle/syringe cannula
Injektionsnadel, Spritzennadel,
Spritzenkanüle
syringe piston/plunger
Spritzenkolben, Stempel, Schieber
syringe pump Spritzenpumpe
systematic systematisch
systematic error/bias
systematischer Fehler, Bias
systematics Systematik
systemic systemisch
systems analysis
Systemanalyse

T-purge (gas purge device)
Spülventil (Inertgas)
table Tisch;
Tabelle, Tafel
➤ **laboratory table/
laboratory workbench/
lab bench**
Labortisch, Labor-Werkbank
➤ **life table** Sterbetafel
➤ **nutrient table/
food composition table**
Nährwert-Tabelle
➤ **periodic table (of the elements)**
Periodensystem (der Elemente)
➤ **timetable**
Stundenplan, Zeitplan,
Fahrplan, Zeittabelle
➤ **weighing table** Wägetisch
➤ **worktable** Arbeitstisch
table salt/common salt/rocksalt (NaCl)
Kochsalz
**table sugar/
cane sugar/beet sugar/sucrose**
Rohrzucker, Rübenzucker,
Saccharose, Sukrose, Sucrose
**table-top centrifuge/
tabletop centrifuge/
benchtop centrifuge
(multipurpose c.)**
Tischzentrifuge
tablet Tablette
tabulate tabellarisieren,
tabellarisch darstellen, tabellieren
tack Stift (Metall~);
(pin) Nadel; (nail) Nagel
➤ **thumb tack** Reißnagel
tackle (pulley)
Flaschenzug, Rollenzug
tag *vb* etikettieren,
markieren, beschildern;
anfügen, anhängen
tag *n* **(for identification)**
Etikett, Plakette, Anstecker,
Abzeichen, Schildchen
➤ **name tag** Namensetikett,
Namensschildchen
tagged molecule
markiertes Molekül

tail (e.g., of a molecule)
Schwanz
tailing(s)/tails *dist/chromat*
Nachlauf, Ablauf;
Schwanzbildung,
Signalnachlauf
take up/take in
einnehmen,
zu sich nehmen
tallow (extracted from animals)
Talg
tally chart Strichliste
tally counter
Zähler (Handzähler),
Zählwerk
tampon/plug/pack *vb*
tamponieren
tan *vb* gerben, beizen;
bräunen
tangential section
Tangentialschnitt;
(wood) Sehnenschnitt
tank (vessel) Tank, Kessel,
großer (Wasser)Behälter,
Becken, Zisterne
**tank car/tank truck
(*Schiene:* rail tank car)**
Kesselwagen
(Chemikalientransport)
tannate (tannic acid)
Tannat (Gerbsäure)
tannic acid (tannate)
Gerbsäure (Tannat)
tanniferous
gerbsäurehaltig,
gerbstoffhaltig
tannin (tanning agent)
Tannin (Gerbstoff)
tanning Gerben
tanning agent/tannin
Gerbstoff
tap Zapfen, Spund, Hahn;
Ausgießhahn;
(tool for forming an internal
screw thread) Gewindebohrer;
med Punktion
tap grease Hahnfett
tap water Leitungswasser

tape
  Band (Klebeband/Messband etc.)
  ➤ **adhesive tape**
    Klebeband
  ➤ **autoclave tape/**
    autoclave indicator tape
    Autoklavier-Indikatorband
  ➤ **barricade tape**
    Absperrband, Markierband
  ➤ **cloth tape**
    Gewebeband,
    Textilband (einfach)
  ➤ **duct tape (polycoated cloth tape)**
    Panzerband, Gewebeband,
    Gewebeklebeband,
    Duct Gewebeklebeband,
    Universalband, Vielzweckband
  ➤ **electric tape**
    (insulating tape/friction tape)
    Elektro-Isolierband
  ➤ **filament tape** Filamentband
  ➤ **insulating tape/duct tape**
    Isolierband
  ➤ **label tape** Etikettierband
  ➤ **magnetic tape** Magnetband
  ➤ **masking tape** Kreppband
  ➤ **packaging tape**
    Verpackungsklebeband
  ➤ **sealing tape**
    Dichtungsband
  ➤ **Teflon tape** Teflonband
  ➤ **thread seal tape**
    Gewindeabdichtungsband
  ➤ **warning tape**
    Signalband, Warnband
tape measure
  Messband, Bandmaß
tape rule/tape measure
  Bandmaß, Messband
taper (tapering/tapered)
  zuspitzen (konisch machen);
  spitz zulaufen, sich verjüngen
tapered joint Kegelschliff,
  Kegelschliffverbindung
tare *n*
  (weight of container/packaging)
  Tara (Gewicht des Behälters/
  der Verpackung)

tare *vb* **(determine weight of container/**
  packaging as to substract from
  gross weight: set reading to zero)
  tarieren, austarieren
  (Waage: Gewicht des Behälters/
  Verpackung auf Null stellen)
target *n* Ziel, Soll
  (Plan/Leistung/Produktion);
  (quota) Quote
target *vb* anvisieren,
  anpeilen, ins Auge fassen, planen
target date
  Stichtag, Termin
target group Zielgruppe
tarnish matt machen, trüben,
  mattieren, anlaufen,
  blind machen beschlagen
tartar Weinstein, Tartarus
  (Kaliumsalz der Weinsäure);
  med Zahnstein
tartaric acid (tartrate)
  Weinsäure, Weinsteinsäure (Tartrat)
taste *n* Geschmack
taste *vb* schmecken
taut straff, gespannt, stramm
  ➤ **clamp taut** *vb polym*
    straff einspannen
  taut wire Zugdraht
tautomeric shift
  tautomere Umlagerung
tautomerism Tautomerie
taxidermist
  Präparator, Tierpräparator
teaching laboratory/
  educational laboratory
  Lernlabor
tear *n* Träne
tear *vb* tränen; reißen, zerren; zerreißen
technical technisch
technical inspection agency/authority
  (technical supervisory association)
  Technischer Überwachungsverein
  (TÜV)
technician Techniker
  ➤ **laboratory technician/**
    technical lab assistant
    Laborassistent(in),
    technische(r) Assistent(in)

technique/technic Technik
(einzelnes Verfahren/Arbeitsweise)
technologic(al)
technologisch, technisch
technology
Technik, Technologie (Wissenschaft)
technology assessment
Technikfolgenabschätzung
Teflon tape Teflonband
teichoic acid Teichonsäure
teichuronic acid Teichuronsäure
tellurium (Te) Tellur
temper tempern; (Glas) verspannen, vorspannen, härten
temperate (moderate) gemäßigt
temperature
Temperatur
➢ **ambient temperature**
Umgebungstemperatur
➢ **body temperature**
Körpertemperatur
➢ **cardinal temperature**
Vorzugstemperatur
➢ **ceiling temperature** *polym*
Ceiling-Temperatur
(meist nicht übersetzt),
Gipfeltemperatur
➢ **disintegration temperature/ decomposition temperature**
Zersetzungstemperatur
➢ **fluctuation of temperature**
Temperaturschwankung
➢ **glass-transition temperature** T_g *polym* Glasübergangstemperatur
➢ **ignition point/ kindling temperature/ flame temperature/ flame point**
Zündpunkt, Zündtemperatur, Entzündungstemperatur
➢ **liquidus temperature**
Liquidustemperatur
➢ **melting temperature**
Schmelztemperatur
➢ **operating temperature**
Arbeitstemperatur
➢ **phase transition temperature**
Phasenübergangstemperatur

➢ **room temperature/ ambient temperature**
Raumtemperatur
➢ **sensitivity to temperature**
Temperaturempfindlichkeit
➢ **spontaneous-ignition temperature (SIT)**
Selbstenzündungstemperatur
temperature controller
Temperaturregler
temperature-dependent
temperaturabhängig
temperature gradient
Temperaturgradient
temperature-gradient apparatus *ecol*
Temperaturorgel
temperature gradient gel electrophoresis
Temperaturgradienten-Gelelektrophorese
temperature sensor
Temperaturfühler
tempered *tech/metal* gehärtet
tempered glass/ resistance glass
Hartglas
tempering beaker (jacketed beaker)/ cooling beaker/ chilling beaker
Temperierbecher, Becher(glas) mit Temperiermantel
template Matrize; Schablone
temporary arrangement
Übergangsregelung, Übergangslösung
temporary storage/interim storage
Zwischenlager
temporary worker (aid/helper/employee/personnel)
Aushilfe, Hilfspersonal
tenacious zäh, hartnäckig; klebrig; reißfest, zugfest
tenacity
Zähigkeit; Klebrigkeit; Reißfestigkeit, Zugfestigkeit
tender/fragile
empfindlich, zerbrechlich

tensile strength
Zugfestigkeit, Zerreißfestigkeit,
Reißfestigkeit (Holz)
tension Spannung; Druck, Spannkraft,
Zugkraft; (suction/pull) Zug, Sog
(z.B. in der physiol. Wasserleitung)
tension spring Zugfeder, Spannfeder
tensioning tool/
  tensioning gun
  (cable ties/wrap-it-ties)
  Spannzange
  (für Kabelbinder/Spannband)
teratogenesis/teratogeny
  Teratogenese,
  Missbildungsentstehung
teratogenic teratogen,
  Missbildungen verursachend
teratology Teratologie
  (Lehre von Missbildungen)
term Ausdruck, Fachausdruck,
  Begriff; Zeit, Dauer;
  (of a contract) Laufzeit (eines
  Vertrages), Termin
terminal *adj/adv* **(terminate)**
  end..., letzt;
  begrenzend, endständig
terminal *n* **electr** Pol (Plus~;Minus~),
  Anschlussklemme,
  Klemmschraube,
  Endstecker, Kabelschuh
  ➢ **luster terminal**
    (insulating screw joint)
    Lüsterklemme
terminal amplifier
  Endverstärker
terminal voltage/terminal potential
  Klemmspannung
terminology
  Fachsprache, Fachterminologie,
  Fachbezeichnungen, Terminologie
terminus Terminus,
  Ende (Molekülende)
territory/range Gebiet, Revier,
  Bereich, Wohnbezirk, Territorium
tertiary structure (proteins)
  Tertiärstruktur
test *n* Test, Prüfung,
  Untersuchung, Probe, Messung

test *vb* testen, prüfen, messen
test data Prüfdaten
test medium Testmedium,
  Prüfmedium (zur Diagnose)
test procedure/testing procedure
  Testverfahren
test results (of an investigation)
  Ermittlungsergebnisse
test run Trockenlauf, Probelauf
test solution (solution to be analyzed)
  Untersuchungslösung, Prüflösung
test tube (glass tube/assay tube)
  Reagensglas
test-tube brush
  Reagensglasbürste
test-tube holder Reagensglashalter
test-tube rack
  Reagensglasständer,
  Reagensglasgestell
testability Prüfbarkeit
testcross *gen* Testkreuzung
  ➢ **three-point testcross**
    Drei-Faktor-Kreuzung
tester/testing device/
  checking instrument
  Prüfgerät, Prüfer, Testvorrichtung;
  gen Testpartner
testing device
  Prüfgerät; Prüfmittel
testing equipment/apparatus
  Untersuchungsgerät
testing procedure/audit procedure
  Prüfverfahren
tether binden, anbinden,
  zusammenbinden
tetrahedral tetraedrisch, vierflächig
tetrahedral intermediate
  tetraedrisches Zwischenprodukt
tetravalent vierwertig
textile fiber Textilfaser
textile finishing Textilveredlung
thaw auftauen
thawing Auftauen
theoretic/theoretical theoretisch
theoretical physics
  Theoretische Physik
theoretical plates *dist/chromat*
  theoretische Böden

theory Theorie
thermal analysis
 Thermoanalyse,
 thermische Analyse
thermal conductance (C)
 Wärmedurchgangszahl
thermal conductivity detector (TCD)
 Wärmeleitfähigkeitsdetektor,
 Wärmeleitfähigkeitsmesszelle
 (WLD)
thermal efficiency
 Wärmewirkungsgrad,
 thermischer Wirkungsgrad
thermal radiation
 Wärmestrahlung
thermic thermisch,
 Wärme..., Hitze...
thermobalance Thermowaage
thermocouple Thermoelement
thermocouple probe
 Thermoelementsonde
thermodynamics
 Thermodynamik
➤ **first/second/third law of thermodynamics**
 1./2./3. Hauptsatz
 (der Thermodynamik)
thermogravimetry (TG)
 (=thermogravimetric analysis)
 Thermogravimetrie (TG)
 (=thermogravimetrische Analyse)
thermolysis Thermolyse
thermometer
 Thermometer
➤ **bimetallic thermometer**
 Bimetallthermometer
➤ **gas thermometer**
 Gasthermometer
➤ **mercury-in-glass thermometer**
 Quecksilberthermometer
➤ **noise thermometer**
 Rauschthermometer
➤ **pyrometer**
 Pyrometer, Hitzemessgerät
➤ **quartz thermometer**
 Quarzthermometer
➤ **vapor pressure thermometer**
 Dampfdruckthermometer

➤ **wet-bulb thermometer**
 Nassthermometer,
 Verdunstungsthermometer
thermophilic
 wärmesuchend, thermophil
thermophobic
 hitzemeidend, thermophob
thermopile
 Thermokette, Thermosäule
thermoplastic Thermoplast
thermoregulation Thermoregulation
thermoregulator Wärmeregler
thermos
 Thermoskanne, Thermosflasche
thermospray Thermospray
thermostat Thermostat
thermowell (for thermocouples)
 Thermoelement-Schutzrohr,
 Thermohülse
thicken eindicken, verdicken;
 verdichten, verstärken
thickener Dickungsmittel,
 Verdickungsmittel, Eindicker
thickening Verdickung,
 Eindickung; Eindickmittel
**thief/thief tube/
 sampling tube (pipet)**
 Stechheber
thimble Fingerhut
thin *vb* ausdünnen
thin section/microsection
 Dünnschnitt
thin-layer chromatography (TLC)
 Dünnschichtchromatographie (DC)
thinner Verdünner, Verdünnungsmittel
thinning Ausdünnen, Ausdünnung
thiocarbonic acids
 Thiocarbonsäuren
thiocyanic acid Thiocyansäure
thiourea Thioharnstoff
**thistle tube funnel/
 thistle top funnel tube**
 Glockentrichter
 (Fülltrichter für Dialyse)
thoracic respiration/costal breathing
 Brustatmung, Thorakalatmung
thoroughness
 Gründlichkeit; Vollkommenheit

thread Faden;
tech/mech Gewinde
(Schrauben, Bolzen etc.)
➤ **British Standard Pipe (BSP) thread/ fittings**
Britisches Standard Gewinde
➤ **external thread/male thread**
Außengewinde
➤ **internal thread/female thread**
Innengewinde
➤ **National Pipe Thread (NPT)**
NPT-Gewinde
(U.S. Standard Gewinde: in Zoll)
➤ **screw thread**
Schraubgewinde, Schraubengewinde
➤ **Unified Fine Thread (UNF)**
UNF-Feingewinde
thread seal tape
Gewindeabdichtungsband
threaded socket (connector/nozzle)
Gewindestutzen
threaded top
Schraubgewindeverschluss
threat/ endangerment Bedrohung
three-dimensional structure/ spatial structure
Raumstruktur, räumliche Struktur
three-finger clamp
Dreifinger-Klemme
three-neck flask
Dreihalskolben
three-prong ... Dreizack...
three-way cock/ T-cock/three-way tap
Dreiweghahn, Dreiwegehahn
three-way connection
Dreiwegverbindung
threshold Schwelle
(z.B. Reizschwelle/ Geschmacksschwelle etc.)
threshold concentration
Schwellenkonzentration
threshold current Schwellenstrom
threshold effect Schwelleneffekt

threshold limit value (TLV) (US: by ACGIH)
maximale Arbeitsplatzkonzentration
(aber nicht identisch mit DFG: MAK)
threshold potential (firing level)
Schwellenpotential
(kritisches Membranpotential)
threshold value
Schwellenwert
thrive/flourish gedeihen, florieren
throttle *n* **(choke)** Drossel
throttle *vb*
(choke/slow down/dampen)
drosseln, herunterfahren, dämpfen
throttle valve/damper
Drosselventil, Drosselklappe
throughput
Durchsatz, Durchsatzmenge;
electr Durchgang
throughput rate
Durchsatzrate
throwaway society
Wegwerfgesellschaft
thrust Vortrieb, Anschub; *aer* Schub
➤ **forward thrust**
Schubkraft, Vortriebkraft
thumbscrew
Daumenschraube, Flügelschraube
tidal volume Atemzugvolumen
tight dicht, fest, eng;
unbeweglich, festsitzend;
(tightly closed/sealed tight)
fest verschlossen
tightness Dichtigkeit
tile Fliese; Kachel
➤ **floor tile** Bodenfliese
tiled gefliest (mit Fliesen ausgelegt); gekachelt
tiled floor/tiling
Fliesenfußboden
timber Nutzholz, Bauholz, Schnittholz; Nutzholzbäume
timber industry
holzverarbeitende Industrie
time averaging
Time-averaging, Zeitmittlung

time frame
zeitlicher Rahmen
time-of-flight mass spectrometry (TOF-MS)
Flugzeit-Massenspektrometrie (FMS)
time-resolved zeitaufgelöst
time-weighted average
zeitgewichtetes Mittel
timer Zeitschaltuhr, Zeitschalter, Schaltuhr
timekeeping Zeitmessung, Zeitkontrolle, Zeitnahme
timetable Zeitplan, Fahrplan, Programm
tin (Sn) Zinn; Weißblech; (*Br*) Blechdose
tincture Tinktur
tinfoil (aluminum foil)
Stanniol (Aluminiumfolie/Alufolie)
tingibility Anfärbbarkeit
tint Farbe, Farbton, Tönung, Schattierung
tinware
Weißblechwaren
tip over
umstoßen, umkippen, umwerfen
tissue
Gewebe, Stoff; Taschentuch
➢ **cleansing tissue**
Reinigungstuch (Papier)
➢ **kitchen tissue (kitchen paper towels)**
Küchenrolle, Haushaltsrolle, Tücherrolle, Küchentücher, Haushaltstücher
➢ **lens tissue** *micros*
Linsenpapier, Linsenreinigungspapier
➢ **paper tissue**
Papiertaschentuch
tissue culture
Gewebekultur
tissue culture flask
Gewebekulturflasche, Zellkulturflasche

tissue forceps
Gewebepinzette
tissue paper
Seidenpapier; Papierhandtuch
titanium (Ti) Titan
titer Titer
titrant Titrationsmittel, Titrant
titrate titrieren
titration Titration
➢ **acid-base titration**
Säure-Basen-Titration, Neutralisationstitration
➢ **amperometric titration**
amperometrische Titration, Amperometrie
➢ **back titration** Rücktitration
➢ **conductometric titration**
konduktometrische Titration, Konduktometrie, Leitfähigkeitstitration
➢ **coulometric titration**
coulometrische Titration, Coulometrie
➢ **end-point dilution technique**
Endpunktverdünnungsmethode (Virustitration)
➢ **flow-injection titration**
Fließinjektions-Titration
➢ **oscillometry/ high-frequency titration**
Oszillometrie, oszillometrische Titration, Hochfrequenztitration
➢ **photometric titration**
photometrische Titration
➢ **precipitation titration**
Fällungstitration
➢ **redox titration**
Redoxtitration
titration curve
Titrationskurve
titration error
Titrationsfehler
titrimetry
Titrimetrie, titrimetrische Analyse (*see* volumetric analysis)
TLC (thin layer chromatography)
DC (Dünnschichtchromatographie)

toggle switch/rocker
    Kippschalter, Hebelschalter
toilet paper
    Toilettenpapier, Klopapier
toiletry Toilettenartikel
tolerance Toleranz,
    Wiederstandsfähigkeit;
    Verträglichkeit; Fehlergrenze,
    zulässige Abweichung, Spielraum
tolerance dose
    Toleranzdosis, zulässige Dosis
tolerance limit Toleranzgrenze
tolerance range/tolerane interval
    Toleranzbereich, Toleranzintervall
tomography Tomographie
tone Tönung, Schattierung;
    med Tonus; Farbgebung;
    (sound) Ton, Klang
tongs Haltezange (Laborzange)
> **beaker tongs** Becherglaszange
> **crucible tongs** Tiegelzange
> **flask tongs** Kolbenzange
tongue depressor
    Mundspatel, Zungenspatel
tonicity Spannkraft
tool (tools) Werkzeug(e)
tool box/tool kit Werkzeugkasten
toolmaker Werkzeugmacher
toothpick Zahnstocher
top-fermenting obergärig
    (Fermentation: Bier)
top up/off bis zum Rand auffüllen
topical topisch, örtlich, lokal
topogenic/topogenous topogen
topographic mapping
    Geländekartierung
topographic survey
    Geländeaufnahme
torch Fackel; *(Br)* Taschenlampe
torpor (hibernation) Torpor, Starre
    (Kältestarre, Winterstarre)
torque wrench
    (torque amplifier handle)
    Drehmomentschlüssel
torsion Drehung, Torsion
total body irradiation
    Ganzkörperbestrahlung
total dose Gesamtdosis

total hardness
    Gesamthärte (Wasser)
total magnification/
    overall magnification *micros*
    Gesamtvergrößerung
total static head (pump)
    Gesamtförderhöhe
tote *n* Last, Traglast, Ladung;
    Tragen, Schleppen
tote *vb* tragen, laden, schleppen,
    mit sich herumschleppen;
    transportieren
tote bag
    Einkaufstasche; Reisetasche
tote box
    Transportkiste, Transportbehälter
tote tray
    Werkstückkasten, Teilekasten
touch/contact berühren
touchstone Probierstein
tough/rigid
    zäh, hart, widerstandsfähig
toughened (glass) gehärtet
toughness Zähigkeit,
    Härte, Robustheit
tourniquet Abschnürbinde,
    Binde, Aderpresse, Tourniquet
towel Handtuch
> **paper towel**
    Papierhandtuch
towel rack
    Handtuchhalter, Handtuchständer
toxic (T)/poisonous giftig (toxisch)
> **cytotoxic**
    cytotoxisch, zellschädigend
> **embryotoxic** embryotoxisch
> **extremely toxic (T+)** sehr giftig
> **highly toxic** hochgiftig
> **fetotoxic** fetotoxisch
> **hepatotoxic**
    leberschädigend, hepatotoxisch
> **moderately toxic** mindergiftig
> **neurotoxic** neurotoxisch
> **phytotoxic**
    pflanzenschädlich, phytotoxisch
> **toxic to reproduction**
    fortpflanzungsgefährdend,
    reproduktionstoxisch

toxic agent Giftstoff
Toxic Substances Control Act (TSCA)
U.S. Gesetz zur Kontrolle toxischer Substanzen (Gefahrstoffe)
toxic waste/poisonous waste Giftmüll
toxicity/poisonousness Toxizität, Giftigkeit
toxicology Toxikologie
toxin Toxin, Gift
➢ **suspected toxin** Verdachtsstoff
toxoid vaccine Toxoidimpfstoff, Toxoidvakzine
trace *n* **(remainder/remains)** Spur, Überrest (meist *pl* Überreste)
trace *vb* verfolgen, nachspühren, ausfindig machen, auffinden; pausen, durchpausen
trace analysis Spurenanalyse
trace element/ microelement/micronutrient Spurenelement, Mikroelement
traceability Rückführbarkeit, Rückverfolgbarkeit
tracer Tracer; Indikator; Leit...
tracer enzyme Leitenzym
tracer nuclide Leitnuklid
track *n* Gleis, Pfad, Spur, Weg, Fährte
track vb nachgehen, nachspühren, folgen, verfolgen
tracing paper Pauspapier
trackability Rückverfolgbarkeit
tracking dye *electrophor* Farbmarker
trade *adj/adv* **(commercial)** **(commonly available)** handelsüblich
trade *vb* tauschen, austauschen; handeln, Handel treiben
trade *n* **(business/occupation)** Handel, Gewerbe
➢ **retail trade** Einzelhandel
➢ **wholesale trade/ wholesale business** Großhandel
trade & industrial supervision (federal agency) Gewerbeaufsicht (staatl. Behörde)

trade cooperative association Berufsgenossenschaft
trade name Warenzeichen, Markenbezeichnung
trade secret Betriebsgeheimnis, Geschäftsgeheimnis
trademark Warenzeichen
➢ **registered trademark** eingetragenes Warenzeichen
training Schulung, Fortbildung
training period Einarbeitungsphase
trait (characteristic/feature) Merkmal; (character) Charakterzug, Eigenschaft
transamination Transaminierung
transducer Wandler, Umwandler
transect (cut through) durchschneiden
transection Durchschnitt (schneiden)
transfer *n* Transfer, Übertragung, Überführung
transfer *vb* transferieren, übertragen; überführen; umfüllen
transfer loop Transferöse
transfer pipet/volumetric pipet Vollpipette, volumetrische Pipette
transferability Übertragbarkeit
transform transformieren, umwandeln
transformation Transformation, Umwandlung
transformation series Transformationsreihe
transformer *electr* Transformator, Trafo, Umwandler
transfuse *med* Blut übertragen, eine Transfusion/Blutübertragung machen
transillumination (transmitted light illumination) Durchlicht, Durchlichtbeleuchtung
transite board (lab bench) Asbestzementplatte (Labortisch)
transition Übergang; (developmental transition) Entwicklungsübergang
transition metal Übergangsmetall (Nebengruppenmetall)
transition phase Übergangsphase
transition state Übergangszustand (Enzymkinetik)

translucent
 (transparent) lichtdurchlässig;
 (pellucid) durchscheinend
transmissible
 (communicable) übertragbar;
 (heritable) vererbbar
transmissible disease/
communicable disease
 übertragbare Krankheit
transmission (transfer) Übertragung
 (z.B. Krankheit); (spreading) *med*
 Verschleppung, Übertragung;
 (of gearing) Getriebe (Motor)
transmission electron microscopy
(TEM) Transmissionselektronenmikroskopie, Durchstrahlungselektronenmikroskopie
transmission of signals/
impulse propagation
 Erregungsleitung
transmit *med/tech/mech*
 (e.g., a disease) übertragen;
 (pass on) vererben
transmitter Überträger,
 Überträgerstoff, Transmitter
transmitter of disease
 Krankheitsüberträger
transparent lichtdurchlässig;
 (pellucid) durchscheinend
transplant/graft *n* Transplantat
transplant/graft *vb med*
 transplantieren, verpflanzen;
 (replant) versetzen, umpflanzen
transplantation
 Verpflanzung, Transplantation
transport *vb* transportieren
transport *n* (transportation) Transport;
 (shipment) Beförderung, Transport
➢ **coupled transport/co-transport**
 gekoppelter Transport
transport of dangerous goods/
transport of hazardous materials
 Gefahrguttransport
transport vehicle Transportfahrzeug
transverse section/cross section
 Hirnschnitt, Querschnitt
trap *n* Falle; *electr* Sperrkreis
trap *vb* fangen, einfangen

trash (*see also:* **waste**)
 Müll, Abfall
➢ **household trash**
 Haushaltsmüll,
 Haushaltsabfälle
trash bag/waste bag
 Müllbeutel, Müllsack
trash can/waste container/litter bin
 Abfallbehälter
trash compactor Abfallpresse
tray Schale, Flachbehälter; Tablett
tray reactor Gärtassenreaktor
treat (treated)
 behandeln (behandelt)
➢ **untreated** unbehandelt
trial Versuch, Probe, Prüfung
trial run ('experimental experiment')
 Probelauf
triangle Dreieck
➢ **clay triangle/pipe clay triangle**
 Tondreieck, Drahtdreieck
tributary *adj* zufließend
 (Rohre/Zuleitungen)
trickle rieseln, tröpfeln
trickling filter (sewage treatment)
 Tropfkörper
 (Tropfkörperreaktor,
 Rieselfilmreaktor)
trickling filter reactor
 Rieselfilmreaktor,
 Tropfkörperreaktor
trigger *n* Auslöser, Drücker; Zünder
trigger *vb* **(elicitate)**
 auslösen (z.B. eine Reaktion)
trigger reaction Auslösereaktion
trigger switch Kippschalter,
 Kipphebelschalter
trigger threshold *med*
 Auslöseschwelle
triggering (elicitation)
 Auslösung (Reaktion)
trim *micros* anspitzen
trimming block *micros* Trimmblock
trimming shears Trimmschere
trinocular head *micros*
 Trinokularaufsatz, Tritubus
trip (fuse/circuit breaker) rausfliegen,
 durchbrennen (Sicherung auslösen)

triple bond Dreifachbindung
triple nosepiece *micros*
  Dreifachrevolver
triple point
  Tripelpunkt, Dreiphasenpunkt
triple-pole dreipolig, Dreipol...
tripod Dreifuß, Dreibein (Stativ)
triturate reiben, zerreiben,
  zermahlen (im Mörser)
trituration Zerreiben,
  Zermahlen (im Mörser)
trivalency Dreiwertigkeit
trivalent dreiwertig
trivet Dreifuß, Untersatz
  (kurzfüßiger Untersetzer zum
  Abstellen von heißen Gefäßen)
troubleshooting Fehlersuche
trough Trog, Wanne
trough-shaped
  wannenförmig, muldenförmig
trowel Kelle, Spachtel
trueness (quality control)
  Richtigkeit (Qualitätskontrolle)
try/attempt probieren, versuchen
tub Wanne, Zuber, Fass, Waschbottich;
  (bathtub) Badewanne
tube Tube; (hose/tubing) Schlauch;
  Rohr, Röhre, Röhrchen;
  (body tube) *micros* Tubus;
  (draw tube) Steckhülse für Okular
➢ **bomb tube/Carius tube/sealing tube**
  Bombenrohr, Schießrohr,
  Einschlussrohr
➢ **bubble tube**
  (slightly bowed glass tube/
  vial in spirit level)
  Libelle (Glasröhrchen der
  Wasserwaage)
➢ **capillary tube** Kapillarrohr
➢ **centrifuge tube**
  Zentrifugenröhrchen
➢ **combustion tube** Glühröhrchen
➢ **culture tube** Kulturröhrchen
➢ **dip tube** Steigrohr
➢ **drift tube (TOF-MS)**
  Driftröhre
➢ **drying tube**
  Trockenrohr, Trockenröhrchen

➢ **ebullition tube**
  Siederöhrchen
➢ **feed tube** Zulaufschlauch
➢ **fermentation tube (bubbler)**
  Gärröhrchen,
  Einhorn-Kölbchen
➢ **fluorescent tube**
  Leuchtstoffröhre,
  Leuchtstofflampe ('Neonröhre')
➢ **funnel tube** Trichterrohr
➢ **fusion tube/melting tube**
  Abschmelzrohr
➢ **gas sampling tube**
  Gassammelrohr
➢ **gas-discharge tube**
  Gasentladungsröhre
➢ **glass tube**
  Glasrohr, Glasröhre, Glasröhrchen
➢ **guard tube**
  Sicherheitsrohr (Laborglas)
➢ **ignition tube**
  Zündröhrchen, Glühröhrchen
➢ **incinerating tube**
  Verbrennungsrohr (Glas)
➢ **rubber tube** Gummischlauch
➢ **spectrophotometer tube (cuvette)**
  Küvette (für Spektrometer)
➢ **tablet tube** Tablettenröhrchen
➢ **test tube**
  (glass tube/assay tube)
  Reagensglas
➢ **thief tube/sampling tube (pipet)**
  Stechheber
➢ **thistle tube/**
  thistle tube funnel/
  thistle top funnel tube
  Glockentrichter (Fülltrichter für
  Dialyse)
➢ **tubule** Röhrchen
➢ **vacuum tube** Vakuumröhre
tube brush (test tube brush)
  Reagensglasbürste
tube clip
  Schlauchschelle
tube furnace Rohrofen
tuberculosis Tuberkulose
tuberculous tuberkulös
tuberous/tuberal tuberös

tubing Rohr, Schlauch, Röhrenmaterial, Rohrleitung, Rohrstück
➤ **capillary tubing** Kapillarrohr
➤ **glass tubing** Glasrohr, Glasröhre, Glasröhrchen
➤ **high-pressure tubing** Hochdruckschlauch
➤ **pressure tubing** Druckschlauch
➤ **rubber tubing** Gummischlauch
tubing attachment socket Schlauchtülle (z.B. am Gasreduzierventil)
tubing clamp/ pinch clamp/pinchcock clamp Schlauchklemme, Quetschhahn
tubing closure *dial* Schlauchverschlussklemme
tubing connection/tube coupling Schlauchkupplung
tubing connector (for connecting tubes)/ tube coupling/fittings Schlauchverbinder, Schlauchverbindung(en)
tubing pinch valve Schlauchventil (Klemmventil)
tubing pump Schlauchpumpe
tubular tubulär
tubular loop reactor Rohrschlaufenreaktor
tubule Röhrchen
Tullgren funnel *ecol* Tullgren-Apparat
tumble/sway/stagger taumeln, torkeln
tumbler (lever) Kipphebel; (tumbling mixer) Fallmischer
tumbler switch/knife switch Kipphebelschalter
tumefacient anschwellend, eine Schwellung verursachend
tumescent geschwollen
tumor (*Br* tumour) Tumor, Wucherung, Geschwulst
tumorous tumorartig, tumorös
tungsten (W) Wolfram
tungstic acid Wolframsäure

tunneling microscopy Tunnelmikroskopie
turbid trübe
turbidimetry Turbidimetrie, Trübungsmessung
turbidity Trübheit, Trübung
turbine Turbine
turbine impeller/turbine stirrer Turbinenrührer
turbulence Turbulenz, Wirbel, Verwirbelung
turbulent flow turbulente Strömung
turgescent prall, schwellend, turgeszent
turgid/swollen (swell) geschwollen (schwellen)
turgidity Turgidität, Geschwollenheit, Schwellungsgrad
turgor (hydrostatic pressure) Turgor (hydrostatischer Druck)
turgor pressure Turgordruck
turn off/switch off/shut off ausschalten, abschalten (computer: power down) herunterfahren
turn on/switch on einschalten, anschalten (computer: power up) hochfahren
turn over *vb* **(become oxygen-deficient/ turn anaerobic)** umkippen (Gewässer)
turn-key delivery schlüsselfertige Lieferung
turnover *n* Umsatz, Umwandlung
turnover number Wechselzahl k_{cat} (katalytische Aktivität)
turnover period Umsatzzeit
turnover rate/rate of turnover Umsatzgeschwindigkeit, Umsatzrate
turntable Drehplatte (z.B. des Mikrowelle-Ofens); Plattenteller
turpentine Terpentin
turret Revolver; (nosepiece turret) *micros* Objektivrevolver
tutor Tutor, Studienleiter; (teaching assistant) Lehrassistent
tutorial Tutorium, Tutorenkurs

tweezers (see also: forceps)
Pinzette
➢ **dissection tweezers**
Präparierpinzette, Sezierpinzette, anatomische Pinzette
➢ **high-precision tweezers**
Präzisionspinzette
➢ **reverse-action tweezers (self-locking tweezers)**
Umkehrpinzette, Klemmpinzette
➢ **sharp-point tweezers/ sharp-pointed tweezers**
Spitzpinzette
➢ **specimen tweezers**
Probennahmepinzette
twin crystals Zwillingskristalle, Doppelkristall, Bikristall
twin electrons Elektronenpaar
twine *n* Zwirn, Garn, starker Bindfaden; Wickelung, Windung; Knäuel
twine *vb*
schlingen, winden, umschlingen

twist *n* Drehung, Verdrehung, Biegung, Krümmung; Drall; (coil/spiral: a series of loops) Windung, Torsion, Spirale
twist *vb* drehen, verdrehen; abschrauben; verbiegen, verkrümmen
twist-grip
Drehgriff
two-neck flask
Zweihalskolben
two-stage impeller
zweistufiger Rührer
two-way cock
Zweiweghahn, Zweiwegehahn
two-way intercom/ two-way radio
Wechselsprechanlage, Gegensprechanlage
type Typus, Standard
typo (typographical error)
Schreibfehler, Druckfehler (typografischer Fehler)

ubiquitous
  (widespread/existing everywhere)
  ubiqitär
  (weitverbreitet/überall verbreitet)
ultracentrifugation
  Ultrazentrifugation
ultracentrifuge Ultrazentrifuge
ultrafiltration Ultrafiltration
ultrahigh vacuum
  Ultrahochvakuum,
  Höchstvakuum
ultramicrotome Ultramikrotom
ultrashort wave Ultrakurzwelle
ultrasonic Ultraschall...,
  den Ultraschall betreffend
ultrasound/ultrasonics Ultraschall
➢ **ultrasonography/sonography**
  Ultraschalldiagnose, Sonographie
ultrastructure Ultrastruktur
ultrathin section *micros*
  Ultradünnschnitt
**ultraviolet spectroscopy/
  UV spectroscopy**
  UV-Spektroskopie
unbalance/unbalanced state
  Unwucht
unbiased *math/stat*
  unverzerrt, unverfälscht
unbiased error Zufallsfehler
unblock (drain)
  frei machen (Abfluss)
unbranched (chain) *chem*
  unverzweigt (Kette)
unbreakable unzerbrechlich
uncertainty Unbestimmtheit,
  Unsicherheit, Ungewissheit;
  phys Unschärfe
uncharged
  ungeladen, ladungsfrei
unconformity
  Abweichung,
  Nichtübereinstimmung
unconscious/unknowing(ly)
  unbewusst
unconsciousness Bewusstlosigkeit
uncontaminated unverschmutzt
uncontrolled
  unkontrolliert, ungesteuert

uncoupler/uncoupling agent
  Entkoppler
undamped ungedämpft
undemanding/modest
  (having low requirements or
  demands)
  anspruchslos
undercool unterkühlen
undercooled liquid
  unterkühlte Flüssigkeit
undernourished (*siehe:* fehlernährt)
  unterernährt
undernourishment Unterernährung
undersaturation
  Untersättigung, Sättigungsdefizit
underside/undersurface
  Unterseite
undersize (sieve)
  Unterkorn (Siebdurchgang)
undersurface Unterseite
undetectable
  nicht feststellbar,
  nicht nachweisbar
undissolved ungelöst
undivided (not divided) ungeteilt
uneatable/inedible
  ungenießbar, unessbar
unequal (different/nonidentical)
  ungleich, nicht identisch, anders
unfertilized unbefruchtet
uniform *adj/adv*
  einheitlich, gleichförmig
uniform rules (standards)
  einheitliche Richtlinie(n)
uniformity Uniformität,
  Gleichförmigkeit, Gleichmäßigkeit
unilateral einseitig, unilateral
unimolecular (monomolecular)
  unimolekular (monomolekular)
unit (measure) Einheit (Maßeinheit);
  (branch) Bereich, Abteilung
unit cell Elementarzelle
unit factor unteilbarer Faktor
unit library Bereichsbibliothek
unit operation Grundoperation
  (Verfahrenstechnik)
unit process Grundverfahren
  (Verfahrenstechnik)

univalence *chem*
Univalenz, Einwertigkeit
univalent/monovalent
univalent, monovalent,
einwertig
unnatural unnatürlich
unpalatable
ungenießbar,
nicht schmackhaft
unpleasant smell
unangenehmer Geruch
unplug/disconnect
usstöpseln,
Stecker herausziehen
unpolluted/
uncontaminated
unverschmutzt
unprovable
nicht nachweisbar,
unbeweisbar
unripe/immature unreif
unsafe unsicher, gefährlich
unsaturated ungesättigt
unsaturation
ungesättigter Zustand
unstable (instable)
instabil, nicht stabil
untreated unbehandelt
up-regulation Heraufregulation
upper phase
Oberphase (flüssig-flüssig)
upperside/upper surface
Oberseite
upstairs die Treppe hoch
uptake/intake/ingestion
Aufnahme, Einnahme
urea (ureide)
Harnstoff (Ureid)
uric acid (urate)
Harnsäure (Urat)
uridylic acid Uridylsäure
urinate/micturate
urinieren, harnen,
harnlassen, miktuieren
urination/micturition
Harnen, Harnlassen,
Urinieren, Miktion

urine
Urin, Harn
➢ **glomerular ultrafiltrate**
Primärharn,
Glomerulusfiltrat
➢ **secondary urine**
Sekundärharn
urocanic acid (urocaninate)
Urocaninsäure (Urocaninat),
Imidazol-4-acrylsäure
uronic acid (urate)
Uronsäure (Urat)
use (usage)
Verwendung, Nutzen
useful energy
Nutzenergie,
nutzbare Energie
user Benutzer, Nutzer;
(consumer) Verbraucher
user-friendly (easy to use)
benutzerfreundlich;
anwenderfreundlich;
bedienungsfreundlich
usnic acid Usninsäure
utensil Utensil, Gerät;
Gebrauchsgegenstand
utilities
Versorgungseinrichtungen;
(public utilities) Leistungen
der öffentlichen
Versorgungsunternehmen:
Gas, Wasser, Strom
utility Nutzen, Nützlichkeit;
nützliche Sache/Einrichtung;
vielseitig verwendbar,
Mehrzweck...
utility company
öffentliches
Versorgungsunternehmen:
Gas, Wasser, Strom
utility pliers
Mehrzweckzange
utilization/use
Verwendung, Nutzung, Ausnutzung;
Verwertung
utilize/use verwenden, nutzen;
metabol/ecol verwerten

vacany Leere, Leerstehen; Vakanz, freie/offene Stelle (Arbeitsstelle)
vacant leer, unbesetzt, offen, frei
vaccinate/immunize impfen, immunisieren
vaccination Vakzination, Vakzinierung, Impfung; (immunization) Immunisierung
vaccine Vakzine, Impfstoff
vacuum Vakuum, Luftleere
vacuum-clean staubsaugen
vacuum cleaner/vacuum Staubsauger
vacuum concentrator/speedy vac Vakuum-Evaporator
vacuum distillation/ reduced-pressure distillation Vakuumdestillation
vacuum filtration/suction filtration Vakuumfiltration
vacuum furnace Vakuumofen
vacuum gauge Vakuummesser (Messgerät)
vacuum line Vakuumleitung
vacuum manifold Vakuumverteiler (mit Hähnen)
vacuum-metallize *micros* aufdampfen, bedampfen
vacuum-proof vakuumfest
vacuum pump Vakuumpumpe
vacuum trap Vakuumfalle
valence/valency Valenz, Wertigkeit
➤ **divalent** zweiwertig, divalent, bivalent
➤ **polyvalent** mehrwertig, polyvalent
➤ **trivalent** dreiwertig, trivalent
valence electron/valency electron Valenzelektron
valeric acid/pentanoic acid (valeriate/pentanoate) Valeriansäure, Baldriansäure, Pentansäure (Valeriat/Pentanat)
valid gültig, bestätigt, richtig; wirksam
validate bestätigen

validation Validierung, Gültigkeit, Gültigkeitserklärung
validity Gültigkeit, Richtigkeit, Stichhaltigkeit
valve (vent) Ventil
➤ **air inlet valve/air bleed** Lufteinlassventil
➤ **backflow prevention/ backstop (valve)** Rückflusssperre, Rücklaufsperre, Rückstauventil
➤ **backstop valve/check valve** Rückschlagventil
➤ **ball valve** Kugelventil
➤ **butterfly valve** Flügelhahnventil
➤ **check valve/backstop valve** Rückschlagventil, Sperrventil
➤ **control valve** Regelventil, Kontrollventil
➤ **cut-off valve** Schlussventil
➤ **delivery valve** Zulaufventil, Beschickungsventil
➤ **diaphragm valve** Membranventil
➤ **drum vent** Fassventil (Entlüftung)
➤ **exhalation valve** Ausatemventil (an Atemschutzgerät)
➤ **injection valve/syringe port** Einspritzventil
➤ **inlet valve** Einlassventil
➤ **limit valve** Begrenzungsventil
➤ **metering valve** Dosierventil
➤ **needle valve** Nadelventil, Nadelreduzierventil (Gasflasche)
➤ **pilot-operated (valve)** hydraulisch vorgesteuert (Ventil)
➤ **pinch valve** Quetschventil
➤ **plug valve** Auslaufventil
➤ **pneumatic valve** Druckluftventil
➤ **positive-displacement valve** Verdrängerventil
➤ **pressure control valve** Druckregelventil

- **pressure-relief valve/
  gas regulator**
  Reduzierventil,
  Druckreduzierventil,
  Druckminderventil,
  Druckminderungsventil
  (für Gasflaschen)
- **pressure valve/pressure relief valve**
  Überdruckventil
- **proportional valve/P valve**
  Proportionalventil
- **purge valve/
  pressure-compensation valve**
  Entlüftungsventil
- **relief valve
  (pressure-maintaining valve)**
  Ausgleichsventil
- **safety valve** Sicherheitsventil
- **security valve/security relief valve**
  Sicherheitsventil
- **shut-off valve**
  Abschaltventil, Absperrventil
- **slide valve**
  Schieberventil, Schieber
- **solenoid valve**
  Magnetventil (Zylinderspule)
- **split valve** *chromat* Abzweigventil
- **throttle valve** Drosselventil
- **T-purge (gas purge device)**
  Spülventil (Inertgas)
- **tubing pinch valve**
  Schlauchventil (Klemmventil)

vanadic *adj* Vanadium...
vane *tech/mech* Flügel,
  Schaufel (Propeller/Turbine/Rotor/
  Ventilator etc.)
vane anemometer
  Flügelradanemometer
vane-type pump
  Propellerpumpe
vanillic acid Vanillinsäure
vapor/vapour *(Br)* Dampf
- **water vapor/steam**
  Wasserdampf

vapor bath Dampfbad
vapor blasting *micros*
  Bedampfung,
  Bedampfen, Aufdampfen

vapor cooling
  Verdunstungskühlung,
  Siedekühlung
vapor density
  Dampfdichte
vapor-deposited
  dampfbeschichtet (bedampft)
vapor pressure
  Dampfdruck
vapor pressure thermometer
  Dampfdruckthermometer
vaporization apparatus *micros*
  Bedampfungsanlage
vaporize verdampfen, verdunsten;
  eindampfen, vergasen
vaporizer (water vaporizer) *dist*
  Dampfentwickler,
  Verdampfungsapparat,
  Zerstäuber (Wasserdampfentwickler)
vaporproof/vaportight
  dampfdicht, dampffest
variability Variabilität,
  Veränderlichkeit, Wandelbarkeit
  (*auch:* Verschiedenartigkeit)
variable *adj/adv* **(variably adjustable)**
  stufenlos
  (regulierbar/regelbar/einstellbar etc.)
variable pitch screw impeller
  Schraubenspindelrührer mit
  unterschiedlicher Steigung
variance/mean square deviation *stat*
  Varianz,
  mittlere quadratische Abweichung,
  mittleres Abweichungsquadrat
**variance ratio distribution/
  F-distribution/Fisher distribution**
  Fisher-Verteilung, F-Verteilung,
  Varianzquotientenverteilung
variate variieren, schwanken
variation
  Variation, Schwankung
- **range of variation/
  range of distribution** *stat*
  Variationsbreite

varnish Lasur, Lack,
  Firnis, Lackfirnis
vat/tub
  Bottich, großes Fass

vector Vektor
- **containment vector** Sicherheitsvektor
- **replacement vector** Substitutionsvektor

vegetable oil Pflanzenöl
vegetation/plant life Vegetation
veined/venulous geädert
velcro Klettverschluss (Haken & Flausch)
vending machine Verkaufsautomat
vendor Verkäufer, Händler (Firma/Lieferant)
- **retail vendor** Einzelhändler
- **wholesaler/wholesale vendor** Großhändler

venom Tiergift
venomous giftig (Tiere)
vent *n* Abzugsöffnung, Luftschlitz
vent *vb* entlüften
ventilate/vent/air ventilieren, belüften, entlüften, durchlüften, Rauch abziehen lassen
ventilating pipe/vent pipe Lüftungsrohr
ventilating shaft/ vent shaft/ ventilating duct/ vent duct Lüftungsschacht, Luftschacht
ventilation Lüftung, Ventilation; (air extraction) Entlüftung; (aeration) Belüftung
ventilation system/vent Lüftungsanlage
ventilation volume Ventilationsvolumen
verdigris/Spanish green (cupric subacetate) Grünspan, Verdigris, Spangrün
verification (control) Bestätigung, Überprüfung, Vergewisserung, Kontrolle
verification assay Bestätigungsprüfung
verify/control bestätigen, überprüfen, kontrollieren
vernier Nonius; Feineinsteller
vertical air flow (clean bench with vertical air curtain) vertikale Luftführung (Vertikalflow-Biobench)
vertical flow workstation/hood/unit Fallstrombank
vertical rotor *centrif* Vertikalrotor
very low density lipoprotein (VLDL) Lipoprotein sehr niedriger Dichte
vesicating/vesicant blasentreibend, blasenziehend
vesicle Vesikel *nt*, Bläschen
vesicular/bladderlike vesikulär, bläschenartig
vessel Gefäß; (container) Behälter
- **agitator vessel** Rührkessel, Rührbehälter
- **Dewar vessel** Dewargefäß
- **glass pressure vessel** Druckbehälter aus Glas
- **pressure vessel** Druckbehälter
- **reaction vessel** Reaktionsgefäß
- **receiving vessel (receiver/collection vessel)** Auffanggefäß
- **safety vessel** Sicherheitsbehälter, Sicherheitskanne

vestibule/vestibulum Vestibül
veterinarian/vet Tierarzt, Veterinär
veterinary medicine/ veterinary science Tiermedizin, Tierheilkunde, Veterinärmedizin
viability Lebensfähigkeit
viable lebensfähig
vial Gläschen, Glasfläschchen, Phiole; (tube) Röhrchen
- **crimp-seal vial** Rollrandgläschen, Rollrandflasche
- **scintillation vial** Szintillationsgläschen

vibration Vibration, Schwingung
➤ **deformation vibration/ bending vibration (IR)**
  Deformationsschwingung
➤ **stretching vibration (IR)**
  Streckschwingung
➤ **wagging vibration (IR)**
  Wippschwingung
vibrational motion
  Schwingungsbewegung
vibrational spectrum
  Schwingungsspektrum
victim Opfer
viewing panel
  Beobachtungsfenster
viewing window Sichtfenster
vigorous heftig (Reaktion etc.)
Vigreux column
  Vigreux-Kolonne
vinegar Essig
violation Missachtung,
  Vergehen (einer Vorschrift)
violent heftig, gewaltig, gewaltsam,
viral infection/virosis
  Viruserkrankung, Virose
viremia Virämie
virology Virologie
virosis Virose, Viruserkrankung
virostatic Virostatikum
virtual image *micros*
  virtuelles Bild
virucidal/viricidal viruzid
virulence (disease-evoking power/ ability of cause disease)
  Virulenz, Infektionskraft,
  Ansteckungskraft
virulent virulent
virus (*pl* viruses) Virus (*pl* Viren)
viscosity/viscousness
  Viskosität, Dickflüssigkeit,
  Zähflüssigkeit
viscous/viscid
  (glutinous consistency)
  viskos, viskös,
  dickflüssig, zähflüssig
vise/vice (*Br*) Schraubstock
Vise-Grip® pliers
  Gripzange, Haltezange

visible sichtbar
➤ **invisible** unsichtbar
vision/sight/eyesight
  Sicht; Sehvermögen; Gesichtssinn
➤ **range of vision/ visual distance**
  Sehweite
➤ **strength of vision** Sehstärke
visor/vizor (*Br*) Schirm, Blende
  (Sicht~), Sichtschutz, Visier;
  (face visor) Gesichtsschutz,
  Sichtschutz
visual acuity Sehschärfe
vital vital, kraftvoll; Lebens...;
  lebenswichtig; wesentlich,
  grundlegend, entscheidend
vital capacity Vitalkapazität
vital dye/vital stain
  Vitalfarbstoff, Lebendfarbstoff
vital red *micros* Brilliantrot
vital staining
  Vitalfärbung, Lebendfärbung
vitality
  Vitalität, Lebenskraft
vitamin deficiency Vitaminmangel
vocational aptitude test
  Berufseignungstest
vocational school Berufsschule
volatile flüchtig
➤ **highly volatile/light**
  leicht flüchtig (niedrig siedend)
➤ **less volatile**
  (boiling/evaporating at higher temp.)
  höhersiedend
➤ **nonvolatile**
  nicht flüchtig; schwerflüchtig
volatility Flüchtigkeit *chem* (von
  Gasen: Neigung zu verdunsten)
volatilization Verflüchtigung
volcanic ash Vulkanasche
voltage Spannung
➤ **high voltage**
  Hochspannung
voltage clamp
  Spannungsklemme
voltage tester screwdriver
  Spannungsprüfer
  (Schraubenzieher)

voltmeter Spannungsmessgerät
voltammetry
Voltammetrie
➢ **cyclic voltammetry**
cyclische Voltammetrie,
Cyclovoltammetrie
➢ **linear scan voltammetry/
linear sweep voltammetry**
lineare Voltammetrie
➢ **stripping analysis/voltammetry**
Stripping-Analyse,
Inversvoltammetrie
volume Volumen, Rauminhalt; Masse, große Menge; (loudness) Lautstärke
volume fraction Volumenanteil
volumetric analysis
Maßanalyse, Volumetrie,
volumetrische Analyse
volumetric flask
Messkolben, Mischzylinder
volumetric solution
(a standard analytical solution)
Maßlösung
voluntary
freiwillig, aus eigenen Stücken, aus eigenem Willen; willkürlich

vomit
brechen (bei Übelkeit),
sich übergeben
vomiting Erbrechen
➢ **induced vomiting**
provoziertes Erbrechen
vortex (*pl* **vortices**)
Wirbel, Strudel; (mixer) Vortex, Mixer, Mixette, Küchenmaschine
**vortex motion/
whirlpool motion (shaker)**
Vortex-Bewegung, Wirbelbewegung, kreisförmig-vibrierende Bewegung
(z.B. Schüttler)
vortex shaker/vortex
Vortexmischer, Vortexschüttler, Vortexer (für Reagensgläser etc.)
voucher
Dokument, Unterlage;
Beleg, Belegzettel, Quittung;
Gutschein, Bon
**voucher specimen/
voucher copy**
Belegexemplar
vulcanize vulkanisieren
vulnerable verletzlich

wad Pfropf, Pfropfen; Wattebausch
wadding Einlage, Füllmaterial; Polsterung; Wattierung, Watte
wafer Wafer, dünne Platte/Scheibe; (e.g., silicon wafer/chip) Halbleiterplatte
wagging vibration (IR) Wippschwingung
waiver Verzicht, Verzichtserklärung
walk-in hood begehbarer Abzug
walkthrough Durchgang
wall cabinet/cupboard Wandschrank
wall chart Wandtafel
wall outlet Wandsteckdose
ware (articles/products/goods) Ware(n)
warehouse Lager, Lagerraum, Warenlager
warming Erwärmung
warmth (heat) Wärme (Hitze)
warning (caution) Warnung
warning label Warnetikett
warning sign/precaution sign Warntafel, Warnzeichen, Warnhinweis
warning tape Signalband, Warnband
warranty Garantie (Hersteller~/Verkäufer~), Haftung
wash waschen
➤ **wash out/rinse out/flush out** auswaschen; (elute) eluieren
wash basin Waschbecken
wash bottle/squirt bottle Spritzflasche
washdown Ganzwäsche
washer Dichtungsring, Dichtungsscheibe, Unterlegscheibe; (washing machine) Waschmaschine
washing bottle Waschflasche
➤ **gas washing bottle** Gaswaschflasche
washing facilities Wascheinrichtung

washroom/lavatory Waschraum, Toilette
washup room Spülküche
waste *vb* verschwenden, vergeuden; verbrauchen; verwüsten, zerstören; nutzlos sein
waste *n* **(trash/rubbish/refuse/garbage)** Müll, Abfall
➤ **chemical waste** Chemieabfälle
➤ **clinical waste** Klinikmüll
➤ **hazardous waste** Sondermüll, Sonderabfall, Problemabfall
➤ **household waste** Haushaltsmüll, Haushaltsabfälle
➤ **industrial waste** Industriemüll, Industrieabfall
➤ **infectious waste** infektiöser Abfall
➤ **municipal solid waste (MSW)** kommunaler Müll
➤ **nuclear waste** Atommüll
➤ **radioactive waste/nuclear waste** radioaktive Abfälle
➤ **toxic waste/poisonous waste** Giftmüll
waste avoidance Müllvermeidung
waste collection Müllabfuhr
waste container/ garbage can/dustbin (*Br*) Mülltonn; (litter bin) Abfallbehälter
waste disposal (waste removal) Abfallentsorgung, Abfallbeseitigung
waste disposal law/ waste disposal act Abfallgesetz, Abfallbeseitigungsgesetz (AbfG)
waste disposal site/waste dump Mülldeponie, Müllplatz, Müllabladeplatz, Müllkippe
waste heat Abwärme
waste incineration plant/ incinerator Müllverbrennungsanlage
waste oil/used oil Altöl
waste paper Altpapier
waste pretreatment Abfallvorbehandlung

waste recycling
  Müllwiederverwertung
waste recycling plant
  Müllverwertungsanlage
waste separation
  Mülltrennung, Abfalltrennung
waste treatment
  Abfallbehandlung, Abfallverwertung
waste treatment facility
  Abfallbehandlunganlage, Abfallverwertungsanlage
wastewater (sewage) Abwasser
wastewater charges
  Abwasserabgaben
wastewater purification plant/ sewage treatment plant
  Kläranlage
watch glass/clock glass
  Uhrglas, Uhrenglas
watchmaker forceps/jeweler's forceps
  Uhrmacherpinzette
watchmaker's screwdriver/ jeweler's screwdriver
  Uhrmacherschraubenzieher
water Wasser
  ➢ **bound water**
    gebundenes Wasser
  ➢ **brackish water (somewhat salty)**
    Brackwasser
  ➢ **capillary water** Kapillarwasser
  ➢ **carbonated water**
    kohlensäurehaltiges Wasser, Sodawasser; Sprudel
  ➢ **cooling water** Kühlwasser
  ➢ **crystal water/ water of crystallization**
    Kristallwasser
  ➢ **deionized water**
    entionisiertes Wasser
  ➢ **dishwater**
    Spülwasser, Abwaschwasser
  ➢ **distilled water**
    destilliertes Wasser
  ➢ **double distilled water**
    Bidest
  ➢ **drinking water/potable water**
    Trinkwasser
  ➢ **film water/retained water**
    Haftwasser
  ➢ **freshwater** Süßwasser
  ➢ **ground water** Grundwasser
  ➢ **hard water** hartes Wasser
  ➢ **heavy water** D_2O schweres Wasser
  ➢ **hot water** Warmwasser
  ➢ **industrial wastewater**
    Industrieabwasser; Fabrikationsabwasser
  ➢ **industrial water**
    Industrie-Brauchwasser
  ➢ **jet of water** Wasserstrahl
  ➢ **mineral water** Mineralwasser
  ➢ **peptone water** Peptonwasser
  ➢ **potable water**
    trinkbares Wasser
  ➢ **process water/ service water/ industrial water (nondrinkable water)**
    Brauchwasser, Betriebswasser (nicht trinkbares Wasser)
  ➢ **purified water**
    gereinigtes Wasser, aufgereinigtes Wasser, aufbereitetes Wasser
  ➢ **saline water** salziges Wasser
  ➢ **salt water** Salzwasser
  ➢ **seawater (salt water)**
    Meerwasser (Salzwasser)
  ➢ **soda water** Selterswasser, Sodawasser, Sprudel
  ➢ **soft water** weiches Wasser
  ➢ **springwater** Quellwasser
  ➢ **tap water** Leitungswasser
  ➢ **wastewater/sewage** Abwasser
  ➢ **well water** Brunnenwasser
water activity
  Wasseraktivität, Hydratur
water analysis
  Wasseruntersuchung, Wasseranalyse
water bath Wasserbad
water column (column of water)
  Wassersäule
water-conducting wasserleitend
water consumption/water usage
  Wasserverbrauch

water content Wassergehalt
water distillation Wasserdestillation
water flow Wasserströmung
water gas Wassergas
water glass/soluble glass $M_2O \times (SiO_2)_x$ Wasserglas
water hardness Wasserhärte
water hazard class Wassergefahrenklasse (WGK)
water jacket Wassermantel (Kühler)
water loss Wasserverlust
water of crystallization Kristallisationswasser
water of hydration Hydratwasser
water outlet Wasserzulauf, Wasserzapfstelle (Wasserhahn)
water pollution Wasserverschmutzung
water potential Wasserpotential, Hydratur, Saugkraft
water pump/ filter pump/vacuum filter pump Wasserstrahlpumpe
water pump pliers/ slip-joint adjustable water pump pliers (adjustable-joint pliers) Wasserpumpenzange, Pumpenzange
water purification Wasseraufbereitung
water purification plant/ water treatment facility Wasseraufbereitungsanlage
water quality Wassergüte, Wasserqualität
water reactive wasserreaktiv
water regime Wasserhaushalt, Wasserregime
water-repellent/water-resistant wasserabstoßend, wasserabweisend
water sample Wasserprobe
water saturation Wassersättigung
water saturation deficit (WSD) Wassersättigungsdefizit
water softener Wasserenthärter
water softening Wasserenthärtung
water solubility Wasserlöslichkeit
water-soluble wasserlöslich
water still Wasserdestillierapparat
water stress Wasserstress
water supply Wasserversorgung, Wasserzufuhr
water tension Zugspannung (Wasserkohäsion); (water suction) Wassersog
water trap/separator Wasserabscheider
water uptake Wasseraufnahme
water vapor Wasserdampf
waterlogged vollgesogen (mit Wasser)
waterlogging Vernässung
waterproof wasserfest, wasserdicht, wasserundurchlässig
waterproofing wasserfest/wasserdicht/ wasserundurchlässig machen ('imprägnieren')
watertight/waterproof wasserdicht, wasserundurchlässig
wave guide Hohlleiter (z.B. an Mikrowelle)
wavelength Wellenlänge
wavenumber (IR) Wellenzahl
wax Wachs
➤ **beeswax** Bienenwachs
➤ **paraffin wax** Paraffinwachs
➤ **sealing wax** Siegelwachs
➤ **synthetic wax** Synthesewachs
➤ **wool wax (wool fat)** Wollwachs (Wollfett)
wax feet (plasticine supports on edges of coverslip) *micros* Wachsfüßchen (Plastilinfüßchen an Deckgläschen)
wax paper Wachspapier
waxy (wax-like/ceraceous) wachsartig
wear *n* Abnutzung, Verschleiß
wear (out) verschleißen, abnutzen, verbrauchen
weather *n* Wetter
weather *vb* verwittern
weathering Verwitterung
weatherproof wetterbeständig
webbing Vernetzung
wedge/peg Keil
weed control Unkrautbekämpfung, Unkrautvernichtung

weigh wiegen
> **weigh in (after setting tare)**
  einwiegen (nach Tara)
> **weigh out** abwiegen
  (eine Teilmenge)
> **weigh out precisely**
  auswiegen (genau wiegen)
weighing Wägung
weighing boat/weighing scoop
  Wägeschiffchen
weighing bottle Wägeglas
weighing paper Wägepapier
weighing spatula Wägespatel
weighing spoon
  Maßlöffel, Wägelöffel
weighing table Wägetisch
weight Gewicht;
  Belastung, Traglast, Last;
  Wägemasse
> **atomic weight**
  Atomgewicht
> **dry weight**
  (*sensu stricto*: **dry mass**)
  Trockengewicht
  (*sensu stricto*: Trockenmasse)
> **fresh weight**
  (*sensu stricto*: **fresh mass**)
  Frischgewicht
  (*sensu stricto*: Frischmasse)
> **gross weight**
  Bruttogewicht
> **live weight**
  Lebendgewicht
weight buret/weighing buret
  Wägebürette
weightless
  ohne Gewicht, schwerelos
weightlessness
  Schwerelosigkeit
weld *vb* schweißen;
  (weld together) verschweißen;
  (welded on/welded to)
  eingeschweißt
welded joint
  Schweißverbindung
welding
  Schweißen, Schweißung;
  Schweißnaht, Schweißstelle

well Brunnen, Quelle;
  (depression: at top of gel)
  electrophor Tasche (Vertiefung:
  Elektrophorese-Gel);
  Rinne (z.B. Pufferrinne,
  Pufferwanne)
well plate *gen/micb* Lochplatte
well water Brunnenwasser
welt (weal) Quaddel
wet *vb* nass machen,
  befeuchten, benetzen
wet blotting Nassblotten
wet cell *electr* Nasselement
wet mount *micros*
  Nasspräparat (Frischpräparat,
  Lebendpräparat, Nativpräparat)
wet rot Nassfäule
wet-bulb thermometer
  Nassthermometer,
  Verdunstungsthermometer
wettability Benetzbarkeit
wettable benetzbar
wetting agent
  (wetter/surfactant/spreader)
  Benetzungsmittel;
  Entspannungsmittel
  (oberflächenaktive Substanz)
wheelchair Rollstuhl
wheelchair accessible
  rollstuhlgerecht
whirl/swirl/eddy strudeln
whisk Schneebesen
whiskers
  (needle-shaped, single crystals)
  Whisker, Haarkristalle
  (Fadenkristalle, Nadelkristalle)
white wash Tünche, Kalkanstrich
whole blood Vollblut
whole-body exposure *rad*
  Ganzkörperbestrahlung
whole mount Totalpräparat
whole mount plastination
  Ganzkörperplastination
wholesale business/
  wholesale trade Großhandel
wholesaler/wholesale vendor
  Großhändler
wick Docht

wide-angle X-ray scattering (WAXS)
  Röntgenweitwinkelstreuung
wide-mouthed
  (widemouthed/
  wide-neck/widenecked)
  Weithals ...
wide-mouthed bottle
  Weithalsflasche
wide-mouthed flask/
  wide-necked flask
  Weithalskolben
wide-neck vat/wide-mouth vat
  Weithalsfass
widefield *micros* Weitwinkel
widespread/ubiquitous
  (existing everywhere)
  weitverbreitet,
  ubiquitär (überall verbreitet)
wilt/wither/fade welken
wilted/withered/faded/limp/flaccid
  welk, schlaff
wilting (withering/fading/flaccid/
  deficient in turgor) welkend
wilting point Welkepunkt
winch Winde, Kurbel;
  (rope winch) Seilwinde
wind *vb* **(twist/coil)** winden
winding/contortion/turn/bend
  Windung, Krümmung, Biegung
window Fenster
window glass Fensterglas
window pane Fensterscheibe
wing nut Flügelmutter
wing-tip (for burner)/
  burner wing top
  Schwalbenschwanzbrenner,
  Schlitzaufsatz für Brenner
wipe *n* Wischtuch; (wiper) Wischer
wipe *vb* wischen
➢ **wipe off/wipe clean**
  abwischen
➢ **wipe up** aufwischen
wire Draht; (cable) Kabel
wire brush Drahtbürste, Stahlbürste
wire cable shears/cable shears
  Drahtseilschere, Kabelschere
wire end sleeve/
  wire end ferrule *electr* Aderendhülse

wire gauze/
  wire gauze screen
  Drahtnetz
wire shears/wire cutters
  Drahtschere
wire stripper Abisolierzange
wiring/electrical wiring
  Verkabelung, Verdrahtung
wither/wilt/fade (shrivel up)
  verwelken
withering/
  wilting/fading/shrivelling/marcescent
  verwelkend
wood drill (bit) Holzbohrer
wood pulp Zellstoff
wood rot Holzfäule
wood spirit/wood alcohol/
  pyroligneous spirit/
  pyroligneous alcohol
  (chiefly: methanol)
  Holzgeist
wood sugar/xylose
  Holzzucker, Xylose
wood tar Holzteer
wood vinegar/
  pyroligneous acid
  Holzessig
wood-wool
  Holzwolle; Zellstoffwatte
wooden hölzern
wool Wolle
➢ **glass wool** Glaswolle
➢ **mineral wool (mineral cotton)**
  Mineralfasern
  (speziell: Schlackenfasern)
➢ **rock wool** Steinwolle
➢ **wood wool**
  Holzwolle, Zellstoffwatte
wool alcohols
  Wollwachsalkohole
wool fat gland Wollfettdrüse
wool wax (wool fat)
  Wollwachs (Wollfett)
work Arbeit;
  (job) Job, Arbeitsstelle
➢ **physical work** körperliche Arbeit
➢ **preparatory work** Vorarbeiten
➢ **shift work** Schichtarbeit

work area
Arbeitsbereich, Arbeitsplatz
work gloves
Arbeitshandschuhe
work hours Arbeitszeit
work of expansion
Ausdehnungsarbeit
work procedure
Arbeitsmethode; Arbeitsvorgang
➤ **prescribed work procedure/**
prescribed operating procedure
Arbeitsvorschrift, Arbeitsanweisung
work surface/working surface
Arbeitsfläche
work up *n* **(working up/processing/**
down-stream processing)
Aufarbeitung
work up *vb* **(process)** aufarbeiten
working conditions (for personnel)
Arbeitsbedingungen (für
BediensteteAngestellte/Arbeiter)
working disability/disablement
Berufsunfähigkeit
working distance
(objective-coverslip) *micros*
Arbeitsabstand
working guideline
Arbeitsrichtlinie
working hypothesis
Arbeitshypothese
working life *tech/mech*
Verwendbarkeitsdauer,
Nutzungsdauer
working order/operating condition
Funktionszustand
working pressure/delivery pressure
Arbeitsdruck: mit Druckausgleich
working procedure
Arbeitsmethode;
Arbeitsvorgang, Arbeitsverfahren
➤ **step in a working procedure**
Arbeitsschritt
working range
Arbeitsbereich
working space
Arbeitsraum
(im Inneren der Werkbank)
workload Arbeitspensum

workman's compensation
Entschädigung~/
Kompensationszahlung bei
Arbeitsunfällen od.
Berufskrankheiten
workpiece
Werkstück, Teil, Stück
workplace Arbeitsplatz
workplace concentration
Arbeitsplatz-Konzentration
➤ **maximum permissible workplace**
concentration/
maximum permissible exposure
MAK-Wert
(maximale Arbeitsplatz-
Konzentration)
workplace safety regulations
Arbeitsschutzverordnung
workshield
Schutzschild, Schutzschirm
workshop/'shop'
Werkstatt
workspace
Arbeitsbereich (räumlich)
worktable Arbeitstisch
worm gear
Schneckengetriebe (DIN)
worm thread
Schneckengewinde
worst-case accident
größter anzunehmender Unfall
worst-case scenario
schlimmster anzunehmender Fall
wort (beer) Würze (Bier)
Woulff bottle Woulff'sche Flasche
wound *n* Wunde
wound healing
Wundheilung
wrap
Folie, Einwickelpapier
➤ **plastic wrap (household wrap)**
Plastikfolie (Frischhaltefolie)
wrap-it tie(s)/
wrap-it tie cable/
cable tie(s)
Kabelbinder, Spannband
wrapfoil heat sealer
Folienschweißgerät

wrapping
 Verpackung(smaterial)
 [mit Folie/Papier]
wrapping paper Einpackpapier
wrench (screw wrench)/
spanner *Br*
 Schlüssel, Schraubenschlüssel,
 Schraubschlüssel
➢ **adjustable wrench**
 'Engländer', Rollgabelschlüssel
➢ **Allen wrench** Inbusschlüssel
➢ **box wrench/box spanner/**
ring spanner (*Br*)
 Ringschlüssel
➢ **hex nutdriver**
 Sechskant-Steckschlüssel
➢ **hex socket wrench**
 Sechskant-Stiftschlüssel
➢ **open-end wrench/**
open-end spanner (*Br*)
 Gabelschlüssel, Maulschlüssel
➢ **ring spanner wrench/**
box wrench/
box spanner/
ring spanner (*Br*)
 Ringschlüssel
➢ **socket wrench/**
box spanner
 Stiftschlüssel, Steckschlüssel
➢ **spider wrench/spider spanner** (*Br*)
 Kreuzschlüssel
➢ **torque wrench**
(torque amplifier handle)
 Drehmomentschlüssel
➢ **wipe wrench**
(rib-lock pliers/
adjustable-joint pliers)
 Rohrzange
wringer (mop)
 Auswringer, Wringer (Mop)
wrought iron
 Schmiedeeisen

xanthan gum
Xanthangummi
**xanthogenic acid/
xanthic acid/
xanthonic acid/
ethoxydithiocarbonic acid**
Xanthogensäure
xenobiotic (*pl* **xenobiotics)**
Xenobiotikum (*pl* Xenobiotika)
X-ray
Röntgenstrahl;
Röntgenaufnahme, Röntgenbild;
vb röntgen,
eine Röntgenaufnahme machen,
bestrahlen
X-ray absorption spectroscopy
Röntgenabsorptionsspektroskopie
X-ray crystallography
Röntgenkristallographie
X-ray diffraction
Röntgenbeugung
X-ray diffraction method
Röntgenbeugungsmethode
X-ray diffraction pattern
Röntgenbeugungsdiagramm,
Röntgenbeugungsaufnahme,
Röntgendiagramm,
Röntgenbeugungsmuster
X-ray emission spectroscopy
Röntgenemissionsspektroskopie
**X-ray fluorescence spectroscopy
(XFS)**
Röntgenfluoreszenzspektroskopie
(RFS)
X-ray microanalysis
Röntgenstrahl-Mikroanalyse
X-ray microscopy Röntgenmikroskopie
**X-ray structural analysis/
X-ray structure analysis**
Röntgenstrukturanalyse
xylene/dimethylbenzene
Xylol, Dimethylbenzol

yarn Garn
yeast Hefe
➢ **baker's yeast**
  Backhefe, Bäckerhefe
➢ **bottom yeast**
  niedrigvergärende Hefe ('Bruchhefe')
➢ **brewers' yeast** Bierhefe, Brauhefe
➢ **distiller's yeast** Brennereihefe
➢ **dried yeast** Trockenhefe
➢ **fission yeast**
  (*Saccharomyces pombe*)
  Spalthefe
➢ **mineral accumulating yeast**
  Mineralhefe
➢ **pitching yeast**
  Stellhefe, Anstellhefe, Impfhefe
➢ **top yeast**
  hochvergärende Hefe ('Staubhefe')
yield Ertrag, Ausbeute, Ergiebigkeit; Gewinn, Ergebnis; Fließen

yield coefficient (Y)
  Ertragskoeffizient,
  Ausbeutekoeffizient,
  ökonomischer Koeffizient
yield increase Ertragssteigerung
yield level/quality class
  Ertragsklasse, Ertragsniveau, Bonität
yield point *tech/mech*
  Fließgrenze,
  Fließspannung, Fließpunkt
yield reduction
  Ertragsminderung
yield strength
  Elastizitätsgrenze, Dehngrenze;
  Fließfestigkeit,
  Verformungsfestigkeit
yield stress
  Streckspannung, Fließspannung
ytterbium (Yb) Ytterbium
yttrium (Y) Yttrium

ze

zero *n* Null
zero *vb* auf Null stellen
zero adjustment Nullabgleich
zero-point adjustment/
  zero-point setting
  Nullpunktseinstellung
zero reading Null-Anzeige
zero-valent/nonvalent nullwertig
zinc (Zn) Zink
zinc blende/blackjack Zinkblende
zinc finger *gen* Zinkfinger
zip seal/zip-lip/zip-lip seal/zipper-top
  Zippverschluss,
  Druckleistenverschluss
zip-lip bag/zipper-top bag/
  zip storage bag/zip-lip storage bag
  Zippverschlussbeutel,
  Druckverschlussbeutel

zipper Reißverschluss
zircon $ZrSiO_4$
  Zirkon
zirconia ZrO_2
  (zirconium oxide/zirconium dioxide)
  Zirconiumdioxid
zirconium (Zr) Zirconium
zonal centrifugation
  Zonenzentrifugation
zonation Zonierung, Stufung
zone electrophoresis
  Zonenelektrophorese
zone refining
  Zonenschmelze(n)
zone sedimentation/
  zonal sedimentation
  Zonensedimentation
zwitterion Zwitterion

Printing: Strauss GmbH, Mörlenbach
Binding: Schäffer, Grünstadt

UMRECHNUNGSTABELLEN / CONVERSION TABLES

VOLUMEN (RAUMINHALT) – *VOLUME (CAPACITY)**

liters	gallons	quarts	pints	fl.oz.
1	0.2642	1.0567	2.1134	33.814
3.7854	1	4	8	128
0.9464	0.25	1	2	32
0.4732	0.125	0.5	1	16
0.0296	0.0078	0.03125	0.0625	1

MASSE – *MASS**

kg/g	pounds	ounces
1kg (1000g)	2.2046	35.274
453.6g	1	16
28.35g	0.0625	1

LÄNGE – *LENGTH**

km/m/cm	miles	yards	feet	inches
1 km	0.62137	1093.61	3280.84	–
1 m	–	1.0936	3.281	39.37
1.61 km (1609 m)	1	1760	5280	63,360
0.9144 m	0.00057	1	3	36
30.48 cm	–	0.333	1	12
2.54 cm	–	0.0278	0.0833	1

DRUCK – *PRESSURE**

N/m^2 (Pa)	torr (mm Hg)	bar	atm (st)	lb/ft^2	lb/in^2 (psi)
1	7.528×10^{-3}	10^{-5}	9.869×10^{-6}	0.02089	0.145×10^{-3}
133.3	1	1.333×10^{-3}	1.3157×10^{-3}	2.784	2.4942×10^{-3}
10^5	750.06	1	0.9869	2116.8	14.7
1.01325×10^5	760	1.0133	1	2116.4	14.6974
47.88	0.3591	4.788×10^{-4}	4.725×10^{-4}	1	6.944×10^{-3}
6894.76	51.7236	0.0689476	0.06804	144	1

T+ = *very toxic*
sehr giftiger
Stoff

T = *toxic*
giftiger Stoff

Xn = *nocent (harmful)*
gesundheits-
schädlicher
Stoff

Xi = *irritant*
reizender
Stoff

C = *corrosive*
ätzender
Stoff